BIOCHEMICAL SOCIETY SYMPOSIA

No. 35

CURRENT TRENDS IN THE BIOCHEMISTRY OF LIPIDS

Indian Institute of Science (main building).

The Department of Biochemistry.

Professor T. W. Goodwin, F.R.S. (Chairman, British Biochemical Society) conveying greetings of the Society at the beginning of the meeting. Seated from left to right:
Professor J. Ganguly, Organizer and Chairman of the Symposium.
Professor D. K. Banerjee, Director, Indian Institute of Science.
Professor H. R. Cama, In-charge, Department of Biochemistry, Indian Institute of Science, Dr G. Ramamanda Rao, Secretary, Society of Biological Chemists, India.

CURRENT TRENDS IN THE BIOCHEMISTRY OF LIPIDS

Biochemical Society, Symposium No. 35
Proceedings of the International Symposium on Lipids held at Bangalore on 3 *and* 4 *December*, 1971

ON THE OCCASION OF THE
GOLDEN JUBILEE OF THE DEPARTMENT OF BIOCHEMISTRY
INDIAN INSTITUTE OF SCIENCE, BANGALORE
INDIA

SYMPOSIUM CHAIRMAN AND ORGANIZER
J. GANGULY

EDITED BY
J. GANGULY AND R. M. S. SMELLIE

1972
ACADEMIC PRESS
LONDON and NEW YORK

ACADEMIC PRESS INC. (LONDON) LTD.
24/28 Oval Road,
London NW1

United States Edition published by
ACADEMIC PRESS INC.
111 Fifth Avenue 7
New York, New York 10003

Library of Congress Catalog Card Number: 72–9778
ISBN: 0–12–274250–8

Text set in 11/12 pt. Monotype Modern, printed by letterpress, and bound in Great Britain at The Pitman Press, Bath

LIST OF CONTRIBUTORS

G. B. Ansell, *Department of Pharmacology (Preclinical), The Medical School, Birmingham, B15 2TJ, England*

K.-I. Arnstadt, *Max-Planck-Institut für Zellchemie, D-8 Munchen 2, Germany*

Y. C. Awasthi, *National Botanic Gardens, Lucknow, India*

E. Baer, *Banting and Best Department Medical Research, University of Toronto, Ontario, Canada*

A. A. Benson, *Scripps Institution of Oceanography, University of California, San Diego, La Jolla, California 92037, U.S.A.*

J. Borensztajn, *Department of Pathology, University of Chicago, Chicago, Illinois 60637, U.S.A.*

P. P. M. Bonsen, *Laboratory of Biochemistry, State University of Utrecht, Vondellaan 26, Utrecht, The Netherlands*

R. O. Brady, *Laboratory of Neurochemistry, National Institute of Neurological Diseases and Stroke, National Institutes of Health, Bethesda, Maryland 20014, U.S.A.*

H. R. Cama, *Department of Biochemistry, India Institute of Science, Bangalore, India*

T. F. Chuang, *Department of Biological Sciences, Purdue University, Lafayette, Indiana, U.S.A.*

J. Clausen, *The Neurochemical Institute, 58 Radmandsgado, 2200 Copenhagen N., Denmark*

M. C. Craig, *Lipid Metabolism Laboratory, Veterans Administration Hospital and the Department of Physiological Chemistry, University of Wisconsin, Madison, Wisconsin 53706, U.S.A.*

F. L. Crane, *Department of Biological Sciences, Purdue University, Lafayette, Indiana, U.S.A.*

A. N. Davison, *Department of Neurochemistry, Institute of Neurology, Queen Square, London, WC1 3BG, England*

R. M. C. Dawson, *Biochemistry Department, Institute of Animal Physiology, Babraham, Cambridge, England*

J. de Gier, *State University of Utrecht, Department of Biochemistry, Vondellaan 26, Utrecht, The Netherlands*

G. H. de Haas, *Laboratory of Biochemistry, State University of Utrecht, Vondellaan 26, Utrecht, The Netherlands*

H. F. DeLuca, *Department of Biochemistry, College of Agricultural and Life Sciences, University of Wisconsin, Madison, Wisconsin 53706, U.S.A.*

R. A. Demel, *State University of Utrecht, Department of Biochemistry, Vondellaan 26, Utrecht, The Netherlands*

K. A. Devor, *Department of Physiological Chemistry, The Johns Hopkins University, School of Medicine, Baltimore, Maryland, U.S.A.*

M. ESFAHANI, *Department of Biochemistry, Baylor College of Medicine, Houston, Texas* 77025, *U.S.A.*

C. G. GAHMBERG, *Department of Biochemistry, University of Helsinki, Helsinki, Finland*

J. GANGULY, *Department of Biochemistry, Indian Institute of Science, Bangalore*-12, *India*

DeW. S. GOODMAN, *Department of Medicine, Columbia University College of Physicians and Surgeons, New York, New York* 10032, *U.S.A.*

T. W. GOODWIN, *Department of Biochemistry, The University, Liverpool, L*69 3*BX, England*

J. L. HARWOOD, *Department of Biochemistry, University of Leeds, Leeds* 2, *England*

T. HASHIMOTO, *Department of Medical Chemistry, Kyoto University Faculty of Medicine, Kyoto, Japan*

J. N. HAWTHORNE, *Department of Chemistry, University of Calgary, Alberta, Canada*

J. HOELZL, *Institut fur Pharmazeutische Arzneimittellehre der Universitat Munchen*, 8 *Munchen, Karlstrasse* 29, *Germany*

A. T. JAMES, *Division of Biochemistry, Biosciences Group, Unilever Research Laboratory, Colworth House, Sharnbrook, Bedfordshire, England*

H. JATZKEWITZ, *Max-Planck-Institut für Psychiatrie, Neurochemische Abteilung, Munich, Germany*

F. B. JUNGALWALA, *Biochemistry Department, Institute of Animal Physiology, Babraham, Cambridge, England*

L. KÄÄRIÄINEN, *Department of Biochemistry, University of Helsinki, Helsinki, Finland*

E. KLENK, *Physiol.-Chem. Institut der Universitat*, 5 *Koln-Lindenthal, Joseph-Stelzmannstr.* 52, *Germany*

M. R. LAKSHMANAN, *Lipid Metabolism Laboratory, Veterans Administration Hospital and the Department of Physiological Chemistry, University of Wisconsin, Madison, Wisconsin* 53706, *U.S.A.*

E. LEDERER, *Institut de Chimie des Substances Naturelles, C.N.R.S.* 91-*Gif-sur-Yvette and Institut de Biochimie, Centre d'Orsay, Université de Paris-Sud*, 91-*Orsay, France*

R. F. LEE, *Scripps Institution of Oceanography, University of California, San Diego, La Jolla, California* 92037, *U.S.A.*

W. J. LENNARZ, *Department of Physiological Chemistry, The Johns Hopkins University School of Medicine, Baltimore, Maryland, U.S.A.*

F. LYNEN, *Max-Planck-Institut für Zellchemie, D*-8 *München* 2, *Germany*

W. C. MCMURRAY, *Biochemistry Department, Institute of Animal Physiology, Babraham, Cambridge, England*

R. A. MAKULA, *Department of Physiological Chemistry, The Johns Hopkins University School of Medicine, Baltimore, Maryland, U.S.A.*

A. K. MALLIA, *Department of Biochemistry, India Institute of Science, Bangalore, India*

E. MILLER, *Biochemistry Department, Institute of Animal Physiology, Babraham, Cambridge, England*

R. A. MORTON, *The University of Liverpool, Liverpool, England*

S. NAKANISHI, *Department of Medical Chemistry, Kyoto University Faculty of Medicine, Kyoto, Japan*

C. M. NEPOKROEFF, *Lipid Metabolism Laboratory, Veterans Administration Hospital and the Department of Physiological Chemistry, University of Wisconsin, Madison, Wisconsin* 53706, *U.S.A.*

J. C. NEVENZEL, *Scripps Institution of Oceanography, University of California, San Diego, La Jolla, California* 92037, *U.S.A.*

S. NUMA, *Department of Medical Chemistry, Kyoto University Faculty of Medicine, Kyoto, Japan.*

T. OKAZAKI, *Department of Medical Chemistry, Kyoto University Faculty of Medicine, Kyoto, Japan*

S. PARTHASARATHY, *Department of Biochemistry, Indian Institute of Science, Bangalore*-12, *India*

W. A. PIETERSON, *Laboratory of Biochemistry, State University of Utrecht, Vondellaan* 26, *Utrecht, The Netherlands*

J. W. PORTER, *Lipid Metabolism Laboratory, Veterans Administration Hospital and the Department of Physiological Chemistry, University of Wisconsin, Madison, Wisconsin* 53706, *U.S.A.*

T. RAMASARMA, *Department of Biochemistry, Indian Institute of Science, Bangalore*-12, *India*

O. RENKONEN, *Department of Biochemistry, University of Helsinki, Helsinki, Finland*

D. S. ROBINSON, *Department of Biochemistry, University of Leeds, Leeds 2, England*

V. W. RODWELL, *Department of Biochemistry, Purdue University, Lafayette, Indiana* 47907, *U.S.A.*

M. P. ROGERS, *Department of Biochemistry, University of Leeds, Leeds 2, England*

H. RUDNEY, *Department of Biological Chemistry, College of Medicine, University of Cincinnati, Cincinnati, Ohio* 45219, *U.S.A.*

B. SAMUELSSON, *Department of Medical Chemistry, Royal Veterinary College, S*-104 05 *Stockholm, Sweden*

K. SIMONS, *Department of Biochemistry, University of Helsinki, Helsinki, Finland*

H. SINGH, *Department of Biochemistry, Indian Institute of Science, Bangalore, India*

S. SPANNER, *Department of Pharmacology (Preclinical), The Medical School, Birmingham, B*15 *2TJ, England*

K. SRIVASTAVA, *The Neurochemical Institute*, 58 *Radmandsgado*, 2200 *Copenhagen N., Denmark*

P. K. STUMPF, *Department of Biochemistry and Biophysics, University of California, Davis, California* 95616, *U.S.A.*

P. V. SUBBAIAH, *Department of Biochemistry, Indian Institute of Science, Bangalore, India*

M. SUMPER, *Max-Planck-Institut für Zellchemie, D-8 München 2, Germany*

J. R. TATA, *National Institute for Medical Research, Mill Hill, London, NW7 1AA, England*

R. H. S. THOMPSON, *Courtauld Institute of Biochemistry, The Middlesex Hospital Medical School, London, W1P 5PR, England*

L. L. M. VAN DEENEN, *State University of Utrecht, Department of Biochemistry, Vondellaan 26, Utrecht, The Netherlands*

I. VIJAY, *Department of Biochemistry, University of Michigan, Ann Arbor, Michigan, U.S.A.*

J. J. VOLWERK, *Laboratory of Biochemistry, State University of Utrecht, Vondellaan 26, Utrecht, The Netherlands*

S. J. WAKIL, *Department of Biochemistry, Baylor College of Medicine, Houston, Texas 77025, U.S.A.*

K.-U. WEITHMANN, *Max-Planck-Institut für Zellchemie, D-8 München 2, Germany*

W. WINNEWISSER, *Max-Planck-Institut für Zellchemie, D-8 München 2, Germany*

J. ZIEGENHORN, *Max-Planck-Institut für Zellchemie, D-8 München 2, Germany*

PREFACE

THE Indian Institute of Science was founded at Bangalore in the year 1909 by Jamshetji Nussarwanji Tata, one of the greatest sons of India and a great captain of industry. When started it had only a few departments. The Department of Biochemistry was added on in the year 1921 with a small number of staff and students. But it is significant that it was the first chair in biochemistry in India, in the Orient, and probably fourth in the Commonwealth.

From such a modest beginning the Department has steadily grown to its present stature of one of the largest and foremost centres of biochemistry. In appreciation of such a position the University Grants Commission of India has recognized it as a Centre of Advanced Study in Biochemistry.

This Department completed fifty years of its eventful life in 1971. The International Symposium on Lipids was a part of the programme for the observance of the Golden Jubilee of the Department. This Symposium has been cosponsored by the Society of Biological Chemists (India), the Biochemical Society (U.K.) and the International Union of Biochemistry (IUB). My special thanks are due to Professor T. W. Goodwin, F.R.S. (Chairman, Biochemical Society, U.K.) and Professor A. P. Mathias (International Secretary of the Biochemical Society) for their personal interest and generous assistance in the organization of the Symposium and subsequent publication of the proceedings. I am equally indebted to Professor P. N. Campbell for his co-operation in having it sponsored by the IUB.

It is a pleasure to record with thanks the generous financial contributions made by the following national and international agencies:

The Commonwealth Foundation (London), the International Union of Biochemistry, the Wellcome Trust (London), the Atomic Energy Establishment, Government of India (Bombay), the Council of Scientific and Industrial Research (New Delhi), the Indian Council of Agricultural Research, (New Delhi), the Indian Council of Medical Research (New Delhi), the Indian Institute of Science (Bangalore), the Indian National Science Academy (New Delhi), the Ministry of Defence, Government of India (New Delhi), the J. R. D. Tata Trust (Bombay), the Sir Dorab Tata Trust (Bombay), the Sir Rattan Tata Trust (Bombay), the University Grants Commission (New Delhi), the Beckman International (U.S.A.) and the Nattermann International (West Germany).

I wish to thank all my colleagues of this Department and also of this Institute who have cheerfully put in so much hard work for the success of the Symposium. I am most grateful to all participants and Chairmen of the various sessions for their co-operation and contributions. With great pleasure I record the most valuable and constant assistance of my colleagues Professor T. Ramasarma, Dr P. S. Sastry and Dr S. Krishnamurthy in the planning and execution of this Symposium and also in the heavy editorial work. Miss Emily Stephens has given valuable secretarial assistance in the work connected with the entire Golden Jubilee programme and also in the publication of this book. Finally, I wish to thank Professors R. Choksi (Chairman of the Governing Council of our Institute) and S. Dhawan (Director of the Institute) for their sustained and keen interest in the entire Golden Jubilee programme.

J. Ganguly

Symposium Chairman and Organizer
Department of Biochemistry,
Indian Institute of Science,
Bangalore,
India

CONTENTS

CHAIRMAN'S INTRODUCTION

By J. GANGULY

Department of Biochemistry, Indian Institute of Science, Bangalore, India

DURING the eventful life of the past fifty years of this Department there have been a variety of research activities, applied as well as basic, reflecting the changing conditions in the country as well as in the world, and also growth of different phases of biochemistry in general. Interest in lipids began in this Department very early and has been sustained throughout with increasing vigour. But, as happened contemporarily in the rest of the world, lipids were considered to be rather messy materials and the work here was confined to analysis of different kinds of lipids, especially ghee and milk fat, and to a certain extent carotenoids, vitamin A and fish-liver oils, and occasionally plant lipids. It might be of interest to recall here that two galactolipids were isolated in this Department as early as in 1930; one of them was later shown to be a sphingolipid, and the other one a galactolipid containing sulphur. This finding is of particular significance in that the plant sulpholipid was actually identified only in the late fifties.

The early lipid work in this Department eventually began to grow vigorously during the mid-fifties, coinciding with the rapid strides lipid biochemistry began to make in the rest of the world. It was at about this time that quite a few of us started to make significant contributions in the areas of vitamin A, carotenoids, ubiquinones and other lipids. At present there are several distinct groups working on different aspects of lipids, namely absorption of lipids, and metabolism of vitamin A, carotenoids, cholesterol, ubiquinones and brain lipids.

Meetings on different aspects of lipids are held almost every year in some part of the world or the other. While planning this Symposium, naturally we kept our local interest in mind and chose to restrict ourselves to a few areas like fatty acid synthesis, isoprenoid metabolism, physiology of lipids, brain biochemistry and membrane biochemistry, and to seek co-operation of the recognized authorities in these areas. We are indeed very happy that our colleagues responded enthusiastically to our invitation, so that we have contributions from most of the stalwarts in these fields. As always happens, a few of them could not come to the meeting due to last minute developments and we missed their participation. But they were kind enough to send their manuscripts, and they do appear in print along with the others.

In this country we feel rather isolated and cut off from the main centres of activity in the rest of the world. Meetings, such as the one this book records, serve a very useful purpose in bringing us closer to our colleagues in other parts of the world. We made special efforts to attract workers at all levels from different parts of our country. It is our sincere hope that the deliberations of this symposium will not only generate new ideas and thoughts, but will also make significant impact on the Indian workers regarding the latest trends in lipid biochemistry. I also hope that such meetings will be held in our country frequently in the future.

I

BIOSYNTHESIS OF FATTY ACIDS

ENZYME SYSTEMS FOR FATTY ACID SYNTHESIS

By F. LYNEN*

With experimental collaboration of K.-I. Arnstadt, M. Sumper, K.-U. Weithmann, W. Winnewisser and J. Ziegenhorn

Max-Planck-Institut für Zellchemie, München, Germany

INTRODUCTION

BIOSYNTHESIS of fatty acids takes place in two steps. In the first acetyl CoA, the common building block for numerous natural products, is carboxylated to form malonyl CoA. In this process protein-bound biotin acts as a CO_2 carrier with the intermediary formation of CO_2-biotin.

$$\text{ATP} + \text{HCO}_3^- + \text{biotin enzyme} \rightleftharpoons \text{CO}_2\text{-biotin enzyme} + \text{ADP} + \text{Pi}$$

$$\text{CO}_2\text{-biotin enzyme} + \text{acetyl CoA} \rightleftharpoons \text{biotin enzyme} + \text{malonyl CoA.}$$

The second step of fatty acid synthesis represents the formation of the long chains of fatty acids, starting from acetyl CoA, malonyl CoA and TPNH.

$$\text{RCO—SCoA} + n\ \text{HOOCCH}_2\text{—CO—SCoA} + 2n\ \text{TPNH} + 2n\ \text{H}^+ \rightarrow$$
$$\text{R(CH}_2\text{—CH}_2)_n\text{—CO—SCoA} + n\ \text{CO}_2 + n\ \text{H}_2\text{O} + n\ \text{HSCoA} + 2n\ \text{TPN}^+$$

In this process two carbon units derived from malonyl CoA are successively added to the priming acyl CoA derivative which is represented principally by acetyl CoA. The carbon dioxide previously fixed in the carboxylation reaction is ejected during the formation of fatty acids and this is quite important from the thermodynamic point of view.

GROUPS OF FATTY ACID SYNTHETASES

The ability to synthesize fatty acids from malonyl CoA is widespread in nature and enzyme systems required have been isolated from various unicellular organisms (Butterworth & Bloch, 1970), including *E. coli* (Goldman, Alberts & Vagelos, 1963; Lennarz, Light & Bloch, 1962; Pugh *et al.*, 1966) and yeast (Lynen, 1961; Klein, 1966) as well as from many plants (Overath & Stumpf, 1964; Brooks & Stumpf, 1966; Simoni, Criddle & Stumpf, 1967; Huang & Stumpf, 1971) and animal tissues (Bressler & Wakil, 1961; Hsu, Wasson & Porter, 1965; Yang *et al.*, 1967; Brady, Bradley & Trams, 1960;

* Not presented at the symposium.

Burton, Haavik & Porter, 1968; Brady, 1960; Martin, Horning & Vagelos, 1961; Larrabee *et al.*, 1965; Smith & Abraham, 1969). All enzyme systems require, in addition to malonyl CoA, TPNH as a reducing agent and an acyl CoA as a "priming" substrate of the synthetic process. A deeper insight into the detailed mechanism of fatty acid synthesis was first obtained in our studies on the purified enzyme system from yeast (Lynen, 1961, 1967) which led to the following scheme (Fig. 1). An essential feature of the biosynthetic

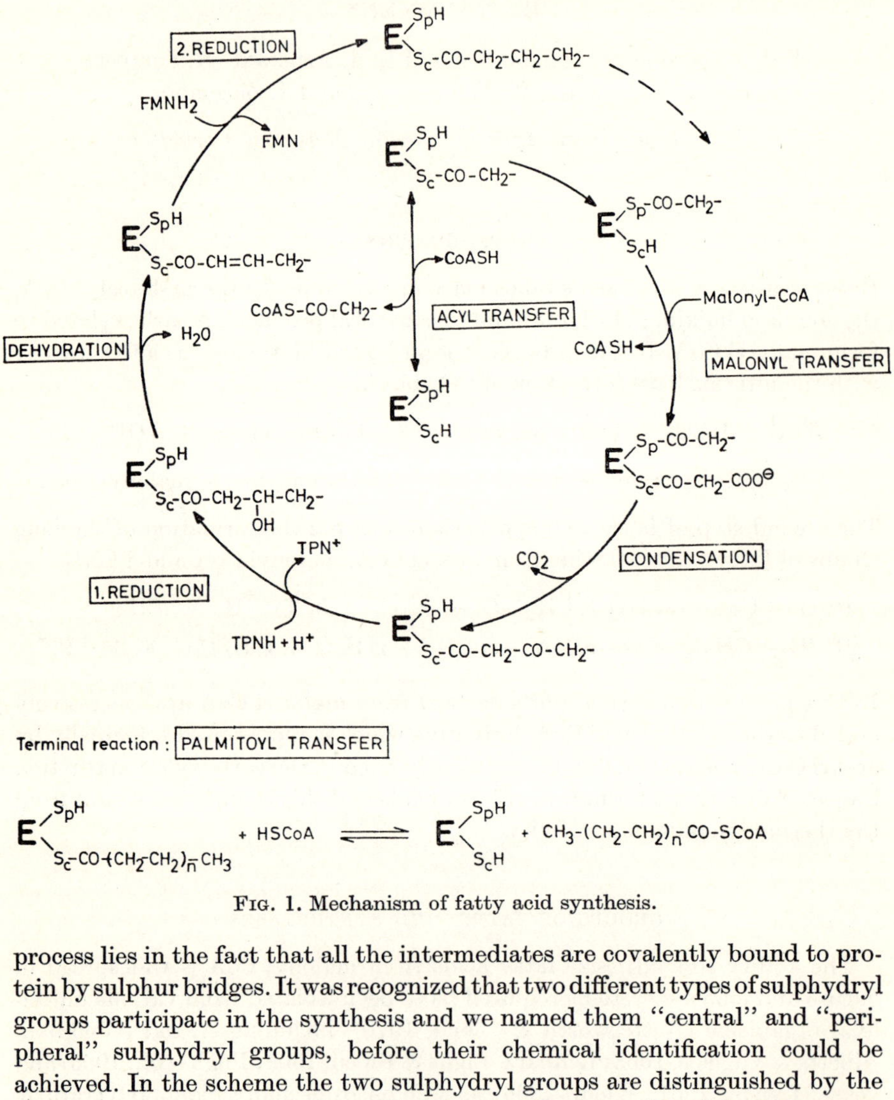

FIG. 1. Mechanism of fatty acid synthesis.

process lies in the fact that all the intermediates are covalently bound to protein by sulphur bridges. It was recognized that two different types of sulphydryl groups participate in the synthesis and we named them "central" and "peripheral" sulphydryl groups, before their chemical identification could be achieved. In the scheme the two sulphydryl groups are distinguished by the prefix "c" for the "central" and "p" for the "peripheral" sulphydryl groups and they carry the two substrates malonate and acetate, respectively. In these positions both acyl groups condense to form enzyme-bound acetoacetate with concomitant release of CO_2. The next steps consist of reduction of the enzyme-bound acetoacetate to β-hydroxybutyrate, which is dehydrated to crotonyl enzyme, which in turn is reduced to butyryl enzyme. The butyrate is not released but enters a corresponding reaction sequence instead of acetate. This cycle is repeated until a final carbon chain of 16–18 C atoms has been synthesized.

In the yeast system the process is terminated by a transfer of palmitic or stearic acid from the "central" sulphydryl group to free coenzyme A. The acids are thus released from the enzyme system. In the mammalian fatty acid synthetases the terminal reaction involves hydrolysis and in this case free fatty acids are produced instead of the CoA derivatives (Porter & Tietz, 1967; Burton *et al.*, 1968).

It is possible to arrange the fatty acid synthetases into two groups. One group includes stable multienzyme complexes having molecular weights between 450,000 and 2·5 million, in which all the individual enzymes of the pathways are assembled into a compact unit. To this group belong the fatty acid synthetases listed in Table 1. They are classified as multienzyme complexes because functionally and operationally they behave as single units. The fatty acid synthetases listed in Table 2 belong to the second category. They are systems only in the sense that they jointly catalyse a multistep metabolic pathway. At the cell-free level, however, the constituent enzymes fail to show

Table 1. *Fatty acid synthetases existing as multienzyme complexes*

Source	Mol. Weight	Products
Pigeon liver	$4{\cdot}5 \times 10^5$	C_{16}; free
Rat liver	$5{\cdot}4 \times 10^5$	C_{16}; free
Adipose tissue	?	C_{14}–C_{16}; free
Mammary gland	$(S_{20,w} = 12{\cdot}9)$	C_8–C_{18}; free
Yeast	$2{\cdot}3 \times 10^6$	C_{16}, C_{18}; CoA
Penicillium patulum	$\sim 2{\cdot}5 \times 10^6$	C_{16}, C_{18}; CoA
Euglena gracilis (etiolated)	$\sim 1 \times 10^6$	C_{16}; CoA
Mycobacterium phlei	$1{\cdot}7 \times 10^6$	C_{14}–C_{26}; esterified

Table 2. *Fatty acid synthetases existing as individual enzymes*

Source	Products
E. coli	C_{16}, C_{18}, $C_{16:1}$, $C_{18:1}$ β-OH-C_{10}–C_{14}; ACP
Clostridium species	β-OH-C_{10}–C_{14}; ACP
Pseudomonas species	β-OH-C_{10}–C_{14}; ACP
Bacillus subtilis	C_{15} and C_{17} iso- and anteiso
Avocado mesocarp	C_{16}, C_{18}
Lettuce chloroplasts	C_{16}, C_{18}
Spinach chloroplasts	C_{16}, C_{18}
Euglena gracilis (photoauxotropic)	C_{18}; ACP

any signs of physical interaction. The best studied example of this category is the enzyme system of *E. coli*. As the experiments in the laboratories of Vagelos (1964) and Wakil (Toomey *et al.*, 1965) have shown, all the individual enzymes of the pathway exist as monofunctional units and they readily separate on conventional protein fractionation.

ACYL CARRIER PROTEIN

Both groups of fatty acid synthetases contain protein-bound 4′-phosphopantetheine (acyl carrier protein = ACP) which represents the "central" SH-group and serves as a carrier of acyl groups during chain elongation (Fig. 2). In the first group of fatty acid synthetases the acyl carrier protein component is integrated into the multienzyme complexes (Larrabee *et al.*, 1965;

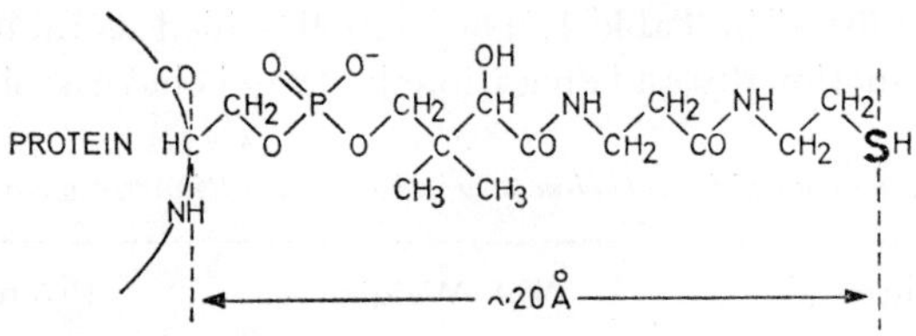

FIG. 2. 4′-Phosphopantetheine, the prosthetic group of acyl carrier protein.

Lynen *et al.*, 1968; Willecke, Ritter & Lynen, 1969; Chesterton *et al.*, 1968). In contrast, the bacterial and plant acyl carrier proteins (Table 2) are not firmly attached to the individual enzymes involved in fatty acid synthesis (Majerus, Alberts & Vagelos, 1964; Wakil, Pugh & Sauer, 1964; Brindley, Matsumara & Bloch, 1969; Overath & Stumpf, 1964; Brooks & Stumpf, 1966; Simoni *et al.*, 1967). Consequently, the acyl carrier protein derivatives themselves can migrate and serve as substrates in subsequent reactions (Nagai & Bloch, 1968; Goldfine, Ailhaud & Vagelos, 1967). In the stable multienzyme complexes, the fatty acids must be released from the complex-associated acyl carrier protein component either by transfer to coenzyme A or by hydrolysis.

1 10
NH_2-Ser-Thr-Ile-Glu-Glu-Arg-Val-Lys-Lys-Ile-Ile-Gly-Glu-

20
Gln-Leu-Gly-Val-Lys-Gln-Glu-Glu-Val-Thr-Asp-Asn-Ala-Ser-

30 * 40
Phe-Val-Glu-Asp-Leu-Gly-Ala-Asp-Ser-Leu-Asp-Thr-Val-Glu-

50
Leu-Val-Met-Ala-Leu-Glu-Glu-Glu-Phe-Asp-Thr-Glu-Ile-Pro-

60
Asp-Glu-Glu-Ala-Glu-Lys-Ile-Thr-Thr-Val-Gln-Ala-Ala-Ile-

70 77
Asp-Tyr-Ile-Asn-Gly-His-Gln-Ala-COOH

FIG. 3. Chemical structure of the *E. coli* acyl carrier protein.

Due to the work of Hill in Wakil's laboratory (Vanaman, Wakil & Hill, 1968) the amino-acid sequence of the acyl carrier protein of *E. coli* is now known (Fig. 3). The polypeptide contains 77 amino-acids and three serine residues. Serine in the position 36 carries the functional 4′-phosphopantetheine in phospho-diester linkage. Similar acyl carrier proteins were isolated from *M. phlei* (Matsumara, Brindley & Bloch, 1969) and from plant materials (Simoni *et al.*, 1967). The various proteins are heat-stable and contain abundant amounts of acidic amino-acids. Their total amino-acid compositions are remarkably similar (Simoni *et al.*, 1967). Further observations indicated that portions of the amino-acids around the 4′-phosphopantetheine may be similar in *E. coli*, Arthrobacter viscosus, and spinach (Matsumara & Stumpf, 1968).

Using a ^{14}C-pantothenate-labelled fatty acid synthetase from yeast Wells, Schultz & Lynen (1967) and Willecke *et al.* (1969) could demonstrate that 4′-phosphopantetheine also belongs to an individual structural element in the stable multienzyme complexes listed in Table 1. It turned out to be smaller than the other proteins aggregated in the complex and to be rather firmly attached to the complex. Dissociation of the complex with 3·5 M guanidine hydrochloride was necessary for the release of the C^{14}-labelled ACP-component.

Winnewisser in our laboratory recently studied the incorporation of ^{14}C-labelled pantothenic acid into various cell constituents of yeast. The yeast cells were aerobically grown in a medium containing pantothenic acid labelled in the β-alanine part and after harvest they were crushed in the cell homogenizer of Merkenschlager, Schlossmann & Kurz (1957). The crude extract was then passed through a column of Sephadex G 75. Figure 4 shows the radioactivity of the various fractions eluted from the column. Three radioactive fractions were obtained. The first fraction represents fatty acid synthetase with a molecular weight of 2·3 million. The last fraction, which contained most of the radioactivity, represents the low molecular weight coenzyme A and its precursors. The radioactive compound in the middle fraction could be identified as ACP. Winnewisser found that this yeast ACP could replace *E. coli* ACP in the CO_2-malonyl CoA-exchange reaction catalysed by *E. coli* enzymes, developed by Vagelos as routine assay for ACP (Majerus, Alberts & Vagelos, 1969). Using this assay Winnewisser embarked on the purification of yeast ACP, following the procedure elaborated for the purification of *E. coli* ACP. The most purified fraction obtained so far contains 0·136 ACP units/mg protein whereas according to Vagelos pure *E. coli* ACP contains 2·4 units. The lower specific activity of yeast ACP may be explained by the fact that its action was measured with *E. coli* enzymes. In this context it may be mentioned that spinach ACP, purified by Simoni *et al.* (1967) and measured with *E. coli* enzymes also contains only 0·14 ACP units/mg for the same reason.

The occurrence of free ACP in the crude yeast extract indicates that the synthesis of ACP and that of the various enzyme proteins of fatty acid synthetase are not fully coordinated. It is true that, although during the growth of yeast cells most of the ACP is incorporated into the multienzyme complex, small amounts of it remain free. This observation raises the interesting question regarding the regulation of synthesis of various components of fatty acid synthetase. At the present moment it is an open question. But it is probable

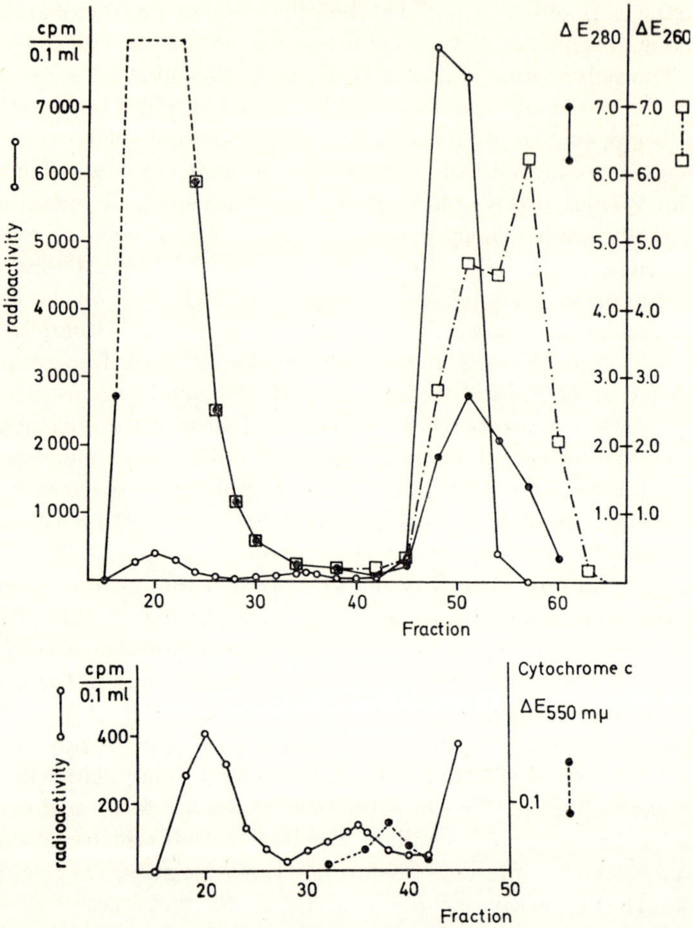

FIG. 4. Distribution of ^{14}C-pantothenate in various fractions of the crude yeast extract. 30 ml of crude extract prepared by breaking yeast cells grown in a medium containing ^{14}C-labelled pantothenic acid was chromatographed over Sephadex G-75 ($57 \times 3{\cdot}3$ cm) after the addition of 6 mg of cytochrome c. Fractions of 11·5 ml were collected and their radioactivity as well as cytochrome content was measured.

that genetic analysis of yeast fatty acid synthetase, recently started by Schweizer & Bolling (1970), will provide more insight into the control process.

The attachment of the carboxylic acid intermediate through 4′-phosphopantetheine seems to be very important for the functioning of the stable multienzyme complexes. In this way the system gains a flexible arm of about 20 Å length (Fig. 2) and this makes it possible to bring the carboxylic acid intermediates covalently bound to the sulphydryl group in contact with the active site of each component enzyme which might have only limited freedom of movement in the rather rigid structure of a multienzyme complex.

In order to build the carbon chain of stearic acid from malonyl CoA and the primer acetyl CoA at least 32 intermediates have to be passed through. All these intermediates are covalently bound to the enzyme system. If we assume for simplicity that the acyl carrier protein is present only once in the complex it follows that at any given moment only one of the 32 intermediates can be present (per multienzyme complex). From the known volume of the fatty acid synthetase the concentration of any intermediate was calculated and found to be 5×10^{-4} M. This locally high concentration of each intermediate facilitates the formation of the Michaelis-complexes and is reflected in the kinetics of the overall process.

The multienzyme complex may be visualized as strict compartmentalization of a metabolic multi-step reaction sequence in a small space. As a result the covalently bound intermediates within the complex are not accessible to competing processes, e.g. the enzymes of fatty acid degradation. According to this concept the multienzyme complexes represent intracellular factories in which the building blocks are assembled piece by piece and only the finished product is released from the complex. With this picture in mind it is tempting to draw an analogy between fatty acid synthetase of type I and the mitochondrion. The function of the mitochondrial membrane, which serves as a selective barrier, in

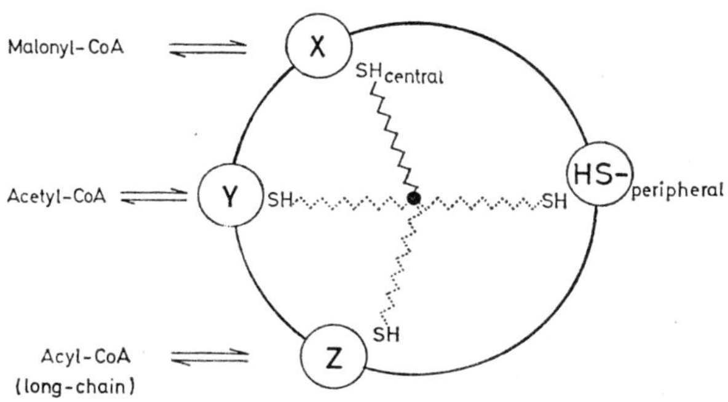

FIG. 5. Schematic representation of the functions of acyl-transferases in fatty acid synthetase. $SH_{central}$: ACP component. $SH_{peripheral}$: Functional group of the condensing enzyme. X, Y and Z: Active sites of acyltransferases.

the fatty acid synthetase is taken over by the acyl transferase enzyme components (see Fig. 5), which deliver acetyl and malonyl fragments to the complex and release the long-chain fatty acids from it. In fatty acid synthesis the transferases play the critical role in compartmentalization.

THE "PRIMING" REACTION

This immediately leads us to the specificity of the "priming" reaction in fatty acid synthesis and the reaction terminating the elongation of the carbon chain. The initial reaction effects a transfer of the "primer" acyl group from the SH-group of acyl CoA to the SH-group of the acyl carrier protein ("central" SH-group). In most of the fatty acid synthetases the enzyme which catalyses this acyl transfer has a fairly narrow specificity, the activity being high with acetyl CoA and decreasing with increasing chain length of the acid. In all these cases the products are mostly saturated straight-chain fatty acids (see Tables 1 and 2). Some microbial fatty acid synthetases produce unsaturated straight-chain fatty acids also (see Table 2). This depends upon the presence of a specific β-hydroxydecanoyl thiolester dehydrase (Brock, Kass & Bloch, 1967), which transforms β-hydroxy decanoyl-ACP into Δ^3-decenoyl-ACP. On the other hand *B. subtilis* synthesizes predominantly branched-chain fatty acids (Saito, 1960) and uses branched-chain acyl CoA esters such as isobutyryl CoA, isovaleryl CoA or 3-methyl butyryl CoA as "priming" substrates (Butterworth & Bloch, 1970). The products of synthesis therefore are C_{15} and C_{19} *iso*- or *anteiso* acids. The cell-free system from *B. subtilis* was unable to utilize acetyl CoA. However, if acetyl-ACP was used as the "priming" substrate, the initial acyl CoA:ACP acyl transferase was bypassed and acetate was rapidly incorporated into straight-chain fatty acids. These experiments of Butterworth & Bloch (1970) clearly demonstrated the important role the initial acyl-transferase of the various synthetases plays in dictating the nature of the fatty acids formed.

The specificity of yeast fatty acid synthetase in regard to the "priming" substrate was extensively studied by Pirson (1970) in our laboratory. Figure 6

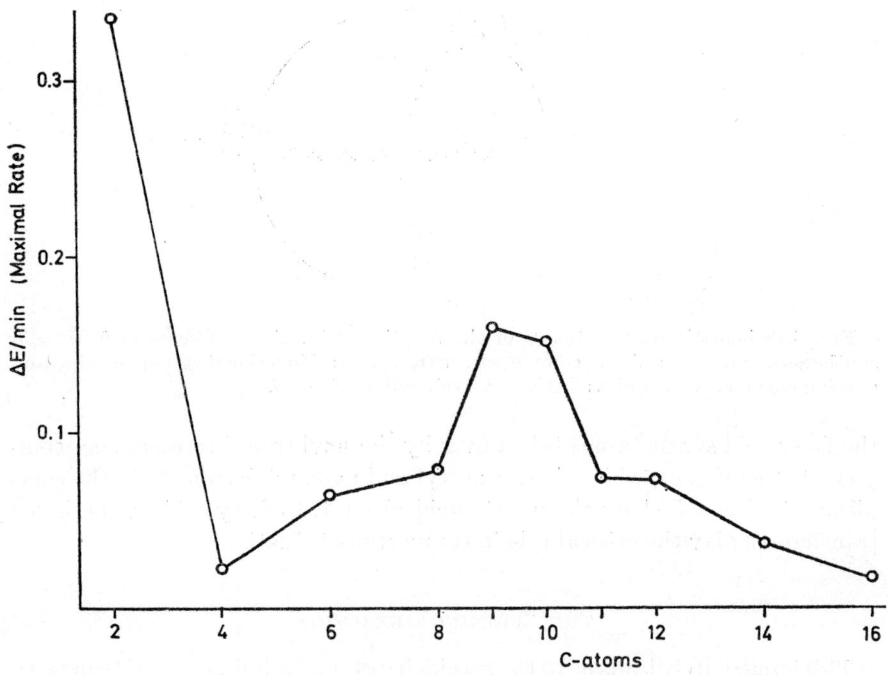

FIG. 6. Influence of the chain length of the "priming" acyl CoA on the activity of fatty acid synthetase. Standard incubation; ΔE/min indicates the rate of TPNH consumption measured at 334 nm and at optimal concentrations of acyl CoA.

summarizes his results. It can be seen that acetyl CoA is a far superior "priming" substrate. The capacity to prime fatty acid synthesis falls abruptly with the next higher homologues of acetyl CoA but then increases again and reaches a new maximum at decanoyl CoA indicating the presence of at least two acyl transferase components with different chain-length specificity. While studying the concentration dependence of the priming reaction with respect to long-chain acyl CoA derivatives an inhibitory effect at higher substrate concentrations became apparent (Fig. 7). It could be shown by kinetic analysis that this inhibition was competitive with respect to malonyl CoA (Lust & Lynen, 1968) indicating a competition between acyl and malonyl residues for the SH-carrier group of the acyl carrier protein component.

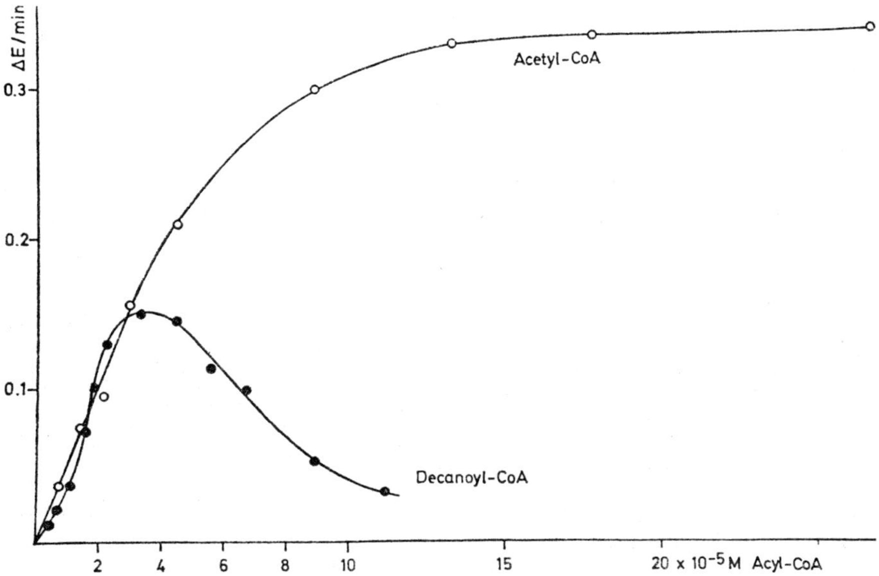

FIG. 7. Activity of acetyl CoA and decanoyl CoA as "priming" substrates. The cuvette (d = 1 cm) contained in a total volume of 2 ml: 200 μmoles of phosphate buffer, pH 6·5; 20 μmoles of cysteine; 0·6 mg of serum albumin; 0·3 μmoles of TPNH; 24·8 μg of fatty acid synthetase (2·2 enzyme units/mg) and the amount of acyl CoA, as given on the abscissa. The reaction was started by the addition of 0·13 μmoles of malonyl CoA. The rate of TPNH consumption (ΔE/min) was measured at T = 25° and λ = 334 nm and is plotted on the ordinate.

Table 3. *Relative capacity of decanoyl CoA and its derivatives to serve as "priming" substrates*

Priming substrate	Relative V_{Max}	Products
Decanoyl CoA	100	C_{12}–C_{18}; CoA
4-Oxodecanoyl CoA	40	Oxo-C_{12}–C_{18}; CoA (Hydroxy-C_{12}–C_{18})
7-Oxodecanoyl CoA	24	—
Δ^3-Decenoyl CoA	34	$C_{12:1}$–$C_{18:1}$; CoA (C_{12}–C_{18})
Δ^9-Decenoyl CoA	127	—
10-Hydroxydecanoyl CoA	8	—
D, L-3-Hydroxydecanoyl CoA	2	C_{12}–C_{18}; CoA
Δ^2-Decenoyl CoA	4	C_{12}–C_{18}; CoA

At the level of C_{10} straight-chain acids Pirson (1970) also studied the influence of functional groups of the hydrocarbon chain on the capacity to serve as "priming" substrates. It can be seen from Table 3 in which the relative rates at optimal substrate concentrations are summarized, that introduction of a hydroxy group or a double bond in the vicinity of the carboxyl group strongly depresses the "priming" activity. Substitutions at positions more distant from the carboxyl group however result in much less effects. The reaction products obtained with various priming substrates are listed in the last column of Table 3. It was found that functional groups further away from the thiolester group were mostly retained during chain elongation whereas the conjugated double bond or the β-hydroxy group such as in Δ^2-decenoic acid or 3-hydroxydecanoic acid were eliminated. This was to be expected, because the latter acids when bound to the acyl carrier protein component of the complex are known to be normal intermediates of fatty acid synthesis. The low "priming" activity of the corresponding CoA-derivatives could be explained by the slow transfer of the acyl residues from coenzyme A to the acyl carrier protein catalysed by the acyl transferase component. In passing it may be remarked that yeast fatty acid synthetase can also reduce the double bond in Δ^3-decenoyl CoA or the carbonyl group in 4-oxodecanoyl CoA at very slow rates. These slow side reactions therefore became apparent only when the "priming" substrate was preincubated with the enzyme complex in presence of TPNH before chain elongation was initiated by the addition of malonyl CoA. The enzyme components of yeast fatty acid synthetase responsible for these side reactions have not yet been identified.

TERMINATION OF CHAIN ELONGATION

Let us now turn to the rather puzzling problem as to why in the experiments, in which acetyl CoA was used as the "priming" substrate, all even numbered fatty acids from 4 to 18 C-atoms exist as enzyme-bound intermediates, whereas mainly palmitic and stearic acids are released from the complex. When propionyl CoA is substituted for acetyl CoA odd numbered fatty acids are produced. In this case a derivative with 17 carbon atoms is the principal product amounting to about 75% of the fatty acids formed (Fig. 8). Apparently the information to stop the synthesis specifically at the C_{17}-level must in some way be incorporated in the multienzyme complex. There are several explanations available for this phenomenon. In one of them it was assumed that the long-chain fatty acids were expelled from the enzyme complex as a result of lipophobic repulsion (Lynen, 1967). Sumper *et al.* (1969) have transformed this rather vague hypothesis into a model which considers some of the thermodynamic problems involved. In the model one considers the saturated acyl group of the elongating fatty acid to be attached to the "central" SH-group of the complex. In this situation the acyl group has two alternatives for further reaction. It can be reversibly transferred to the "peripheral" SH-group and can be elongated there by a C_2-unit derived from malonyl CoA, or it can be transferred via a transferase site from the "central" SH-group to coenzyme A and in this way it leaves the multienzyme complex.

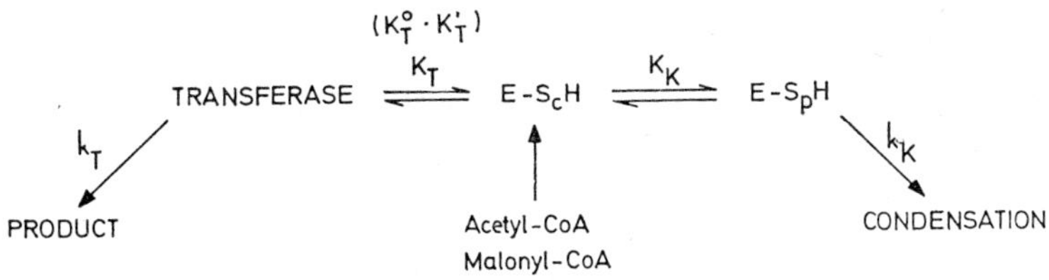

$$W = \frac{V_T}{V_T + V_K} = \frac{1}{1 + \frac{V_K}{V_T}} = \frac{1}{1 + c \cdot \frac{1}{K_T'}} = \frac{1}{1 + c \cdot e^{\frac{m\Delta F}{RT}}}$$

$$c \sim \frac{k_K \cdot K_K}{k_T \cdot K_T^o} \cdot \frac{[\text{Mal CoA}]}{[\text{Mal} \cdot \text{CoA} + \text{Ac-CoA}]}$$

This model is based on two assumptions that are supported by experimental evidence:

(a) the probability of any covalently enzyme-bound saturated acyl residue forming a product by transfer to coenzyme A is determined by the relative velocities of the condensing and transferring reactions;

(b) the elongating carbon chain interacts with the enzyme protein only after a chain length of 13 C-atoms has been attained; this interaction changes the relative velocities in favour of product formation by an energy increment of —0·9 Kcal for each additional methylene group beyond the thirteenth.

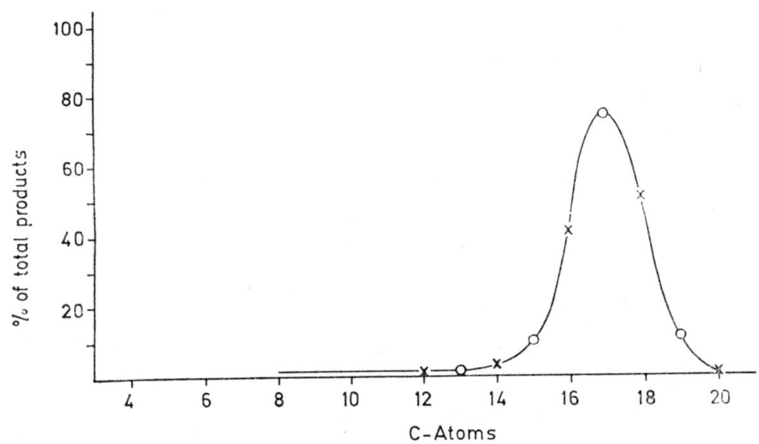

FIG. 8. Percentage distribution of products formed by fatty acid synthetase (Sumper *et al.*, 1969). The reaction mixture contained in a volume of 2 ml: 200 μmoles of phosphate buffer, pH 6·5; 0·8 mg of serum albumin; 1·2 μmoles of TPNH; 200 μg of fatty acid synthetase (1·5–2·5 enzyme units/mg) and either 0·25 μmoles of -acetyl CoA or 0·25 μmoles of propionyl CoA. $T = 25°$. The reaction was started by the addition of 0·5 μmoles of ^{14}C-malonyl-CoA and the rate of reaction was followed optically at $\lambda =$ 366 nm. The reaction was stopped at the linear region by the addition of 1 ml of 10% KOH in methanol. Separation of fatty acid methyl esters was by gas chromatography.

To calculate the probability of the product released at a particular chain length (P_n) the following equation was derived from the model describing quantitatively the observed product distribution.

$$P_n = \frac{1}{1 + \frac{k_K}{k_T} \cdot \frac{K_K}{K_T^0} \cdot \frac{[\text{Mal CoA}]}{[\text{Mal CoA} + \text{Ac CoA}]} \cdot e^{\frac{m \,.\, \Delta F}{RT}}}$$

where k_K and k_T are rate constants of condensation and transfer respectively, K_K and K_T^0 are equilibrium constants of acyl transfer between the "central" SH-group and the active sites of the condensing enzyme or transferase, respectively. ΔF represents the energy increment of —0·9 Kcal, that is the energy of interaction per methylene group. The variable m is the number of C-atoms in the elongating acyl chain minus 13.

The expression

$$\frac{[\text{Mal CoA}]}{[\text{Mal CoA} + \text{Ac CoA}]}$$

takes into consideration the fact that both malonyl and acetyl residues compete for the "central" SH-group but condensation can only occur if malonic acid is present there.

The formula suggests conditions under which either short-chain acyl CoA derivatives or exclusively stearyl CoA can be produced. Synthesis under these conditions was examined experimentally and results indicated that the formula can be applied to a wide range of experimental conditions (Sumper *et al.*, 1969).

In order to explain that the growing hydrocarbon chain interacts with the enzyme protein only after a chain length of 13 C-atoms has been attained the following hypothesis was put forward (Fig. 9). It was assumed that the transferase possesses two active sites, one of which binds to the thiolester group and must have the same affinity for all the saturated acyl intermediates. The other one, presumably an area of apolar amino-acid side chains, is so far away from the first site that it comes into play only after a critical length of the fatty acyl

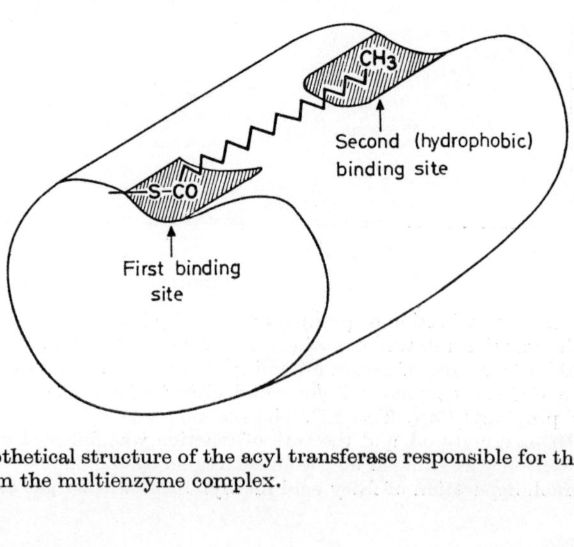

FIG. 9. Hypothetical structure of the acyl transferase responsible for the release of the fatty acids from the multienzyme complex.

chain has been exceeded. When this stage has been reached each additional methylene group in the fatty acid would add an energy increment to the hydrophobic interaction and thereby increase the affinity of the transferase for the acyl residue. A ΔF value of $-0{\cdot}9$ Kcal which permits maximal agreement between theory and experiment lies in the range of hydrophobic interactions as calculated by Nemethy & Scheraga (1962).

Most of the fatty acid synthetases behave like the yeast enzyme and with sufficient supply of malonyl CoA and TPNH preferentially produce C_{16} and C_{18} acids (see Tables 1 and 2). In the *E. coli* system unsaturated and β-hydroxy acids are also formed as acyl carrier protein derivatives, and this can be related to the fact that this enzyme system contains the specific β-hydroxydecanoyl thiolester dehydrase of Bloch (Brock *et al.*, 1967) and does not exist as a stable multienzyme complex. In the group of stable multienzyme complexes the enzyme system of *M. phlei* is exceptional in so far as it gives products with a biphasic chain-length distribution, with the maxima occurring at C_{18} and C_{24} (Brindley *et al.*, 1969). It is probable therefore that in this case two chain-length specific enzymes are involved in the termination of the elongating process.

SUBSTRATE BINDING SITES OF THE MULTIENZYME COMPLEX

In view of the deciding role the specific acyl transferases play in fatty acid synthesis by stable multienzyme complexes it was tempting to obtain further insight into the chemical make-up of their active sites. When we studied the transfer of the acyl moieties of ^{14}C-labelled acetyl CoA, malonyl CoA or palmityl CoA to the yeast enzyme complex where they remain in covalent linkage, it was found that only a portion of the acids was bound to the enzyme complex in thiolester linkage. The residual portion was bound to the enzyme in some other linkage, differing from thiolesters by their resistance towards cleavage on oxidation with performic acid.

By subjecting ^{14}C-malonyl-enzyme to pepsin digestion Schweizer *et al.* (1970*a*) could isolate two malonyl peptides, which proved to be stable against performic acid oxidation. One of them proved to be a heptapeptide containing the amino acids His, Ser, Glu, Gly, Gly, Ala and Leu. The other was apparently derived from the first peptide by further digestion, since it was short of two amino-acid residues (Glu and Gly). Leucine was the C-terminal amino-acid in this peptide. A detailed study of these two peptides indicated that most probably malonate is bound to the hydroxyl group of serine.

A similar study of the non-thiol acetate-binding site of the complex led Ziegenhorn (1970), after trypsin digestion of ^{14}C-acetyl enzyme at pH 6·9, to the isolation of two performic acid stable ^{14}C-acetyl peptides. By stepwise degradation with carboxypeptidase A the following amino-acid sequences were established:

N-^{14}C-acetyl-*N*-Ser-Gln-Gly-Leu-Val-Thr-Ala-Val

and *N*-^{14}C-acetyl-*N*-Ser-Gln-Gly-Leu-Val-Thr-Ala.

It is apparent that the second peptide is derived from the first peptide by enzymic release of the terminal valine residue. Since it is known that the acetate bound to the hydroxyl group of the amino-terminal serine very easily moves to the serine amino group, it is conceivable that in the ^{14}C-acetyl enzyme the acyl residue was originally bound to the hydroxyl group of serine. Experimental evidence for this assumption was obtained from kinetic studies in which the alkaline release of ^{14}C-acetate from acetyl-enzyme, from purified acetyl-peptides (obtained by peptic digestion of the acetyl-enzyme), and from the model of O-acetyl-pyrrolidonyl-seryl-glycine was measured and was found to be identical.

The absence of basic amino acids in the C-terminal position of the tryptic peptides now raises the question whether these peptides represent the amino-acid sequence at the carboxyl end of the acetyltransferase polypeptide chain. Since pretreatment of the acetyl-enzyme with carboxy-peptidase before digestion with trypsin leads to the same acetyl peptides this possibility seems to be excluded and therefore we may assume that the trypsin preparation used in these experiments still contained some other protease(s) as impurity which was (were) responsible for the hydrolysis of the peptide bonds at positions away from basic amino-acids. This could also explain as to why two acetyl-peptides so closely related were isolated from the tryptic digest of acetyl-enzyme.

When we measured the number of acetyl carrier sites in the yeast enzyme by titration with Ellman's reagent* or by simultaneous radioactive assay, some

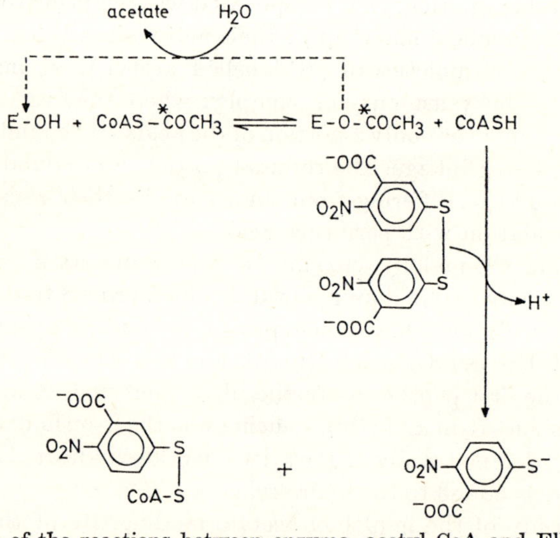

FIG. 10. Scheme of the reactions between enzyme, acetyl CoA and Ellman's reagent.

conflicting results were obtained (Fig. 10). According to radioactivity measurements five ^{14}C-acetate residues were transferred from ^{14}C-acetyl CoA to the enzyme complex in presence of Ellman's reagent. At the same time optical measurements indicated that about eight sulphydryl groups had interacted with the reagent (Sumper, 1970). This discrepancy could be explained if we assumed that about three of these sulphydryl groups belong to protein and are

* 5,5′-dithio-bis(nitrobenzoate).

exposed only as a result of some conformational changes induced by the binding of acetate. Earlier titration experiments with Ellman's reagent had shown that out of 140 cysteine residues present in native fatty acid synthetase only between five and ten react immediately with the reagent; others seem to be buried in the interior of the protein complex and become accessible only after denaturation of the polypeptide chains following the addition of urea or desoxycholate.

We thus come to the conclusion that about five non-sulphur acceptor sites are present per molecule of yeast fatty acid synthetase (molecular weight of 2·3 million). This value compares favourably with the presence of 5 ACP units and 5 FMN molecules per molecule of fatty acid synthetase as determined by microbiological analysis of bound pantothenic acid (Winnewisser, unpublished; see also Schweizer *et al.*, 1970*b*) and by spectrophotometric measurements, respectively (Oesterhelt, Bauer & Lynen, 1969).

One may therefore ask whether the synthetase as a whole could be a pentamer. At this moment we must leave this question open because of lack of more analytical data. Nevertheless it is worth mentioning in this context that fatty acid synthetase isolated from animal tissues has one fifth of the molecular weight of the yeast enzyme (500,000) and only one ACP unit per molecule (Yang *et al.*, 1967).

On the other hand, *E. coli* pyruvate dehydrogenase, another intensively studied multienzyme system, which comprises three different enzyme components, contains these enzymes in widely differing molecular amounts (Reed & Cox, 1966) and this may also be true for the composition of fatty acid synthetase.

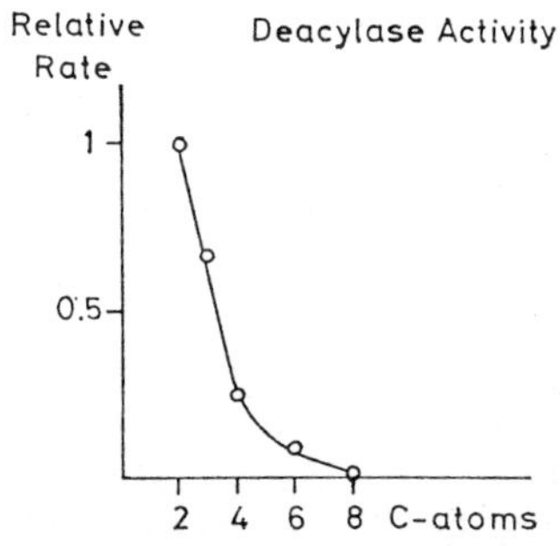

FIG. 11. Dependence of deacylase activity on the chain length of the carboxylic acid.

When Sumper (1970) studied the stoichiometry of acetyl transfer between acetyl CoA and the nonsulphur acceptor sites of the complex using Ellman's reagent [5,5′-dithio-bis(nitrobenzoate)] as an indicator, he found that the acetyl-enzyme produced is slowly hydrolysed to form acetate and the free enzyme. The regenerated enzyme is then re-charged with acetate from acetyl CoA (see Fig. 10). As a result the reagent continues to be used up until the acetyl CoA added is completely hydrolysed. Using similar experimental techniques Sumper (1970) also measured the rates of hydrolysis of propionyl-enzyme, butyryl-enzyme, etc. His findings are summarized in Fig. 11, which shows that the reaction rate gradually slows down with increasing chain length

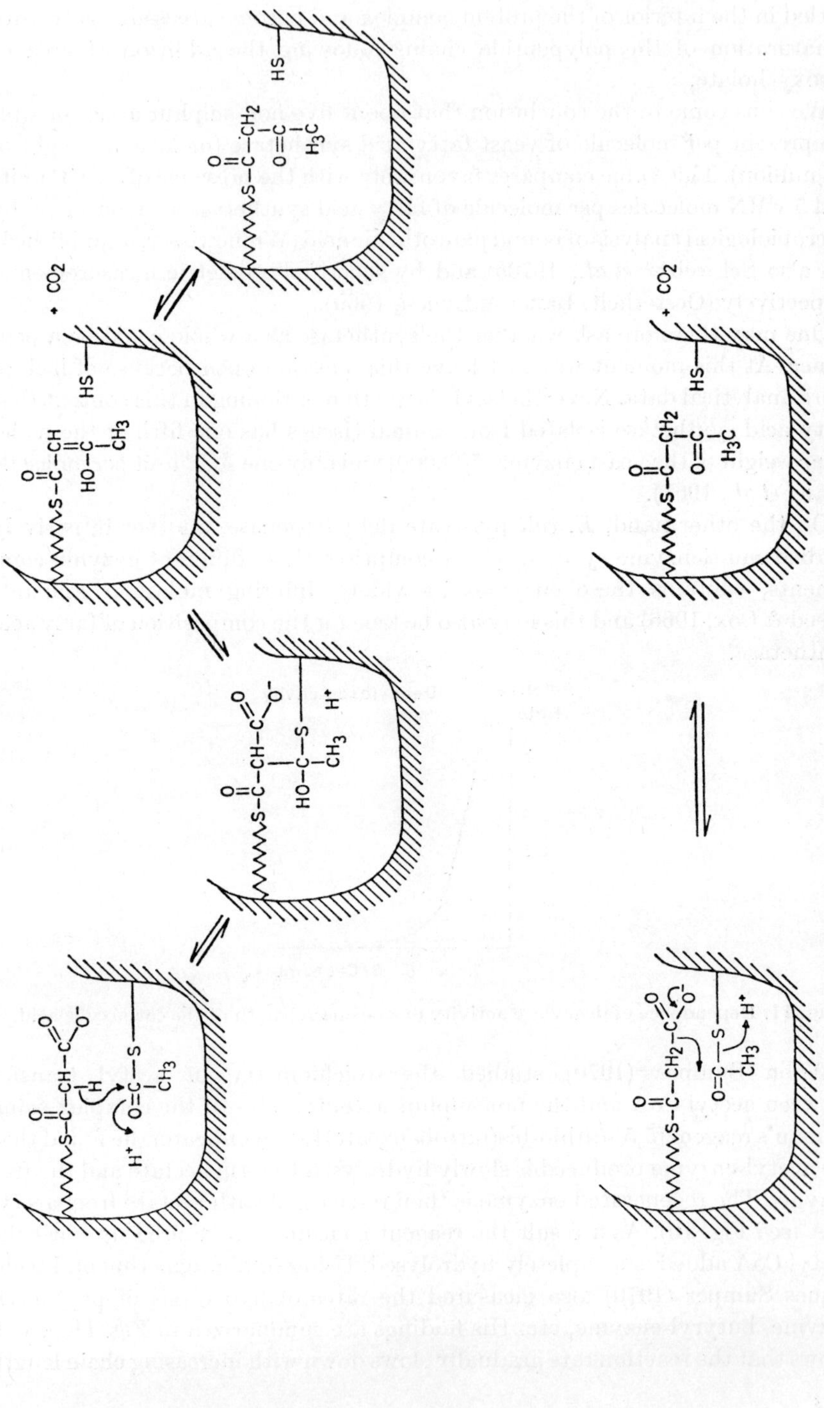

FIG. 12. Mechanism of the formation of acetoacetic acid by condensation of acetyl and malonyl thiolesters.

of the carboxylic acid and becomes negligible at the stage of the octanoyl derivative.

The participation of serine at the active sites of the acyl-transferase components of fatty acid synthetase was also demonstrated in experiments with the multienzyme complex from avian liver (Phillips *et al.*, 1970; Joshi, Plate & Wakil, 1970) and with the purified malonyl transferase of *E. coli* (Joshi & Wakil, 1971).

CHEMICAL MECHANISM OF THE CONDENSING REACTION

From a chemical standpoint the condensation reaction between acetate and malonate may be considered to be acylation of a malonic ester. The methylene anion of the malonyl thiolester, which is known to be more nucleophilic than the methyl anion of an acetyl thiolester, adds to the electrophilic carbonyl carbon of the sulphur-bound carboxylic acid, as shown in the upper half of Fig. 12. The intermediate formed is converted into the β-oxoacetyl derivative by a subsequent elimination of thiol and carbon dioxide. An alternative reaction mechanism might be the concerted process shown in the lower half of Fig. 12.

Recently we embarked upon experiments in order to establish which of the two mechanisms operates. Because the first mechanism involves a deprotonation of the malonic acid to form an anion for the subsequent nucleophilic attack on the carbonyl carbon, Arnstadt (unpublished work) in our laboratory studied the effect of exchanging the hydrogen atoms in malonyl CoA by deuterium atoms on the kinetics of fatty acid synthesis. Such experiments should be informative because we had learned from many previous studies of fatty acid synthesis with the purified multienzyme complex that the condensation reaction is the rate-limiting step in the overall process (Oesterhelt, 1967; Sumper *et al.*, 1969). If the upper mechanism were correct, one would expect a kinetic isotope effect of about 1·4, as measured by comparing maximum reaction velocities with acetyl CoA and trideutero-acetyl CoA in the malate synthase reaction, where the deprotonation is rate-limiting (see Cornforth *et al.*, 1969). Arnstadt prepared dideutero-malonyl CoA by dissolving malonyl CoA in heavy water and by incubating the mixture for several hours. Using NMR-spectroscopy it was found that the half-life of the deuterium exchange with the methylene group of thiolbound malonic acid at pH 6·5–7·5(25°) was in the range of 10 min. Therefore, in the experiments with the enzyme, where initial rates were measured, the loss of deuterium from dideutero-malonyl CoA was negligible. When he compared the rates of fatty acid synthesis with malonyl CoA and dideutero-malonyl CoA he did not find any difference (Arnstadt, unpublished work). These results exclude a primary isotope effect and suggest very strongly that the condensation reaction may follow the concerted process assuming deprotonation is rate-determining.

Further evidence for the concerted process was obtained from experiments with purified enzymes of *E. coli*. In these experiments Arnstadt studied the formation of β-hydroxybutyryl-ACP starting from acetyl-ACP, malonyl-ACP and TPNH in the presence of rate-limiting amounts of the condensing enzyme and an excess of β-ketoacyl-ACP reductase, the enzyme responsible for the first reduction (see Fig. 1). Under these conditions acetoacetyl-ACP formed by condensation of acetyl-ACP and malonyl-ACP cannot accumulate, but is immediately reduced by TPNH to form β-hydroxybutyryl-ACP. In the presence of tritiated water incorporation of tritium into the β-hydroxybutyric acid is to be expected if the upper reaction mechanism proves right, whereas no tritium should be incorporated if condensation occurs in the concerted manner (Fig. 12).

Arnstadt isolated the β-hydroxybutyric acid by gas chromatography after alkaline hydrolysis of the ACP-derivative and found it to be practically free of

Table 4. *Formation of β-hydroxybutyryl-ACP in the presence of tritiated water*

Experiment No.	nmoles of acetyl-ACP	nmoles of malonyl-ACP	radioactivity of β-hydroxybutyrate (d.p.m.)
1	—	—	200–400
2	10	10	280
3	20	20	303
4	40	40	620
5	40	40	23170

The reaction mixture contained in a total volume of 0·5 ml: 50 μmoles of potassium phosphate buffer, pH 7·0; 5 μmoles of thioglycol; 0·1 μmole of ethylenediamine tetraacetate, 1 μmole of TPNH; 170 milliunits of purified condensing enzyme from *E. coli*; 1000 milliunits of β-ketoacyl-ACP reductase from *E. coli*, tritiated water corresponding to $10{\cdot}4 \times 10^{10}$ d.p.m., and acetyl-ACP and malonyl-ACP, as indicated in the table. In experiments 1–4, the mixture was incubated for 2 min at 25°. The incubation was then stopped and the β-hydroxybutyrate was isolated by gaschromatography after alkaline hydrolysis and addition of carrier material. In experiment 5 the reaction mixture was incubated for 2 min in the absence of reductase, and then for an additional 2 min after the addition of reductase.

radioactivity (Table 4). However, if the formation of β-hydroxybutyryl-ACP took place in two separate steps with the intermediate accumulation of actoacetyl-ACP the isolated product should contain much tritium (Table 4). In these experiments the mixture of acetyl-ACP, malonyl-ACP and TPNH was preincubated for 2 min with the condensing enzyme alone before the reductase was added. The uptake of tritium in this case can be explained by the rapid exchange of the methylene hydrogens of the acetoacetyl-ACP accumulated in the first phase with the tritium atoms of the solvent.

SUBUNIT STRUCTURE OF FATTY ACID SYNTHETASES

The two extremes of fatty acid synthetase, those that represent tightly-aggregated multienzyme complexes and those that lack any higher order of structural organization might represent two stages in the evolution of the enzyme system. According to this concept the synthesis of fatty acids from malonyl CoA, TPNH and acetyl CoA might originally have been carried out by separate enzymes with an easily dissociable acyl carrier protein as in the case of the *E. coli* and related systems (Table 2). In the course of evolution the various component enzymes and the acyl carrier protein came to be associated in the form of fatty acid synthetase in which the components are firmly bound to each other as in the multienzyme complexes (Table 1). Therefore, the various protein components in the complexes must possess both: (1) catalytic sites responsible for the enzyme activity, and (2) binding or recognition sites responsible for the interaction with other proteins to form a specific spatial structure.

In this context the work of Delo, Ernst-Fonberg & Bloch (1971) on both etiolated and green cells of *Euglena gracilis* should be mentioned. Fatty acid synthetase isolated from dark-grown cells is a complex of high molecular weight that does not require the addition of acyl carrier protein. In contrast, cells grown in the light yield two distinct synthetases. One of them has the same properties as the synthetase from the dark-grown cells. The second, which appears to be dependent on the protein synthesis machinery of the chloroplast (Ernst-Fonberg & Bloch, 1971), requires the addition of acyl carrier protein and generally resembles the corresponding *E. coli* and plant enzymes.

The stability of the various synthetase complexes varies. As for example successful isolation of the complex from *M. phlei* requires an environment of

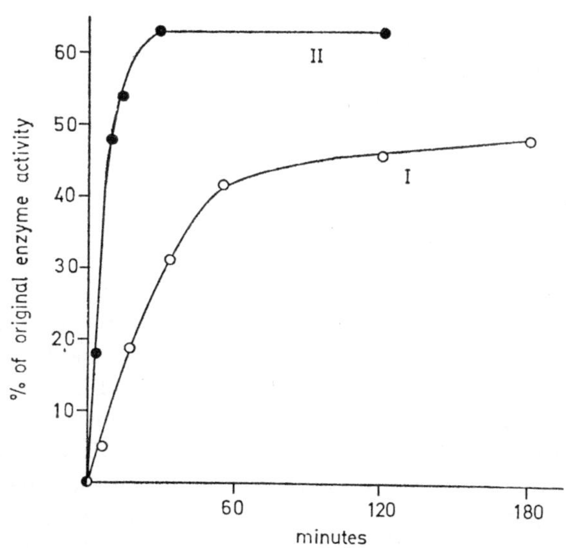

FIG. 13. Kinetics of reconstitution of fatty acid synthetase and reductase activities (Sumper, Riepertinger and Lynen, 1969). Fatty acid synthetase, dissolved in 1·0 M LiCl, was dissociated by two freezings and thawings. The inactive enzyme solution was diluted five-fold by the addition of 0·1 M phosphate, pH 7·5, containing 0·04 M cysteine, and incubated at 20°. Samples were taken at several time intervals, as given in the abscissa, and their enzymic activities were measured.

(I) Fatty acid synthetase activity
(II) Reductase activity

high ionic strength. When the buffer concentration is lowered from 0·1 M to 0·01 M phosphate the activity is rapidly lost and, as indicated by sucrose density centrifugation, is probably due to fragmentation of the complex into units of smaller sizes (Brindley *et al.*, 1969). The pigeon-liver synthetase also looses its activity and dissociates into subunits in dilute buffer solutions (Hsu, Wasson & Porter, 1965; Yang *et al.*, 1965). The yeast multienzyme complex appears to be much more stable and requires more drastic treatment for its dissociation (Sumper, Riepertinger & Lynen, 1969). The native yeast fatty acid synthetase can be dissolved in 1·0 M NaCl or LiCl without loss of its enzymatic activity or changes in its sedimentation and electrophoretic behaviours. Freezing of such solutions two or more times for 15 min at —70° followed by thawing at room temperature results in a drastic alteration of enzymatic and physical properties. After a single freezing and thawing the appearance of more slowly sedimenting species can be noted. After two freezings the native fatty acid synthetase is no longer present. It was found that the dissociation of fatty acid synthetase into subunits occurs stepwise, first to a species having a molecular weight of about 500,000 and then to a species of molecular weight about 250,000.

When the native fatty acid synthetase is no longer present enzymatic activity disappears. Reactivation occurs on decreasing ionic strength by dilution or by dialysis (Sumper, Riepertinger & Lynen, 1969; Sumper, 1970). The kinetics of reactivation are shown in Fig. 13. It was found that the reactivation process is strongly temperature dependent and is also influenced by the protein concentration, as expected. When the reconstituted enzyme was compared with the native fatty acid synthetase no differences in the sedimentation behaviour could be observed (Sumper, Riepertinger & Lynen, 1969).

Since synthesis depends on the proper functioning of all component enzymes, loss of total synthetic activity does not imply loss of all partial reactions. Indeed it was found that acetyl and malonyl transferase activities and the activities of the enzymes responsible for the formation and reduction of unsaturated acids, as measured with model substrates (Lynen, 1961; 1969), were not impaired by the freezing and thawing procedure (Weithmann, unpublished). On the other hand, the activity of the enzyme responsible for the reduction of the β-keto acid was completely lost after such treatment. The kinetics of reconstitution of this enzyme are also shown in Fig. 13. As can be seen the partial activity reappears more quickly than total synthetase activity. This observation proves that the loss of the reductase activity alone cannot be responsible for the loss of the overall activity. The dissociation of the complex must also be responsible. But why did the reductase activity sink to zero after dissociation of the complex ? One of the explanations could be that the enzymatically active conformation of the reductase depends on specific protein-protein interactions with neighbouring enzymes in the complex which become physically separated during dissociation.

REFERENCES

Brady, R. O., (1960). *J. biol. Chem.* **235**, 3099.

Brady, R. O., Bradley, R. M. & Trams, E. G. (1960). *J. biol. Chem.* **235**, 3093.

Bressler, R. & Wakil, S. J. (1961). *J. biol. Chem.* **236**, 1643.

Brindley, D. N., Matsumara, S. & Bloch, K. (1969). *Nature* **224**, 666.

Brock, D. J. H., Kass, L. R. & Bloch, K. (1967). *J. biol. Chem.* **242**, 4432.

Brooks, J. L. & Stumpf, P. K. (1966). *Archs. Biochem. Biophys.* **116**, 108.

Burton, D. N., Haavick, A. G. & Porter, J. W. (1968). *Archs. Biochem. Biophys.* **126**, 141.

Butterworth, P. H. W. & Bloch, K. (1970). *European J. Biochem.* **12**, 496.

Chesterton, C. J., Butterworth, P. H. W., Abramovitz, A. S., Jacob, E. J. & Porter, J. W. (1968). *Archs. Biochem. Biophys.* **124**, 386.

Cornforth, J. W., Redmond, J. W., Eggerer, H., Buckel, W. & Gutschow, C. (1969). *Nature* **221**, 1212.

Delo, J., Ernst-Fonberg, M. L. & Bloch, K. (1971). *Archs. Biochem. Biophys.* **143**, 384.

Ernst-Fonberg, M. L. & Bloch, K. (1971). *Archs. Biochem. Biophys.* **143**, 392.

Goldfine, H., Ailhaud, G. P. & Vagelos, P. R. (1967). *J. biol. Chem.* **242**, 4466.

Goldman, P., Alberts, A. W. & Vagelos, P. R. (1963). *J. biol. Chem.* **238**, 1255.

Huang, K. P. & Stumpf, P. K. (1971). *Archs. Biochem. Biophys.* **143**, 412.

Hsu, R. Y., Wasson, G. & Porter, J. W. (1965). *J. biol. Chem.* **240**, 3736.

Joshi, V. C., Plate, C. A. & Wakil, S. J. (1970). *J. biol. Chem.* **245**, 2857.

Joshi, V. C. & Wakil, S. J. (1971). *Archs. Biochem. Biophys.* **143**, 493.

Klein, H. P. (1966). *J. Bacteriol.* **92**, 130.

Larrabee, A. R., McDaniel, E. G., Backerman, H. A. & Vagelos, P. R. (1965). *Proc. natn. Acad. Sci. U.S.A.* **54**, 267.

Lennarz, W. J., Light, R. J. & Bloch, K. (1962). *Proc. natn. Acad. Sci. U.S.A.* **48**, 840.

Lust, G. & Lynen, F. (1968). *European J. Biochem.* **7**, 68.

Lynen, F. (1961). *Fedn. Proc. Fedn. Am. Socs. exp. Biol.* **20**, 941.

Lynen, F. (1967). *Biochem. J.* **102**, 381.

Lynen, F. (1969). In *Methods in Enzymology*, Vol. 14, p. 17 (Ed. by Colowick, S. P. and Kaplan, N. O.). New York and London: Academic Press Inc.

Lynen, F., Oesterhelt, D., Schweizer, E. & Willecke, K. (1968). In *Cellular Compartmentalization and Control of Fatty Acid Metabolism*. Oslo: Universitetsforlaget.

Majerus, P. W., Alberts, A. W. & Vagelos, P. R. (1964). *Proc. natn. Acad. Sci. U.S.A.* **51**, 1231.

Majerus, P. W., Alberts, A. W. & Vagelos, P. R. (1969). In *Methods of Enzymology*, Vol. 14, p. 43 (Ed. by Colowick, S. P. and Kaplan, N. O.), New York and London: Academic Press Inc.

Martin, D. M., Horning, M. G. & Vagelos, P. R. (1961). *J. biol. Chem.* **236**, 663.

Matsumara, S., Brindley, D. N. & Bloch, K. (1969). *Biochem. biophys. Res. Commn.* **38**, 369.

Matsumara, S. & Stumpf, P. K. (1968). *Archs. Biochem. Biophys.* **125**, 932.

Merkenschlager, M., Schlossmann, K. & Kurz, W. (1957). *Biochem. Z.* **329**, 332.

Nagai, J. & Bloch, K. (1968). *J. biol. Chem.* **243**, 4626.

Nemethy, G. & Scheraga, H. A. (1962). *J. chem. Phys.* **36**, 3401.

Oesterhelt, D. (1967). Ph.D. Thesis, University of Munich.

Oesterhelt, D., Bauer, H. & Lynen, F. (1969). *Proc. natn. Acad. Sci. U.S.A.* **63**, 1377.

Overath, P. & Stumpf, P. K. (1964). *J. biol. Chem.* **239**, 4103.

Phillips, G. T., Nixon, J. E., Dorsey, J. A., Butterworth, P. H. W., Chesterton, C. J. & Porter, J. W. (1970). *Archs. Biochem. Biophys.* **138**, 380.

Pirson, W. (1970). Ph.D. Thesis, University of Munich.

Porter, J. W. & Tietz, A. (1957). *Biochim. biophys. Acta* **25**, 41.

Pugh, E. L., Sauer, F., Waite, M., Toomey, E. R. & Wakil, S. J. (1966). *J. biol. Chem.* **241**, 2635.

Reed, L. J. & Cox, D. J. (1966). *Ann Rev. Biochem.* **35**, 57.

Saito, K. (1960). *J. Biochem.* (*Tokyo*) **47**, 710.

Schweizer, E. & Bolling, H. (1970). *Proc. natn. Acad. Sci. U.S.A.* **67**, 660.

Schweizer, E., Willecke, K., Winnewisser, W. & Lynen, F. (1970*a*). *Vit. Horm.* **28**, 329.

Schweizer, E., Piccinini, F., Duba, C., Günther, S., Ritter, E. & Lynen, F. (1970*b*). *Eur. J. Biochem.* **15**, 483.

Simoni, R. D., Criddle, R. S. & Stumpf, P. K. (1967). *J. biol. Chem.* **242**, 573.

Smith, S. & Abraham, S. (1969). *Fedn. Proc. Fedn. Am. Socs. exp. Biol.* **28**, 537.

Sumper, M. (1970). Ph.D. Thesis, University of Munich.

Sumper, M., Riepertinger, C. & Lynen, F. (1969). *FEBS Lett.* **5**, 45.

Sumper, M., Oesterhelt, D., Riepertinger, C. & Lynen, F. (1969). *Eur. J. Biochem.* **10**, 377.

Toomey, R. E., Waite, M., Williamson, I. P. & Wakil, S. J. (1965). *Fedn. Proc. Fedn. Am. Socs. exp. Biol.* **24**, 290.

Vagelos, P. R. (1964). *Ann. Rev. Biochem.* **33**, 139.

Vanaman, T. C., Wakil, S. J. & Hill, R. L. (1968). *J. biol. Chem.* **243**, 6420.

Wakil, S. J., Pugh, E. L. & Sauer, F. (1964). *Proc. natn. Acad. Sci. U.S.A.* **52**, 106.

Wells, W. W., Schultz, F. & Lynen, F. (1967). *Biochem. J.* **346**, 474.

Willecke, K., Ritter, E. & Lynen, F. (1969). *Eur. J. Biochem.* **8**, 503.

Yang, P. C., Bock, R. M., Hsu, R. Y. & Porter, J. W. (1965). *Biochim. biophys. Acta* **110**, 608.

Yang, P. C., Butterworth, P. H. W., Bock, R. M. & Porter, J. W. (1967). *J. biol. Chem.* **242**, 3501.

Ziegenhorn, J. (1970), Ph.D. Thesis, University of Munich.

REGULATORY MECHANISMS FOR LIVER ACETYL COENZYME A CARBOXYLASE

By SHOSAKU NUMA, TAKASHI HASHIMOTO, SHIGETADA NAKANISHI AND TAKESHI OKAZAKI

Department of Medical Chemistry, Kyoto University Faculty of Medicine, Kyoto, Japan

ACETYL-CoA carboxylase (acetyl-CoA:CO_2 ligase (ADP), EC 6.4.1.2), which catalyses the formation of malonyl-CoA, plays a critical role in the control of long-chain fatty acid synthesis (see Numa *et al.*, 1970). There is accumulating evidence which indicates that the regulation at this carboxylation step is achieved both by changes in the quantity of the carboxylase and by changes in the catalytic efficiency per carboxylase molecule (see Numa *et al.*, 1970). It appears that the former mechanisms are involved mainly in the long-term control, while the latter mechanisms are more responsible for the short-term control. The present paper deals with our recent findings on the regulatory mechanisms for the quantity as well as the catalytic efficiency of liver acetyl-CoA carboxylase.

REGULATION OF THE QUANTITY OF ACETYL-COA CARBOXYLASE

In order to study the regulatory mechanisms for the quantity of liver acetyl-CoA carboxylase, we used animals that were subjected to different metabolic conditions associated with either increased or decreased lipogenesis. These were normal rats and mice fed a balanced diet, rats fasted for 48 hr or more, rats fasted for 48 hr and subsequently refed a fat-free diet for 72 hr or more, rats suffering from alloxan-diabetes, and genetically obese hyperglycemic mice (C57BL/6J-*ob*) carrying a single recessive mutant gene for obesity (Ingalls, Dickie & Snell, 1950). It is well established that the level of acetyl-CoA carboxylase activity in liver cytosol is lowered in fasted and in alloxan-diabetic rats as compared with that in normal animals (Numa, Matsuhashi & Lynen, 1961; Wieland *et al.*, 1963), whereas the level is elevated in refed rats and in obese mice (Allmann, Hubbard & Gibson, 1965; Chang *et al.*, 1967). In the work described in the present paper, we investigated the questions whether the changing levels of the carboxylase activity are actually determined by changing quantities of the enzyme protein, and if so, whether these changes are due to variations in the rate of enzyme synthesis or in that of enzyme degradation. (Numa, Nakanishi & Iritani, 1969; Nakanishi & Numa, 1970; Nakanishi & Numa, 1971). For this purpose, we employed immunochemical techniques with the use of antibody prepared against rat liver acetyl-CoA carboxylase which was purified to homogeneity; this antibody was found to cross-react with the carboxylase from mouse liver also.

Figure 1 represents immunochemical titration of acetyl-CoA carboxylase from livers of obese and normal mice. Despite the fact that the specific activity

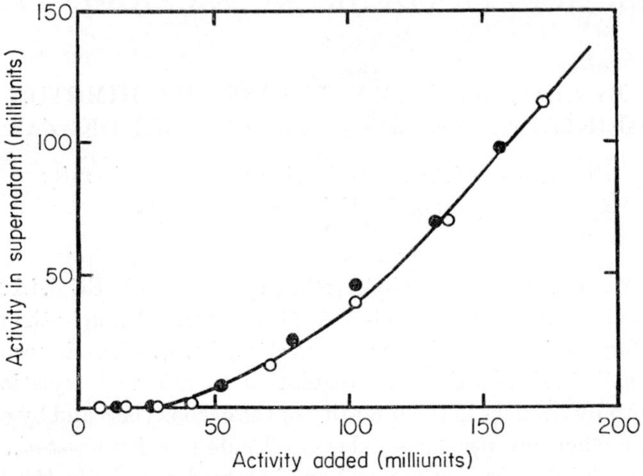

FIG. 1. Immunochemical titration of acetyl-CoA carboxylase from livers of obese and normal mice. To 90 μg of antiacetyl-CoA carboxylase γ-globulin were added increasing amounts of liver cytosol. Following completion of precipitation, the supernatant fluids were assayed for carboxylase activity. ○, obese mice; ●, normal mice. From Nakanishi & Numa (1971).

(or total activity) of the carboxylase in liver cytosol from obese mice was 3- to 4-fold (or 6- to 10-fold) higher than the normal value, the equivalence point was the same for both types of liver cytosol when based on the amount of enzyme activity added. This showed that the change in the level of carboxylase activity was accompanied by a proportionate change in the quantity of the immunochemically reactive protein. Similar results were also obtained with liver cytosol preparations derived from normal, fasted, refed and alloxan-diabetic rats. These findings, together with the results of Ouchterlony double diffusion analyses, indicated that the changes in the level of carboxylase activity in liver cytosol observed under the aberrant metabolic conditions studied can be ascribed to changes in the quantity of the enzyme protein, which is indistinguishable from that derived from normal animals. Such a conclusion based on immunological evidence was further supported by the results of kinetic and heat inactivation studies carried out with partially purified carboxylase preparations from obese and normal mice (Table 1). Thus, it is suggested that the mutation causing obesity does not affect the structure of the carboxylase, but alters the enzyme concentration in the liver.

The quantity of an enzyme is affected by changes in the rates of its synthesis and/or degradation. Under steady-state conditions, the content of an enzyme is related to these rates as follows: $E = k_s/k_d$, where E is the content of enzyme per mass, k_s is a zero-order rate constant of synthesis per mass, and k_d is a first-order rate constant of degradation expressed as the reciprocal of time (Schimke

Table 1. *Some properties of liver acetyl-CoA carboxylase from obese and normal mice*

Property		Obese	Normal
Michaelis constant	ATP	12 μM	12 μM
	Acetyl-CoA	32 μM	19 μM
	Citrate	4·3 mM	5·0 mM
Palmityl-CoA concentration required for 50% inhibition		5 μM	6 μM
$t_{\frac{1}{2}}$ of heat inactivation at 45°C		6 min	6 min

Data taken from Nakanishi & Numa (1971).

& Doyle, 1970). In order to explore the mechanisms underlying the observed changes in the quantity of acetyl-CoA carboxylase, we estimated the rates of its synthesis and degradation in livers of the various animals. As a measure of the rate of synthesis, the extent of [^{3}H]leucine incorporation into the protein precipitated by antiacetyl-CoA carboxylase after pulse-labelling was determined. As a measure of the rate of degradation, the rate of loss of radioactivity from the prelabelled carboxylase was determined. Table 2 summarizes the

Table 2. *Synthesis and degradation of acetyl-CoA carboxylase in livers of animals under different metabolic conditions*

Animal	Conditions	Enzyme content per liver (E) (%)	Rate of synthesis per liver (k_s) (%)	Rate constant of degradation (k_d) (%)	k_s/k_d (%)
Rats	Normal	100*	100	100‡	100
	Refed	376	405	107	378
	Alloxan-diabetic	53	54	100	54
	Fasted	28	59	190	31
Mice	Normal	100†	100	100§	100
	Obese	1 020	775	58	1 340

* 4·18 units.
† 0·36 unit.
‡ $t_{\frac{1}{2}}$ = 59 hr.
§ $t_{\frac{1}{2}}$ = 67 hr.

Data taken from Nakanishi & Numa (1970, 1971).

results of these studies. It is evident that the rate constant of degradation is essentially the same in normal, refed and alloxan-diabetic rats, and that the increase or decrease in the carboxylase content in refed or alloxan-diabetic rats is attributed to accelerated or retarded synthesis. The increase in the carboxylase content in the genetically obese mice is due mainly to a rise in the rate of synthesis, and to a minor extent, to a decrease in the rate of degradation.

On the other hand, the decrease in the carboxylase content in fasted rats is ascribed both to a diminished rate of synthesis and to an increased rate of degradation. In refed and alloxan-diabetic rats as well as in obese mice, all of which, like normal animals, can be assumed to be in steady states, the values for the enzyme content predicted from the ratio k_s/k_d agree fairly well with the actually found values. Majerus & Kilburn (1969), using antibody prepared against homogeneous chicken liver acetyl-CoA carboxylase, independently carried out similar studies with regard to rats under different dietary conditions and obtained results that were in general agreement with ours.

It is of interest to note that the synthesis and degradation of the carboxylase are controlled independently, and that under steady-state conditions the carboxylase content is regulated mainly by changes in the rate of synthesis, while under nonsteady-state conditions like starvation both the rates of synthesis and degradation are altered leading to changes in the carboxylase content. These findings, together with those of Schimke (1964) on liver arginase, support the concept that the control of the enzyme content by changes in the rate of degradation may play an important role only when the animal deviates from a steady state for adjusting itself to a new environment.

MECHANISM OF ACTIVATION OF ACETYL-COA CARBOXYLASE BY CITRATE

The foregoing results revealed that the changes in the level of acetyl-CoA carboxylase activity in liver cytosol observed under different metabolic conditions are determined by the changing quantities of the carboxylase protein. However, several lines of evidence indicate that the rate of the carboxylase reaction *in vivo* is also controlled by changes in the catalytic efficiency per carboxylase molecule, especially when the rate of fatty acid synthesis has to be adjusted promptly (see Numa *et al.*, 1970). A number of positive and negative allosteric effectors, including citrate as an activator and long-chain acyl-CoA derivatives as inhibitors, are known to change the catalytic efficiency of this enzyme (see Numa *et al.*, 1970). In the work described in the present paper, we analysed the stimulatory effect of citrate by kinetic experiments and studied the interaction of the carboxylase with citrate, in order to elucidate the mode of action of this allosteric effector (Hashimoto & Numa, 1971; Hashimoto *et al.*, 1971*a*, *b*).

Acetyl-CoA carboxylase contains biotin as a prosthetic group (Wakil, Titchener & Gibson, 1958), which is bound covalently to the ε-amino group of a lysyl residue of the apoenzyme (Numa, Ringelmann & Lynen, 1964). Exchange experiments with labelled substrates (Lynen *et al.*, 1963; Matsuhashi, Matsuhashi & Lynen, 1964), as well as isolation of the carboxylated enzyme as an intermediate (Numa *et al.*, 1964) indicated the following reaction mechanism involving two partial reactions:

$$E\text{-biotin} + \text{ATP} + \text{HCO}_3^- \overset{\text{Mg}^{2+}}{\rightleftharpoons} E\text{-biotin} \sim \text{CO}_2 + \text{ADP} + \text{P}_i \quad (1)$$

$$E\text{-biotin} \sim \text{CO}_2 + \text{acetyl-CoA} \rightleftharpoons E\text{-biotin} + \text{malonyl-CoA} \quad (2)$$

Overall:

$$\text{ATP} + \text{HCO}_3^- + \text{acetyl-CoA} \overset{\text{Mg}^{2+}}{\rightleftharpoons} \text{ADP} + \text{P}_i + \text{malonyl-CoA} \quad (3)$$

In order to corroborate this two-step mechanism, we carried out kinetic experiments, using purified rat liver acetyl-CoA carboxylase. The results of initial velocity studies, and product and dead-end inhibition studies are summarized in Tables 3, 4 and 5. Kynurenate, which we found to be a novel dead-end

Table 3. *Initial velocity studies on the overall reactions*

Reaction	Pair of substrates varied	Pattern of double reciprocal plots
Forward	ATP-HCO_3^-	Intersecting
	ATP-acetyl-CoA	Parallel
	HCO_3^--acetyl-CoA	Parallel
Reverse	Malonyl-CoA-P_i	Parallel
	Malonyl-CoA-ADP	Parallel
	P_i-ADP	Intersecting

Citrate was added in the optimal concentration (10 mM). Data taken from Hashimoto & Numa (1971).

Table 4. *Product inhibition patterns for the forward reaction*

Product	Substrate varied	Inhibition
Malonyl-CoA	ATP*	Noncompetitive
	HCO_3^-	Noncompetitive
	Acetyl-CoA	Noncompetitive
ADP	ATP*	Noncompetitive
	Acetyl-CoA	Uncompetitive

* At lower concentrations.

Data taken from Hashimoto & Numa (1971).

Table 5. *Patterns of dead-end inhibition by kynurenate*

Reaction	Component varied	Inhibition*
Forward	ATP	Competitive
	HCO_3^-	Noncompetitive
	Acetyl-CoA	Uncompetitive
	Citrate	Uncompetitive
Reverse	Malonyl-CoA	Competitive

* Kynurenate is a parabolic inhibitor with respect to both slope and intercept.

Data taken from Hashimoto *et al.* (1971*b*).

inhibitor, bound to the uncarboxylated form of the enzyme, was suitable for kinetic studies, because it inhibits the carboxylase instantaneously when it comes in contact with the enzyme, and the inhibition is readily removed upon dilution of the inhibitor. From these kinetic data we conclude that the acetyl-CoA carboxylase reaction proceeds through an ordered bi bi uni uni ping pong mechanism; the order of addition of substrates to the enzyme is ATP, HCO_3^- and acetyl-CoA in the forward reaction, and malonyl-CoA, P_i and ADP in the reverse reaction (see Scheme 1). Thus the two-step reaction mechanism with

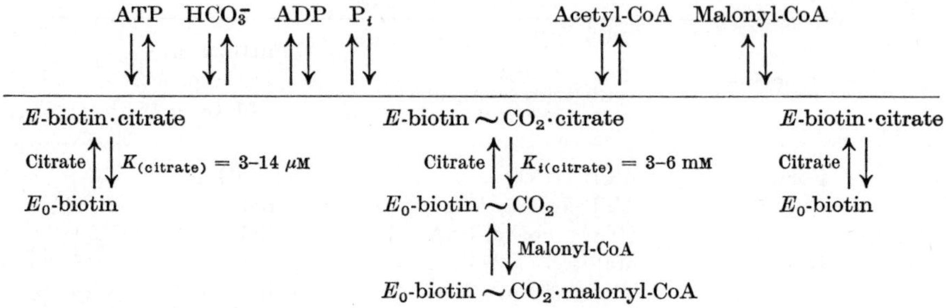

Scheme 1. The mechanism of the acetyl-CoA carboxylase reaction and the mode of citrate activation. $K_{(citrate)}$ and $K_{i(citrate)}$ denote the dissociation constants of the uncarboxylated and carboxylated forms of the enzyme for citrate, respectively.

intermediate formation of the carboxylated enzyme [Reactions (1) and (2)] is substantiated by kinetic studies. Noncompetitive inhibition by malonyl-CoA with respect to ATP, which would be inconsistent with this mechanism, is accounted for by mixed product and dead-end inhibition, because malonyl-CoA-[^{14}C]acetyl-CoA exchange studies revealed that malonyl-CoA forms a dead-end complex with the inactive (or less active) species of the carboxylated form of the enzyme (see Scheme 1).

Based on this reaction mechanism, we investigated then which of the enzyme forms involved is dependent on the allosteric activator, citrate. Table 6

Table 6. *Patterns of citrate activation*

Reaction	Substrate varied	Citrate effect on double reciprocal plots
Forward	ATP	Intercept
	HCO_3^-	Intercept
	Acetyl-CoA	Slope
Reverse	Malonyl-CoA	Intercept
	P_i	Slope
	ADP	Slope and intercept
Malonyl-CoA-[^{14}C]acetyl-CoA exchange	Acetyl-CoA	Slope

Data taken from Hashimoto & Numa (1971) and Hashimoto *et al.* (1971*b*).

summarizes the results of kinetic studies of the citrate effect on the overall and partial reactions. In order to activate rat liver acetyl-CoA carboxylase to the full extent, citrate must be present not only during the carboxylase reaction but also during prior incubation (Numa & Ringelmann, 1965). In these experiments we used the enzyme which had been preincubated with 10 mM citrate for a sufficiently long period of time, so that only the citrate effect during the carboxylase catalysis was investigated. The data in Table 6 indicate that, of the obligatory enzyme forms, only the carboxylated form of the enzyme is dependent on the presence of citrate. The dissociation constant of the carboxylated form of the enzyme for citrate was calculated from these data to be 3–6 mM.

On the other hand, it is well established that the citrate effect is also exerted on the uncarboxylated form of the enzyme for inducing polymerization of protomeric carboxylase molecules (see Numa *et al.*, 1970). The apparent citrate-independence of the uncarboxylated form of the enzyme during the carboxylase catalysis, as found by the kinetic studies, suggests therefore that this enzyme form is converted completely to the active polymeric form under the conditions employed for the kinetic experiments, in which the citrate concentration was varied from 1 mM to 5 mM. In order to verify this concept, we investigated the interaction of the uncarboxylated enzyme with citrate by means of different methods. As summarized in Table 7, the dissociation constant of the uncarboxylated

Table 7. *Interaction of uncarboxylated enzyme with citrate*

Criterion	Dissociation constant (μM)
[^{14}C]Citrate binding	3·8
Protection from inactivation by avidin	14
Protection from activation by trypsin	10
Polymerization of protomers	5

Data taken from Hashimoto, *et al.* (1971*a*,*b*).

enzyme for citrate was estimated to be 3–14 μM. This value is about 10^3-fold lower than the dissociation constant of the carboxylated form of the enzyme for citrate, as found by the kinetic studies, i.e. 3–6 mM, so that the effect of citrate on the uncarboxylated form of the enzyme does not manifest itself upon kinetic studies of the carboxylase-catalysed reactions.

The above-mentioned findings indicate that carboxylation of the biotinyl prosthetic group gives rise to a conformational change at the citrate site, resulting in a marked decrease in the affinity of the enzyme for the activator. This, in conjunction with the protective effect of citrate against avidin inactivation (Ryder *et al.*, 1967; Hashimoto *et al.*, 1971*b*), suggests that there is a close conformational relationship between the catalytic and regulatory sites on the enzyme. This concept is also supported by the earlier results with the chicken liver enzyme showing that citrate increases the rate of decarboxylation of the carboxylated enzyme (Ryder *et al.*, 1967; Ohtsu *et al.*, 1968). Furthermore, the carboxylated chicken liver enzyme exhibits a greater tendency to dissociate than the uncarboxylated enzyme, and citrate prevents this dissociation or reassociates the dissociated protomers into polymers (Gregolin *et al.*, 1968; Ohtsu *et al.*, 1968). We conclude therefore that the main role of citrate in the carboxylase catalysis is to keep the carboxylated form of the enzyme in active conformation by shifting the equilibrium between the active species (E-biotin $\sim CO_2 \cdot$citrate) and the inactive (or less active) species (E_0-biotin $\sim CO_2$) of this enzyme form. Scheme 1 outlines the mechanism of the acetyl-CoA carboxylase reaction and the mode of activation by citrate. The kinetic constants for acetyl-CoA carboxylase are listed in Table 8.

Table 8. *Kinetic constants for acetyl-CoA carboxylase*

Compound	Michaelis constant* (M)	Dissociation constant (M)
ATP	$1{\cdot}5 \times 10^{-5}$	
HCO_3^-	$2{\cdot}5 \times 10^{-3}$	
Acetyl-CoA	$2{\cdot}5 \times 10^{-5}$	5×10^{-6}†
Malonyl-CoA	$1{\cdot}6 \times 10^{-5}$	1×10^{-5}†
P_i	$7{\cdot}0 \times 10^{-3}$	
ADP	$1{\cdot}0 \times 10^{-5}$	
Citrate		$3–6 \times 10^{-3}$‡

* Estimated from the results of initial velocity studies on the overall reactions.

† Estimated from the results of malonyl-CoA-[^{14}C]acetyl-CoA exchange studies.

‡ Dissociation constant of the E-biotin $\sim CO_2 \cdot$citrate complex estimated from the results of both initial velocity studies on the overall reactions and malonyl-CoA-[^{14}C]-acetyl-CoA exchange studies.

Data taken from Hashimoto & Numa (1971) and Hashimoto *et al.* (1971*b*).

Information on the subunit structure of acetyl-CoA carboxylase is of great importance for the elucidation of its allosteric characters. Studies of Vagelos' and Lane's groups (Alberts & Vagelos, 1968; Alberts, Nervi & Vagelos, 1969; Dimroth *et al.*, 1970; Guchhait *et al.*, 1971; Nervi, Alberts & Vagelos, 1971; Alberts, Gordon & Vagelos, 1971) demonstrated that acetyl-CoA carboxylase from *E. coli* dissociates into three functionally dissimilar proteins, i.e. biotin carboxylase, carboxyl transferase and biotin carboxyl carrier protein (BCCP). It appears reasonable to assume that acetyl-CoA carboxylase from animal sources is a complex of tightly bound subunits which correspond to the three proteins constituting the bacterial enzyme system. Moreover, in view of the fact that the *E. coli* system is not activated by citrate, the allosteric nature of the animal enzyme may be based on its complex character. Figure 2 shows the sedimentation patterns of rat liver acetyl-CoA carboxylase in different forms. The polymer (A) with active conformation and the protomer (B) with inactive (or less active) conformation exhibited sedimentation coefficients ($s_{20,w}$) of 44 S and 16 S respectively, and were thus analogous to the corresponding molecular species of the chicken liver enzyme (Numa, Ringelmann & Riecel, 1966; Gregolin *et al.*, 1966*a,b*). The molecular weight of the protomer (containing one molecule of biotin) of the rat liver enzyme was estimated from its biotin

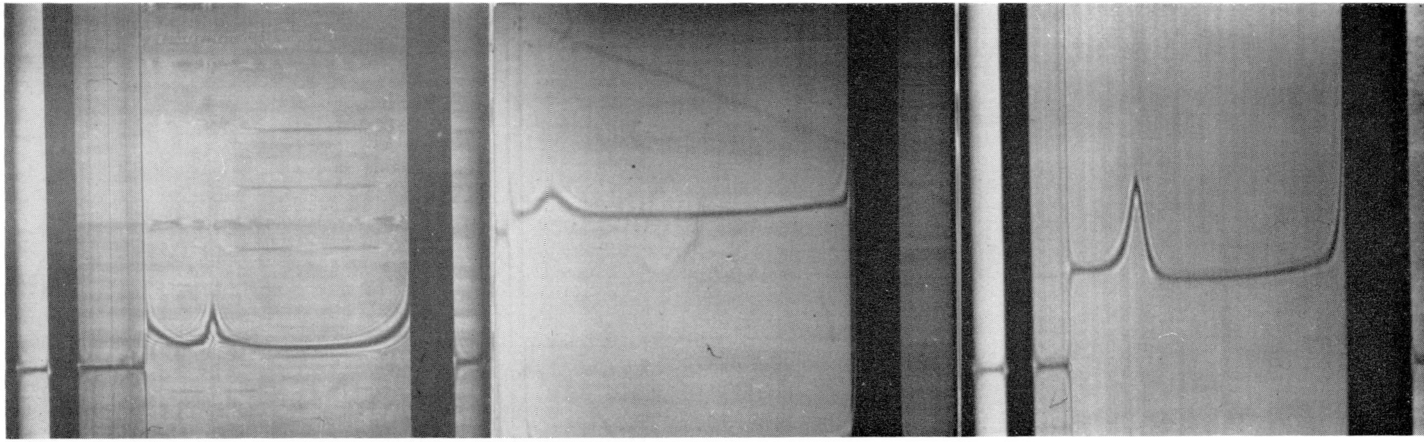

FIG. 2. Sedimentation patterns of rat liver acetyl-CoA carboxylase in different forms; (left), polymer; (middle), protomer; (right), subunits. Solvents used: (A) 0·1 M potassium phosphate buffer, pH 7·5, 0·01 M potassium citrate, 5 mM 2-mercaptoethanol, 1 mM EDTA; (B) 0·05 M *tris*-HCl buffer, pH 8·0, 0·5 M NaCl, 5 mM 2-mercaptoethanol, 1 mM EDTA; (C) 0·2 M *tris*-HCl buffer, pH 8·5, 6·7 M guanidine-HCl, 0·14 M 2-mercaptoethanol. Centrifugation was in a Spinco model E ultracentrifuge at 31 410 rev/min and 21·5°C (A), at 31 410 rev/min and 20·0°C (B), and at 59 780 rev/min and 22·7°C (C); protein concentration, 2·5 mg/ml (A), 1·3 mg/ml (B) and 5·8 mg/ml (C); photographs were taken at 13 min (A), 24 min (B) and 358 min (C) after reaching speed.

content to be 540000. When the carboxylase was treated with 6·7 M guanidine-HCl, the protomer dissociated further into an apparently homogeneous species (C) exhibiting an $s^0_{20,w}$ of 3·3 S. In an attempt to characterize the biotin-containing subunit, rat liver acetyl-CoA carboxylase was labelled by injecting biotin-deficient animals with [^{14}C]biotin, and was then isolated in the homogeneous form (Okazaki & Numa, 1971). The labelled enzyme was treated with 6·7 M guanidine-HCl and centrifuged in a sucrose density gradient containing the denaturant. The sedimentation pattern is illustrated in Fig. 3. Protein and

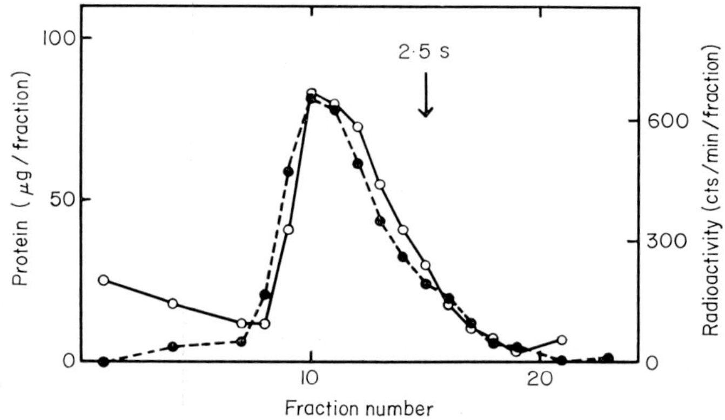

FIG. 3. Sucrose density gradient centrifugation of rat liver acetyl-CoA carboxylase in the presence of 6·7 M guanidine-HCl. The 5–20% (w/v) sucrose gradient (5 ml) contained 0·2 M *tris*-HCl buffer, pH 8·5, 6·7 M guanidine-HCl and 0·14 M 2-mercaptoethanol. *D*-[*carbonyl*-^{14}C]biotin-labelled acetyl-CoA carboxylase from rat liver was dissolved in the same solution (minus sucrose) and carboxymethylated with iodoacetamide. After addition of excess 2-mercaptoethanol, 0·2 ml of the enzyme solution (0·6 mg) was layered over the gradient and centrifuged at 48 500 rev/min and at 20°C for 44 hr in a SW-50 rotor in a Spinco model L2 ultracentrifuge. Twenty-four fractions were collected from the bottom, and aliquots were assayed for protein (○——○) and radioactivity (●- - -●). Trypsin was used as an external marker. Direction of centrifugation, right to left.

^{14}C activity migrated nearly concomitantly with an estimated $s_{20,w}$ of 3·6 S. Although attempts to separate this biotin subunit from other component proteins have thus far been unsuccessful, these results support the above-mentioned concept that the animal enzyme is a complex of tightly bound subunits which are similar to the constituent proteins of the bacterial enzyme system.

The mechanism of carboxylation of acetyl-CoA by the *E. coli* system must involve intersubunit translocation of the biotinyl prosthetic group, which is bound covalently to BCCP. Accumulating evidence also indicates that other biotin enzymes similarly possess dual sites for the bicyclic ring of biotin which are located adjacent to the two substrate sites responsible for the two partial reactions [Reactions (1) and (2)] (*see* Hashimoto *et al.*, 1971*b*). The biocytin (ε-*N*-D-biotinyl-lysine) side chain can reach a distance of 1·4 nm when fully extended (Mildvan, Scrutton & Utter, 1966). Considerations of the structure of biotin enzymes, together with the observed kinetics of the oxalacetate transcarboxylase reaction, led Northrop (1969) to propose a novel mechanism for this enzymic reaction. This mechanism is designated as hybrid ping pong to indicate that it involves intermediate formation of a substituted form of the enzyme and allows independent binding of substrates to two distinct sites on the enzyme. McClure *et al.* (1971) also reported that the pyruvate carboxylase reaction likewise proceeds through a similar two-site ping pong mechanism. The observed kinetics of the liver acetyl-CoA carboxylase reaction described in the present paper can be interpreted by a standard ping pong mechanism, provided that activation by citrate and dead-end inhibition by malonyl-CoA are taken into consideration as outlined in Scheme 1. However, there are several findings which indicate the hybrid character of the mechanism of the acetyl-CoA carboxylase reaction. First, we found that the initial rate of the ATP-[^{32}P]P_i exchange reaction, which represents the first partial reaction [Reaction (1)], was increased 2- to 3-fold by addition of 10–200 μM acetyl-CoA, a substrate for the second partial reaction [Reaction (2)] (Fig. 4). At the same

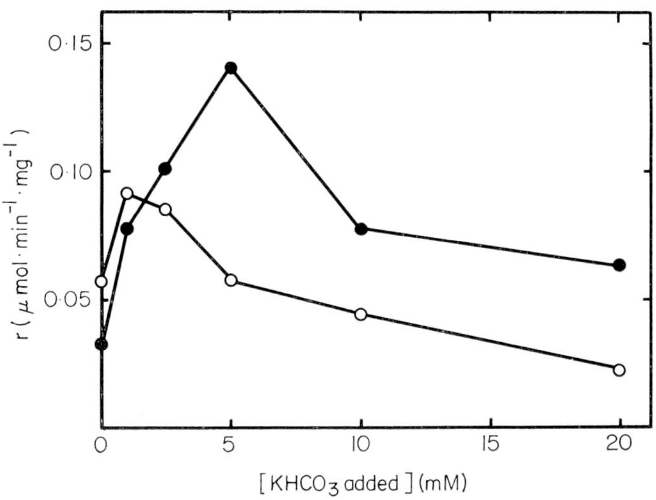

FIG. 4. ATP-[^{32}P]P_i exchange studies in the presence (●——●) and absence (○——○) of acetyl-CoA. The initial rate of isotope exchange (r) was determined in a reaction mixture (final volume, 1·0 ml) containing 50 μmol of *tris*-HCl buffer, pH 7·5, 3 μmol of GSH, 0·6 mg of bovine serum albumin, 10 μmol of potassium citrate, 10 μmol of $MgCl_2$, 1·0 μmol of ATP, 0·5 μmol of ADP, 10 μmol of [^{32}P]P_i ($1{\cdot}0 \times 10^5$ counts per min), indicated amounts of $KHCO_3$, 37 μg of rat liver acetyl-CoA carboxylase (specific activity, 2·7 units/mg) and, when indicated, 0·05 μmol of acetyl-CoA. The reaction was initiated by addition of the enzyme which had been preincubated in the presence of 10 mM citrate at 37°C for 30 min. Following incubation at 37°C for 10 min, the reaction was terminated by heating in a boiling water bath for 1 min. The mixture was diluted by addition of 4 ml of water, and then applied on a Dowex-1 (Cl^-) column (bed volume, 1 ml). After complete removal of P_i by washing the column with 120 ml of a solution containing 0·02 M NaCl and 0·01 N HCl, ATP was eluted with 10 ml of 1 N HCl. The radioactivity of [^{32}P]ATP was determined by the Cherenkov effect in a liquid scintillation spectrometer.

time, the optimal concentration of bicarbonate was increased. Secondly, citrate and acetyl-CoA exert a synergistic effect on the uncarboxylated form of the chicken liver enzyme in terms of protection from avidin inactivation and conversion of the protomer to the polymer (Ryder *et al.*, 1967; Gregolin *et al.*, 1968). Thirdly, binding experiments revealed that the uncarboxylated enzyme has one tight binding site per protomer for acetyl-CoA, the dissociation constant being very close to that of the carboxylated enzyme for acetyl-CoA estimated by kinetic experiments (Gregolin *et al.*, 1968; Hashimoto *et al.*, 1971*b*).

SUMMARY

Immunochemical and enzymological studies indicate that the wide variations in the level of acetyl-CoA carboxylase activity in liver cytosol observed in rats and mice under different metabolic conditions can be ascribed to changes in the quantity of the carboxylase protein. The mechanisms underlying the changes in the carboxylase content were investigated by combined immunochemical and isotopic techniques in terms of the rates of synthesis and degradation of the enzyme. The results of these studies indicate that under steady-state conditions the carboxylase content is regulated mainly by changes in the rate of synthesis, while under nonsteady-state conditions both the rates of synthesis and of degradation are altered to change the carboxylase content.

The acetyl-CoA carboxylase reaction was shown by kinetic studies to proceed through an ordered bi bi uni uni ping pong mechanism; the order of addition of substrates to the enzyme is ATP, HCO_3^- and acetyl-CoA in the forward reaction, and malonyl-CoA, P_i and ADP in the reverse reaction. Evidence is also presented which indicates the hybrid (or two-site) ping pong character of the mechanism of the acetyl-CoA carboxylase reaction. The mode of allosteric activation of the carboxylase by citrate was investigated by kinetic experiments and by studies on the interaction between the enzyme and the effector. The results of these investigations lead to the conclusion that the main role of citrate in the carboxylase catalysis is to keep the carboxylated form of the enzyme in active conformation by shifting the equilibrium between the active and inactive (or less active) species of this enzyme form.

With the use of rat liver acetyl-CoA carboxylase labelled with [^{14}C]biotin, the biotin-containing subunit of this enzyme was shown to exhibit a sedimentation coefficient ($s^0_{20,w}$) of 3·3 S.

ACKNOWLEDGEMENTS

We are indebted to Professor K. Ogata and Dr. Y. Izumi for biotin determinations. This investigation was supported in part by research grants from the Ministry of Education of Japan, the Toray Science Foundation, the Japanese Association for Study of Metabolism and Disease, the Naito Foundation and the Tanabe Amino Acid Research Foundation.

REFERENCES

Alberts, A. W., Gordon, S. G. & Vagelos, P. R. (1971). *Proc. natn. Acad. Sci. U.S.A.* **68,** 1259.

Alberts, A. W., Nervi, A. M. & Vagelos, P. R. (1969). *Proc. natn. Acad. Sci. U.S.A.* **63,** 1319.

Alberts, A. W. & Vagelos, P. R. (1968). *Proc. natn. Acad. Sci. U.S.A.* **59,** 561.

Allmann, D. W., Hubbard, D. D. & Gibson, D. M. (1965). *J. Lipid Res.* **6,** 63.

Chang, H. C., Seidman, I., Teebor, G. & Lane, M. D. (1967). *Biochem. biophys. Res. Commun.* **28,** 682.

Dimroth, P., Guchhait, R. B., Stoll, E. & Lane, M. D. (1970). *Proc. natn. Acad. Sci. U.S.A.* **67**, 1353.
Gregolin, C., Ryder, E., Kleinschmidt, A. K., Warner, R. C. & Lane, M. D. (1966*a*). *Proc. natn. Acad. Sci. U.S.A.* **56**, 148.
Gregolin, C., Ryder, E., Warner, R. C., Kleinschmidt, A. K. & Lane, M. D. (1966*b*). *Proc. natn. Acad. Sci. U.S.A.* **56**, 1751.
Gregolin, C., Ryder, E., Warner, R. C., Kleinschmidt, A. K., Chang, H. C. & Lane, M. D. (1968). *J. biol. Chem.* **243**, 4236.
Guchhait, R. B., Moss, J., Sokolski, W. & Lane, M. D. (1971). *Proc. natn. Acad. Sci. U.S.A.* **68**, 653.
Hashimoto, T. & Numa, S. (1971). *Eur. J. Biochem.* **18**, 319.
Hashimoto, T., Isano, H., Iritani, N. & Numa, S. (1971*a*). *Proc. Jap. Conf. Biochem. Lipids*, **13**, 151.
Hashimoto, T., Isano, H., Iritani, N. & Numa, S. (1971*b*). *Eur. J. Biochem.* **24**, 128.
Ingalls, A. M., Dickie, M. M. & Snell, G. D. (1950). *J. Hered.* **41**, 317.
Lynen, F., Matsuhashi, M., Numa, S. & Schweizer, E. (1963). In *The Control of Lipid Metabolism*, p. 43. (Ed. by Grant, J. K.). New York: Academic Press Inc.
Majerus, P. W. & Kilburn, E. (1969). *J. biol. Chem.* **244**, 6254.
Matsuhashi, M., Matsuhashi, S. & Lynen, F. (1964). *Biochem. Z.* **340**, 263.
McClure, W. R., Lardy, H. A., Wagner, M. & Cleland, W. W. (1971). *J. biol. Chem.* **246**, 3579.
Mildvan, A. S., Scrutton, M. C. & Utter, M. F. (1966). *J. biol. Chem.* **241**, 3488.
Nakanishi, S. & Numa, S. (1970). *Eur. J. Biochem.* **16**, 161.
Nakanishi, S. & Numa, S. (1971). *Proc. natn. Acad. Sci. U.S.A.* **68**, 2288.
Nervi, A. M., Alberts, A. W. & Vagelos, P. R. (1971). *Archs. Biochem. Biophys.* **143**, 401.
Northrop, D. B. (1969). *J. biol. Chem.* **244**, 5808.
Numa, S., Nakanishi, S. & Iritani, N. (1969). *Proc. Jap. Conf. Biochem. Lipids*, **11**, 235.
Numa, S., Matsuhashi, M. & Lynen, F. (1961). *Biochem. Z.* **334**, 203.
Numa, S. & Ringelmann, E. (1965). *Biochem. Z.* **343**, 258.
Numa, S., Ringelmann, E. & Lynen, F. (1964). *Biochem. Z.* **340**, 228.
Numa, S., Ringelmann, E. & Riedel, B. (1966). *Biochem. biophys. Res. Commun.* **24**, 750.
Numa, S., Nakanishi, S., Hashimoto, T., Iritani, N. & Okazaki, T. (1970). *Vit. Horm.* **28**, 213.
Ohtsu, E., Akagami, H., Okazaki, T. & Numa, S. (1968). *Seikagaku*, **40**, 278.
Okazaki, T. & Numa, S. (1971). *Proc. Jap. Conf. Biochem. Lipids*, **13**, 147.
Ryder, E., Gregolin, C., Chang, H. C. & Lane, M. D. (1967). *Proc. natn. Acad. Sci. U.S.A.* **57**, 1455.
Schimke, R. T. (1964). *J. biol. Chem.* **239**, 3808.
Schimke, R. T. & Doyle, D. (1970). *Annu. Rev. Biochem.* **39**, 929.
Wakil, S. J., Titchener, E. B. & Gibson, D. M. (1958). *Biochim. biophys. Acta* **29**, 225.
Wieland, O., Eger-Neufeldt, I., Numa, S. & Lynen, F. (1963). *Biochem. Z.* **336**, 455.

THE POLYENOIC FATTY ACID AND THE PROBLEM OF ESSENTIAL FATTY ACIDS

By E. KLENK†*

Physiol.-Chem. Institut der Universitat Koln-Lindenthal, Germany

In animal tissues polyenoic fatty acids are common constituents of phospholipids, like phosphatidyl choline and phosphatidyl ethanolamine. When our investigations were started more than 40 years ago, only 3 polyenoic fatty acids were known to be present in these phospholipids: linoleic, linolenic and arachidonic acids (Table 1).

Table 1

Linoleic acid: all-*cis*-$\Delta^{9,12}$-C_{18}-dienoic acid	18:2 w6
Linolenic acid: all-*cis*-$\Delta^{9,12,15}$-C_{18}-trienoic acid	18:3 w3
Arachidonic acid: all-*cis*-$\Delta^{5,8,11,14}$-C_{20}-tetraenoic acid	20:4 w6

It was originally known that arachidonic acid is a C_{20}-tetraenoic acid. Later on, during the 1930's, the positions of the double bonds were shown by Nunn & Smedley-MacLean (1938), Dolbey, Nunn & Smedley-MacLean (1940) and by Arens & Smedley-MacLean (1943). The exact positions of only the first and last double bonds were established, while those at the positions 8 and 11 were tentatively indicated. It may be assumed that the sequence of the double bonds in each of the three fatty acids follows adivinyl-methane rhythm. The third column shows the abbreviations now commonly used in the nomenclature of polyenoic fatty acids: 18:2 means C_{18}-dienoic acid, 18:3, C_{18}-trienoic acid and 20:4, C_{20}-tetranoic acid. The number after w indicates the double bond next to the terminal methyl group. This nomenclature is not like the ordinary one starting from the carboxyl group and it can be used only if all the double bonds follow the divynyl-methane rhythm.

By analysing phospholipids from liver and brain we were able to detect, in addition to the polyenoic acids mentioned above, considerable amounts of higher unsaturated C_{22}-polyenoic acids. They are more unsaturated than arachidonic acid and are present in particularly large amounts in the phospholipids of the brain.

† Not presented at the symposium.
* Deceased on 29 December, 1971.

In the phospholipids of brain as well as of the liver *ca* 30% of the total fatty acids are C_{20}- and C_{22}-polyenoic acids. Since in brain phospholipids the amount of C_{22}-acids is more than twice that of C_{20}-acids it was difficult to detect the C_{20}-acids with the crude methods available at that time. In contrast, liver phospholipids contain almost twice as much C_{20}-acids as C_{22}-acids (Table 2).

Table 2. *Amounts of polyenoic fatty acids in phosphatides of human brain, human and beef liver*

	Brain phosphatides† (man) (%*)	Liver phosphatides‡ (cow) (%*)
C_{20}-Polyenoic fatty acids	8·7	18
C_{22}-Polyenoic fatty acids	19·9	10·5

* % of total fatty acids.
† Data taken from Klenk (1931).
‡ Data taken from Klenk & Schönebeck (1932).

These investigations on polyenoic fatty acids of phospholipids were interrupted by the war at a premature stage and after 15 years we started where we had left off. It was possible to show that we were dealing with a rather complex mixture of fatty acids. By using a modification of the old Harries method of ozonide cleavage we (Klenk & Bongard, 1952*a,b*; Klenk & Dreike, 1955) were able to demonstrate that the double bonds in the C_{20}- and C_{22}-polyenoic acids of brain and liver phospholipids show the pattern of a divinyl-methane rhythm. Starting from the methyl end the first double bond has the same position as in either oleic, linoleic or linolenic acid, i.e. positions w9, w6 or w3. We can thus differentiate the polyenoic acids of the oleic, linoleic and linolenic acid type.

We have isolated and characterized the various components from a rather complex mixture of acids and the results are shown in Tables 3 and 4. In agreement with previous findings only polyenoic acids of the linolenic, linoleic and oleic type could be detected, the acids of the linolenic and linoleic type being

Table 3. *Polyenoic fatty acids of brain phosphatides* (*lecithin and cephalin*) (*Klenk & Lindlar*, 1955; *Klenk & Montag*, 1958)

C_{20}	$\Delta^{5,8,11,14}$	C_{20}-tetraenoic acid	20:4 w6
	$\Delta^{5,8,11}$	C_{20}-trienoic acid	20:3 w9*
	$\Delta^{8,11,14}$	C_{20}-trienoic acid	20:3 w6
	$\Delta^{11,14}$	C_{20}-dienoic acid	20:2 w6*
C_{22}	$\Delta^{4,7,10,13,16,19}$	C_{22}-hexaenoic acid	22:6 w3
	$\Delta^{4,7,10,13,16}$	C_{22}-pentaenoic acid	22:5 w6*
	$\Delta^{7,10,13,16}$	C_{22}-tetraenoic acid	22:4 w6
C_{24}	$\Delta^{9,12,15,18}$	C_{24}-tetraenoic acid	24:4 w6*

* Small amounts.

Table 4. *Polyenoic fatty acids of phosphatides from cows' liver (lecithin and cephalin)*

(*Klenk & Tomuschat*, 1957; *Klenk & Montag*, 1957)

C_{20}	$\Delta^{5,8,11,14,17}$	C_{20}-pentaenoic acid	20:5 w3
	$\Delta^{5,8,11,14}$	C_{20}-tetraenoic acid	20:4 w6
	$\Delta^{5,8,11}$	C_{20}-trienoic acid	20:3 w9*
	$\Delta^{8,11,14}$	C_{20}-trienoic acid	22:3 w6
	$\Delta^{8,11}$	C_{20}-dienoic acid	20:2 w9*
	$\Delta^{11,14}$	C_{20}-dienoic acid	20:2 w6*
C_{22}	$\Delta^{4,7,10,13,16,19}$	C_{22}-hexaenoic acid	22:6 w3
	$\Delta^{7,10,13,16,19}$	C_{22}-pentaenoic acid	22:5 w3
	$\Delta^{7,10,13,16}$	C_{22}-tetraenoic acid	22:4 w6

* Small amounts.

predominant. The oleic acid type was present only in small amounts in the phospholipids of the brain as well as of the liver and for the first time we were able to find a C_{20}-trienoic acid of the oleic acid type. We will come back to this compound later on.

These investigations should throw new light on the field of essential fatty acids, which, despite a lot of efforts, is still full of problems. The disovery of essential fatty acids was due to Burr & Burr (1929, 1930), who found pathological changes in the skin, particularly of the tail and the legs of rats given a fat-free diet. The animals were retarded in growth and eventually died. These malfunctions could be reversed by the administration of linoleic or linolenic acid, linoleic acid showing better effect, while oleic and saturated acids were ineffective. These results have repeatedly been confirmed. Turpeinen (1938) has shown that arachidonic acid also belongs to the group of essential fatty acids and that it is the most potent of all of the polyenoic acids.

Further progress in this field took place after the work of Smedley-MacLean (Dolbey, Nunn & Smedley-MacLean, 1940; Nunn & Smedley-MacLean, 1938; Arens & Smedley-MacLean, 1943), almost completely forgotten today. On the basis of her investigations on the structure of arachidonic acid, in the thirties she suggested that this fatty acid is derived in the animal from linoleic acid. She showed that the arachidonic acid content of the liver phospholipids of rats receiving a fat-free diet rapidly decreases and that there is a simultaneous increase in the C_{20}-trienoic acid. These results proved to be fundamental for all future work in this field. Later on, by using isotope techniques Mead, Steinberg & Howton (1953) were able to prove that linoleic acid is converted into arachidonic acid. At the same time (Klenk, 1954, 1955) could show that C_{20}- and C_{22}-polyenoic acids of the linoleic acid type are synthesized from linoleic acid and those of the linolenic acid type from linolenic acid by chain elongation and dehydrogenation towards the carboxyl end. Clear cut results could be obtained by simple feeding experiments and without the use of isotopes. This is shown in Table 5.

It can be seen that on a fat-free diet arachidonic acid is replaced by a C_{20}-trienoic acid, which confirmed the results of Smedley-MacLean. According

Table 5. *Amounts of C_{20}- and C_{22}-polyenoic acids in the phosphatides of rat liver in mol/100 mol of total polyenoic acid fraction.* (*Klenk & Oette*, 1960)

		Main component	Fat-free basal/diet (I)	Basal diet + linoleic acid (II)	Basal diet + linolenic acid (III)
C_{20}	dienoic acid	—*	trace	trace	trace
	trienoic acid	w9	28	5	10
	tetraenoic acid	w6	15	40	7
	pentaenoic acid	w3	trace	trace	18
C_{22}	trienoic acid	—*	0·5	—	trace
	tetraenoic acid	—*	1·5	6	—
	pentaenoic acid	—*	—	—	2
	hexaenoic acid	w3	2·5	6	21

* Main component not identified.

to Mead & Slaton (1956) this fatty acid is identical with a C_{20}-trienoic acid of the oleic acid type, which had previously been isolated by us in small amounts from the phospholipids of brain and liver. During linoleic acid deficiency it is synthesized *de novo* via oleic acid (Fulco & Mead, 1959; Klenk & Pfluger, 1963). A linoleic acid diet induces a decrease in the trienoic and an increase in arachidonic acid.

On a linolenic acid diet, however, there is an accumulation of C_{20}-pentaenoic and C_{22}-hexaenoic acids, both of which are of the linolenic acid type. It is interesting that with this diet the C_{20}-trienoic acid content does not decrease as extensively as with the linoleic acid diet.

In another series of experiments (Klenk, 1965) diets containing labelled oleic, linoleic and linolenic acids were used. These fatty acids were obtained by total synthesis and were labelled at specific positions with ^{14}C. It could be shown that each of these fatty acids forms C_{20}- and C_{22}-polyenoic acids of the respective type only and that interconversion from one type to another does not take place. Especially the linolenic acid type could not be converted into linoleic acid type and later into the oleic acid type and *vice versa*. In other words, the double bonds at w6 or w9 position cannot be hydrogenated. Insertion of double bonds towards the terminal methyl group is also not possible.

To a certain extent the situation in plants is just the opposite. Work of James and of Stumpf (James, 1962, 1963; Stumpf & James, 1963; Harris & James, 1965; Nicols, James & Breuer, 1967) has shown that plant polyenoic acids, which consist mainly of linoleic and linolenic acids, are derived from oleic acid. In this case these new double bonds are inserted towards the terminal methyl group leading to polyenoic acids of the divinyl-methane rhythm; oleic acid becomes linoleic and linoleic probably linolenic acid. In the latter case, however, a slightly different mechanism might operate. We (Klenk & Pflüger, unpublished observations) have also done some work in this field. Our experiments did not provide new fundamental knowledge, but they might be of some interest because of the simplicity and the clear cut results obtained. When lettuce leaves are incubated in light in a buffer solution containing potassium salts of unsaturated fatty acids, the fatty acids are absorbed by the leaves and are incorporated mainly into the glycerophospholipids. Using suitable ^{14}C-labelled fatty acids the conversion of oleic into linoleic and of linoleic into linolenic acid could be shown to occur as follows.

Scheme for the transformation of labelled oleic acid into labelled linoleic acid by lettuce leaves

CH_3—$(CH_2)_6$—*CH_2—CH=CH—$(CH_2)_7$—COOH (18:1 w9)

↓ (11-^{14}C-*cis*Δ9-octadecenoic acid)

CH_3—$(CH_2)_4$—CH=CH—*CH_2—CH=CH—$(CH_2)_7$—COOH (18:2 w6)

↓ oxidative ozonization

CH_3—$(CH_2)_4$—COOH + HOOC—*CH_2—COOH + COOH—$(CH_2)_7$—COOH

(malonic acid) (azelainic acid)

Scheme for the transformation of labelled linoleic acid into labelled linolenic acid by lettuce leaves

*CH_3—$(CH_2)_4$—CH=CH—CH_2—CH=CH—$(CH_2)_7$—COOH (18:2 w6)

↓ (18-^{14}C-all-*cis*-$\Delta^{9,12}$-octadecadienoic acid)

*CH_3—CH_2—CH=CH—CH_2—CH=CH—CH_2—CH=CH—$(CH_2)_7$—COOH (18:2 w3)

↓ reductive ozonization ↓ Oxidative ozonization

CH_3—CH_2—C=OH 2 HOOC—CH_2—COOH HOOC—$(CH_2)_7$—COOH

(Propanal) (malonic acid) (azelainic acid)

Table 6 proves that oleic acid is converted into linoleic acid. Incubation of leaves in buffer solutions containing oleic acid-11-^{14}C leads to its conversion

Table 6. *Specific activities of linoleic acid and their degradation products in dpm/μmole*

Linoleic acid	Malonic acid	Azelainic acid
5800	5450	35

into linoleic acid labelled at position 11. Linoleic acid isolated after the incubation contains considerable amounts of the radioactivity incorporated into phospholipids. After oxidative cleavage of the linoleic acid with ozone, the radioactivity can be recovered quantitatively in malonic acid. The rest of the cleavage products are not radioactive.

In order to prove the conversion of linoleic into linolenic acid we used linoleic acid labelled at position 18 with ^{14}C as demonstrated in Table 7. This experiment was carried out in a similar fashion as the previous one and the isolated linolenic acid was considerably less radioactive than linoleic acid derived from oleic acid. There are two reasons for this: (1) lettuce leaves contain large amounts of linoleic acid resulting in considerable dilution effect; (2) the conversion of linoleic into linolenic acid seems to be much slower than from oleic

Table 7. *Specific activities of linolenic acid and their degradation products in dpm/µmoles*

	Linolenic acid	Propanol (dinitrophenyl hydrozone)	Malonic acid	Azelainic acid
1.	117	121	0	
2.	247	250	2	3·6
3.	270	263		

to linoleic acid and this confirms previous results. However from the cleavage experiments with ozone it is quite clear that linolenic acid is derived from linoleic acid. All the radioactivity was located at the methyl end, i.e. it was present as propanal. The other cleavage products of the ozonide degradation, malonic and azelainic acids were inactive. This shows that the enzyme systems for the synthesis of polyenoic acids are different in animals and plants.

For the formation of their specific polyenoic acids, animals preferentially utilize polyenoic acids preformed in plants, namely linoleic and linolenic acids. With the aid of this enzymatic machinery they are also able to synthesize polyenoic acids from monoenoic acids, for instance the already mentioned C_{20}-trienoic acid from oleic acid. However this seems to happen only when

Table 8. *Behaviour of different polyenoic fatty acids of divinyl-methane rhythm in rats. Chain elongation and incorporation into liver lipids*

Suitable		Not suitable
<u>18:1 w9 → 20:3 w9</u>		16:1 w7
16:2 w6 →		18:2 w7
	20:4 w6	18:9 w5
18:2 w6 →		
16:3 w3 →		
16:4 w3 →	20:5 w3	16:4 w1
18:3 w3 →	↓	18:3 w4
	22:6 w3	
<u>17:2 w5 → 19:4 w5</u>		17:2 w6

plant polyenoic acids are not present in the diet. According to our *in vivo* experiments the characteristic substrates for this enzyme system in higher animals are polyenoic acids of the linoleic and linolenic acid type, particularly linoleic and linolenic acids themselves. Table 8 shows a summary of our results obtained during the last 10 years (Klenk, 1965; Geib, 1967).

A special mention may be made of the two unsaturated acids underlined in Table 8. As already mentioned, oleic acid can be converted to C_{20}-trienoic acid which is incorporated in high amounts into liver phospholipids. However this conversion takes place only when linoleic acid and other polyenoic acids of this type are not present. To this group also belongs a C_{17}-dienoic acid similar to linoleic acid. According to the findings of Schlenk & Sand (1967), which we could fully confirm (unpublished observations), this acid is also incorporated into liver phospholipids after chain elongation and dehydrogenation. However this fatty acid represents only about 3% of the total fatty acids, whereas the other acids in this column are 10% or more.

The fatty acids in the second column are also incorporated into phospholipids to a very small extent after chain elongation and dehydrogenation. They however represent only about 1% of the total fatty acids if the same experimental conditions are used. Thus the enzyme system seems to be rather specific for the polyenoic acids of the linoleic and linolenic acid type.

It can be assumed that the polyenoic acids of phospholipids are important for the structure and function of cellular membranes or organelles consisting of membranes. In this respect polyenoic acids of the linoleic and linolenic acid type should be rather similar. Thus both types should serve as essential fatty acids. On the other hand, it is mainly the deficiency of linoleic and arachidonic acids, which induces considerable pathological changes in rats. Therefore only polyenoic acids of the linoleic acid type are frequently considered to be essential fatty acids (Thomasson, 1953).

However polyenoic acids of the linolenic acid type are also of some biological importance. It has been repeatedly shown that under a fat-free diet linolenic acid has an additional growth-promoting effect in rats, which however is considerably less pronounced than that of linoleic acid. In growing animals particularly the mass of the muscles increases. Muscle phospholipids contain higher amounts of arachidonic acid than polyenoic acids of the linolenic acid type. Thus in the growing animals the requirements for linoleic acid will be greater than for linolenic acid. The prevalence of arachidonic acid in the phospholipids of heart and spleen (Klenk & Dittmer, 1936) could have the same effect. Therefore in the ordinary dietary experiments difficiency of linoleic acid should have more pronounced effects than that of linolenic acid. On the other hand, it should be kept in mind that under certain circumstances the polyenoic acids of the linolenic acid type might as well be of biological importance. For instance, it is known that in hypercholesterolemia the cholesterol levels can be lowered by polyenoic fatty acids regardless of whether they belong to the linoleic or linolenic acid type. Apparently the acids of the linolenic acid type which have the highest degree of unsaturation have the most effect (Ahrens *et al.*, 1959; Ahrens *et al.*, 1957).

In this context it is interesting that in some organs where membranes play an important role, i.e. the brain and nerves, polyenoic acids of the linolenic acid type prevail over arachidonic acid.

The function of the polyenoic acids in membrane lipids is not known. It seems unlikely that there are significant differences between both types.

It is not the purpose of this contribution to put forth another hypothesis. My intention is only to show how complex are the problems of essential fatty acids today.

REFERENCES

Ahrens, E. H., Hirch, J., Insull, W., Th. Tsaldas, Bloomstrand, R. & Peterson, M. L. (1957). *Lancet* (1957/1) 943.

Ahrens, E. H. Jr., Insull, W. Jr., Hirch, J., Stofell, W., Peterson, M. S., Faquar, J. W., Miller, Th. & Thomasson, H. J. (1959). *Lancet* (1959/1) 115.

Arens, C. L. & Smedley-MacLean, I. (1943). *Biochem. J.* **37,** 1.

Burr, G. O. & Burr, M. M. (1929). *J. biol. Chem.* **86,** 345.

Burr, G. O. & Burr, M. M. (1930). *J. biol. Chem.* **82,** 587.

Dolbey, D. E., Nunn, L. C. A. & Smedley-MacLean, I. (1940). *Biochem. J.* **34,** 1422.

Fulco, A. J. & Mead, J. F. (1959). *J. biol. Chem.* **234,** 1411.

Geib, K. H. (1967). Dissertation, Univ. Cologne.

Harris, R. V. & James, A. T. (1965). *Biochim. biophys. Acta* **106,** 456.

James, A. T. (1962). *Biochim. biophys. Acta* **57,** 167.

James, A. T. (1963). *Biochim. biophys. Acta* **70,** 9.

Klenk, E. (1931). *Z. Physiol. Chemie* **200,** 51.

Klenk, E. (1954). *Naturwiss.* **41,** 68.

Klenk, E. (1955). *Z. physiol. Chem.* **302,** 268.

Klenk, E. (1965). In *Advances in Lipid Research,* New York, **3,** 1.

Klenk, E. & Bongard, W. (1952*a*). *Z. physiol. Chem.* **290,** 181.

Klenk, E. & Bongard, W. (1952*b*). *Z. physiol. Chem.* **291,** 104.

Klenk, E. & Dittmer, J. (1936). *Z. physiol. Chem.* **244,** 203.

Klenk, E. & Dreike, A. (1955). *Z. physiol. Chem.* **300,** 113.

Klenk, E. & Lindlar, F. (1955). *Z. physiol. Chem.* **299,** 74; **301,** 156.

Klenk, E. & Montag, W. (1957). *Leibigs Annalen Chem.* **604,** 4.

Klenk, E. & Montag, W. (1958). *J. Neurochemistry* **2,** 226.

Klenk, E. & Oette, K. (1960). *Z. physiol. Chem.* **318,** 86.

Klenk, E. & Pfluger, H. (1963). *Z. physiol. Chem.* **335,** 53.

Klenk, E. & Schönebeck, O. V. (1932). *Z. physiol. Chem.* **209,** 112.

Klenk, E. & Tomuschat, J. H. (1957). *Z. physiol. Chem.* **308,** 165.

Mead, J. F. & Slaton, W. H. Jr. (1956). *J. biol. Chem.* **219,** 705.

Mead, J. F., Steinberg, G. & Howton, D. H. (1953). *J. biol. Chem.* **205,** 683.

Nicols, B. W., James, A. T. & Breuer, J. (1967). *Biochem. J.* **104,** 486.

Nunn, L. C. A. & Smedly-MacLean, I. (1938). *Biochem. J.* **32,** 2187.

Schlenk, H. & Sand, D. M. (1967). *Biochim. biophys. Acta* **144,** 305.

Stumpf, P. K. & James, A. T. (1963). *Biochim. biophys. Acta* **70,** 20.

Thomasson, H. J. (1953). *Rev. Vit. Res.* **25,** 1.

Turpeinen, O. (1938). *J. Nutrit.* **15,** 351.

THE SPECIFICITY OF THE FATTY ACID DESATURASES

By A. T. JAMES

Division of Biochemistry, Biosciences Group, Unilever Research Laboratory, Colworth House, Sharnbrook, Bedfordshire, England

INTRODUCTION

THE widespread occurrence of long chain mono- and poly-unsaturated acids of a variety of chain lengths and positions of the double bonds has led to many speculations as to their chemical relationships. It was early pointed out (Thomasson, 1953) that some families of unsaturated fatty acids with related biological activity could be formed if double bond positions in the chain were specified with respect to the terminal methyl group rather than, as in the normal chemical nomenclature, with respect to the carboxyl group.

During the last ten years a number of specific enzymic reactions involving dehydrogenation have been discovered and some of their characteristics outlined, e.g. stearic acid → oleic acid (Bloomfield & Bloch, 1960; James & Marsh, 1962; Fulco & Bloch, 1964); oleic → linoleic → α-linolenic (Yuan & Bloch, 1961; James, 1962; McMahon & Stumpf, 1964; Harris & James, 1965; Harris, Harris & James, 1965); and linoleic → γ-linolenic → homo-γ-linolenic → arachidonic (Mead & Howton, 1957; Howton & Mead, 1960; Nugteren, 1962, 1963; Stoffel & Ach, 1964). The difficulty of isolating single pure enzymes continues to be a major disadvantage in absolute determination of enzymic specificity. Despite this it has been shown for the conversion of stearic to oleic acid that only the D–H atoms are removed at C atoms 9 and 10 (Schroepfer & Bloch, 1965; Morris *et al.*, 1967; Bloch, 1969; Morris, 1970). Similar stereospecificity is shown in the conversion of oleic to linoleic and thence to α-linolenic acid (Morris *et al.*, 1967). Such absolute stereospecificity argues for a rigid holding of the hydrocarbon chain of the fatty acid substrate.

These results together with more recent work on substrate chain length specificity and requirements of double bond position for some of these enzymes now allow some generalization to be made as to the range of substrates that a given enzyme will accept. This we now propose as a basis for classification of some of the fatty acid desaturases.

A SUGGESTED CLASSIFICATION OF THE FATTY ACID DESATURASES

General

Based on published data on crude enzyme specificities we recently suggested a system of nomenclature (Howling *et al.*, 1971). A number I, II or III was used to denote the number of double bonds in the products, and the symbols Δ or W were added to show whether the double bond was introduced in a fixed position with respect either to the carboxyl group or to the terminal methyl group. We now suggest adding a term in brackets to denote the position and configuration of the bond introduced, e.g. (6 *cis*).

Type I. Desaturase enzymes normally converting a saturated to a monoenoic fatty acid

We and others have described the substrate specificities, in terms of chain length and position of double bond, introduced by crude enzyme preparation or by whole cells of a variety of systems (Howling, Morris & James, 1968; Johnson *et al.*, 1969; Brett *et al.*, 1971; Paulsrud, Stewart & Holman, 1970). The relationship between chain length and percentage conversion to the Δ9 monoenoic acid is shown in Fig. 1 for the simplest system (whole cells of *Torulopsis bombicola*). The Δ9 desaturases from other sources have varying abilities

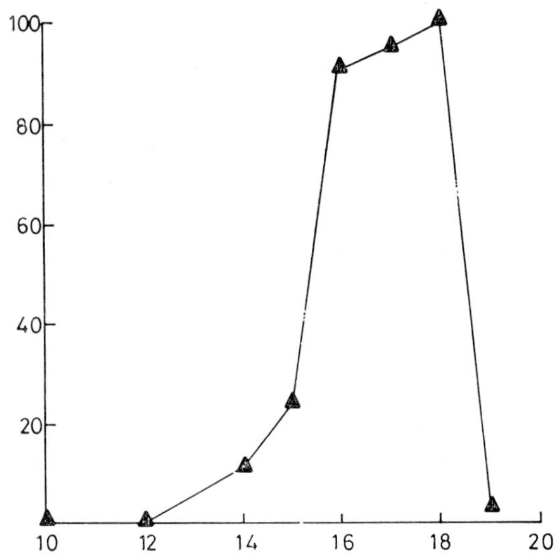

FIG. 1. The relationship between precursor chain length and percentage conversion to the corresponding Δ9-monoenoic fatty acid (normalized to stearic acid = 100%) for whole cells of *Torulopsis bombicola*.

of formation of the monoenoic acid from the shorter chain acids; this we suggested was due to varying mixtures of two enzymes, one with maximal activity at C_{14} and the other at C_{18}.

Further studies were carried out using as substrates most of the isomeric methyl stearic acids, i.e. with the methyl group at varying positions in the chain. The failure of the enzymic reaction when the substrate possessed side chain methyl groups at positions 5, 6, 8, 11, 13, 14 and 15 suggested that the fatty acid chain between C atoms 5 and 15 was severely constrained. This enabled us to produce a schematic model of the enzyme-cleft into which the substrate fitted (Fig. 2). Such a model with its cleft having a length equal to

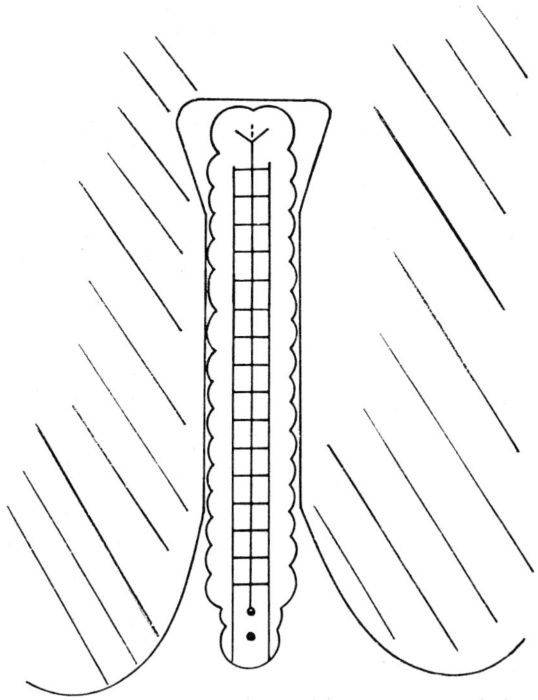

FIG. 2. Suggested relationship between cleft size and substrate chain (stearic acid) in the enzyme-substrate complex of the desaturase converting stearic to oleic acid.

that of the extended chain of stearic acid would explain the limited range of chain lengths that can be accommodated efficiently.

A number of other enzymes that introduce double bonds in positions other than Δ9 have been positively identified and these are summarized in Table 1 together with the proposed classification. In very few cases has the chain length specificity been defined; hopefully this review will encourage this work to be done.

The fact that saturated acids are the normal substrates for these enzymes does not mean that these are the only possible substrates. The existence of Δ5:10-hexadecadienoic acid in *B. Licheniformis* was attributed by Fulco to desaturation of Δ10-hexadecenoic acid by the IΔ(5 *cis*) desaturase. This suggestion is supported by our observation that Δ12-octadecenoic acid is efficiently converted to Δ9:12 octadecadienoic acid by a variety of systems (some not normally able to synthesize linoleic acid) (Brett *et al.*, 1971; Howling *et al.*, 1971). Presumably the active enzyme is either the desaturase IΔ(9 *cis*)$_{M_{14}}$ or IΔ(9 *cis*)$_{M_{18}}$. In the former case the bend in the molecule produced by the existing double bond could be sufficient to lift the chain from C atoms 13 to 18 out of the enzyme cleft, and so mimic a C_{12} saturated acid.

Similarly 18-nonadecenoic acid is converted to 9:18-octadecadienoic acid by hen liver and *Chlorella vulgaris* more efficiently than nonadecanoic acid is converted to Δ9-nonadecenoic acid. This again could be due to lifting of the 19C atom of the chain out of the cleft (this is presumed to terminate at a length

Table 1. *Partially characterized enzymes converting saturated to monounsaturated acids*

Substrates	Position and configuration of double bond introduced	System	Author	Suggested classification
C_{16}	Δ3 *trans*	*Chlorella vulgaris*	Nichols *et al.* (1965)	Desaturase IΔ (3 *trans*)
C_{16}, C_{18}	Δ5 *cis*	*B. megatherium*	Fulco *et al.* (1964)	Desaturase IΔ (5 *cis*)
(C_{14}, C_{12} ineffective)		*B. licheniformis subtilis*, etc.	Fulco (1967)	
C_{16}	Δ8 *cis*	*B. licheniformis macerans*, etc.	Fulco (1969)	Desaturase IΔ (8 *cis*)
C_{10} to C_{15}	Δ9 *cis*	Hen liver particularly	Howling *et al.* (1971)	Desaturase IΔ (9 *cis*)M_{14}
C_{14} to C_{19}	Δ9 *cis*	Widespread, probably most aerobic systems	Howling *et al.* (1971)	Desaturase IΔ (9 *cis*)M_{18}
C_{16}	Δ10 *cis*	*B. licheniformis*	Fulco (1969)	Desaturase IΔ (10 *cis*)

Subscripts denote chain length at which maximum activity is observed.

equal to C_{18} chain) (Fig. 2), by the terminal double bond. Neither Δ7-, Δ8-, Δ10-, nor Δ11-octadecenoic acids were converted to dienes by this system. This could be due either to the wrong orientation of the carboxyl to C9 part of the chain because of the Δ7 or Δ8 double bond, or to the closeness of the Δ8, Δ10 or Δ11 double bonds to the reaction centre.

The classification suggested here may need to be extended in order to cover the nature of the enzymic substrates since these may be either the acyl-S-ACP or acyl-S-CoA thiol esters, depending on the source of the enzyme (Schultz & Lynen, 1971; Nagai & Bloch, 1966; Vijay & Stumpf, 1971). Furthermore there is some evidence that 2-palmitoyl-phosphatidyl glycerol is the substrate for the IΔ(3 *trans*)desaturase since Δ3 *trans* hexadecenoic acid is specifically bound to this lipid (Nichols, Harris & James, 1967). This further classification has not been attempted, however, since more data are required.

Type II. Enzymes converting monoenoic to dienoic fatty acids

Such enzymes are widely distributed especially in plants, and in the latter give rise to Δ7:10 or Δ9:12 dienoic acids; mammalian systems on the other hand form only Δ6:9 dienoic acids (Brenner, 1971). Our investigations of the chain length and double bond specificity of the enzymes present in whole cells are subject to obvious reservations until pure enzymes are available. These results are summarized in Tables 2 and 3 and complement our previous data on the formation of the Δ7 and Δ7:10 acids by *Chlorella vulgaris* (Howling *et al.*, 1968).

Table 2. *The conversion of monoenoic to dienoic fatty acids by* Chlorella vulgaris

Precursor	direct conversion (%)	Double bond position
cis-9-hexadecenoic (w7)	21·0	Δ9-10, 12-13 (w7, w4)
cis-9-heptadecenoic (w8)	37·4	Δ9-10, 12-13 (w8, w5)
cis-9-octadecenoic (w9)	79·0	Δ9-10, 12-13 (w9, w6)
cis-9-nonadecenoic (w10)	25·0	Δ9-10, 12-13 (w10, w7)
cis-7-hexadecenoic (w9)	36·0	Δ7-8, 10-11 (w6, w9)
cis-8-heptadecenoic (w9)	26·0	Δ8-9, 11-12 (w6, w9)
cis-9-octadecenoic (w9)	79·0	Δ9-10, 12-13 (w6, w9)
cis-10-nonadecenoic (w9)	24·1	Δ10-11, 13-14 (w6, w9)
cis-8-octadecenoic (w10)	<1	—
cis-10-octadecenoic (w8)	<1	—
cis-11-octadecenoic (w7)	<1	—
cis-12-octadecenoic (w6)	48·0	Δ9-10, 12-13 (probably due to IΔ (9 *cis*)desaturase since this conversion also occurs in hen liver microsomes)
trans-9-octadecenoic (w9)	<1	—
9-octadecynoic	<1	—
cis, cis-9,18-nonadecadienoic (w9)	72·7	Δ9-10, 12-13, 18-19

Table 3. *Conversion of monoenoic fatty acids to dienoic fatty acids by* Ricinus communis *seeds*

Precursor	Direct conversion into dienoic acid (%)	Double bond position in product
cis-9-hexadecenoic (w7)	6·7	9-10, 12-13
cis-9-heptadecenoic (w8)	6·8	9-10, 12-13
cis-9-octadecenoic (w9)	9·2	9-10, 12-13
cis-9-nonadecenoic (w10)	4·5	9-10, 12-13
cis-8-heptadecenoic (w9)	2·8	8-9, 11-12 (w6, w9)
cis-9-octadecenoic (w9)	9·2	9-10, 12-13 (w6, w9)
cis-10-nonadecenoic (w9)	2·0	10-11, 13-14 (w6, w9)
cis-8-octadecenoic (w10)	<1	—
cis-10-octadecenoic (w8)	<1	—

Our results suggested that two enzymes exist, one recognizing only a double bond distant 9C atoms from the carboxyl group (i.e. Δ9) and the other recognizing only a double bond distant 9C atoms from the terminal methyl group (i.e. w9). This theory accounts for our previous finding (Howling *et al.*, 1968) that only Δ7-hexadecenoic acid (w9) was converted to a Δ7:10 dienoic acid, despite the fact that both the C_{14} and C_{15} saturated acids were readily converted to the Δ7 monoenes. The C_{14} Δ7 acid is w7 and the C_{15}Δ7 acid is w8; neither could be utilized by enzymes specifically requiring either an existing Δ9 or an w9 acid.

In Table 4 these enzymes are listed together with their proposed classification, and indicate clearly their high specificity. However, reference to Table 3

Table 4. *Chain length and double bond position specificity for enzymes converting monoenoic to dienoic fatty acids*

Effective substrate	Position of double bond introduced	Occurrence	Proposed classification
Δ9, C_{16}, C_{17}, C_{18} and C_{19} monoenoic acids No clear maximum	Δ12	Probably all higher plants	II Δ9 (12 *cis*)
w9, C_{16}, C_{17}, C_{18} and C_{19} monoenoic No clear maximum	w6	Probably all higher plants	II w9 (6 *cis*)
Δ9-octadecenoic acid	Δ6	Mammalian liver	Possibly II Δ9 (6 *cis*) or II w9 (12 *cis*)

shows that provided an existing double bond is further towards the methyl group than the double bond to be introduced then a dienoic acid can be converted to a trienoic acid (Δ9:18 octadecadienoic acid to Δ9:12:18-octadecatrienoic acid).

Desaturases converting dienoic to trienoic and trienoic to tetraenoic fatty acids

There have been no detailed studies of enzyme specificity for this group but I hope to develop in a later paper a theory to cover these transformations also.

REFERENCES

Bartels, C. T., James, A. T. & Nichols, B. W. (1967). *Eur. J. Biochem.* **3**, 7.
Bloch, K. (1969). *Acc. Chem. Res.* **2**, 193.
Bloomfield, D. K. & Bloch, K. (1960). *J. biol. Chem.* **235**, 337.
Brenner, R. R. (1971). *Lipids* **6**, 567.
Brett, D., Howling, D., Morris, L. J. & James, A. T. (1971). *Archs Biochem. Biophys.* **143**, 535.
Fulco, A. J. (1967). *Biochim. biophys. Acta* **144**, 701.
Fulco, A. J. (1969*a*). *J. biol. Chem.* **244**, 889.

Fulco, A. J. (1969*b*). *Biochim. biophys. Acta* **187,** 169.
Fulco, A. J. & Bloch, K. (1964). *J. biol. Chem.* **239,** 993.
Fulco, A. J., Levy, R. & Bloch, K. (1964). *J. biol. Chem.* **239,** 993.
Harris, R. V. & James, A. T. (1965). *Biochim. biophys. Acta* **106,** 456.
Harris, R. V., Harris, P. & James, A. T. (1965). *Biochim. biophys. Acta* **106,** 465.
Howling, D., Morris, L. J. & James, A. T. (1968). *Biochim. biophys. Acta* **152,** 224.
Howling, D., Morris, L. J., Gurr, M. I. & James, A. T. (1971). *Biochim. biophys. Acta* (in press).
Howton, D. R. & Mead, J. F. (1960). *J. biol. Chem.* **235,** 3386.
James, A. T. (1962). *Biochim. biophys. Acta* **57,** 167.
Johnson, A. R. Fogerty, A. C., Pearson, J. A., Shenstone, F. S. & Bersten, A. M. (1969). *Lipids* **4,** 265.
Marsh, J. B. & James, A. T. (1962). *Biochim. biophys. Acta* **60,** 320.
McMahon, V. & Stumpf, P. K. (1964). *Biochim. biophys. Acta* **84,** 359.
Mead, J. F. & Howton, D. R. (1967). *J. biol. Chem.* **229,** 575.
Morris, L. J. (1970). *Biochem. J.* **118,** 681.
Morris, L. J., Harris, R. V., Kelly, W. & James, A. T. (1967). *Biochem. biophys. Res. Commun.* **28,** 904.
Nagai, J. & Bloch, K. (1966). *J. biol. Chem.* **241,** 1925.
Nichols, B. W., Harris, P. & James, A. T. (1965). *Biochim. biophys. Res. Commun.* **21,** 473.
Nugteren, D. H. (1962). *Biochim. biophys. Acta* **60,** 655.
Nugteren, D. H. (1963). *Biochem. J.* **89,** 28.
Paulsrud, J. R., Stewart, S. E. & Holman, R. T. (1970). *Lipids* **5,** 611.
Schroepfer, G. J. & Bloch, K. (1965). *J. biol. Chem.* **240,** 54.
Schultz, J. & Lynen, F. (1971). *European J. Biochem.* **21,** 48.
Stoffel, W. & Ach, K. L. (1964). *Z. Physiol. Chem.* **337,** 123.
Thomasson, H. J. (1953). *Int. Rev. Vit. Res.* **25,** 1.
Vijay, I. K. & Stumpf, P. K. (1971). *J. biol. Chem.* **246,** 2910.
Yuan, C. & Bloch, K. (1961). *J. biol. Chem.* **236,** 1277.

BIOSYNTHESIS OF UNSATURATED FATTY ACIDS BY HIGHER PLANT SYSTEMS*

By P. K. STUMPF, INDER VIJAY** and J. L. HARWOOD‡

Department of Biochemistry and Biophysics,
University of California,
Davis, California, USA

INTRODUCTION

One of the unique differences between prokaryotic and eukaryotic organisms is the total absence of non-conjugated polyunsaturated fatty acids in prokaryotic organisms and their presence in significant amounts in eukaryotic organisms. Moreover, in animal cells, additional double bonds are always introduced towards the carboxyl end of the fatty acid, whereas in plant cells, double bonds are introduced towards the methyl end. These reactions are

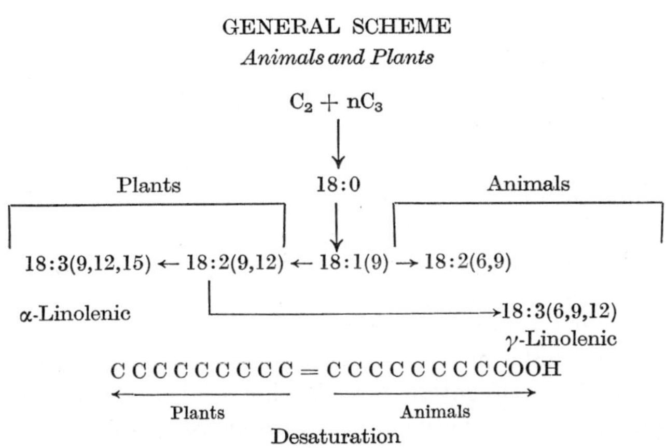

Fig. 1. Summary of desaturation reactions of eukaryotic organisms (plants and animals).

summarized in Fig. 1. All animal cells lack the enzyme which catalyses desaturation of oleic acid to linoleic acid, while plant cells readily carry out this conversion. This specific desaturation reaction is therefore of prime importance for the synthesis of linoleic acid, an essential fatty acid in the animal diet.

* Supported by a NSF Grant 19732X.

** I.V., at present, Department of Biological Chemistry, University of California, Davis, California, U.S.A.

‡ J.L.H., at present at Department of Biochemistry, University of Leeds, Leeds 2, England.

Prokaryotic organisms readily synthesize monoenoic fatty acids by the anaerobic pathway elucidated by Bloch and his coworkers. However, a limited number of prokaryotic organisms including the *Bacilli, Mycobacteria* and *Corynebacteria* have evolved an aerobic pathway with the total exclusion of the anaerobic pathway for the synthesis of monoenoic fatty acids. In eukaryotic organisms, however, there is no evidence to suggest the presence of an anaerobic pathway. All unsaturated fatty acids are formed by a series of aerobic desaturation steps, each step being catalysed by a specific desaturase (Brenner, 1971). The general sequence for this reaction requires an input of electrons into an iron-containing protein which in turn reacts with molecular oxygen and a desaturase to extract two hydrogen atoms to form a *cis* double bond:

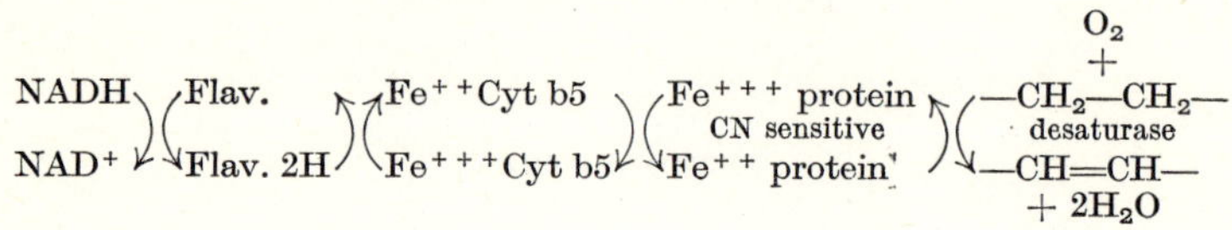

While NADH is normally considered a source of electrons in desaturation reactions, evidence will be presented which suggests that NADH is not an obligatory reductant. Although considerable work has been expended in elucidating desaturation reactions in prokaryotic and eukaryotic organisms from the laboratories of Bloch, Lynen, Sato, Wakil, Brenner and James, we shall limit our discussion to recent developments in our laboratory at the University of California at Davis.

In 1964, McMahon and Stumpf observed that a cell free system obtained from maturing seeds of *Carthamus tinctorius* readily synthesized linoleic acid from oleyl CoA in the presence of molecular oxygen and either NADPH or NADH. Maximum activity for this system manifested itself 14–18 days after flowering. The desaturation system was essentially absent prior to or after this critical time period. This system was re-examined in more detail recently (Vijay & Stumpf, 1971, 1972). The following observations were made.

(1) The microsomal fraction had the highest oleyl CoA desaturase activity. The soluble protein fraction obtained by centrifugation was completely devoid of activity. Although the mitochondrial fraction contained desaturase activity, a sucrose gradient centrifugation of a mitochondrial enriched pellet indicated that the activity was distributed among a dispersed population of membrane fragments rather than with a fraction highly enriched with mitochondrial marker enzymes.

(2) As summarized in Table 1, the cofactors for the system consisted of NADH and molecular oxygen, although NADPH could also serve as a reductant. The Km for NADH was 0·18 mM while that for NADPH was 2·5 mM.

(3) As I have indicated earlier, it was of interest to determine if NADH was a direct obligatory reductant for the oxygen dependent desaturation or merely served as a readily available source of electrons which were required for the reduction of an intermediate electron carrier. Because of the absolute requirement for molecular oxygen it was obvious that reduced reagents which readily undergo auto-oxidation under the assay conditions could not be

Table 1. *Requirement of reduced pyridine nucleotide and oxygen for oleyl-CoA desaturation*

Cofactor	nmoles [1-^{14}C]-linoleate	Conversion (%) from [1-^{14}C]-oleyl CoA
EXPERIMENT 1		
None	0·0	0·0
DPN^+ (1 μmole)	0·0	0·0
TPN^+ (1 μmole)	0·0	0·0
DPNH (1 μmole)	1·7	33·0
TPNH (1 μmole)	1·4	28·0
DPNH + TPNH (1 μmole of each)	1·8	36·0
EXPERIMENT 2		
Aerobic incubation	2·6	52·0
Anaerobic incubation	0·0	0·0

employed. We however decided to use the photoreducing system of spinach grana which, by photo-oxidizing water, can supply electrons at a redox potential of —0·42 eV to the desaturase system in the presence of spinach ferredoxin. As shown in Fig. 2, since photochemically reduced ferredoxin

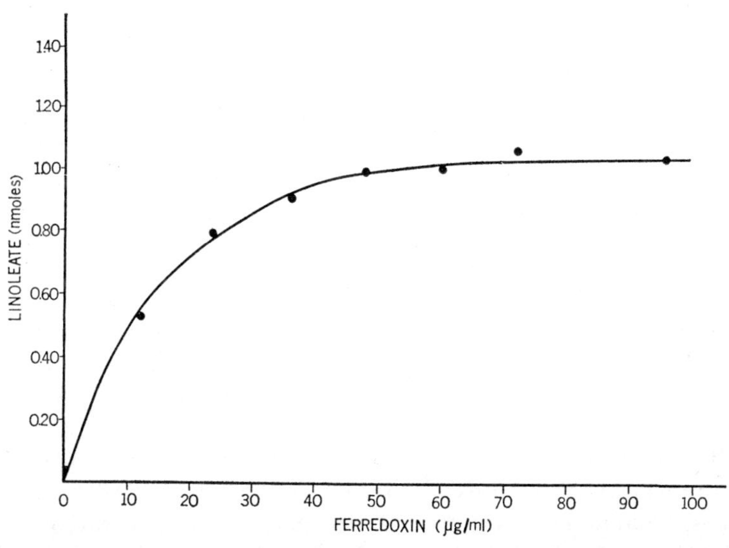

FIG. 2. Effect of increasing concentrations of spinach ferredoxin on oleyl CoA desaturation by the *Carthamus* microsomal desaturase.

readily replaced NADH as a reductant, NADH was not an obligatory reductant involved directly in desaturation but served only as one of the several possible reductants. Ferredoxin did not appear to be the iron containing carrier protein in the *Carthamus* membrane system since repeatedly washed particles never showed a response to the addition of ferredoxin.

(4) A number of inhibitors markedly affected the activity of the desaturase. While metal ion chelators such as *p*-hydroxyquinoline and sodium diethyldithiocarbamate inhibited 100% at 1 mM concentration, cyanide (1 mM) inhibited only 58% and azide was ineffective. Carbon monoxide in a number of concentrations was completely ineffective. The desaturase was also very sensitive to sulfhydryl reagents such as *p*-chloromercuribenzoate.

(5) The *Carthamus* desaturase had remarkable specificity. Thus while oleyl CoA was readily desaturated to linoleyl CoA, *cis*-vaccenyl CoA, elaidyl CoA, palmitoleyl CoA, linoleyl CoA, palmityl CoA and stearyl CoA as well as oleyl ACP were all completely inactive. The substrate requirements for desaturation would therefore include (a) a thioester; (b) a *cis* double bond between the C_9 and C_{10} position and (c) a C_{18} rather than a C_{16} chain length fatty acid.

(6) Although oleyl CoA was the only substrate which was active in the *Carthamus* system, the possibility existed by which the oleyl component was transferred to an endogeneous acceptor to form phosphatidyl choline which in turn was the actual substrate for desaturation. In order to examine this possibility, a sensitive, quick and quantitative procedure was necessary to discriminate between the thioesters and oxygen esters. Fortunately the thioester procedure of Barron & Mooney (1968) made possible the development of a rapid assay system which could indeed discriminate between thioesters and oxygen esters. In essence, acyl thioesters were quantitatively reduced to alkyl alcohols by sodium borohydride and then oxygen esters were saponified, extracted and converted to methyl esters by diazomethane. Injection of the mixture of alkyl alcohols and methyl esters into a radio-GLC system yielded peaks, one set of which were the alkyl alcohols equivalent to acyl thioesters and the others were the methyl esters equivalent to oxygen esters and free fatty acids. By applying this technique to the elucidation of the reaction sequence catalyzed by the membrane bound desaturase of *Carthamus tinctories*, we were able to come to the following conclusions:

(a) The microsomal particle contained acyl transferases which in the absence of NADH and O_2 rapidly transferred the acyl moiety (stearyl, oleyl and linoleyl) to an endogeneous acceptor to form β-^{14}C-acyl phosphatidyl choline.

(b) In the presence of NADH and molecular oxygen a rapid conversion of oleyl CoA to linoleyl CoA occurred with a simultaneous acyl transfer of the substrate and product to an endogeneous acceptor to form β-oleyl and β-linoleyl phosphatidyl choline.

(c) To test the hypothesis that the true substrate for desaturation was actually a β oleyl phosphatidyl choline rather than oleyl CoA (Gurr, Robinson & James, 1969) we allowed the *Carthamus* microsomal preparation to form *in situ* the hypothetical substrate, namely β oleyl phosphatidyl choline by preincubating the microsomal enzyme with oleyl CoA in the absence of NADH, and then removing the unreacted oleyl CoA by the addition of glycerylphosphoryl choline. The desaturation reaction was initiated by the addition of NADH. Analysis of the products at the end of the reaction revealed no conversion of the oleyl component to the linoleyl component.

(d) To determine if oleyl CoA was first converted to either an oleyl-*S*-En or an oleyl-*S*-ACP complex, unlabelled oleyl CoA was incubated with the microsomal enzyme in the presence of ^{3}H-CoA. After 30 min of incubation the reaction products were separated by paper chromatography. The results indicated no exchange of ^{3}H-CoA into oleyl CoA in the absence or presence of ACP, thus excluding any acyl exchange reactions which would be a prelude to desaturation.

These results are summarized in Fig. 3.

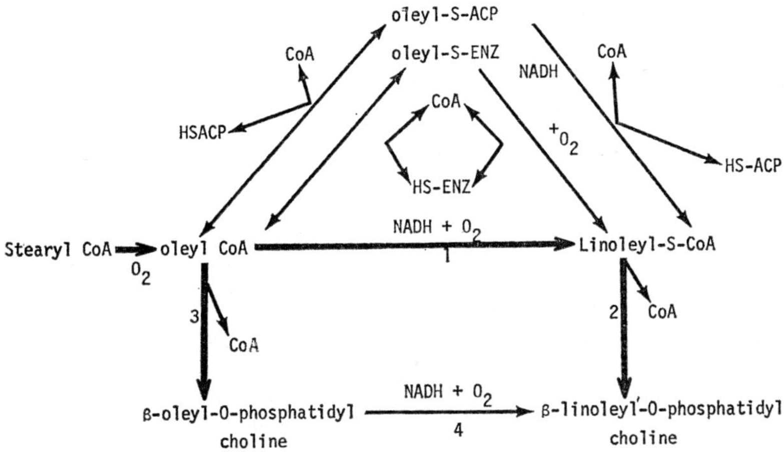

FIG. 3. Summary of reactions involved in the desaturation of oleyl CoA to linoleyl CoA by microsomal preparations from *Carthamus tinctorius*. The heavy arrows indicate experimentally proven reactions and the light arrows indicate those reactions which did not occur experimentally.

The interesting question arises as to the nature of the synthesis of oleyl CoA in plant tissues. No completely defined system has as yet been described although Nagai & Bloch (1968) have described a chloroplast preparation which converted stearyl ACP to oleyl ACP, and Inkpen & Quackenbush (1969) have described a soybean cotyledon system which converted stearic acid to oleic acid.

As we have already indicated, *Carthamus* microsomal preparations were completely inactive with stearyl CoA as a substrate. If, however, ^{14}C-malonyl CoA was added as a substrate to a number of fractions obtained from maturing

Table 2. *Fatty acids produced from* (^{14}C) *malonyl CoA by* Carthamus tinctorius *microsomal and supernatant fractions*

Fraction	Fatty acids (% total) 14:0	16:0	18:0	18:1	20:0	Sp. activity pM/min/mg prot
Microsomal	1	57	16	26	0	72 ± 11
105 000 g Supernatant	1	19	49	30	1	127 ± 24
105 000 g Supernatant (anaerobic incubation)	3	26	71	0	0	114 ± 13
Acetone powder of 105 000 g Supernatant	0	34	61	0	5	141 ± 34

Carthamus seeds, the results summarized in Table 2 were obtained. The principal incorporation site was the soluble protein fraction which under aerobic conditions synthesized as much as 30% of the total ^{14}C fatty acid as oleic acid. Under anaerobic conditions no oleic acid was formed, but a concomitant amount of stearic acid accumulated. It was interesting that, while this system synthesized significant amounts of oleic acid from ^{14}C-malonyl CoA, the addition of ^{14}C-stearic acid, ^{14}C-stearyl CoA, and ^{14}C-stearyl ACP to the system did not lead to ^{14}C-oleic acid. These preliminary results would therefore suggest that the site of saturated and monoenoic acid syntheses was in the soluble protein fraction while conversion of oleyl CoA to linoleyl CoA occurred in a membrane bound enzyme of the endoplasmic reticulum. While the *Carthamus* system is inactive with a number of thioester derivatives of stearic acid, the precursor of oleic acid is probably a specific protein bound stearyl moiety. Purification and characterization of this system is now underway.

We have recently examined a spinach chloroplast system which readily incorporated ^{14}C-acetate into ^{14}C-oleate, ^{14}C-linoleate and ^{14}C-linolenate in the presence of oxygen (Kannangara, Gamini & Stumpf, 1972). As indicated in Table 3 the addition of cyanide at the low level of 0·1 mM had no effect on the

Table 3. *Cyanide inhibition of polyunsaturation by isolated immature spinach chloroplasts incorporating* ^{14}C*-acetate*

KCN	Total lipid	14:0	16:0	18:0 (% ^{14}C)	18:1	18:2	18:3
(mM)	(cpm)						
0	484 000	1	27	2·5	56	4·8	8·7
0·1	529 000	1	28·5	8	54	0	8·6

total incorporation of ^{14}C-acetate into ^{14}C-lipid but had a rather profound effect on the synthesis of linoleic acid. Thus, the biosynthesis of linoleic acid was completely inhibited whereas that of oleic and linolenic acids were not altered. Azide at similar concentrations had no inhibitory effects. These results are of considerable interest since they are not compatible with current thoughts concerning the biosynthesis of polyunsaturated fatty acids; namely: 18:0 → 18:1 → 18:2 → 18:3. Several possible routes such as those given below can be suggested:

(1)

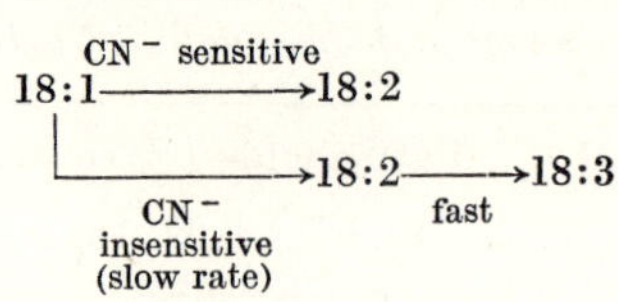

(2)

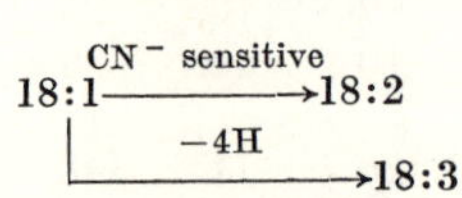

(3)

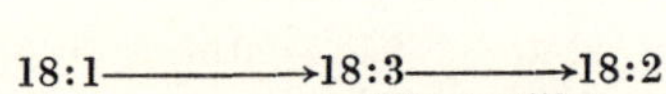

Reaction (1) would postulate two different reactions for the formation of 18:2, a cyanide sensitive step and a cyanide resistant step which is then coupled to the desaturase responsible for the formation of 18:3. In reaction (2) the biosynthesis of 18:3 would require a concerted removal of 4H from 18:1 to form 18:3 while in reaction (3), the formation of 18:3 from 18:1 would first take place and would then be followed by a reduction of the C_{15-16} double bond to form 18:2; this step would presumably be cyanide sensitive. However the concerted removal of 4H and reduction of a double bond in polyunsaturated fatty acids have not been reported in higher plants. Further interpretation of these results must await isolation and characterization of the various chloroplast desaturases.

SUMMARY

The general aspects of desaturation by prokaryotic and eukaryotic organisms are discussed.

The properties of oleyl CoA desaturase from *Carthamus tinctorius* are described. The system is a membrane bound enzyme complex probably associated with the endoplasmic reticulum, and requires NADH and molecular oxygen. Reduced ferredoxin replaces NADH in the system. Presumably an iron containing protein links the reductant to the substrate and the desaturase based on inhibitor studies.

The enzyme has a very stringent substrate specificity in that a *trans* double bond in the C_9–C_{10} position, a shift in the position of the *cis* double bond, chain length, and the substitution of ACP for CoA as the thioester moiety result in complete loss of activity. A careful analysis of the nature of the substrate and product indicates that the correct substrate is oleyl CoA and the product linoleyl CoA. Both thioesters are rapidly and specifically transferred to phosphatidyl choline.

A soluble system which rapidly forms oleic acid is described. The effect of cyanide on desaturation in a chloroplast system is discussed. The results are of interest since in the absence of cyanide, ^{14}C-acetate is rapidly converted to oleic, linoleic and linolenic acids but in the presence of cyanide only oleic and linolenic acids are synthesized.

REFERENCES

Barron, E. J. & Mooney, L. A. (1968). *Anal. Chem.* **40,** 1742.
Brenner, R. (1971). *Lipids* **6,** 567.
Gurr, M. L., Robinson, M. P. & James, A. T. (1969). *Eur. J. Biochem.* **9,** 70.
Inkpen, J. A. & Quackenbush, F. W. (1969). *Lipids* **4,** 539.
Kannangara, C. Gamini & Stumpf, P. K. (1972). *Archs Biochem. Biophys.* (in press).
McMahon, V. & Stumpf, P. K. (1964). *Biochim. biophys. Acta* **84,** 359.
Nagai, J. & Bloch, K. (1968). *J. biol. Chem.* **243,** 4626.
Vijay, I. K. & Stumpf, P. K. (1971). *J. biol. Chem.* **247,** 2910.
Vijay, I. K. & Stumpf, P. K. (1972). *J. biol. Chem.* (in press).

II

CHEMICAL AND PHYSIOLOGICAL ASPECTS OF LIPIDS

INTESTINAL ABSORPTION OF LIPIDS

By J. GANGULY, P. V. SUBBAIAH AND S. PARTHASARATHY

Department of Biochemistry, Indian Institute of Science, Bangalore, India

THE question of intestinal absorption of lipids has been a subject of lively interest for more than a century. Looking back, it now appears that the attempts of earlier workers, though heroic, were handicapped for several reasons. Thus, modern tools like the electron microscope, ultracentrifuge, isotope techniques, etc. were not available to them. Also, biochemistry in general, and cell biology and lipid biochemistry in particular have made rapid strides only in recent years. Naturally, the results obtained were subject to the limitations of the techniques available, and the conclusions reached were limited by the knowledge prevailing at that time. It is no wonder therefore that the entire field remained confused and that a clearer picture regarding fat absorption began to emerge only after advances were made in these other areas.

Thus at the time when Varzar & McDougall (1936) and Frazer (1946) put forward their respective theories of fat absorption the mechanism of action of lipase andthe steps involved in the biosynthesis of TG* had not been elucidated. Only during early fifties did we begin to understand that the pancreatic lipase attacks the primary ester linkages of the TG molecule so that the end products of lipase action are mostly 2-MG and FFA (Mattson *et al.*, 1952; Borgström, 1953, 1954; Mattson & Beck, 1956; Savary & Desnuelle, 1956). At about the same time the pathway (at present known as the α-glycerophosphate (α-GP) pathway) for the biosynthesis of TG was worked out by Kennedy (1957). However, not until early sixties was it known that this is not the major pathway by which TG is synthesized in the mucosal cell, and that the enzymes involved in the resynthesis of TG in the mucosa are different in that, they carry out direct acylation of 2-MG with fatty acyl CoA and are mainly present in the smooth endoplasmic reticulum of the mucosal cell (Cardell, Badenhausen & Porter, 1967).

Yet another point that was overlooked by the early pioneers in this field was the question of solubility of lipids in an aqueous environment. Lipids are inherently insoluble in water and in order that enzymes and cells in general can deal with them effectively they have to be rendered water soluble. An answer to this question was provided by Hofmann & Borgström (1962) who demonstrated that during the course of digestion of TG in human intestine the MG

* Abbreviations used:
TG: Triglyceride; 2-MG: 2-Monoglyceride; FFA: Free fatty acids;
DG: dyglyceride; GP: Glycerophosphate; GPC: Glyceryl phosphoryl choline.

and FFA appear along with bile salts in a water-clear micellar solution. They also suggested that the diameter of the micellar aggregates is less than 100 Å.

Similarly the technique of electron microscopy, which was destined to become a powerful tool in the investigations on absorption of lipids, has been developed only very recently and has revealed that the mucosal cell is fully covered with finger like projections called microvilli, the average distance between two adjacent villi being around 500 to 1000 Å. It is now generally accepted that most of the absorptive surfaces of the mucosal cell are on the membranes of the microvilli, which makes it obvious that for any compound to be absorbed it has to reach these surfaces and its diameter should therefore be less than 500 Å. The formation of micellar aggregates of about 100 Å diameter meets such requirements.

The picture that has emerged from all this recent work can be summarized as follows: During digestion TG are hydrolysed in the lumen of the small intestine by the pancreatic lipase to 2-MG and FFA, which together with bile salts form a water-clear micellar solution. These micellar aggregates readily fit into the space between the microvilli, after which 2-MG and FFA diffuse into the mucosal cell and are re-assembled to TG on the smooth endoplasmic reticulum of the cell.

The possibility will be discussed here that similar mechanisms of micellar solubilization of several other lipids like phospholipids, and cholesteryl and retinyl esters operate during their absorption.

TRIGLYCERIDES

This elegant picture of absorption of TG as summarized above has been supported by fairly good evidence from many directions. However, one of the crucial points that has to be placed on a very firm foundation is an unequivocal demonstration that during the processes of digestion and resynthesis the fatty acid at the 2-position of the triglyceride molecule remains unaffected. Several years ago Mattson & Volpenhein (1964) attempted to provide such information. In their experiments they fed lymph-fistula rats, TG containing labelled fatty acids at different positions on the glycerol and collected the lymph for 24 hr. Analysis of the lymph TG showed that about 70–75% of the fatty acid at the 2-position remained unchanged. However the data on the lymph TG give only overall information on the whole process of absorption and do not reveal details regarding the events that take place during the digestive and cellular phases of absorption.

In order to obtain simultaneous and quantitative information on these two phases of absorption Raghavan & Ganguly (1969) injected into the duodenum of rats, glycerides containing (1-^{14}C) palmitic acid at specific positions and after 1 hr analysed the net amounts and radioactivity of the lipid components in the intestinal lumen and mucosa. It is shown in Table 1 that following injections of glycerol 1,3-dioleate 2(1-^{14}C) palmitate very small proportions of the labelled fatty acid were split off in the intestinal lumen, while more than 80% of the label of the hydrolysed TG was present in the MG and DG. In contrast, when glycerol 1,2-dioleate 3(1-^{14}C) palmitate was administered, no

Table 1. *Appearance of* (1-^{14}C) *palmitic acid in the FFA and glycerides at* 30 *min after an intraduodenal injection of the given compound. Details are described by Raghavan & Ganguly* (1969)

Compound administered	Intestinal contents: Amount of labelled component (μmoles)				Intestinal mucosa: Amount of labelled component (μmoles)			
	MG	DG	FFA	TG	MG	DG	FFA	TG
Glycerol 1,3-dioleate 2(1-^{14}C) palmitate	3·0	4·4	1·2	5·5	1·1	0·8	0·2	12·1
Glycerol 1,2-dioleate 3(1-^{14}C) palmitate	—	1·2	4·3	5·6	—	0·33	0·85	9·3
Mixture of (1-^{14}C) palmitic acid and unlabelled triolein	0·29	0·36	6·8	0·45	—	0·38	2·10	6·1
Mixture of (1-^{14}C) palmitic acid and unlabelled oleic acid	0·24	0·63	9·2	0·81	—	0·51	5·5	2·7

Adapted from Raghavan & Ganguly (1969).

labelled MG accumulated in the lumen, whereas considerable amounts of radioactivity were present in the DG and FFA. These results thus established that only the primary ester linkages of TG are attacked by the pancreatic lipase during digestion, which agrees with the currently accepted concepts of the mechanism of action of lipase on TG.

In these experiments TG was the main labelled component in the mucosal cell thereby indicating that the products of lipolysis are rapidly re-assembled to TG inside the cell. This was further supported by the experiment where labelled palmitic acid was injected in combination with unlabelled triolein. In this case, in spite of markedly high concentrations of the labelled fatty acid in the intestinal lumen, TG constituted the major portion of the labelled lipids in the mucosa. As against these results, when the exogenous glycerol backbone was not provided, e.g. when a mixture of labelled palmitic acid and unlabelled oleic acid was administered, the synthesis of TG in the mucosal cell was limited, which led to an accumulation of labelled FFA in the cell.

The extent to which the fatty acid at the 2-position of the administered TG was retained at the same position of the TG resynthesized in the mucosal cell was then investigated by subjecting the TG isolated from the cells to the action of pancreatic lipase. It will be seen from Table 2 that, when the labelled fatty acid was at the 2-position of the administered TG, 75–80% of the total radioactivity of the mucosal TG was still present at the same position. In contrast, when the labelled fatty acid was at the 1-position of the injected TG or when it

Table 2. *Relative incorporation of* (1-^{14}C) *palmitic acid into the β-position of mucosal TG in the experiments described in Table* 1. *Values of three experiments for each type are presented separately*

Types of compounds used	(1-^{14}C) Palmitic acid remaining at the β-position (%)		
	I	II	III
Glycerol 1,3-dioleate 2(1-^{14}C) palmitate	80	79	75
Glycerol 1,2-dioleate 3 (1-^{14}C) palmitate	16	15	20
(1-^{14}C) Palmitic acid + unlabelled triolein	16	12	20
(1-^{14}C) Palmitic acid + unlabelled free oleic acid	38	32	35

Adapted from Raghavan & Ganguly (1969).

was given as FFA together with unlabelled triolein, only about 20% of the total radioactivity was present at the 2-position. On the other hand, when the exogenous glyceride backbone was not available (following administration of FFA only), the radioactivity was randomly distributed in all the three positions of the TG resynthesized in the cell, which obviously produced evidence for the presence of the α-glycerophosphate pathway for the resynthesis of TG in such cases.

It can be calculated from the amounts of radioactivity present in the lower glycerides and in the FFA of the intestinal lumen (Table 1), that only 14% of the hydrolysed TG undergoes complete hydrolysis to glycerol and free fatty acids, the rest being partially hydrolysed to the MG-stage. It will also be seen from the same table that, following administration of TG alone there was no significant change in the amounts of FFA in the mucosa, while increased amounts of TG accumulated there. In contrast, administration of FFA, alone or together with triolein, resulted in an accumulation of FFA in the cell, obviously due to the limited availability of the glycerol backbone, which had to come from endogenous sources, for the esterification of the excess fatty acids. Thus it is reasonable to conclude that in the absorption of TG, 2-MG is the major end-product of lipolysis and that the 2-MG pathway is the natural and quantitatively more important pathway for the resynthesis of TG.

PHOSPHOLIPIDS

In contrast to such extensive investigations on the absorption of TG the question of absorption of phospholipids has received very little attention. On the basis of experiments with a mixture of phospholipids earlier investigators had come to the conclusion that most of the dietary phospholipids are completely broken down in the intestinal tract. But later workers who used ^{32}P and ^{14}C-fatty acid labelled phospholipids were led to the general conclusion that, while the bulk of the phospholipids are broken down to water-soluble products, small portions of them may be absorbed intact, with respect to their fatty acid moieties. More recent *in vitro* experiments of Nilsson & Borgstörm (1967) with everted intestinal sacs, and *in vivo* experiments of Scow, Stein & Stein (1967), who fed doubly or triply labelled lecithin to lymph-fistula rats, have produced evidence suggesting that during absorption lecithin is hydrolysed to 1-acyl GPC, which is re-acylated to the parent compound, before it is incorporated into chylomicra. Just as in the case of absorption of TG, here also all this earlier work had provided information only on the overall process of absorption and gave little details regarding the digestive and cellular phases of absorption.

Before it can be accepted that such a mechanism operates, it is obviously essential that the presence of enzymes that carry out both hydrolysis and resynthesis of lecithin is demonstrated in the intestine. A powerful phospholipase A that can rapidly hydrolyse lecithin to lysolecithin has been reported in rat pancreas by Gallai-Hatchard & Thompson (1965), and our work has revealed the presence of three types of phospholipases in the homogenates of rat intestinal mucosa namely, phospholipase A, phospholipase B and lysolecithinase. While the phospholipase A was distributed between the brush border and brush border-free particulate fractions, the phospholipase B and lysolecithinase activities were distributed between the brush border-free particulate and soluble fractions (Table 3) (Subbaiah & Ganguly, 1970). Our very recent

Table 3. *Intracellular distribution of the phospholipase and lysolecithin acylase activities in rat intestinal mucosa*

(The specific activity of phospholipase A is expressed as μmoles of lysolecithin formed/hr/mg of protein and those of phospholipase B and lysolecithinase are expressed as μ equiv. of ester bonds hydrolysed/hr/mg of protein. The specific activity of lysolecithin acylase is expressed in nmoles of oleic acid incorporated/hr/mg of protein)

Tissue fraction	Phospholipase A		Phospholipase B		Lysolecithinase		Lysolecithin acylase	
	Specific activity	% of total activity	Specific activity	% of total activity	Specific activity	% of total activity	Specific activity	% of total activity
Homogenate	0·35	100	1·10	100	1·60	100	2080	100
Brush border	0·90	43·2	0·40	19	1·50	17	2060	20
Brush border-free particles	0·51	42·0	1·80	25	3·20	40	5540	65
Soluble	0·20	23·0	0·80	37	2·00	45	114	4

Adapted from Subbaiah & Ganguly (1970) and Subbaiah, Sastry & Ganguly (1970).

work has shown yet another enzyme system in the same tissue that can convert two molecules of lysolecithin into lecithin and GPC by a mechanism of trans-esterification (Subbaiah & Ganguly, 1971).

Regarding acylation of lysolecithin, past attempts to demonstrate such an enzyme in the intestinal mucosal preparations had been unsuccessful (Webster, 1965). But more recently Subbaiah, Sastry & Ganguly (1970) were able to demonstrate such enzyme activity and it was possible to do so by incubating mucosal homogenates with lysolecithin and oleoylCoA for a very short period of time. The enzyme was found to be concentrated in the brush border-free particulate fraction of rat mucosa (Table 3) and its specific activity was highest of all the tissues studied (Table 4).

Table 4. *Tissue distribution of the lysolecithin transacylase in rats*

	nmoles of oleic acid incorporated/hr/mg of protein
Intestinal Mucosa	
Homogenate	1364
Brush border-free particles	4800
Liver	
Homogenate	595
Microsomes	2252
Kidneys (homogenate)	576
Testes (homogenate)	995
Brain (homogenate)	169
Pancreas (homogenate)	285
Epididymal fat pads	303
Spleen (homogenate)	435

Adapted from Subbaiah, Sastry & Ganguly (1970).

Attempts were also made to obtain information on the several phases of absorption of lecithin. For this purpose lecithin containing ^{32}P, and ^{14}C-labelled fatty acids at different positions was prepared and introduced into rat intestine separately after which the intestinal contents and mucosa were analysed for different lipid components after 1 hr; it was found that administration of $\beta(1\text{-}^{14}C)$ linoleoyl lecithin-^{32}P gave rise to an accumulation of lysolecithin in the lumen. Table 5 shows that the $^{14}C/^{32}P$ ratio of the lysolecithin

Table 5. $^{14}C/^{32}P$ *Ratio of lysolecithin and lecithin in the intestinal contents and mucosa of rats at 60 min after an intraduodenal injection of the given compounds*

Compounds administered	Intestinal Contents		Intestinal Mucosa	
	Lysolecithin ($^{14}C/^{32}P$)	Lecithin ($^{14}C/^{32}P$)	Lysolecithin ($^{14}C/^{32}P$)	Lecithin ($^{14}C/^{32}P$)
$\beta(1\text{-}^{14}C)$ linoleoyl lecithin-^{32}P	0·19	1·1	0·19	1·05
$\beta(1\text{-}^{14}C)$ linoleoyl lecithin-^{32}P + unlabelled linoleic acid (1:3)	0·17	0·99	0·00	0·33
$\alpha\text{-}^{14}C$-acyl lecithin-^{32}P	1·30	1·10	1·20	1·10
$\alpha\text{-}^{14}C$-acyl lysolecithin-^{32}P	1·05	—	1·2	1·00

was only about 1/5 of that of the injected lecithin, which indicated that lecithin is hydrolysed at the β-position producing 1-acyl GPC. On the other hand, the $^{14}C/^{32}P$ ratios of the lysolecithin and lecithin of the intestinal contents and mucosa were comparable, which suggested that either the lysolecithin is absorbed into the mucosal cell and is re-acylated with the same labelled fatty acid which was split off, or the lecithin molecule is absorbed unhydrolysed. But there was a considerable decrease in the $^{14}C/^{32}P$ ratio of the mucosal lecithin when the labelled lecithin was given along with unlabelled linoleic acid, which supported the first possibility. Similar decrease in the ratio was observed with unlabelled linolenic acid also, but not with unlabelled palmitic or oleic acid (results not included here), indicating fatty acid specificity for re-acylation.

In those experiments, where the labelled fatty acid was at the α-position of the lecithin or lysolecithin, no changes in the $^{14}C/^{32}P$ ratio of the mucosal lecithin or lysolecithin could be found. At the same time it was also noticed in these experiments that the $^{14}C/^{32}P$ ratio of lecithin of intestinal muscles, blood and liver remained unchanged, which meant that there is little turnover of the fatty acid at the α-position of lecithin (Parthasarathy, Subbaiah & Ganguly, unpublished observations). Sufficient evidence has thus been gathered in support of the concept that the β-ester linkage of lecithin is hydrolysed in the lumen and is re-esterified during absorption.

RETINOL AND CHOLESTEROL

Over the past several years our laboratory has been extensively working on the enzymes and mechanisms involved in the absorption of vitamin A and cholesterol, and we have shown that rat pancreas contains powerful retinyl ester hydrolase and synthetase activities. We have further shown that rat intestinal mucosa also contains these activities and that the pancreatic enzymes are more powerful (Murthy & Ganguly, 1962). By using the procedure of Clark & Porteous (1965) for the fractionation of rat intestinal mucosa and also on the basis of *in vitro* experiments with everted intestinal sacs we had suggested that much of the retinyl acetate hydrolase of rat intestine is situated on the mucosal cell membrane, while the synthetase is mostly intracellular (Ganguly, 1967; 1969). We had also shown that, following feeding of cholesteryl oleate no esterified sterol can be found either in the micellar phase of the intestinal contents or in the brush border, while considerable amounts of unesterified sterol were present in both these fractions. In contrast, the fat phase of the intestinal contents contained considerable amounts of both esterified and unesterified cholesterol. When retinyl acetate was fed the distribution of retinol and its esters was quite different in that, both the alcohol and its esters were present in different proportions in all the fractions, with the fat phase showing a tendency to have more of the esters and the micelle considerably more of the alcohol. It was therefore suggested by us that during absorption retinyl esters are partially hydrolysed in the intestinal lumen, after which retinol and its esters are distributed between the micellar and fat phases. The hydrolysis of the micellar esters is then completed by the hydrolase of the brush border, after which retinol is re-esterified intracellularly. The hydrolysis of the cholesteryl esters, on the other hand, is completed in the intestinal lumen, while it is in the fat phase. (Ganguly, 1969; Ganguly & Murthy, 1970.)

More recently we raised the question as to whether the retinyl ester is inherently incapable of entering the mucosal cell or whether the powerful hydrolase of the brush border does not allow it to enter the cell unchanged. By incubating everted intestinal sacs of rats with retinyl acetate at 0° (a condition under which the hydrolase and synthetase activities are considerably reduced) we could demonstrate that small proportions of the ester can enter the mucosal cell under such conditions (Mathur *et al.*, 1970). At the same time we re-investigated the intracellular distribution of retinyl acetate hydrolase in rat mucosa by following the procedure of Miller & Crane (1961), when the brush border showed about 20% of the total hydrolase activity of the whole homogenate. On successive washing of the brush border fraction with 5 mM EDTA, while the total and the specific activities of the retinyl acetate hydrolase progressively decreased, the specific activities of the marker enzymes, namely, invertase and alkaline phosphatase progressively increased (Table 6) (Mathur, Murthy & Ganguly, 1972). Therefore it now appears that this hydrolase is not a brush border enzyme, and that the terminal hydrolysis of retinyl acetate takes place intracellularly. In sharp contrast to such low recovery of the retinyl acetate hydrolase in the brush border isolated by the procedure of Miller & Crane (1961) it contained almost all of the cholesteryl ester hydrolase activity of the mucosa (David, Malathi & Ganguly, 1965).

Table 6. *Effect of repeated washing of the isolated brush border of rat intestine on the three enzymes*

	Invertase	Alkaline Phosphatase	Retinyl Acetate Hydrolase
Homogenate	6·9*	6·4	0·210
	(100)†	(100)	(100)
Crude brush border	24·0	16·2	0·273
	(60·5)	(54·0)	(22·4)
1st Washing	35·4	19·8	0·153
	(45·5)	(27·4)	(6·4)
2nd Washing	40·5	27·0	0·109
	(35·0)	(25·2)	(3·1)

* Specific activity.
† Per cent of total activity recovered in the fraction.
Adapted from Mathur, Murthy & Ganguly (1972).

RETINOIC ACID

Retinoic acid has had a rather interesting history. It was first synthesized by the two Dutch chemists, Arens & van Dorp (1946). On feeding it to vitamin A-deficient rats they found it supported growth, but they could detect neither retinoic acid, nor retinol or its esters in the rats. Therefore it was assumed that it was neither absorbed nor metabolized to other forms of vitamin A. Such an assumption received full support from Dowling & Wald (1960), who also failed to detect retinoic acid or other derivatives of retinol in the tissues of rats raised on a vitamin A-deficient diet and supplemented with retinoic acid. But in our laboratory we noticed that, following a single dose of large amounts of retinoic acid vitamin A-deficient rats continued to grow for about 4 weeks, after which they started to lose weight (Malathi *et al.*, 1963).

For the first time therefore clear evidence was obtained to show that this compound is not only absorbed by the rat, but also that its effects persisted in the animal for a long time. Soon after this we conclusively showed that retinoic acid is rapidly absorbed by rats and that it quickly disappears from its tissues (Deshmukh *et al.*, 1964). In the meantime Dunagin, Meadows & Olson (1965) demonstrated the presence of retinoic acid in the form of glucuronide in rat bile. It now became clear that retinoic acid is rapidly absorbed but is probably quickly excreted through the bile, and soon in collaboration with Professor Goodman of Columbia University it was demonstrated by using rats with lymph and bile cannulae that it is absorbed entirely through the portal route and is rapidly excreted through the bile (Ganguly, 1969; Ganguly & Murthy, 1970).

DISCUSSION

The question of intestinal absorption of lipids in animals has been a challenging problem for a very long time. Until very recently it was not well understood as to how the animal organism, which operates its metabolic activities in an aqueous environment, can deal with lipids, which are inherently insoluble in water, and yet are required for its survival. Even though the concept of lipoproteins had partly answered this point, the baffling question still remained in that, lipo-proteins cannot exist during digestion of lipids in the intestine. Eventually we learnt that in order to circumvent such difficulties higher organisms have evolved two important secretions, namely, the pancreatic and biliary secretions. The importance of these two secretions in the absorption of lipids was clearly established by Hofmann & Borgström (1962), when they demonstrated that the MG and FFA produced in the intestinal lumen by the action of pancreatic lipase on TG appear along with bile salts in a water-clear micellar solution. Ultimately this observation led to the concept of micellar solubilization of lipids, according to which lipids are absorbed from a finely dispersed and molecularly defined micellar state. It is now generally assumed that from such micellar solutions lipids penetrate the microvillus membranes and enter the mucosal cell, presumably by a process of diffusion (Johnston & Borgström, 1964; Cardell *et al.*, 1967). The concept of micellar solubilization, though mainly developed for understanding the absorption of TG, seems to be important in the absorption of other classes of lipids also, like cholesteryl and retinyl esters and phospholipids.

Although it is now widely accepted that during absorption TG are broken down to 2-MG and FFA, which in turn are re-acylated to TG in the cell, an important missing link in this picture was an unequivocal demonstration that the 2-MG is absorbed intact during this process. We have attempted to provide such information and our results have demonstrated that, while 86% of the hydrolysed TG is present as 2-MG in the lumen, only the remaining 14% is hydrolysed to glycerol and FFA. We could further demonstrate that about 80% of the absorbed 2-MG maintains its positional integrity during re-acylation to TG in the mucosal cell.

Yet another important class of dietary lipids is phospholipids. The small intestine receives considerable amounts of these compounds from the biliary and dietary sources. No doubt these compounds are relatively more soluble in water than TG, but the solubility of their lyso-derivatives is even more pronounced. While the pancreatic secretion has been known to contain a very powerful phospholipase A which can rapidly hydrolyse lecithin to lysolecithin, attempts of earlier workers to demonstrate direct acylation of lysolecithin to lecithin with cell-free preparations from rat intestinal mucosa had failed. But, very recently by using a brush border-free particulate fraction of rat intestinal mucosa Subbaiah *et al.* (1970) succeeded in demonstrating the presence of a highly active enzyme in this tissue. Further, our studies on the mechanism of absorption of phospholipids using lecithin labelled with ^{32}P and ^{14}C-fatty acid esterified at specific positions have clearly shown that dietary lecithin is hydrolysed to its lyso-derivative in the intestinal lumen which is then taken up by the mucosal cell and is re-acylated to lecithin. Therefore it would appear that lysolecithin is the preferred form for micellar solubilization.

Even though lysolecithin might be the preferred species, one should not lose sight of the fact that it is a strong cytolytic agent. Following an intake of large amounts of lecithin a situation might arise where the mucosa may have to face a large concentration of its lyso-derivative. But the mucosal cell is rather unique in that, nature seems to have endowed it with a suitable machinery (in the form of three enzyme systems) for effectively dealing with such a situation. Thus lysolecithin can be converted to lecithin by direct acylation with fatty acyl CoA. It can also be hydrolysed to GPC and free fatty acid. Finally, it can be converted to GPC and lecithin by intermolecular transesterification. While all the three enzymes are active in this tissue, the extent to which lysolecithin is metabolized through any of the three pathways may be determined by several factors. One such controlling factor is the concentration of lysolecithin itself, and the Km values and the specific activities of the three enzymes with respect to lysolecithin would give an indication of their relative importance and significance with respect to the normal physiology of the cell. The results given in Table 7 show that under those conditions where the concentration of

Table 7. *The Km values and specific activities of the three enzymes of lysolecithin metabolism in rat mucosal homogenates. Details are given in Subbaiah & Ganguly* (1970; 1971)

	Acyl transferase	Lysolecithinase	Dismutase
Km	0·25 mM	2·2 mM	3·6 mM
Specific activity	1·36	4·0	0·8

lysolecithin in the cell is low, the acyltransferase reaction would predominate, while the lysolecithinase and dismutase would become more effective when lysolecithin concentration becomes higher. On the other hand, their relative specific activities indicate that, of all the three reactions lysolecithinase is the most active, thereby suggesting that under normal physiological conditions hydrolysis of lysolecithin is the most predominant pathway in this tissue.

Though both retinyl and cholesteryl esters have only one ester group, they differ markedly in their mechanisms of absorption in that, the retinyl esters are partially hydrolysed in the lumen with the result that both the ester and the free alcohol are found in the micelle. Probably small amounts of the micellar esters are taken up by the mucosal cell, and in fact we have been able to demonstrate that under certain conditions unhydrolysed retinyl acetate can enter the mucosal cell. Cholesteryl esters on the other hand are completely hydrolysed during the digestive phase so that only free cholesterol could be detected in the micelle as well as in the brush-border during its absorption (Ganguly, 1969).

One common feature in the absorption of all these lipids thus appears to be their partial or complete hydrolysis in the lumen so as to make the products more soluble in the aqueous environment. Another common feature is re-esterification of such products inside the cell. Such seemingly wasteful processes of hydrolysis followed by resynthesis to the parent compounds appear to serve two important purposes. Firstly, the hydrolytic products are more soluble than the parent compounds. Secondly, through such processes the mucosal cell can exert its selectivity, so that only a particular molecular species is synthesized and transported through the lymph. The latter is well supported by the observations that, while the fatty acid composition of the lymph TG closely resembles that of the dietary TG, the fatty acid composition of the lymph retinyl esters is similar to that at the α-position of the lymph lecithin (Huang & Goodman, 1965). It was therefore suggested by these authors that the fatty acids at the α-position of lymph lecithin and of retinyl esters originate from the same acyl donor.

There are certain differences also in the overall process of absorption of these lipids, e.g. in re-esterification in the cell. Thus resynthesis of retinyl and cholesteryl esters is independent of ATP and CoA, and therefore it seems possible that they are formed by means of transesterification; probably the fatty acids are derived from the α- and β-positions of the lecithin molecule, respectively. The acylation of 2-MG and lysolecithin on the other hand is dependent on ATP and CoA and takes place in the smooth endoplasmic reticulum of the cell. The overall mechanisms of absorption of these lipids are summarized in Fig. 1.

Finally, absorption of retinoic acid is rather unusual in that, it is absorbed through the portal system and is quickly excreted through the bile, in contrast to the behaviour of the common long-chain fatty acids which are absorbed through the lymphatic system. The reasons for such behaviour of retinoic acid are not quite clear. It is possible that it is conjugated in the intestine with glucuronic acid, which prevents its esterification with glycerol or other alcohols and consequently leads to its transport through the portal system.

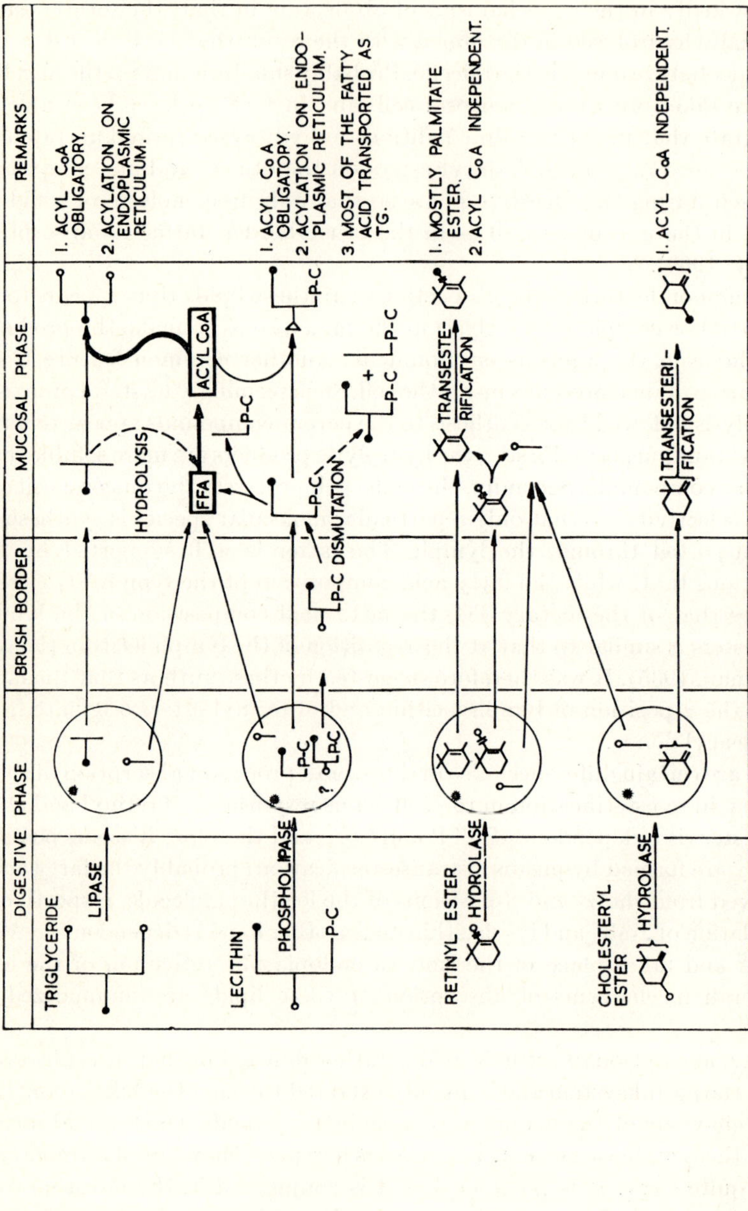

FIG. 1. Summary of the mechanisms of hydrolysis and re-esterification during absorption of triglycerides, lecithin, and retinyl and cholesteryl esters.

SUMMARY

The general mechanisms of absorption of several lipids like triglycerides, lecithin, and cholesteryl and retinyl esters were reviewed. It was discussed that the problems posed by the inherent insolubility of lipids in the aqueous environment during digestion in the intestinal lumen are circumvented by their partial or complete hydrolysis leading to the formation of micellar aggregates. Thus TG is converted to 2-MG, lecithin to its lysoderivative, cholesteryl esters to cholesterol and retinyl esters mostly to retinol. From such micellar state they readily enter the mucosal cell where they are re-esterified to the parent compounds. But the re-esterification mechanism is different for different lipids. Thus 2-MG and lysolecithin are re-esterified through acyl CoA, while retinol and cholesterol are esterified by means of transesterification. Presumably the fatty acids for the esterification of retinol and cholesterol originate from the α and β positions, respectively of the lecithin of the mucosal cell.

ACKNOWLEDGEMENTS

Most of this work was supported by continuous grants from the Indian Council of Medical Research, New Delhi, and by a PL–480 Project No. 314305 from the National Institutes of Health (U.S.A.). Grants for equipment from the Rockefeller Foundation, New York and the Wellcome Trust, London are also acknowledged.

REFERENCES

Arens, J. F. & Dorp, van D. A. (1946). *Nature, Lond.* **158,** 622.

Borgström, B. (1953). *Acta Chim. Scand.* **7,** 557.

Borgström, B. (1954). *Biochim. biophys. Acta* **13,** 491.

Cardell, R. R., Badenhausen, S. & Porter, K. R. (1967). *J. Cell Biol.* **34,** 123.

Clark, B. & Porteous, J. W. (1965). *Biochem. J.* **96,** 539.

David, J. S. K., Malathi, P. & Ganguly, J. (1965). *Biochem. J.* **98,** 662.

Deshmukh, D. S., Malathi, P., Subba Rao, K. & Ganguly, J. (1964). *Indian J. Biochem.* **1,** 164.

Dowling, J. E. & Wald, G. (1960). *Vit. Horm.* **18,** 515.

Dunagin, P. E. Jr., Meadows, E. H. Jr. & Olson, J. A. (1965). *Science,* **148,** 86.

Frazer, A. C. (1946). *Phys. Rev.* **26,** 103.

Gallai-Hatchard, J. J. & Thompson, R. H. S. (1965). *Biochim. biophys. Acta* **98,** 128.

Ganguly, J. (1967). *J. scient. ind. Res.* **26,** 110.

Ganguly, J. (1969). *Amer. J. clin. Nutr.* **22,** 923.

Ganguly, J. & Murthy, S. K. (1970). *J. scient. ind. Res.* **29,** *Suppl.,* pp. S8–S12.

Hofmann, A. F. & Borgström, B. (1962). *Fedn. Proc. Fedn. Am. Socs. exp. Biol.* **21,** 43.

Huang, H. S. & Goodman, D. S. (1965). *J. biol. Chem.* **240,** 2839.

Johnston, J. M. & Borgström, B. (1964). *Biochim. biophys. Acta* **84,** 412.

Kennedy, E. P. (1957) *Ann. Rev. Biochem.* **26,** 119.

Malathi, P., Subba Rao, K., Sastry, P. S. & Ganguly, J. (1963). *Biochem. J.* **87,** 305.

Mathur, S. N., Murthy, S. K. & Ganguly, J. (1972). *Int. J. Vit. Nutr. Res.* **42,** 115.

Mathur, S. N., David, J. S. K., Murthy, S. K. & Ganguly, J. (1970). *Indian J. Biochem,* **7,** 104.

Mattson, F. H. & Beck, L. W. (1956). *J. biol. Chem.* **219,** 735.

Mattson, F. H. & Volpenheim, R. A. (1964). *J. biol. Chem.* **239,** 2772.

Mattson, F. H., Benedict, J. H., Martin, J. B. & Beck, L. W. (1952). *J. Nutr.* **48,** 335.

Miller, D. & Crane, R. K. (1961). *Biochim. biophys. Acta* **52,** 293.

Murthy, S. K. & Ganguly, J. (1962). *Biochem. J.* **83**, 460.
Nilsson, A. & Borgström, B. (1967). *Biochim. biophys. Acta* **137**, 240.
Raghavan, S. S. & Ganguly, J. (1969). *Biochem. J.* **113**, 81.
Savary, P. & Desnuelle, P. (1956). *Biochim. biophys. Acta* **21**, 349.
Scow, R. O., Stein, Y. & Stein, O. (1967). *J. biol. Chem.* **242**, 4919.
Subbaiah, P. V. & Ganguly, J. (1970). *Biochem. J.* **118**, 233.
Subbaiah, P. V. & Ganguly, J. (1971). *Indian J. Biochem. Biophys.* **8**, 197.
Subbaiah, P. V., Sastry, P. S. & Ganguly, J. (1970). *Biochem. J.* **118**, 241.
Varzar, F. & McDougall, E. J. (1936). In *Absorption from the Intestine*. London, New York & Toronto: Longmans, Green & Co.
Webster, G. R. (1965). *Biochim. biophys. Acta* **98**, 512.

STUDIES ON THE EXCRETION OF A $^{14}C/^{32}P$–POLYENE PHOSPHATIDYLCHOLINE IN THE BILE

By J. Hoelzl*

Institut fur Pharmazeutische Arzneimittellehre der Universitat Munchen, Munchen, Germany

It has been reported that the phospholipids of plasma and bile are produced by the liver (Baxter, 1966). While these lipids become water soluble in plasma through binding with specific proteins, they occur in the bile as micelles together with cholesterol and bile acids. The stability of these micelles is influenced by the type of the fatty acids, as for example, it increases on increasing the polyene fatty acids of the phosphatidylcholine. By using a double labelled $^{14}C/^{32}P$-phosphatidylcholine it has now been proved that an exogenous phospholipid can be excreted from the liver into the bile. This labelled phospholipid was prepared by using soybeans (Hoelzl & Wagner, 1971). The ^{14}C-radioactivity in the phosphatidylcholine molecule was localized in the fatty acids with an equal distribution between the 1- and 2-positions. The content of polyene fatty acids was over 70%, and thus this phosphatidylcholine was identical with the so-called EPL (essential phospholipids of Nattermann and Cl). A phospholipid suspension prepared with an ultrasonic disintegrator was injected into rats (20 mg $^{14}C/^{32}P$-phosphatidylcholine per 250 g rat). It could be shown by electrophoresis that the phosphatidylcholine was immediately bound to α-lipo-protein. The injected phospholipid disappeared from the plasma rather rapidly and was incorporated into the organs; highest concentration was detected in the liver. To prove whether an injected phosphatidylcholine can be excreted from the liver into the bile, bile was collected at regular time intervals after cannulating the bile duct. The labelled phospholipid could be detected in the bile as early as 30 min after the injection and maximum radioactivity was registered after 3 hr. One per cent of the injected phospholipid was excreted in the bile during a 6-hr experiment. It could be shown that a part of the injected phosphatidylcholine is excreted unaltered within a period of 1 to 2 hr, after which the $^{14}C/^{32}P$-ratio of the phosphatidylcholine in the bile decreased. Evidently this loss in ^{14}C-radioactivity was due to the deacylation-reacylation cycle of the phosphatidylcholine in the liver (Van Golde, Van den Bosch, Slotboom & Scherpho, 1969). After an injection of $^{14}C/^{32}P$-phosphatidylcholine into the intestine a phosphatidylcholine was detected in the liver with the same $^{14}C/^{32}P$-ratio and

* Not presented at the symposium.

labelled phosphatidylcholine was excreted in the bile also. These experiments with $^{14}C/^{32}P$-phosphatidylcholine therefore show that an exogenous phospholipid, which differs significantly in its high content of the polyene fatty acids from the phosphatidylcholine of the bile, can be excreted from the liver into bile.

REFERENCES

Baxter, J. H. (1966). *J. Lipid Res.* **7**, 158.

Hoelzl, J. & Wagner, H. (1971). *Z. Naturf.* **26b**, 425.

Van Golde, L. M. G., Van den Bosch, H., Slotboom, A. J. & Scherpho, G. L. (1969). *Progr. biochem. Pharmacol.*, Vol. 5, p. 129. Basel/New York: Karger.

THE EFFECTS OF STARVATION AND OF COLD STRESS ON CLEARING FACTOR LIPASE ACTIVITY AND TRIGLYCERIDE FATTY ACID UTILIZATION IN RAT HEART

M. PERENNA ROGERS, J. BORENSZTAJN* AND D. S. ROBINSON

Department of Biochemistry, University of Leeds, Leeds, England

THE main physiological function of the enzyme, clearing factor lipase or lipoprotein lipase, is to facilitate the uptake of triglyceride fatty acids (TGFA) from blood by extrahepatic tissues (see Robinson, 1970). This function is thought to be exercised through an initiation by the enzyme of the hydrolysis of chylomicron and very low density lipoprotein triglycerides that are sequestered at the luminal surface of the endothelial cells of blood capillaries.

Changes in activity of the enzyme that occur in particular tissues with changes in physiological status are also believed to be responsible for corresponding changes in the uptake of plasma TGFA by those tissues. Thus, the enzyme probably performs an additional directive function in determining the *pattern* of TGFA removal from the bloodstream. Convincing evidence in support of this view is provided by the correlations between the activity of the enzyme and the capacity for TGFA removal that have been reported for adipose tissue and the mammary gland in different physiological states (see Robinson, 1970). Similar correlations have been sought with respect to the enzyme in heart muscle, but generally results in this case have not been so clear cut. This could be explained if only a part of the total clearing factor lipase activity of heart was functional in TGFA uptake and there is, in fact, evidence to support this. Thus, when heparin is perfused through the heart for a very short time, only a part of the total activity of the tissue is released into the perfusate (Robinson & Jennings, 1965). However, once this enzyme has been released, the capacity of the heart to utilize chylomicron TGFA perfused through it is subsequently reduced to a very low level indeed (Enser *et al.*, 1967; Borensztajn & Robinson, 1970).

Heparin is believed to liberate specifically the proportion of the total enzyme that is located at the endothelial cell surface (Robinson, 1963) and these findings suggest that correlations between TGFA utilization and clearing factor lipase activity in the heart should properly be sought only with this portion of the total activity of the tissue. We summarize here results of studies which support this view for the enzyme in the hearts of rats that have been starved

* Present address: Department of Pathology, University of Chicago, Chicago, Illinois, U.S.A.

and/or exposed to acute cold stress. Some of the findings have been reported previously (Borensztajn, Otway & Robinson, 1970; Borensztajn & Robinson, 1970).

EXPERIMENTAL

The clearing factor lipase activity of the heart was measured in aqueous homogenates and in perfusates of the organ by methods presented elsewhere (Robinson & Jennings, 1965; Borensztajn *et al.*, 1970). The techniques of heart perfusion in a non-recirculatory system and of measurement of ^{14}C-labelled chylomicron TGFA utilization by the heart in terms of $^{14}CO_2$ output and of ^{14}C incorporation into the tissue lipids, as well as the analytical methods used, have also been described previously (Enser *et al.*, 1967; Borensztajn & Robinson, 1970).

RESULTS

Effects of starvation

On the capacity of the perfused rat heart to utilize chylomicron triglyceride fatty acids. The capacity of the rat heart to take up and utilize chylomicron TGFA perfused through it has been amply demonstrated (Delcher, Fried & Shipp, 1965; Kreisberg, 1966; Crass & Meng, 1966; Enser *et al.*, 1967). The effect of a short period of starvation on this capacity is shown in Table 1. Perfusion of

Table 1. *The effect of starvation for 10 hr on the utilization of ^{14}C-labelled chylomicron lipid by the perfused rat heart*

State of animals	TGFA oxidation (μeq. oxidized/g tissue fresh weight/30 min perfusion)	TGFA incorporation into heart lipids (μeq. incorporated/g tissue fresh weight/30 min perfusion)
Fed	1·25 ± 0·60	0·42 ± 0·20
Starved for 10 hr	2·66 ± 0·40	3·03 ± 1·70

Groups of six hearts, taken from rats on their normal diets which were killed between 8 a.m. and 10 a.m. (fed rats) or from rats that had been starved for 10 hr from 8 a.m. on the day of the experiment, were perfused for 2 min with Krebs-Henseleit bicarbonate buffer (pH 7·4) in a non-recirculatory system and then for 30 min in the same system but with buffer containing albumin-washed ^{14}C-labelled chylomicrons (0·9 μeq. TGFA/ml). The perfusates were collected and their $^{14}CO_2$ contents were determined. At the end of each perfusion, the heart was again perfused with buffer alone for 2 min. Portions of the heart ventricles were then taken for the measurement of ^{14}C incorporated into the tissue lipids. Results are expressed as the means ± S.D.

^{14}C-labelled chylomicron lipids for 30 min through hearts from rats that had been starved for 10 hr resulted in a markedly higher rate of $^{14}CO_2$ production and of ^{14}C incorporation into the heart lipids than was observed with hearts from fed rats.* The time course of the $^{14}CO_2$ production is not shown in Table 1 but it was, in fact, unchanged by starvation (see Borensztajn & Robinson, 1970).

Some 98% of the radioactivity in the ^{14}C-labelled chylomicrons perfused through the heart in this experiment was in the glyceride fraction and no significant changes in this proportion occurred during the perfusion (Enser *et al.*, 1967). Moreover, the chylomicrons used were pretreated to reduce their content of ^{14}C-labelled FFA to between 0·1 and 0·3% of the total lipid radioactivity (Borensztajn & Robinson, 1970). Thus, the $^{14}CO_2$ produced was undoubtedly derived largely from the ^{14}C-labelled TGFA in the chylomicrons.

The possibility that the increased $^{14}CO_2$ production by the hearts from the starved animals might be accounted for by an increase in the capacity of the hearts to oxidize FFA that had been produced from the chylomicron TGFA was excluded by perfusing groups of hearts from fed and from starved animals with ^{14}C-labelled palmitic acid (0·2 mM). The production of $^{14}CO_2$ was, in fact, significantly *reduced* from 0·63 ± 0·12 to 0·39 ± 0·12 μmoles fatty acid oxidized/g fresh weight/15 min with the hearts from the starved animals. Similar findings have been reported by Vahouny, Katzen & Entenman (1967).

On the clearing factor lipase activity of rat heart. The effect of starvation on the total clearing factor lipase activity of rat heart is shown in Fig. 1. The activity increased rapidly during the first few hours of starvation to reach a maximum value after 10 to 24 hr. Thereafter, there was a slow decline so that, after 72 hr, it was not significantly higher than in fed animals.

Table 2. *The effect of starvation for* 10 *hr on the release of clearing factor lipase from the perfused rat heart by heparin*

State of animals	Clearing factor lipase activity (units/g tissue fresh weight) of: Perfusate samples 0–1 min	1–3 min	3–5 min	5–15 min	Hearts at the end of perfusion
Fed	22	5	2	9	47
Starved for 10 hr	153	19	9	34	99

Groups of five hearts, taken from rats on their normal diets which were killed between 8 a.m. and 9 a.m. (fed rats) or from rats that had been starved for 10 hr from 8 a.m. on the day of the experiment, were perfused with 5% (v/v) serum in Krebs-Henseleit bicarbonate buffer containing 5 IU heparin/ml. The perfusate from each heart was collected in a series of samples over the time intervals shown. The samples from each time interval were combined and clearing factor lipase activity was measured in duplicate samples of aqueous homogenates of the combined ventricles of each group. Mean values are reported.

* The rate of $^{14}CO_2$ production by the hearts of rats starved for 10 hr in this experiment is lower than that reported in Table 6. This may be accounted for by the fact that the hearts were perfused at a higher TGFA concentration. Thus, it has been found that the rate of oxidation reaches a maximum at a concentration of between 0·2 and 0·5 μeq. TGFA/ml of perfusion medium and declines somewhat at concentrations above this.

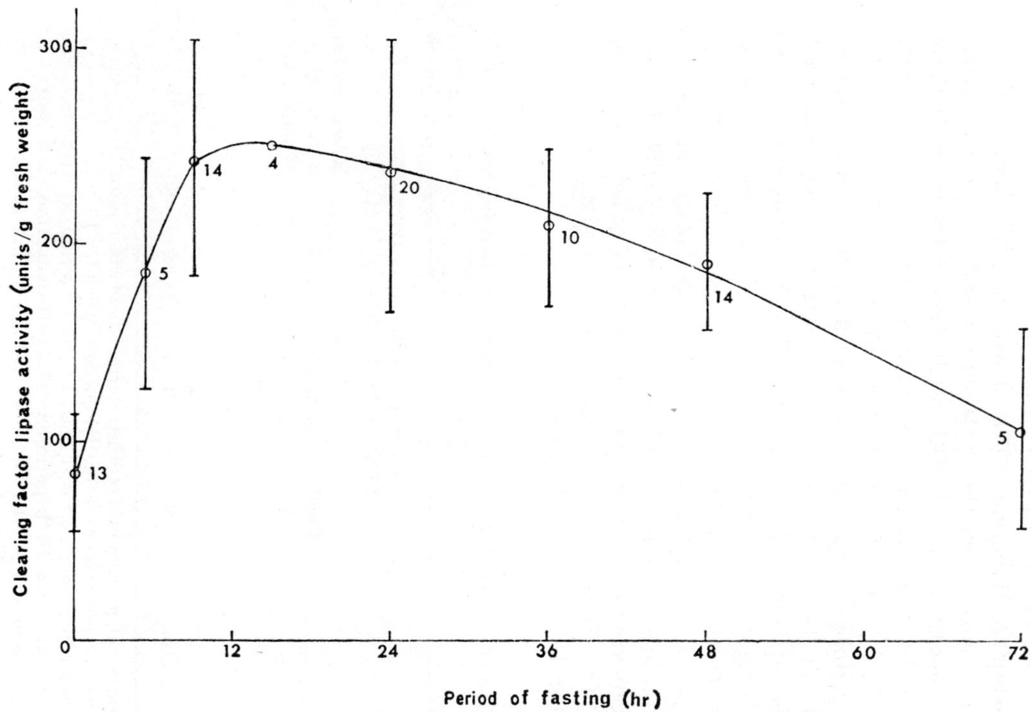

FIG. 1. Time-course of the effect of starvation on the clearing factor lipase activity of rat heart. Hearts were taken between 8 a.m. and 9.30 a.m. from rats which had been maintained on their normal diet, and at appropriate later times from rats which had been starved from 8 a.m. on the first day of the experiment. Clearing factor lipase was assayed in duplicate samples of homogenates of the hearts of each rat. The mean activities at each time interval are recorded. The vertical bars represent SD, and the number of animals in each group is shown. All values for starved animals are significantly higher than those for fed animals ($P < 0{\cdot}001$), except for those for animals starved for 72 hr. The values for animals starved for 36 hr or longer are significantly lower than those for animals starved for 10 hr ($P < 0{\cdot}05$).

The increase in the total enzyme activity occurring during the first 10 hr of starvation was mainly accounted for by a rise in the activity of the enzyme which can be rapidly released from the perfused heart by heparin. Thus, the results in Table 2 show that the activity released during the first minute of a 15 min-heparin perfusion was 153 units/g fresh weight, using hearts from animals starved for 10 hr, as compared to 22 units/g fresh weight using hearts from fed animals.

The enzyme activity remaining in the heart at the end of the perfusion for 15 min was also higher in the starved than in the fed rat, however (99 *vs* 47 units/g fresh weight). Although the absolute difference between these activities was not so marked as that between the heparin releasable enzyme activities, it is clear that the increased TGFA utilization by the hearts from starved animals shown in Table 1 cannot in these circumstances be unequivocally attributed to the rise in the heparin releasable enzyme activity.

Effects of cold exposure

On the clearing factor lipase activity of rat heart. Grafnetter *et al.* (1965) have reported briefly that exposure of rats to a low temperature for short periods increases the total clearing factor lipase activity of the heart. The results of a more detailed investigation of this effect are described here.

The effect of exposure at 4° for 10 hr of rats that were either starved or fed glucose by stomach tube at intervals during this period are shown in Table 3.

Table 3. *The effect of cold exposure for* 10 *hr on the total clearing factor lipase activity of rat heart. Studies on fed rats and on rats starved for* 10 *hr*

State of animals	Number in group	Clearing factor lipase activity (units/g tissue fresh weight)	P value
Kept at 25°. Fed glucose	5	91 ± 23	<0·001
Kept at 4°. Fed glucose	5	189 ± 29	
Kept at 25°. Starved	10	299 ± 46	<0·01
Kept at 4°. Starved	10	370 ± 43	

Food was removed from the cages of four groups of rats at 9 a.m. on the day of the experiment and the animals in two of the groups were put into a cold room at 4° in separate cages, while the other two groups remained at room temperature (25°). At 12 noon and at 4 p.m. the animals in one of the groups at 4° and in one of the groups at 25° were each given 4 ml of a 60% (w/v) glucose solution in water by stomach intubation while under light ether anaesthesia. The animals were killed between 7 and 8 p.m. and clearing factor lipase was assayed in aqueous homogenates of the heart ventricles. The results are expressed as means ± S.D.

It is clear that cold exposure leads to a marked increase in the total clearing factor lipase activity of the heart in both starved and glucose-fed animals, and that, in the starved rats, this rise is superimposed on the increase due to the starvation *per se*.

Further studies were carried out to see how far the increases due to cold exposure could be attributed to changes in the heparin releasable activity of the tissue, this activity being equated as in the experiment described in Table 2 with that which was released into the perfusate by perfusion of the hearts for 1 min with heparin. Table 4 shows that, in animals given glucose during the

Table 4. *The effect of cold exposure for* 10 *hr on the release of clearing factor lipase from the perfused rat heart by heparin. Studies on fed rats and on rats starved for* 10 *hr*

State of animals	Number in group	Clearing factor lipase activity (units/g tissue fresh weight)			P values (4° *v* 25°)		
		Heparin perfusate	Heart at the end of perfusion	Total	Perfusate	Heart	Total
Kept at 25°. Fed glucose	5	13 ± 10	79 ± 9	92 ± 19			
Kept at 4°. Fed glucose	5	47 ± 25	158 ± 48	205 ± 46	<0·05	<0·01	<0·001
Kept at 25°. Starved	5	152 ± 18	173 ± 23	325 ± 40			
Kept at 4°. Starved	5	130 ± 5	253 ± 29	383 ± 27	<0·05	<0·01	<0·05

The experimental procedure was as described in the legend for Table 3, except that, after the rats had been killed between 7 and 8 p.m., their hearts were perfused for 1 min with Krebs-Henseleit bicarbonate buffer (pH 7·4) containing heparin (5 IU/ml) and 5% (w/v) serum. Clearing factor lipase was measured in each 0–1 min perfusate sample and in aqueous homogenates of the heart ventricles prepared at the end of the perfusion. The results are expressed as means ± S.D.

period of cold exposure, the rise in the heparin releasable activity, though significant, accounted for only a small part of the increase in the total activity. And, in rats that were starved during the period of exposure at 4°, there was, in fact, a significant fall in the heparin releasable activity, the rise in total activity being wholly attributable in these circumstances to the enzyme that remained associated with the heart muscle after a 1 min-perfusion with heparin.

In rats starved at room temperature for longer periods than 24 hr there was a slow fall in the total clearing factor lipase activity of the heart (see Fig. 1) and it was of interest to see whether cold exposure in such animals was able to arrest this decline and produce a rise in the activity of the enzyme similar to that observed in animals that had not been starved prior to exposure at 4°. Accordingly groups of rats which had already been starved at room temperature for 48 hr were starved for a further period of 3 or 10 hr at 4°. Their hearts were then perfused for 1 min with heparin, after which clearing factor lipase was measured in the perfusates and in the hearts at the end of the perfusion. Table 5 shows that cold exposure of such animals did indeed cause a marked rise in the total clearing factor lipase activity of the heart. In fact, the activity in the hearts of animals starved for 48 hr at room temperature and then for 10 hr at 4° was as high as that in the hearts of rats starved for only 10 hr at 4° (cf., Table 4).

However, only a very small part of this rise in total activity was accounted for by an increase in the heparin releasable activity. In fact, though the

Table 5. *The effect of cold exposure for 3 hr or 10 hr on the release of clearing factor lipase from the perfused rat heart by heparin. Studies on rats starved for 48 hr*

State of animals	Clearing factor lipase activity (units/g tissue fresh weight)		
	Heparin perfusate	Heart at the end of perfusion	Total
Starved 48 hr at 25°	60 ± 20	80 ± 15	140 ± 15
Starved 48 hr at 25° + 3 hr at 4°	115 ± 15	150 ± 30	265 ± 20
Starved 48 hr at 25° + 10 hr at 4°	95 ± 10	285 ± 35	380 ± 30

Three groups of six rats were starved for 48 hr at room temperature (25°). At 9 a.m. on the day of the experiment, the animals in two of the groups were put into a cold room at 4° in separate cages. The rats in the other group were killed and their hearts were perfused with heparin for 1 min as described in the legend for Table 4. The hearts of the rats in the cold-exposed groups were perfused similarly after the animals had been at 4° for either 3 or 10 hr. Clearing factor lipase was measured in each 0–1 min perfusate sample and in aqueous homogenates of the heart ventricles prepared at the end of the perfusion. Results are expressed as means ± S.D.

activity of this fraction increased significantly ($P < 0{\cdot}01$) after 3 hr of cold exposure it then declined slightly. The continuing increase in the total activity from 3 to 10 hr was accounted for, therefore, by the fact that the fraction of the total activity which was not released by heparin increased markedly throughout this period.

On the capacity of the perfused rat heart to utilize chylomicron triglyceride fatty acids. The various changes in heart clearing factor lipase activity brought about by exposure of rats at 4° reported in the previous section suggested that a study of the utilization of chylomicron TGFA by such hearts would be of interest. Thus, the activities of the total and the heparin non-releasable enzyme in the heart were higher in rats which were starved for 10 hr at 4° than they were in those starved at room temperature, whereas that of the heparin releasable enzyme was slightly lower (Table 4). If total, or heparin non-releasable, enzyme determined the capacity of the heart for TGFA utilization it should clearly be higher in the animals exposed at 4°. The results of an experiment to test this are shown in Table 6. In fact, both $^{14}CO_2$ release and ^{14}C incorporation into heart lipid were lower in the animals kept at 4°. The pattern of $^{14}CO_2$ production, though not shown here, did not change on cold exposure and was similar to that described previously (Borensztajn & Robinson, 1970). These results therefore show that neither the total nor the heparin non-releasable clearing factor lipase activity controls the rate of chylomicron TGFA utilization by the heart and suggest rather that the activity of the heparin releasable enzyme is the determining factor.

A similar conclusion can be drawn from a study of chylomicron TGFA utilization by the hearts of rats starved for 48 hr at room temperature and then kept either at 4° or at room temperature. During the first 3 hr of cold exposure, the activity of both heparin releasable and non-releasable clearing factor lipase

Table 6. *The effect of cold exposure and starvation for* 10 *hr on the utilization of* ^{14}C-*labelled chylomicron TGFA by the perfused rat heart*

State of animals	TGFA oxidation (μeq. oxidized/g tissue fresh weight/20 min perfusion)	TGFA incorporation into heart lipids (μeq. incorporated/g tissue fresh weight/20 min perfusion)
10 hr starved at 25°	3·38 ± 0·48	3·00 ± 0·40
10 hr starved at 4°	2·70 ± 0·42	2·17 ± 0·50
P value	$<0·05$	$<0·05$

Two groups of five rats were used. On the day of the experiment at 9 a.m. food was removed and the animals in one group were put into a cold room at 4° in separate cages while the animals in the other group remained at room temperature (25°). Between 7 p.m. and 9 p.m. the hearts were perfused for 20 min with Krebs-Henseleit bicarbonate buffer (pH 7·4) containing albumin-washed ^{14}C-labelled chylomicrons (0·45 μeq. TGFA/ml). The perfusates were collected and their $^{14}CO_2$ contents were determined. At the end of each perfusion, the heart was again perfused with buffer alone for 2 min. Portions of the heart ventricles were then taken for the measurement of ^{14}C incorporated into the tissue lipids. Results are expressed as the means ± S.D.

rose in the heart (Table 5) and an increase in TGFA oxidation would be expected whichever fraction was concerned in its utilization. However, between 3 and 10 hr of cold exposure further increase in total clearing factor lipase was entirely attributable to a rise in the non-releasable fraction and there was, in fact, a slight decline in heparin releasable activity. If there were increased TGFA utilization over this period then the non-releasable enzyme

Table 7. *The effect of cold exposure for* 3 *hr and* 8 *hr on the utilization of* ^{14}C-*labelled chylomicron TGFA by the perfused hearts of rats starved for* 48 *hr*

State of animals	TGFA oxidation (μeq. oxidized/g tissue fresh weight/20 min perfusion)	TGFA incorporation into heart lipids (μeq. incorporated/g tissue fresh weight/20 min perfusion)
48 hr starved at 25° + 3 hr starved at 25°	1·43 ± 0·31	1·11 ± 0·20
48 hr starved at 25° + 3 hr starved at 4°	2·40 ± 0·10	1·89 ± 0·39
48 hr starved at 25° + 8 hr starved at 25°	1·55 ± 0·10	1·29 ± 0·10
48 hr starved at 25° + 8 hr starved at 4°	2·04 ± 0·40	1·41 ± 0·21

Four groups of rats (six animals per group) were used. Two days before the experiment, all animals were starved at room temperature (25°). On the day of the experiment the rats in two of the groups were put in separate cages at 4°. Heart perfusions were carried out when the rats at 4° had been exposed for 3 hr or for 8 hr, the hearts of the rats kept at 25° being perfused for the same period of time as those kept at 4°. Perfusions and subsequent procedures were carried out as described in the legend for Table 6. Results are expressed as means ± S.D.

would be implicated. If not, then it would seem reasonable to conclude that the increased utilization during the first 3 hr of exposure was due to the increase in the heparin releasable enzyme activity over this period. The results of an experiment to test these various possibilities are shown in Table 7. As anticipated both $^{14}CO_2$ production and ^{14}C incorporation into the heart lipids were increased significantly ($P < 0{\cdot}01$) in animals exposed to 4° for 3 hr. However, after 8 hr at 4°, both these parameters were reduced below the levels at 3 hr. Although the decline in the utilization of TGFA with increasing time of exposure at 4° was not significant in this experiment it has always been observed in other experiments not reported here. Moreover, it should be noted that the fall in the heparin releasable enzyme activity with increasing time of cold exposure is itself not very marked (Table 5). The lack of any significant change in TGFA utilization by the hearts exposed throughout at 25° was expected since the heart clearing factor lipase activity does not change significantly in 48 hr-starved animals kept at room temperature for such a time (Fig. 1).

DISCUSSION

The definition in this study of the heparin releasable enzyme of rat heart as that proportion of the total that is released during a 1 min-perfusion with heparin is clearly a somewhat arbitrary one. Nevertheless, its distinction from the remainder of the enzyme in the heart that is not released by heparin under these conditions does seem to be valid in view of the different effects of starvation and cold stress on these two fractions that have been observed in the present work. Moreover, it is clear that there is an excellent relationship in the various conditions studied between the heparin releasable enzyme activity and the utilization of chylomicron TGFA by the perfused heart, as expressed both in terms of $^{14}CO_2$ production and of ^{14}C-incorporation into the heart lipids. On the other hand, there was no corresponding relationship between $^{14}CO_2$ production or ^{14}C-incorporation and either the total or the heparin non-releasable enzyme activities. These findings then provide strong evidence for the view that, when correlations are sought between changes in heart TGFA utilization and clearing factor lipase activity, it is only the heparin releasable fraction of the latter which should be considered.

Consideration of the mechanism of the changes in heart clearing factor lipase activity observed on starvation and cold stress is outside the scope of this paper. However, the rapid and substantial increase in the activity of the heparin releasable enzyme that occurs in animals starved for only a few hours is of considerable interest since such periods of starvation produce a marked fall in the activity of the enzyme in adipose tissue (Wing & Robinson, 1968). Moreover, similar proportional increases to those found in the total activity of heart are observed in skeletal muscle (Borensztajn & Robinson, 1970). It seems reasonable to conclude, therefore, that these changes may be responsible for a direction of the plasma TGFA away from adipose tissue and towards the muscular tissues during the early stages of starvation. A similar redirection may occur in the cold-stressed animal though this cannot be regarded as established until the effects of cold exposure on the heparin releasable enzyme activity and on TGFA utilization in skeletal muscle and adipose tissue have been carried out. Moreover, it will clearly be necessary to establish in such studies that any changes observed are due to cold stress *per se* rather than to any incidental changes in nutritional status.

The rapidity and extent of the changes in the activity of the heart enzyme due to starvation and cold stress are of interest from another stand point. Thus, the rat is normally a nocturnal feeder and, when animals with free access to food are killed late in the day, the activity of the enzyme in the heart may be expected to be appreciably higher than that in the hearts of animals killed several hours earlier. Moreover, it seems possible that the increase in activity due to cold stress may be one example of a relatively non-specific response to stress (M. P. Rogers, unpublished observations). It is clearly important, therefore, in experiments on the activity of the enzyme in the heart, to pay particular attention to the status of the animal with respect to nutrition and stress. These factors may indeed have been responsible for some of the changes that have hitherto been attributed to specific experimental procedures (see Robinson, 1970).

ACKNOWLEDGEMENT

Dr. M. P. Rogers was in receipt of an M.R.C. Scholarship for training in research methods when this work was carried out and Dr. J. Borensztajn was a Technical Assistance Trainee supported by the Ministry of Overseas Development. Overall support for the study was provided by the Medical Research Council.

REFERENCES

Borensztajn, J., Otway, S. & Robinson, D. S. (1970). *J. Lipid Res.* **11**, 102.

Borensztajn, J. & Robinson, D. S. (1970). *J. Lipid Res.* **11**, 111.

Crass, M. F. & Meng, H. C. (1966). *Am. J. Physiol.* **206**, 610.

Delcher, H. K., Fried, M. & Shipp, J. C. (1965). *Biochim. biophys. Acta* **106**, 10.

Enser, M. B., Kunz, F., Borensztajn, J., Opie, L. H. & Robinson, D. S. (1967). *Biochem. J.* **104**, 306.

Graftnetter, D., Grafnetterova, J., Grossi, E. & Morganti, P. (1965). *Med. Pharmacol. Exptl.* **12**, 266.

Kreisberg, R. A. (1966). *Am. J. Physiol.* **210**, 379.

Robinson, D. S. (1963). *Advanc. Lip. Res.* **1**, 133.

Robinson, D. S. (1970). *Comp. Biochem.* **18**, 51.

Robinson, D. S. & Jennings, M. A. (1965). *J. Lipid. Res.* **6**, 222.

Vahouny, G. V., Katzen, R. & Entenman, C. (1967). *Biochim. biophys. Acta* **137**, 181.

Wing, D. R. & Robinson, D. S. (1968). *Biochem. J.* **106**, 667.

THE SYNTHESIS OF PROSTAGLANDINS IN HUMAN PLATELETS

By J. CLAUSEN AND K. SRIVASTAVA

The Neurochemical Institute, Copenhagen, N. Denmark

PLATELET aggregation in human, rat and pig plasma is induced by ADP (Born, 1962) and inhibited by PGE_1 (Kloeze, 1966). Aggregation caused by epinephrine or collagen, the increased adhesiveness of platelets to glass, and the reduced electrophoretic mobility of the platelets effected by ADP is also reduced by PGE_1 (Emmons *et al.*, 1967). Since the half-life of the prostaglandins in the blood stream is only about 2 min due to inactivation in the lung tissue (Piper, Vane & Myllie, 1970), the role of prostaglandins as inhibitors of agglutination of the platelets can be of biological importance only if the blood platelets possess the whole enzymic system for prostaglandin synthesis. The present communication has been designed to elucidate this point.

MATERIALS AND METHODS

Preparation of platelets

Blood was collected from normal human donors using trisodium citrate (3·8%, adjusted to pH 7·4 with citric acid) as anticoagulant in the ratio of 9:1 (v/v). All glassware was siliconized. Platelet-rich plasma (PRP) was obtained by centrifugation at 275 g for 15 min at room temperature (Deykin & Desser, 1968). The PRP was pipetted into another centrifuge tube and centrifuged at 2500 g for 30 min. The platelet-pellet thus obtained was suspended in the Krebs-Ringer-phosphate buffer (KRP) to a final concentration of about $2{\cdot}0 \times 10^9$ platelets/ml. The platelets were counted in a Bürker-Türk's counting chamber. The suspension used was 99·9% pure with practically no admixture of leucocytes and erythrocytes (Brecher, Schmierderman & Cronkite, 1953).

Chemicals

Unless otherwise stated, the chemicals were of the highest purity obtainable from the British Drug Houses, (BDH) Poole, Dorset, England.

Preparation of Acetate-1-^{14}C

The concentration of the labelled sodium acetate (Radiochemical Centre, Amersham, Buckinghamshire, England, specific activity 59 mCi/mmole, total activity 250 μCi) was adjusted to 50 μmoles of acetate by adding non-radioactive acetate (AR, BDH) and made up to a volume of 1000 μl.

Glutathione (reduced form, GSH, Biochemical, BDH)

Eleven mg of glutathione was dissolved in 2 ml redistilled water, stored in a refrigerator and used within a week.

Prostaglandin E

This was prepared from human seminal plasma as described by Hamberg & Samuelsson (1966). Reference prostaglandin E_1 was a gift from the Unilever Research Laboratories, Vlaardingen, The Netherlands.

Metabolic studies

Incubations were done in siliconized pyrex glass test tubes at 37° with a continuous passage of oxygen (11 ml/min in water-saturated O_2). In order to elucidate the experimental conditions, two kinds of incubations were done.

(a) *Incubations with varying concentrations of the acetate*
The incubation mixture consisted of 1 ml of the platelet suspension, 1 ml of the buffer (KRP), 0·36 μmoles of glutathione and an appropriate amount of the radioactive acetate. The incubation was carried out for 2 hr with a continuous supply of oxygen as described above.

(b) *Incubations with varying time*
The incubation mixture contained 1 ml of platelet suspension, 1 ml of the buffer, 0·36 μmoles of glutathione and the acetate in a concentration of 2·4 m moles/litre.

The reactions were initiated by adding sodium acetate and terminated by shaking the reaction mixture with chloroform-ethanol mixture (7:3 v/v).

Extraction of lipids

500 μl of clear plasma was added to the incubation mixture. The mixture was extracted overnight with 20 ml chloroform-methanol in the cold at 4°. The clear upper aqueous phase was carefully removed and 1 ml of distilled water was added to the extraction mixture in such a way that it did not mix with the organic phase (as judged from no emulsion resulting on addition of water). The water phase was again pipetted out immediately. This procedure was repeated twice. The organic phase was filtered through a phase separating filter (Whatman No. 1) paper previously soaked with the extraction mixture. The filter paper was washed twice with 2 ml of the solvent. The filtrate was concentrated to dryness under N_2 at 65°, followed by further drying under vacuum at 80°. This step was essential, because sodium acetate is soluble both in methanol and in methanol-water mixture, but is insoluble in chloroform. The dry residue was extracted with 10 ml chloroform and again filtered through a phase separating paper (vide supra). Prior to filtration this paper was soaked in chloroform in a dessicator, so that the water vapour which might condense due to the cooling produced by evaporation of chloroform during filtration did not dissolve the acetate and thereby transfer it through the paper. The filter paper was thrice washed with 3 ml chloroform and the filtrate then was evaporated to a residue under N_2 at 60°.

Experiments controlling the filtration process were done by using 25 μl of radioactive sodium acetate (sp. activity 59 mCi/m mole; total activity 250 μCi in 1000 μl) and by equilibrating this between 2 ml of water and 15 ml of the extraction mixture. The residue obtained by the process described above showed less than 170 cpm.

Thin-layer chromatography (TLC)

TLC was carried out with silica gel G (0·25 m) (Merck, Darmstadt, W. Germany) using the equipment of Desega Werke (W. Germany). The plates were activated at 110° for 30 min and stored in a dessicator until use. The PGE mixture (ca. 500 μg) was added in ethereal solution to the residue mentioned above and the material was applied with a Hamilton syringe (50 μl) on the TLC-plate. The plate was developed with ethylacetate-isooctane-acetic acid-water (110:20:10:100 (v/v) equilibrated for 1 hr before using the organic phase). Reference spots of the PGE mixture were placed on the lateral parts of the plate. A total run of 12 cm was maintained in all the experiments. The lipid fractions were visualized by iodine vapour.

Assay of radioactivity

The radioactivity of the various zones from the plates was measured in a Beckman liquid scintillation counter (model 230) with a fluor prepared by mixing 8·0 g of butyl PBD and 0·5 g of PBBO in 1 litre of toluene. Counting efficiency was 72%. Corrections for quenching were made with an external standard. Total acetate incorporated in all the experiments was calculated on the basis of the sum of all radioactivity from the entire plate which was divided into five zones as follows: (i) application line zone—representing an area 0·5 cm above and below the application line. This area contained most of the polar lipids, (ii) zone between (i) and (iii), the PGE zone (Rf 0·49 to 0·50), (iv) between the PGE zone and the solvent front and (v) the solvent front which contained most of the neutral lipids.

The major part of the radioactivity was present in zones (ii) and (iv), which in the present solvent system might contain other prostaglandins as well.

ISOLATION OF CRUDE PGE COMPOUNDS (PGES)

Thin layer chromatography

500 μl of clear plasma was added after completion of incubation and prior to extraction of the incubation mixture with $CHCl_3$-MeOH (7:3 v/v). The material obtained on extraction of the incubation mixture followed by other treatments described above was taken up in 200–300 μl of chloroform to which was added 500 μg of PGE mixture. The resulting mixture was resolved by TLC. The plate (20 × 20 cm) was coated with a 0·25 mm thick layer of silica gel G and was activated at 110° for 30 min. A margin of 3 cm wide was marked on either side while the rest of the plate was divided into two halves. On the margin was applied the mixture of PGE markers. The developing solvent consisted of ethylacetate-2,2,4-trimethylpentane-acetic acid-water (100:20:10:100 v/v equilibrated for 1 hr before using the organic phase) (Bygdeman,

Svanborg & Samuelsson, 1969). Visualization was achieved by exposure to iodine vapour. The zone due to PGE compounds (R_p 0·48–0·49) was marked. The polar lipids (phospholipids) remained at the line of application and a zone 0·5 cm above and below the line of application was marked. The solvent front zone which mainly contained the neutral lipids was also marked. The plate was kept in the open until such time that the iodine was completely evaporated, as judged by the disappearance of the stain due to it. The marked areas were scraped off, eluted with methanol (*vide infra*) and counted.

Fractionation of PGE compounds

In another similar experiment the zone due to PGE compounds was scraped off and three times extracted with 3 ml of methanol. After centrifugation the clear supernatant was pooled together and methanol evaporated under N_2 at 60°. The material thus obtained was taken up in 200 μl methanol and resolved by TLC on $AgNO_3$-impregnated silica gel (1:25 w/w, 0·25 mm layer, activated at 110° for 30 min) (Hamberg & Samuelsson, 1966). The solvent system consisted of ethyl acetate-acetic acid-methanol-2,2,4-trimethylpentane-water (110:10:15:10:100, v/v equilibrated for one hour before using the organic layer). The zones due to PGE_1, PGE_2 and PGE_3 were visualized by water spray and their respective Rf values were 0·74, 0·54 and 0·30. The zones were scraped off and used for radioactivity counting.

RESULTS

Figure 1 shows the incorporation of C^{14} acetate into neutral lipids, polar lipids (phospholipids) and prostaglandins as function of acetate concentration. At lower acetate concentrations (up to 0·9 mM of acetate) the labelling of neutral

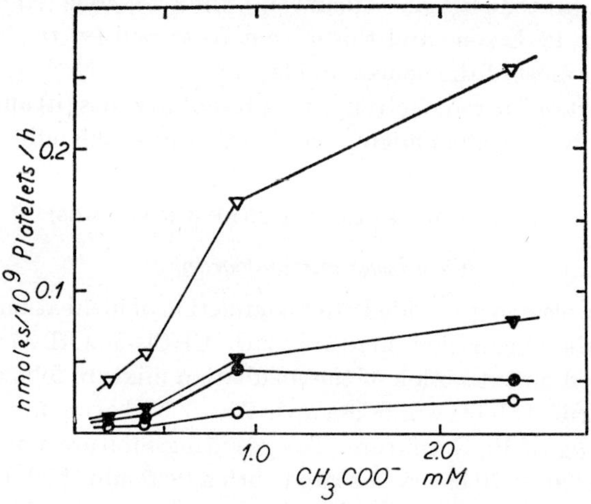

FIG. 1. The incorporation of C^{14} labelled acetate into lipid fractions of human platelets. *Conditions of incubation:* Krebs-Ringer-Phosphate buffer (pH = 6·9); Glutathione, 55 μg/ml; Platelets, 1·0 × 10^9/ml; Incubation time, 120 min. Ordinate: Acetate incorporated (nmoles acetate/10^9 cells/h). Abscissa: mM acetate. (▽——▽——▽) Total acetate incorporated; (▼——▼——▼) Polar lipids; (○——○——○) Prostaglandin E; (●——●——●) Neutral lipids.

lipids and phospholipids increased proportionately. Thereafter, only the phospholipid labelling increased slightly and the labelling of neutral lipids showed a slightly lower value (0·039 nmoles acetate/10^9 platelets/h) at 2·4 mM acetate, as compared with 0·9 mM, where the incorporation was 0·047 nmoles acetate/10^9 cells/h. Incorporation of the label into PGEs steadily increased as a function of acetate concentration during the whole experimental period. The same was the case with the labelling of the total lipids, of which only about 60% was accounted for by the lipids mentioned.

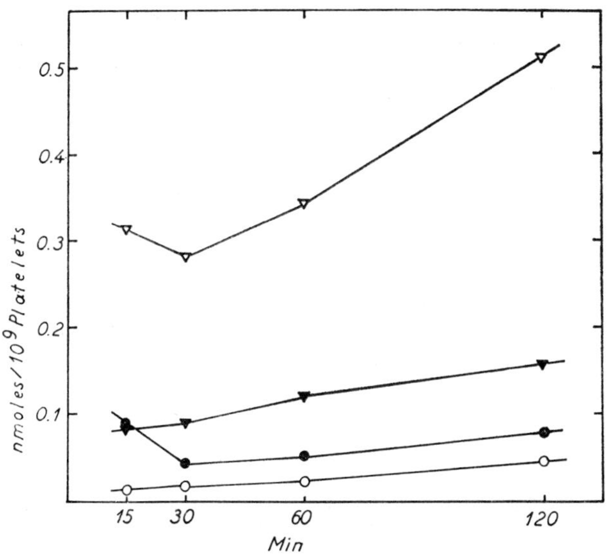

FIG. 2. The incorporation of ^{14}C labelled acetate into lipid fraction of human platelets. *Conditions of incubation:* KRP buffer, pH = 6·9; Glutathione, 55 μg/ml; Platelets, 1·0 × 10^9/ml; Acetate, 2·4 mmoles/l or (2·4 μmoles/ml). Ordinate: nmoles acetate incorporated/10^9 cells/h. Abscissa: time (min). (▼——▼——▼) Polar lipids; (●——●——●) Neutral lipids; (○——○——○) Prostaglandin E; (▽——▽——▽) Total acetate incorporated.

Figure 2 demonstrates the labelling as a function of time. The labelling of phospholipids and PGEs steadily increased during the period of 2 hr. The total label and the labelling of neutral lipids initially (within the first half hour) showed a slight fall followed by a steady increase during the next 90 min.

Furthermore, it is obvious from Figs. 1 and 2 that only about 10% of the total label was found in PGEs.

Figure 3 shows the percentage distribution of the radioactivity on the TLC plate. It will be seen that the highest label was present at the point of application, where polar lipids are localized (33%). High label (22%) was also present at the interzone between the application area and PGE zone (29%). This was also the case with the interzone between PGE and the solvent front where the neutral lipids are found (23%). These intermediate radioactive compounds have not yet been identified with known labelled compounds.

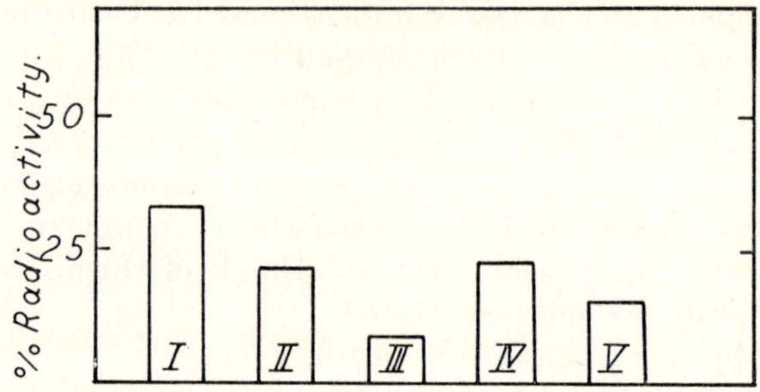

FIG. 3. Distribution of radioactivity among lipid classes after incubation with acetate-1-^{14}C. The incubation mixture consisted of 1 ml of the platelet suspension ($2 \cdot 0 \times 10^9$ ml), 1 ml of KRP buffer, glutathione (55 μg/ml) and acetate (2·4 mmoles/l) in a total volume of 2 ml (cf. the text). Ordinate: per cent of radioactivity on TLC-plates. (i) Application line zone containing highly polar lipids; (ii) Zone between (i) and (iii) containing some of the less polar lipids; (iii) PGE zone; (iv) Zone between PGE and solvent front; (v) Solvent front—containing neutral lipids (NL).

Table 1. *Composition of PGE compounds formed in human platelets*

PGE	% radioactivity
PGE_1	48·6
PGE_2	43·4
PGE_3	7·2*

* PGE_3 and the line of application share the 7·2% of the total activity in almost equal proportions.

Table 1 demonstrates the separation of the PGEs on TLC. Although PGE_1 possessed the highest percentage of label (48·6%), the PGE_2 had nearly equal amounts of the label (43·4%). On the other hand, the PGE_3 and a fraction at the application line together shared 7·2% of the label in equal proportions.

DISCUSSION

The concept for the function of essential fatty acids is related to their transformation into prostaglandins (Samuelsson, Granström & Hamberg, 1966; Nugteren, Beerthuis & van Dorp, 1966). These compounds act on the hormone-stimulated cyclic-AMP system (Butcher & Baird, 1970) and the present paper may support this concept because it has shown that PGEs are enzymically formed in platelets. Thus, as the agglutination of circulating thrombocytes is enhanced by adrenalin and ADP, probably through the cyclic AMP system (Marquis, Vigdahl & Tavormina, 1969), the same is inhibited by prostaglandin E (Hornstra, 1970). This may explain the enhanced formation of thrombi in the arteriosclerotic subjects (Hornstra, 1970) having a relatively low supply of linoleic acid since this acid, through chain elongation and desaturation reactions, can be transformed into dihomo-γ-linoleic and arachidonic acids which give rise respectively to PGE_1 and PGE_2 ($PGF_{2\alpha}$) (Samuelsson *et al.*, 1966; Nugteren *et al.*, 1966). These data thus explain the reasons as to why experiments have shown that a diet rich in butter or stearic acid markedly predisposes to ADP and endotoxin-initiated thrombosis, reactions which however, are reversed by feeding a diet rich in linoleic acid (Nordöy *et al.*, 1968; Renaud *et al.*, 1970). The blood platelets can thus (as shown by the present studies) be added to a series of cells synthesizing PGs, i.e. the epithelial cells of the vesicular gland and the collecting tubules of the Kidney (Janszen & Nugteren, 1971). Our present results showing active incorporation of radioactive acetate by the platelets into different lipid fractions, have confirmed previous studies (Hennes, Awai & Hammarstrand, 1964; Deykin & Desser, 1968; Cohen *et al.*, 1970; Deykin, 1971), which demonstrated that both acetate and fatty acid are easily incorporated into the platelet lipids, and have argued for an active metabolic role of these corpuscular elements of the blood stream.

SUMMARY

Human blood platelets were found to possess the whole biological system for the synthesis of prostaglandins from C^{14}-labelled acetate and endogenous derivatives of essential fatty acids. The incorporation of radioactive acetate into prostaglandins, polar lipids and neutral lipids was followed as a function of acetate concentration and time. Of the fractions mentioned, about 10% of the label was found in PGEs. Among the PGEs, PGE_1 and PGE_2 accounted for 48·6% and 43·4% of the radioactivity. The results obtained support the theory of inhibitory function of prostaglandins in platelet agglutination.

ACKNOWLEDGEMENT

These studies have been supported by a grant from the M.I.F.U. Research Fund, Copenhagen, Denmark. Gratitude is expressed to Unilever Research Laboratories, Vlaardingen, The Netherlands, for samples of the PGE marker.

REFERENCES

Born, G. V. R. (1962). *Nature, Lond.* **194**, 927–929.

Brecher, G., Schmiederman, M. & Cronkite, E. G. (1953). *Am. J. clin. Path.* **23**, 14.

Butcher, R. W. & Baird, C. E. (1970). *Proc. int. Congr. Pharmacol.* **4**, 42–48.

Bygdeman, M., Svanborg, K. & Samuelsson, B. (1969). *Clin. Chim. Acta* **26**, 373–379.

Cohen, P., Derksen, A. & Van den Bosch, H. (1970). *J. Lipid Res.* **49**, 128–139.

Deykin, D. (1971). *J. Lipid Res.* **12**, 9–11.

Deykin, D. & Desser, R. K. (1968). *J. clin. Invest.* **47**, 1590–1602.

Emmons, P. R., Hampton, J. R., Harrison, M. J. E., Honour, A. J. & Mitchell, J. R. A. (1967). *Brit. Med. J.* **20**, 468.

Hamberg, M. & Samuelsson, B. (1966). *J. biol. Chem.* **241**, 257–263.

Hennes, A. R., Awai, K. & Hammarstrand, K. (1964). *Biochim. biophys. Acta* **84**, 613–614.

Hornstra, G. (1970). *Fette. Seifen, Anstrichmittel* **72**, 960–964.

Janszen, F. H. A. & Nugteren, D. H. (1971). *Histochemie* **27**, 159–164.

Kloeze, J. (1966). *Proc. Nobel Symp. II*, Alqvist & Wiksell, Stockholm, pp. 241–252.

Marquis, N. R., Vigdahl, R. L. & Tavormina, P. A. (1969). *Biochem. biophys. Res. Commun.* **36**, 965–972.

Nordöy, A., Hamlin, J. T., Chandler, A. B. & Newland, H. (1968). *Scand. J. Haematol.* **5**, 458–473.

Nugteren, D. H., Beerthuis, R. K. & van Dorp, D. A. (1966). *Proc. Nobel Symp. II*. Alqvist & Wiksell, Stockholm, pp. 45–50.

Piper, P. S., Vane, J. R. & Myllie, J. A. (1970). *Nature, Lond.* **225**, 600–604.

Renaud, S., Kuba, K., Goulet, C., Lemire, Y. & Allard, C. (1970). *Circulat. Res.* **26**, 553–564.

Samuelsson, B., Granström, E. & Hamberg, M. (1966). *Proc. Nobel Symp. II*, Alqvist & Wiksell, Stockholm, pp. 31–44.

THE SYNTHESIS AND BIOLOGICAL ROLE OF PROSTAGLANDINS

By BENGT SAMUELSSON

Department of Medical Chemistry, Royal Veterinary College, Stockholm, Sweden

THE prostaglandins have been shown to have a broad spectrum of pharmacological actions (for references see Bergström, Carlson & Weeks, 1968; Weeks, 1972). These include effects on, e.g. smooth muscle organs, cardiovascular systems, systems involving cyclic AMP formation, nervous systems and reproductive systems.

Previous studies have led to information on the mechanism of the enzymatic formation of prostaglandins. This work has been described in previous reviews (Samuelsson, Granström & Hamberg, 1967; Samuelsson, 1969; van Dorp, 1969; Samuelsson, 1972).

Prostaglandins are rapidly metabolized by prostaglandin dehydrogenase, Δ^{13}-reductase, β-oxidation and ω-hydroxylation and oxidation (for references see, Samuelsson, 1969; Granström, 1971*a,b*; Hamberg, Israelsson & Samuelsson, 1971; Samuelsson *et al.*, 1971). The structures of a large number of metabolites in guinea-pigs, rats and man have been determined in our laboratory (Samuelsson, 1969; Granström & Samuelsson, 1971*a,b*; Gréen, 1971*a,b*; Samuelsson *et al.*, 1971) and studies on the endogenous synthesis of prostaglandins in man (Hamberg & Samuelsson, 1971) and the rat (Gréen & Samuelsson, 1971) have also been described.

We have recently applied the biochemical information on enzymatic synthesis and catabolism of prostaglandins referred to above to studies of some biological systems where control mechanisms involving prostaglandins might be involved.

The endogenous excretion in rats of two metabolites of prostaglandins E_1 and E_2, namely, tetranor-prostaglandin E_1 and tetranor-prostaglandin B_1, was found to be $2{\cdot}90 \pm 0{\cdot}5$ μg/kg/24 hr in females. When animals were subjected to acute cold stress a 2 to 2·5-fold increase in the excretion of the two metabolites was observed (Gréen & Samuelsson, 1971). It is of interest in this context that prostaglandins of the E series (prostaglandins E_1 and E_2) have been shown to inhibit the release of noradrenaline in response to nerve stimulation in various tissues and species (Hedqvist, 1970; Wennmalm & Hedqvist, 1970). We have found that inhibition of nerve-stimulation-induced synthesis of prostaglandins E_1 and E_2 results in increased release of noradrenaline (Samuelsson & Wennmalm, 1971). These studies provide strong evidence for the occurrence of a prostaglandin mediated mechanism for control of release of the adrenergic transmitter.

Our laboratory has recently been involved in the development of sensitive and specific methods for prostaglandin analysis. One promising approach has been the application of combined gas chromatography-mass spectrometry using synthetic deuterium labelled prostaglandins as carriers (Axen *et al.*, 1971). With this method prostaglandins of both the E and F series can be measured accurately with a lower limit of less than one picomole. This method has been used to show that prostaglandin $F_{2\alpha}$ appears in concentrations of 25–40 ng/ml plasma in uterine venous blood or sheep at the time of normal luteal regression (McCracken *et al.*, 1972). A unique transport mechanism involving direct transfer of prostaglandin $F_{2\alpha}$ from the uterine vein to the adhering ovarian artery through the walls of the vessels seem to operate.

ACKNOWLEDGEMENT

The work from the author's laboratory was supported by grants from the Swedish Medical Research Council (proj. no. 13X-217), Knut and Alice Wallenbergs stiftelse and Stiftelsen Riksbankens Jubileumsfond.

REFERENCES

Axen, U., Gréen, K., Hörlin, D. & Samuelsson, B. (1971). *Biochem. biophys. Res. Commun.* **45**, 519.
Bergström, S., Carlson, L. A. & Weeks, J. R. (1968). *Pharmacol. Rev.* **20**, 1892.
Dorp, van, D. A. (1969). *Naturwissenschaften* **56**, 124.
Granström, E. (1971*a*). *Eur. J. Biochem.* **20**, 451.
Granström, E. (1971*b*). *Biochim. biophys. Acta* **239**, 120.
Granström, E. & Samuelsson, B. (1971*a*). *J. biol. Chem.* **246**, 5254.
Granström, E. & Samuelsson, B. (1971*b*). *J. biol. Chem.* **246**, 7470.
Gréen, K. (1971*a*). *Biochim. biophys. Acta* **231**, 419.
Gréen, K. (1971*b*). *Biochemistry* **10**, 1072.
Gréen, K. & Samuelsson, B. (1971). *Eur. J. Biochem.* **22**, 391.
Hamberg, M., Israelsson, U. & Samuelsson, B. (1971). *Ann. N.Y. Acad. Sci.* **180**, 164.
Hedqvist, P. (1970). *Life Sci.* **9**, 269.
McCracken, J. A., Carlson, J. C., Glew, M. E., Goding, J. R., Baird, D. T., Gréen, K. & Samuelsson, B. (1972). *Nature New Biol.* **238**, 129.
Samuelsson, B. (1969). *Proc. 4th Int. Congr. Pharmacol.* (*Basel, July* 14–18). p. 12.
Samuelsson, B., (1972). *Fed. Proc* **31**, (5), 1442.
Samuelsson, B. & Wennmalm, Å. (1971). *Acta physiol. scand.* **83**, 163.
Samuelsson, B., Granström, E. & Hamberg, M. (1967). *Nobel Symposium* **2**, 31 (ed. S. Bergström and B. Samuelsson). Almqvist & Wiksell AB, Stockholm.
Samuelsson, B., Granström, E., Gréen, K. & Hamberg, M. (1971). *Ann. N.Y. Acad. Sci.* **180**, 138.
Weeks, J. R. (1972). *Ann. Rev. Physiol.* **12**, 317.
Wennmalm, Å. & Hedqvist, P. (1970). *Life Sci.* **9**, 931.

FATTY ACID METABOLISM IN MULTIPLE SCLEROSIS

By R. H. S. THOMPSON

Courtauld Institute of Biochemistry,
The Middlesex Hospital Medical School,
London, England

WORK in this field began some 10 years ago with a comparison of the fatty acids in the lecithin faction of cerebral white matter taken from normal brains and from brains affected by multiple sclerosis.

Our reasons for turning to this approach were twofold. First, it had been suggested on clinical and epidemiological grounds by Swank (1950), Sinclair (1956) and others that the development of multiple sclerosis might depend on the level of dietary intake of particular types of fat; and secondly it had been shown that in the rat it is possible to alter the fatty acid composition of the brain lipids by varying the levels of different fatty acids in the diet (Witting *et al.*, 1961).

Since in multiple sclerosis there is a breakdown of myelin it is not surprising that there are marked alterations in the levels of various lipids in the demyelinating plaques, as had indeed been shown by Cumings and numerous other workers. But, in addition, some evidence had been obtained that even in apparently unaffected areas of the brain in multiple sclerosis lipid abnormalities are present (Cumings, 1953), although the fatty acids themselves had not been examined.

In 1962, therefore, we extracted the lecithin fraction from white matter obtained from the brains of 6 persons dying from non-neurological causes and from 9 patients with multiple sclerosis (Baker, Thompson & Zilkha, 1963). Comparable samples of white matter were dissected from the internal capsule and subcortical areas of the white matter from each of the 2 groups of brains, and in the case of the multiple sclerosis brains care was taken to dissect only those areas of white matter which were not in the immediate vicinity of a recognizable plaque, and which appeared normal on visual inspection. The proportions of the different fatty acids derived from these lecithin fractions were then determined by gas chromatography.

Table 1 shows that there appears to be a shift in the proportions of saturated and unsaturated fatty acids in the abnormal tissue, the saturated acids being increased and the unsaturated ones decreased, and it is important to stress again that these changes are found not in the plaques, but in apparently normal areas of white matter.

The fatty acid composition of lecithin from the white matter of normal and multiple sclerosis brains has also more recently been compared by three other

Table 1. *Percentage composition of fatty acid mixture from lecithins of white matter (mean values, expressed as moles individual fatty acids/100 moles total fatty acids)*

	16:0	16:1	18:0	18:1	20:4	Total saturated acids (16:1 + 18:0)	Total unsaturated acids (16:1 + 18:1 +20:4)
Normal (6)	24·6	4·9	13·3	51·1	3·3	37·9	59·3
Multiple sclerosis (9)	33·6	2·3	10·4	44·4	1·7	44·0	48·4

(Number of brains used indicated in brackets.)

groups of workers, Cumings and his colleagues in England (Cumings, Shortman & Skrbic, 1965), Arnetoli, Pazzagli & Amaducci (1969) in Italy, and Gerstl and his colleagues (1970) in America. It is unfortunately not easy to compare the reports from these other groups of workers since the levels of fatty acids present in only small amounts are not given by all of them. It is however possible to compare the levels of the four fatty acids which have been estimated by each group, i.e. two of the saturated acids (palmitic and stearic) and two of the unsaturated ones, oleic and arachidonic acid.

Table 2. *Percentage fatty acid composition of lecithin fraction from brain white matter*

Authors	No. of brains	Saturated Control	Saturated M.S.	Saturated Δ	Unsaturated Control	Unsaturated M.S.	Unsaturated Δ
Baker *et al.* (1963)	6C, 9MS	37·9	44·0	+6·1	54·4	46·1	−8·3
Cumings *et al.* (1965)	4C, 5MS	36·5	39·7	+3·2	47·3	40·6	−6·7
Arnetoli *et al.* (1969)	4C, 5MS	46·4	50·5	+4·1	42·3	39·7	−2·6
Gerstl *et al.* (1970)	4C, 4MS	43·4	45·7	+2·3	42·6	42·4	−0·2

Table 2 gives a comparison of the results obtained by the four groups who have so far carried out these studies. It will be seen that each of the four groups has reported an admittedly small increase in the multiple sclerosis brains in the mean percentage of the sum of the two saturated acids (16:0 + 18:0), while three of the four groups have described a small reduction in the mean percentage of the two unsaturated acids (18:1 + 20:4), although in the four brains studied by Gerstl and his colleagues no significant difference was found.

Both Arnetoli and Gerstl and their colleagues have each also examined the fatty acid composition of the ethanolamine-containing glycerophospholipids extracted from the white matter of normal brain and from areas of multiple sclerosis brains which appeared normal on visual inspection, and they have each found differences in the percentages of two mono-unsaturated fatty acids in the multiple sclerosis brains, oleic acid and eicosamonoenoic acid (Table 3).

Table 3. *Percentage of* 18:1 *and* 20:1 *in the ethanolamine-containing phospholipids extracted from white matter*

Authors	No. of brains	18:1 Control	18:1 M.S.	18:1 Δ	20:1 Control	20:1 M.S.	20:1 Δ
Arnetoli *et al.*	4C, 5MS	22·5	14·2	−8·2	6·4	2·8	−3·6
Gerstl *et al.*	3C, 4MS	45·8	39·7	−6·1	9·5	6·9	−2·6

Clausen & Hansen (1970) have recently carried out a somewhat similar and very detailed examination of fatty acid patterns. They studied fractions prepared from purified myelin obtained from twelve normal brains and from apparently normal areas of white matter dissected from the brains of twenty-two patients with multiple sclerosis, and they also found significant differences in the fatty acid patterns of the two groups.

Thus, in agreement with our own findings for lecithin extracted from whole white matter, they have shown that the lecithin fraction of myelin contains significantly lower amounts of 20:4 in multiple sclerosis brain (controls 2·85%, multiple sclerosis 1·14%), although they did not observe the changes in the levels of oleate found by us.

Their ethanolamine phosphoglyceride fractions differed more markedly than any other fraction examined, and in agreement with Arnetoli and Gerstl and their colleagues they found low levels of 18:1 in multiple sclerosis brains, but high levels of arachidonic and docosahexaenoic acids.

Owing to the difficulty of obtaining suitable post-mortem specimens of brain tissue from patients who have died with multiple sclerosis the numbers of brains that have been examined by most of these different groups of workers are small, and the differences between the control and the diseased tissues are also small. One must be cautious therefore in reaching any conclusions. But when we consider all these results which I have outlined together, there is, I think, a sufficient measure of agreement between the various groups to indicate that, both for the choline-containing and perhaps more strikingly in the case of the ethanolamine-containing phospholipids extracted from apparently normal areas of white matter from multiple sclerosis brains, there is evidence of an alteration in the levels of certain of the unsaturated fatty acids in this disease.

Without commenting for the moment on the possible significance of this I want to consider mechanisms that might account for these differences.

They could be due to some purely local cause such as an abnormality in the action of one of the enzymes concerned in phospholipid synthesis, fatty acid incorporation or degradation, or they could be the result of some more general cause, such as a change in the relative amounts of saturated and unsaturated fatty acids reaching the brain.

We therefore next studied the blood lipids to see whether any change could be observed in their fatty acid composition in multiple sclerosis.

Specimens of fasting blood were obtained from thirty-eight controls, about equally divided between healthy persons and patients with non-demyelinating neurological conditions, and from forty-seven patients with multiple sclerosis. Estimation of the various fatty acids derived from the total lipid extract obtained from these specimens showed that there were changes in the levels of linoleic acid (Baker, Thompson & Zilkha, 1964).

Table 4 shows the levels of linoleic acid (18:2) expressed as a percentage of the total fatty acids contained in the serum lipids.

Table 4. *Percentage of linoleic acid* (18:2) *in the total lipid extract of serum*

	No. of subjects	18:2 as % of total fatty acids (±SEM)	
Controls	38	25·6 ± 0·5	P < 0·0001
M.S.	47	22·3 ± 0·5	
M.S. Grade I	10	25·8 ± 0·9	
M.S. Grade II	11	23·6 ± 0·6	
M.S. Grade III	16	21·6 ± 0·6	
M.S. Grade IV	10	18·6 ± 0·4	

It will be seen that there is a small but highly significant lowering of the percentage of linoleate in the serum lipids in the multiple sclerosis group as a whole. The forty-seven sclerosis patients were then divided into four groups, depending on the clinical assessment of the severity of the disease (Group I showing no evidence of clinical deterioration over the preceding month, Groups II, III and IV showing increasing evidence of recent deterioration), and it would seem that the percentage of linoleate falls progressively with increasing evidence of recent deterioration.

When we take into account our more recent results, and we have now studied a very large number of both controls and multiple sclerosis patients, a statistical evaluation of our more recently studied patients confirms this finding of a small but significant fall in the level of 18:2 in the serum lipids obtained from fasting blood, and our impression continues to be that the most marked falls are found in those patients who are showing active and rapid deterioration.

Furthermore, our results in serum have been confirmed by a group in Prague (Tichy, Vymazal & Michalec, 1969) who have published their findings on the levels of cholesteryl linoleate and of monounsaturated and saturated cholesteryl esters in the α- and β-lipoproteins present in the serum of twelve controls and forty-two multiple sclerosis patients (Table 5).

Table 5. *Cholesteryl esters in serum lipoproteins*
(*Mean percentage* ± *S.D.*)

	Linoleate	Monounsaturated	Saturated
α-Lipoproteins			
Controls	54·5 ± 4·0	30·5 ± 4·1	15·0 ± 4·0
M.S.	45·6 ± 6·7*	32·3 ± 3·9	22·6 ± 5·5
β-Lipoproteins			
Controls	57·8 ± 5·9	30·2 ± 3·5	12·0 ± 3·2
M.S.	45·7 ± 6·8*	33·3 ± 4·6†	20·7 ± 5·6*

* Significant at 0·01 level.
† Significant at 0·05 level.
(From Tichy *et al.*, 1969.)

Further confirmation of this finding in serum has recently been provided by Mahler, unpublished observations (1971), who has also found a lowering of the percentage of linoleate in specimens of fasting blood obtained from patients with multiple sclerosis.

A different picture, however, appears to emerge if the blood is drawn in the non-fasting state: thus Cumings *et al.* (1965), using blood from fourteen patients with multiple sclerosis and from eleven control subjects found no significant differences in the levels of 18:2. The groups of subjects studied were, however, small, and as no clinical details are given as to the state or extent of the disease it is hard to evaluate their results. And secondly, Karlsson, Alling & Svennerholm (1971) have been unable to detect any difference in linoleate levels in non-fasting samples obtained from even smaller groups—eight controls and eight patients with multiple sclerosis, in at least two of whom the disease does not appear to have been active.

In 1952 it was reported that the blood platelets from patients with multiple sclerosis were abnormally sticky (Nathanson & Savitsky, 1952). We have also found increased platelet adhesiveness in multiple sclerosis, and this finding has now been amply confirmed by several other groups of workers (Fog, Kristensen & Helweg-Larsen, 1955; Wright, Thompson & Zilkha, 1965; Caspary *et al.*, 1965; Millar, Merrett & Dalby, 1966). Furthermore, we have shown that in multiple sclerosis there is a highly significant inverse correlation ($P < 0{\cdot}001$) between platelet adhesiveness and the serum cholesteryl linoleate levels (Sanders *et al.*, 1968).

In additon to increased platelet stickiness, changes in the red blood cells of multiple sclerosis patients have also been described. Plum & Fog (1959) showed that the mean erythrocyte diameter is increased in multiple sclerosis, a finding which has since been confirmed by Prineas (1968), who found that the increase correlated with the activity of the disease. And secondly, Laszlo in 1964 reported a highly significant increase in osmotic fragility of the red blood cells. This finding has been confirmed by Caspary, Sewell & Field (1967), who also correlated the extent of this change with the activity of the disease.

Because of these findings of abnormalities both in platelets and in red cells, we studied the fatty acid composition of platelets and erythrocytes from a series of patients with multiple sclerosis (Gul *et al.*, 1970).

Table 6. *Linoleate content of platelet and erythrocyte phospholipids* (18:2 *as* % *of total fatty acids; Mean* ± *S.E.M.*)

	Controls	M.S.	P
Platelets	6·3 ± 0·27	4·9 ± 0·12	0·0001
Erythrocytes	14·0 ± 0·42	11·0 ± 0·44	0·001

Table 6 shows that the percentage of linoleate in the phospholipids is significantly lower in both the platelets and the erythrocytes derived from multiple sclerosis patients than in those obtained from healthy controls.

Mahler (unpublished, 1971) has made similar observations on leucocytes obtained from multiple sclerosis patients, and again has found that the percentage of linoleate is reduced, while that of oleate is increased.

In view therefore of this growing evidence in multiple sclerosis of what would appear to be a widespread abnormality in the handling of certain fatty acids in the body, and since the view has been put forward that the high prevalence of multiple sclerosis in certain parts of the world might be related to a relative dietary deficiency of unsaturated fatty acids (Swank, 1950; Sinclair, 1956; Allison, 1963), it seemed important to determine whether any difference could be revealed between healthy subjects and patients with multiple sclerosis in their ability to incorporate orally administered linoleate into the plasma lipids.

It is of course well known that the feeding of diets with a high content of linoleate normally causes an increase of linoleate in the serum lipids, together with a decrease of oleate and sometimes of palmitate, palmitoleate and stearate as well. In comparatively few instances have these changes been studied over short time-intervals, but Farquhar & Ahrens (1963) obtained indications that an increase of linoleate in the β-lipoproteins can be observed within hours after a single dose of corn oil.

In a series of healthy subjects and of patients with multiple sclerosis we have therefore measured the fatty acid patterns in the serum lipids in blood samples, drawn after an overnight fast, both before and during a five-day period when the normal diet was supplemented with linoleate, and during the two days after cessation of these supplements (Belin *et al.*, 1971). Two doses of the linoleate supplements were taken each day, at 9.0 a.m. and 5.0 p.m., each dose consisting of 50 ml of a sunflower seed oil emulsion, that is, an intake of 14·3 g linoleic acid per dose. Fourteen members of the laboratory staff (eight males and six females) were used as controls; their ages ranged from 19 to 59 years, with a mean of 36·3. The fourteen patients with multiple sclerosis (seven males and seven females) who were used for this investigation had ages ranging from 20 to 53 years, with a mean of 36·6 years; they were hospitalized throughout the ten-day period, but were all on the full, normal hospital diet, and all but one were fully ambulant.

The levels of linoleate and of oleate in the hydrolysates of the total lipid extracts prepared from the various serum samples are shown in Fig. 1.

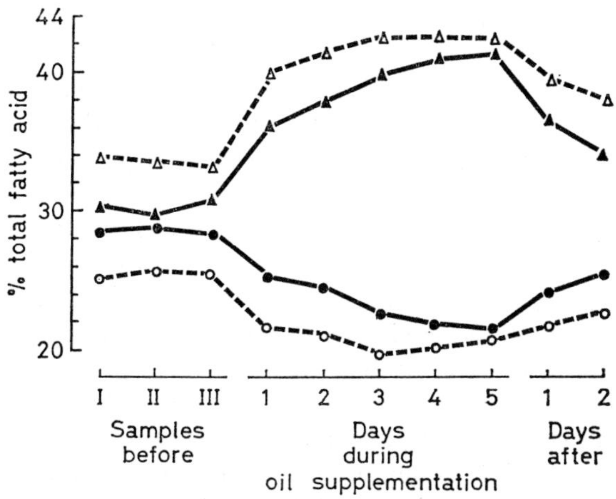

FIG. 1. Mean levels of oleate and linoleate in the serum lipids of controls and M.S. patients, expressed as a percentage of the total fatty acids, before, during and after supplementation of the diet with sunflower seed oil emulsion.

○ = oleate in controls. ● = oleate in M.S. patients.
△ = linoleate in controls. ▲ = linoleate in M.S. patients.

In confirmation of our earlier work, in the pre-supplementation period the mean level of linoleate for the multiple sclerosis patients (29·7%) was significantly lower ($P < 0{\cdot}01$) than that of the control subjects (33·7%). The oleate levels in the pre-supplementation period showed a change in the reverse direction, the mean level for the multiple sclerosis patients (28·9%) being significantly higher ($P < 0{\cdot}005$) than that of the controls (25·4%), so that the 18:2/18:1 ratio fell from 1·35 for the controls to 1·05 for the patients ($P < 0{\cdot}001$).

However, it will be noted that when large amounts of linoleate are available for absorption from the gut the patients appear to absorb this as efficiently as the controls, since, as will be seen from Fig. 1, the rise in 18:2 in the multiple sclerosis patients during the supplementation period is as great as in the controls. The levels for the two groups are not significantly different after four days' supplementation, although when the subjects returned to their normal unsupplemented diet the levels of 18:2 seemed to fall more steeply in the multiple sclerosis patients than in the controls.

This experiment has therefore certainly provided us with no evidence of defective alimentary absorption of linoleate when this is present in relatively large amounts in the diet, inasmuch as the increase in the linoleate content of the plasma lipids resulting from the extra dietary intake does not appear to differ in the multiple sclerosis patients from that occurring in normal subjects. On the other hand, we still do not know how the utilization of the much smaller amounts of linoleate ingested in the normal Western diet is carried out in multiple sclerosis. Such evidence as we have, however, leads to believe that the abnormality in unsaturated fatty acids in multiple sclerosis is more likely to concern the uptake and incorporation of 18:2 into the lipids of cell membranes, or the further metabolism or degradation of the 18:2 present in membrane lipids, rather than any general defect in the absorption, storage or transport of 18:2.

We have therefore recently embarked on a study of the uptake of ^{14}C-labelled linoleate and palmitate into the lipids of red cell membranes obtained from blood samples from healthy subjects and from patients with multiple sclerosis.

The significance of this work on fatty acid metabolism with regard to the aetiology and pathogenesis of multiple sclerosis is, at the present stage of our knowledge, hard to assess. But I must stress that these are findings in patients with established multiple sclerosis, and that the degree of change of certain of these parameters appears to bear a relation to the severity of the disease, and in particular to the degree of deterioration of the patient at the time the investigation is made. They may therefore be more in the nature of a reflection of the disease process rather than of any primary aetiological significance, although we do not think that they merely reflect changes in the state of nutrition of the patients since care was taken at the clinical level to ensure that the patients were eating well, and were on a full diet.

However, the fact that the fatty acid composition of the brain can be altered by changes in the dietary intake of various different fatty acids (Witting *et al.*, 1961; Mohrhauer & Holman, 1963; Biran *et al.*, 1964; Rathbone, 1965; Galli, White & Paoletti, 1970), together with the bizarre geographical pathology of the disease, and the evidence obtained by Swank (1950), Allison (1963) and others suggesting a high incidence in areas where there is a low dietary intake of unsaturated fatty acids, is all consistent with the view that these fatty acid changes may be of aetiological significance. In this connection it is worth bearing in mind the report from Clausen & Møller (1967) that rats bred and raised on a diet deficient in polyunsaturated fatty acids show an increased susceptibility to the development of allergic encephalomyelitis following the injection of a brain antigen.

Of the evidence which is at present available, relating to the small abnormalities of fatty acid levels and metabolism in patients with multiple sclerosis, perhaps the most interesting outcome is the indication that even normal, apparently unaffected areas of the central nervous system show changes in chemical composition, and also that these small changes in fatty acid levels are not restricted to the CNS, but appear to be widespread throughout the body, although manifesting themselves symptomatically primarily in disturbances of nervous function.

Evidence already exists pointing to a familial tendency in this disease, and it would seem possible that there may be some inborn abnormality in fatty acid metabolism, which allows some other environmental factor—possibly a virus infection, nutritional imbalance, or some other cause—to precipitate the pathological changes in the nervous system.

ACKNOWLEDGEMENTS

The permission of the Editor of the Journal of Neurology, Neurosurgery and Psychiatry to reproduce Fig. 1 is gratefully acknowledged.

REFERENCES

Allison, R. S. (1963). *Proc. R. Soc. Med.* **56,** 71.

Arnetoli, G., Pazzagli, A. & Amaducci, L. (1969). *J. Neurochem.* **16,** 461.

Baker, R. W. R., Thompson, R. H. S. & Zilkha, K. J. (1963). *Lancet,* **1,** 26.

Baker, R. W. R., Thompson, R. H. S. & Zilkha, K. J. (1964). *J. Neurol. Neurosurg. Psychiat.* **27,** 408.

Belin, J., Pettet, N., Smith, A. D., Thompson, R. H. S. & Zilkha, K. J. (1971). *J. Neurol. Neurosurg. Psychiat.* **34,** 35.

Biran, L. A., Bartley, W., Carter, C. W. & Renshaw, A. (1964). *Biochem. J.* **93,** 492.

Caspary, E. A., Sewell, F. & Field, E. J. (1967). *Brit. med. J.* **2,** 610.

Caspary, E. A., Prineas, J., Miller, H. & Field, E. J. (1965). *Lancet* **2,** 1108.

Clausen, J. & Hansen, I. B. (1970). *Acta Neurol. Scand.* **46,** 1.

Clausen, J. & Møller, J. (1967). *Acta Neurol. Scand.* **43,** 375.

Cumings, J. N. (1953). *Brain,* **76,** 551.

Cumings, J. N., Shortman, R. C. & Skrbic, T. (1965). *J. clin. Path.* **18,** 641.

Farquhar, J. W. & Ahrens, E. H. (1963). *J. clin. Invest.* **42,** 675.

Fog, T., Kristensen, I. & Helweg-Larsen, H. F. (1955). *Arch. Neurol. (Chic.)* **73,** 267.

Galli, C., White, H. B. & Paoletti, R. (1970). *J. Neurochem.* **17,** 347.

Gerstl, B., Eng, L. F., Tavaststjerna, M., Smith, J. K. & Kruse, S. L. (1970). *J. Neurochem.* **17,** 677.

Gul, S., Smith, A. D., Thompson, R. H. S., Wright, H. P. & Zilkha, K. J. (1970). *J. Neurol. Neurosurg. Psychiat.* **33,** 506.

Karlsson, I., Alling, C. & Svennerholm, L. (1971). *Acta neurol. scand.* **47,** 403.

Millar, J. H. D., Merrett, J. D. & Dalby, A. M. (1966). *J. Neurol. Neurosurg. Psychiat.* **29,** 187.

Mohrhauer, H. & Holman, R. T. (1963). *J. Neurochem.* **10,** 523.

Nathanson, M. & Savitsky, J. P. (1952). *Bull. N.Y. Acad. Med.* **28,** 462.

Prineas, J. (1968). *Acta neurol. scand.* **44,** 81.

Plum, C. M. & Fog, T. (1959). *Acta psychiat. scand.* **34,** Suppl. No. 128.

Rathbone, L. (1965). *Biochem. J.* **97,** 620.

Sanders, H., Thompson, R. H. S., Wright, H. P. & Zilkha, K. J. (1968). *J. Neurol. Neurosurg. Psychiat.* **31,** 321.

Sinclair, H. M. (1956). *Lancet* **1,** 381.

Swank, R. L. (1950). *Am. J. med. Sci.* **220,** 421.

Tichy, J., Vymazal, J. & Michalec, C. (1969). *Acta neurol. scand.* **45,** 32.

Witting, L. A., Harvey, C. C., Century, B. & Horwitt, M. K. (1961). *J. Lipid Res.* **2,** 412.

Wright, H. P., Thompson, R. H. S. & Zilkha, K. J. (1965). *Lancet* **2,** 1109.

DISORDERS OF LIPID METABOLISM

By ROSCOE O. BRADY*

Developmental and Metabolic Neurology Branch,
National Institute of Neurological Diseases and Stroke
National Institutes of Health
Bethesda, Maryland, USA

THE pathological chemistry of nine heritable lipid storage diseases is now well established. This impressive medical advance has occurred within the short space of six years. The principal experiments which provided this breakthrough devolved about three critical investigations on Gaucher's diseases. The first disclosure was the observation that the lipid which accumulated in patients with Gaucher's disease was not produced because of a defective synthetic pathway, which would bring about the formation of an abnormal tissue component (Thannhauser, 1950). The second series of revealing experiments proved that excessive synthesis of a normally minor tissue lipid is not involved in the pathogenesis of these diseases (Trams & Brady, 1960). Finally, the fundamental demonstration of a deficiency of a normally-required catabolic enzyme was provided through studies in which the metabolism of the accumulating lipid labelled in critical portions of the molecule was investigated in normal human tissues and in similar specimens obtained from patients with Gaucher's disease (Brady, Kanfer & Shapiro, 1965; Brady *et al.*, 1966*a*). These investigations provided clear evidence for the unbroken precedent that all lipid storage diseases are caused by an attenuation or lack of a specific hydrolytic enzyme required for the disposal of a normal tissue metabolite.

In time, additional investigations revealed the clinically important fact that the *degree* of diminution of activity of the various sphingolipid hydrolases is similar in all of the tissues of an affected individual. This statement does not mean that the specific activity of the various enzymes is the same in all tissues, but the remark does clearly imply that the same percentage decrease in enzyme activity occurs in all tissues. This observation is extremely important because it means that homozygotes and heterozygotes can be identified through appropriate enzyme assays performed on easily accessible cells or tissues such as leukocytes or cultured skin fibroblasts. A number of these discoveries and some of the practical benefits derived therefrom have been summarized in recent reviews (Brady, 1970*a*; 1970*b*; 1971). In this paper I wish to outline briefly the salient enzymological features of the lipid storage diseases and emphasize recent conceptual and practical advances with which biochemists should be familiar.

* Not presented at the symposium.

GAUCHER'S DISEASE

Patient's with Gaucher's disease are divided into at least two clinical groups called the infantile and adult forms of this disease, respectively. Patients in both of these categories exhibit hepatosplenomegaly, hip joint and long bone erosion, and large Oil-red-O and periodic acid-Schiff-positive cells in bone marrow. These patients may have mild anemia, thrombocytopenia, and an increase in serum acid phosphatase activity. Patients with the infantile form of Gaucher's disease also have central nervous system damage and progressive mental retardation. The metabolic defect in Gaucher's disease is a deficiency of a β-glucosidase which catalyses the cleavage of glucose from glucocerebroside (Table 1–A) (Reaction 1):

$$\text{Glucocerebroside} + H_2O \xrightarrow{\text{glucocerebrosidase}} \text{glucose} + \text{ceramide } (N\text{-fatty acyl sphingosine}) \quad (1)$$

Patients with the infantile form of Gaucher's disease have extremely little glucocerebrosidase activity in their tissues, whereas patients with the adult form have attenuated but definite residual enzyme activity. Patients with least glucocerebrosidase activity have the most rapid onset and progression of the clinical manifestations of the disease. Generally, the degree of enzymatic deficiency is constant in various members of a family in which Gaucher's disease has appeared. These observations suggest that the disorder is genetically heterogeneous and that specific mutations have occurred which result in variations of amino acid composition of the enzyme which adversely affect its catalytic activity.

Glucocerebroside appears to be derived from lipids which are turned over during the normal metabolism and replacement of cells. Probably the principal source of this glycolipid in peripheral tissues is ceramide lactoside

$$(\text{ceramide} \overset{\beta}{\underset{1\leftarrow 1}{—}} \text{glucose} \overset{\beta}{\underset{4\leftarrow 1}{—}} \text{galactose}),$$

the major neutral glycolipid of polymorphonuclear leukocytes. Although "Gaucher cells" are frequently observed in sections of spleens from patients with chronic myelogenous leukemia and the level of glucocerebroside is slightly elevated in these patients' spleens (Kattlove *et al.*, 1969), the possibility exists that the major accumulating material might be ceramide lactoside. This speculation derives some support from the fact that a detailed examination of electron photomicrographs indicated some histological differences between authentic Gaucher cells and the occasional lipid-laden cells in the tissues of patients with chronic myelogenous leukemia (Brady & King, 1972). Another source of glucocerebroside in systemic tissues are the lipids in membranes of senescent erythrocytes. There are two glyco-lipids in red cell stroma; globoside

$$(\text{ceramide} \overset{\beta}{\underset{1\leftarrow 1}{—}} \text{glucose} \overset{\beta}{\underset{4\leftarrow 1}{—}} \text{galactose} \overset{\alpha}{\underset{4\leftarrow 1}{—}} \text{galactose} \overset{\beta}{\underset{3\leftarrow 1}{—}} N\text{-acetylgalactosamine})$$

and hematoside

$$(\text{ceramide} \overset{\beta}{\underset{1\leftarrow 1}{—}} \text{glucose} \overset{\beta}{\underset{4\leftarrow 1}{—}} \text{galactose} \overset{\alpha}{\underset{3\leftarrow 2}{—}} N\text{-acetylneuraminic acid}).$$

Table 1. *Sphingolipid hydrolase activity in tissues and cells obtained from patients with lipid storage diseases*

Disorder	Enzyme	Tissue	Circulating leukocytes (or serum)	Cultured skin fibroblasts
			Percent of normal)	
A. Gaucher's disease	Glucocebrosidase	Spleen		
Infantile form		3% (0–9%)	3–4%	4–5%
Adult form		15% (12–44%)	16% (13–28%)	17% (10–30%)
B. Niemann-Pick disease	Sphingomyelinase	Liver		
Infantile form		3·5% (0–7%)	0%	6·5%
Adult form		18% (15–20%)	17%	10%
C. Krabbe's disease	Galactocerebrosidase	Brain		
		8% (4–11%)	2%	11%
D. Metachromatic	Sulfatidase	Kidney		
Leukodystrophy		<1%	11%*	2·6%*
E. Ceramide lactoside	Lactosylceramidase	Liver	†	11%
lipidosis		17%		
F. Fabry's disease	Ceramide trihexosidase	Intestine		
		<1%	3%*	10%*
			(Serum)	
G. Tay-Sachs disease	G_{M2}-ganglioside-hexosaminidase	Brain	Hex A + Hex B	Hex A + Hex B
O-variant		5%	2·8%	4·0%
B-variant		0%	15%* 396%*	5·8% 82%*
AB-variant		Liver		
		14%	†	†
H. Generalized gangliosidosis	G_{M1}-ganglioside-galactosidase	Brain	Leukocytes	
		5%	5·6%*	5·6%*
I. Fucosidosis	α-Fucosidase	Liver		
		0%*	†	†

* Measured with artificial chromogenic or fluorogenic substrate.

† No reported values.

In infants with central nervous system difficulties, glucocerebroside accumulates in the neuronal cells and it is probably derived from gangliosides (see pp. 20 and 21).

Several important practical developments have resulted from the elucidation of the nature of the metabolic defect in this condition. Both homozygotes (Kampine *et al.*, 1967*a*) and heterozygotes (Brady, Johnson & Uhlendorf, 1971) for Gaucher's disease may be identified by enzyme assays on washed, sonicated leukocyte preparations (Table 2–A). Similarly, patients and carriers may be detected through determination of glucocerebrosidase activity in extracts of cultured skin fibroblasts with the labelled natural substrate or under carefully adjusted conditions, through the use of the artificial fluorogenic substance 4-methyl-umbelliferyl-β-D-glucopyranoside (Ho *et al.*, 1971). These tests have been successfully employed for the prenatal detection of a foetus affected with Gaucher's disease (Schneider, E. L., Brady, R. O. & Epstein, C. J., manuscript in preparation).

NIEMANN-PICK DISEASE

Patients with the classic infantile form of Niemann-Pick disease have the signs and symptoms of splenomegaly, hepatomegaly, along with cachexia, mental retardation, foam cells in the marrow, and about 30% of these patients have a cherry-red spot in the macula. The skin may have a brownish discoloration, and waxy texture. X-rays of the chest often show evidence of a granular or reticular infiltration in the lungs along with lymphadenopathy. An increasing number of patients with the disorder are being detected beyond the infantile period since the advent of increased diagnostic accuracy for Niemann-Pick disease (Kampine *et al.*, 1967*a*). These individuals have organomegaly without manifestations of nervous system damage. The metabolic defect in Niemann-Pick disease is a deficiency of the enzyme which catalyses the hydrolytic cleavage of sphingomyelin (Brady *et al.*, 1966*b*) (Table 1–B) (Reaction 2):

$$\text{Sphingomyelin} + H_2O \xrightarrow{\text{sphingomyelinase}} \text{ceramide} + \text{phosphorylcholine} \quad (2)$$

Patients with the infantile form have very little sphingomyelinase activity in their tissues while patients with the less rapidly progressing form have a reduction from normal, but some definitely detectable residual sphingomyelinase activity. Sphingomyelin probably arises from the normal metabolic turnover of almost all tissues since it is a constituent of the plasma membrane of all cells. It is also a prominent component of erythrocyte stroma, and it is one of the major lipids of the myelin sheath.

Patients with Niemann-Pick disease may be diagnosed by determining sphingomyelinase activity in specimens of solid tissues (Brady *et al.*, 1966*b*), in circulating leukocytes (Kampine *et al.*, 1967*a*) or in cultured skin fibroblasts (Sloan *et al.*, 1969). The identification of heterozygous carriers is now possible through careful enzymatic assays with leukocyte preparations (Brady, Johnson & Uhlendorf, 1971) and through sphingomyelinase assays performed on extracts of cultured skin fibroblasts (Brady, 1972*a*) (Table 2–B). The determination of sphingomyelinase activity in cultured foetal cells has provided for the correct

Table 2. *Detection of heterozygous carriers of lipid storage diseases*

Disorder	Substrate	Enzyme source	Activity	References
			(percent of normal)	
A. Gaucher's disease	Glucocerebroside-^{14}C	Circulating leukocytes	59%	Brady, Johnson & Uhlendorf, 1971
		Cultured skin fibroblasts	47%	
B. Niemann-Pick disease	Sphingomyelin-^{14}C	Leukocytes	55%	
		Cultured skin fibroblasts	64%	Brady, 1972*a*
C. Krabbe's disease	Galactocerebroside-^{3}H	Serum	26%	Suzuki & Suzuki, 1971
D. Metachromatic leuko-dystrophy	*p*-Nitrocatechol sulphate	Leukocytes	49%	Bass *et al.*, 1970; Taniguichi & Nanba, 1970
		Cultured skin fibroblasts	19%	Kaback & Howell, 1970
E. Ceramide lactoside lipidosis	Ceramide lactoside-^{3}H	Cultured skin fibroblasts	52%	Dawson *et al.*, 1971
F. Fabry's disease	*p*-Nitrophenyl-α-D-galactopyranoside	Leukocytes	31%	Kint, 1970
		Cultured skin fibroblasts	57%	Romeo & Migeon, 1970; Brady, Uhlendorf & Jacobson, 1971
G. Tay-Sachs disease	4-Methylumbelliferyl-β-D-*N*-acetylglucosaminide	Serum	Hex A / Hex B	
1. O-variant			56% / 61%	Brady, 1972*c*
2. B-variant			55% / 84%	O'Brien *et al.*, 1970
H. Generalized ganglio-sidosis	*p*-Nitrophenyl-β-D-galactopyranoside	Leukocytes	48%	Wolfe *et al.*, 1970

prenatal diagnosis of foetuses with Niemann-Pick disease (Epstein *et al.*, 1971). All of these assays were performed with labelled sphingomyelin as substrate. A potentially useful artificial substrate for measuring sphingomyelinase has recently been proposed (Brady, Johnson & Uhlendorf, 1971).

GLOBOID LEUKODYSTROPHY (KRABBE'S DISEASE)

Patients with globoid leukodystrophy exhibit the usual signs of severe mental retardation. There is nearly a total absence of myelin in the nervous system. It is replaced by a severe gliosis and "globoid bodies" appear in the white matter. The metabolic defect in this disorder is a deficiency of a β-galactosidase which catalyses the hydrolysis of galactocerebroside (Suzuki & Suzuki, 1970) (Table 1–C) (Reaction 3):

$$\text{Galactocerebroside} + H_2O \xrightarrow{\text{galactocerebrosidase}} \text{ceramide} + \text{galactose} \quad (3)$$

Galactocerebroside is one of the chief lipid components of myelin and its accumulation is presumed to arise from the turnover and restructuring of the myelin sheath during development of the nervous system. The lipids are engulfed by phagocytic cells, and because of the attenuated catalytic activity of galactocerebrosidase, lipid accumulates and is accompanied by distention and proliferation of these cells. The heterozygous carriers of globoid leukodystrophy are said to be identifiable through galactocerebrosidase assays performed on serum samples (Table 2–C). The prenatal detection of a homozygous foetus with Krabbe's disease has recently been accomplished (Epstein, C. J. & Suzuki, K., personal communication).

METACHROMATIC LEUKODYSTROPHY (MLD)

The primary clinical manifestation in these patients is severe, progressive mental retardation. Again it seems it is possible to make a clinical distinction between patients with early or late onset of the disorder. Patients in the latter category frequently show psychological disturbances before signs of nervous system damage become obvious. There is a decrease in nerve conduction time in these patients, and nerve biopsies show brownish-yellow metachromatic granules when stained with cresyl violet dye. The substance which accumulates in tissues of patients with MLD is sulfatide, the 3′-O-sulfate ester of galactocerebroside. The enzymatic lesion is a deficiency of sulfatidase (Mehl & Jatzkewitz, 1965) (Table 1–D) (Reaction 4):

$$\text{Sulfatide} + H_2O \xrightarrow{\text{sulfatidase}} \text{galactocerebroside} + H_2SO_4 \quad (4)$$

The enzyme may be conveniently assayed using the chromogenic substrate *p*-nitrocatechol sulfate (Austin *et al.*, 1963). This assay procedure has proven very useful for the detection of homozygotes (Percy & Brady, 1968) and heterozygotes with washed leukocytes as well as in extracts of cultured skin fibroblasts (Table 2–D).

In the report of Mehl & Jatzkewitz (1965), two tissue fractions were said to be required for catalytic activity. One was a heat-labile component, and the other a heat-stable factor. All MLD patients are deficient in the heat-labile (presumably enzyme) fraction. The requirement for the heat-stable substance has recently been re-evaluated. This dependency seems to have been due to the particular method employed for the preparation of the enzyme and the conditions used in the incubation, and probably should not be regarded as a general phenomenon (Sandhoff, K. & Jatzkewitz, H., personal communication).

CERAMIDE LACTOSIDE LIPIDOSIS

This is an extremely rare disorder, and to date, only one authenticated case has been reported (Dawson & Stein, 1970). The patient had slowly progressing CNS impairment, hepatosplenomegaly, foam cells in the bone marrow, increased serum acid phosphatase, and mild anemia. The metabolic defect is a deficiency of a β-galactosidase required for the hydrolytic cleavage of lactosylceramide (Table 1–E) (Reaction 5):

$$\text{Ceramide-glucose-galactose} + H_2O \xrightarrow[\beta\text{-galactosidase}]{\text{ceramide lactoside}} \text{ceramide-glucose} + \text{galactose} \qquad (5)$$

Carriers may be detected through enzyme assays performed on extracts of cultured fibroblasts (Table 2–E).

FABRY'S DISEASE

Patients with Fabry's disease have a reddish-purple, maculopapular rash in the umbilical, inguinal and scrotal areas. They also have severe renal impairment, corneal opacities, peripheral neuralgias, and electro-cardiogram changes. Only this sphingolipid storage disease is transmitted as an X-linked genetic mutation. Therefore, in order to produce an affected male offspring, only the female needs to be heterozygote. On a statistical basis half of her sons will be hemizygous and affected; half of her daughters will be carriers and they may have mild or even a moderate degree of the characteristic symptomatology of Fabry's disease. The enzymatic defect is a deficiency of ceramidetrihexoside-α-galactosidase (Brady *et al.*, 1967) (Table 1–F) (Reaction 6):

$$\text{Ceramide-glucose-galactose-galactose} + H_2O \xrightarrow{\alpha\text{-galactosidase}} \text{ceramide-glucose-galactose} + \text{galactose} \qquad (6)$$

The activity of the involved enzyme may be assayed through the use of chromogenic or fluorogenic α-galactopyranosides as substrates (Kint, 1970). Heterozygous carriers can be detected by these assays performed on washed leukocytes or extracts of cultured skin fibroblasts (Table 2–F). The prenatal identification of a Fabry foetus has recently been accomplished (Brady, Uhlendorf & Jacobson, 1971).

The principal source of the accumulating ceramide trihexoside is red cell globoside. The tissue distribution of trihexoside in the tissues of patients with Fabry's disease is quite interesting. It accumulates in the endothelial lining of blood vessels, and it is deposited extensively in kidney glomeruli. The material begins to accumulate prenatally, and in the foetus in which the disease was authenticated, a ten-fold increase in the quantity of ceramide-trihexoside in the liver was already present at the 21st gestational week (Brady, Uhlendorf & Jacobson, 1971). Kidney tissue from this foetus did not appear to contain more ceramidetrihexoside than that found in two age-matched control kidney tissue specimens. These findings suggest that the degradation of globoside occurs primarily in the reticuloendothelial cells of the liver and spleen (and perhaps bone marrow) and then the trihexoside is carried in the blood and deposited in other tissues, especially the kidney.

TAY-SACHS DISEASE

Infants with this disorder suffer from progressive mental deterioration which usually begins to manifest itself around the sixth or seventh month of life. Blindness and convulsions follow and there is a cherry-red spot in the macula. The neuronal cells are distended with "membranous cytoplasmic bodies" which originate from lysosomes (Tallman, Brady & Suzuki, 1971). The pathological biochemistry in patients with Tay-Sachs disease is a deficiency of an enzyme which catalyses the cleavage of the molecule of *N*-acetylgalactosamine from Tay-Sachs ganglioside (G_{M2}) (Kolodny, Brady & Volk, 1969). (Table 1–G) (Reaction 7):

$$\text{Ceramide-glucose-galactose } (N\text{-acetylneuraminic acid})\text{-}N\text{-acetylgalactosamine} + H_2O \xrightarrow{G_{M2}\text{-hexosaminidase}} N\text{-acetylgalactosamine} + \text{Ceramide-glucose-galactose-}N\text{-acetylneuraminic acid} \quad (7)$$

The complete elucidation of the pathogenesis of Tay-Sachs disease is complicated by the fact that there is normal G_{M2}-sialidase activity in brain tissue of patients with Tay-Sachs disease (Brady *et al.*, 1972) (Reaction 8).

$$\text{Ceramide-glucose-galactose-}(N\text{-acetylneuraminic acid})\text{-}N\text{-acetylgalactosamine} + H_2O \xrightarrow{G_{M2}\text{-sialidase}} N\text{-acetylneuraminic acid} + \text{Ceramide-glucose-galactose-}N\text{-acetylgalactosamine} \quad (8)$$

Furthermore, there are three "variant" forms of Tay-Sachs disease based on the presence or absence of hexosaminidase isozymes whose activity may be determined with artificial substrates (Sandhoff *et al.*, 1971). This classification is based on the fact that there are at least two enzymes in normal human tissues which catalyse the hydrolysis of chromogenic or fluorogenic derivatives of *N*-acetylglucosamine (or *N*-acetylgalactosamine) (Robinson & Stirling, 1968). These isozymes can be separated because of differences in electrophoretic migration on starch gels or acrylamide strips. Isozyme A is more negatively charged than isozyme B and it is more heat-labile than isozyme B. Only isozyme A appears to be able to catalyse the hydrolysis of G_{M2}, and the demonstration of this reaction is presently restricted to preparations of lysosomes obtained from fresh brain or muscle tissue specimens (Brady *et al.*, 1972; Kolodny *et al.*, 1969). Extracts of tissues obtained from patients with the three categories of Tay-Sachs disease may show no activity with the artificial substrates (Variant O) (Sandhoff, Andreae & Jatzkewitz, 1968), have hexosaminidase isozyme B only (Variant B) (Okada & O'Brien, 1969), or even show increased activity of both isozymes with artificial substrates, but do not catalyse the cleavage of *N*-acetylgalactosamine from Tay-Sachs ganglioside (Variant AB) (Sandhoff *et al.*, 1971). Detailed resolution of the complexities of the pathogenesis of Tay-Sachs disease is discussed elsewhere (Brady, 1972*b*) and the interested investigator is referred to this article. Homozygotes and heterozygotes of the conventional (B) and (O)-variant forms of Tay-Sachs disease may be readily identified (Tables 1 and 2, item G). These tests also permit the prenatal diagnosis of foetuses with these forms of this disorder (Schneck *et al.*, 1970; O'Brien *et al.*, 1971). It must be kept in mind that the diagnosis of the AB-variant can be substantiated only through appropriate enzyme assays with authentic labelled Tay-Sachs ganglioside (Quirk, Tallman & Brady, 1972). Finally, of particular importance is the accumulation of globoside in tissues of patients with the (O)-variant form of Tay-Sachs disease. It is apparent that the terminal molecule of *N*-acetylgalactosamine cannot be hydrolysed when hexosaminidase activity is totally lacking. This observation suggests that hexosaminidase isozyme B may be largely responsible for the catabolism of globoside; however, this deduction is not yet established. Since globoside does not accumulate in the (AB)-variant form of Tay-Sachs disease, it suggests that a genetic mutation has occurred which involves only the A isozyme in these patients. To date there is no report of patients lacking only the B isozyme. This fact probably has considerable relevance with regard to the underlying genetic abnormalities in Tay-Sachs disease.

GENERALIZED[G_{M1}] GANGLIOSIDOSIS

The clinical manifestations of patients with this disorder include mental retardation, hepatosplenomegaly, foam cells in bone marrow, rarefaction of bones and skeletal deformation. Approximately 50% of these infants have a cherry-red spot in the macula. The disease is caused by a deficiency of G_{M1}-ganglioside-β-galactosidase (Okada & O'Brien, 1969) (Table 1–H) (Reaction 9):

Ceramide-glucose-galactose-(*N*-acetylneuraminic acid)-*N*-acetylgalactosamine-galactose + H_2O $\xrightarrow{G_{M1}\text{-galactosidas}}$ galactose +
Ceramide-glucose-galactose-(*N*-acetylneuraminic acid)-*N*-acetylgalactosamine (9)

Patients with this disorder may be diagnosed through the use of chromogenic or fluorogenic β-galactopyranoside substrates since most of the tissue β-galactosidase activity seems to be related to G_{M1}-ganglioside catabolism (Brady *et al.*, 1970). Galactocerebroside and ceramide lactoside metabolism is not impaired. Artificial substrates may be employed to identify heterozygotes (Table 2–H). The prenatal identification of a foetus with G_{M1}-gangliosidosis has recently been realized (M. M. Kaback & H. R. Sloan, personal communication). The accumulating ganglioside may arise from the turnover of plasma membrane components of various cells since gangliosides are selectively concentrated in these structures.

FUCOSIDOSIS

A discussion on lipid storage diseases should include some very recent findings in a small group of patients with the disorder called fucosidosis. These children show progressive cerebral degeneration, weakness, and spasticity, emaciation, thickening of the skin, and cardiomegaly (Durand *et al.*, 1969). The enzymatic defect is a complete deficiency of α-L-fucosidase activity in the tissue of these patients (Van Hoof & Hers, 1968). To date the natural substrate(s) of this enzyme is not completely settled. However, it is well known that intestinal tissue, red blood cells, and some other tissues such as human adenocarcinoma contain fucogylcolipids with H, Le^a and Le^b isoantigens (Hakomori, 1970). In accordance with the anticipated pathophysiology of such a disorder, a very large increase of a pentahexosylfucoglycolipid has been observed in the liver tissue obtained from a patient with fucosidosis (Dawson & Spranger, 1971). It is presumed that the specific metabolic defect in this disorder will soon be conclusively demonstrated.

REMARKS

The metabolic derangement in lipid storage diseases is well established for nine of the ten now recognized disorders. The last of these, fucosidosis, seems on the verge of resolution. Patients and heterozygous carriers of all of these disorders may be identified through a determination of the activity of the enzyme involved. The prenatal diagnosis of all of these diseases is now feasible and accurate genetic counselling can be provided to prospective parents in which the foetus is at risk for any of these disorders.

What shall be done with the cases which are not diagnosed prenatally? The main concern at present is to try to replace the missing enzyme. This is quite a complicated process. We have purified human hexosaminidase A to homogeneity and expect to undertake enzyme infusion studies, with appropriate precautions, into Tay-Sachs patients. The question is, how effective will such replacement therapy be because of the doubt that the enzyme can pass across the blood brain barrier. If it does not reach the brain, other procedures will have to be taken under consideration (cf. Brady, 1972*d*). Actually, two of the lipid storage diseases probably can be successfully treated at the present time by enzyme replacement therapy. One is the Fabry male patient without central nervous system damage. Such men probably will be helped by enzyme replacement. The other group consists of patients with the juvenile form of Gaucher's disease in which again it may not be necessary to get the enzyme across the blood-brain barrier.

On an entirely different track, Fig. 1 shows a cell which is distended with glycolipid and it probably is glucocerebroside, the same substance that accumulates in Gaucher's disease; however, this is from a patient with chronic myelogenous leukemia. There is something wrong with glycolipid metabolism in this patient, otherwise he would not have such a lipid storing cell. And, it is very much like a Gaucher cell. We wondered if there was some inhibition of glucocerebrosidase activity in such a patient. There is no inhibition; actually there is an increase in glucocerebrosidase activity in leukemic

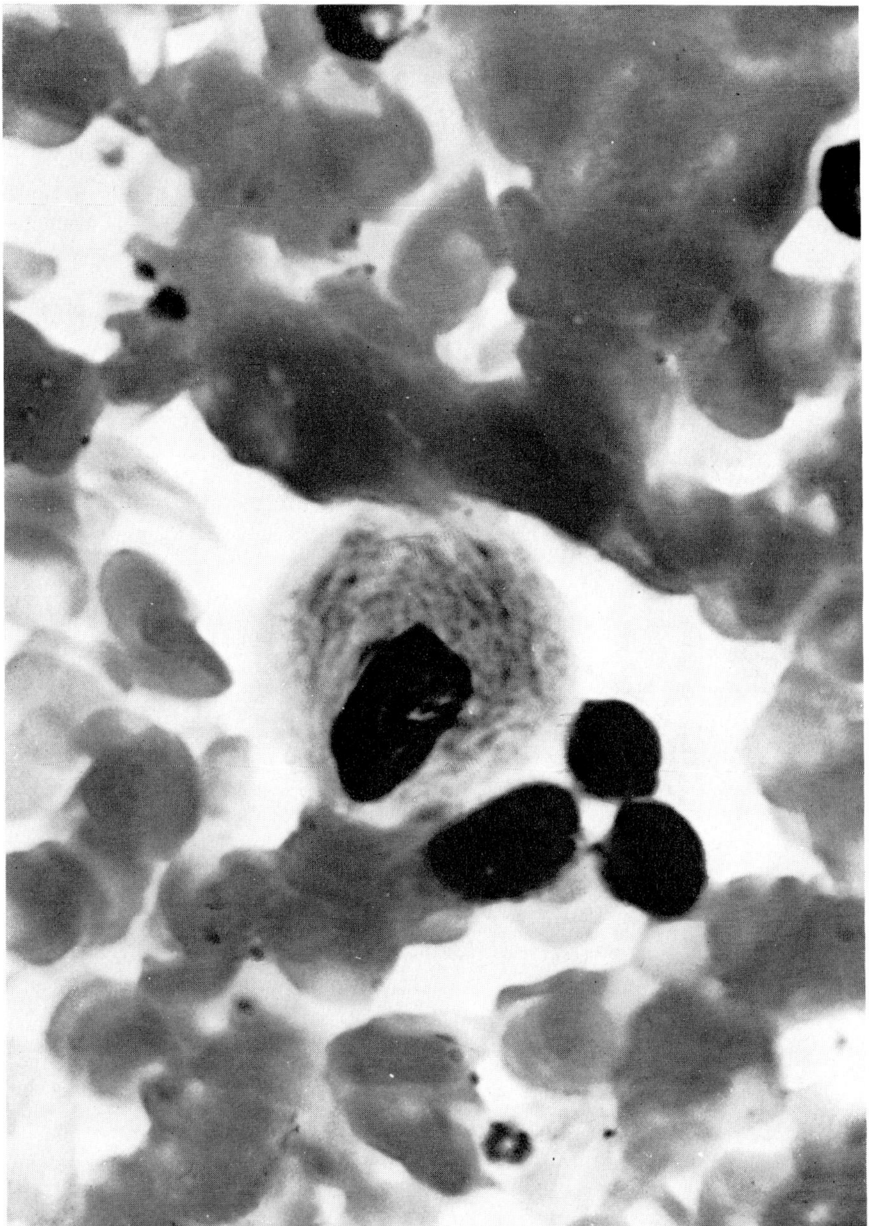

FIG. 1. Glycolipid storage cell in bone marrow of a patient with chronic myelocytic leukemia. (Reproduced with permission from Kattlove *et al.*, 1969.)

leukocytes (Kampine *et al.*, 1967*b*). We had to assume that the glycolipid accumulates because the pathway for leukocyte lipid catabolism is overwhelmed, and we then wondered why the glucocerebrosidase activity was increased in leukemic cells. Was this phenomenon a reflection of the immaturity of the cells or was it a manifestation of an enzymatic alteration in neoplastic processes in general? In order to try to answer these questions, we carried out a large amount of work and did not come up with too much until we decided that it would be helpful if we could get some closely controlled situation in which lipid metabolism could be compared in normal cells with that in transformed tumorigenic cells. We decided that it would be worthwhile looking at characteristic membrane glycolipids because we were concerned about the loss of contact inhibition in the tumorigenic cells. Since gangliosides are concentrated in the plasma membrane of cells, we focused our attention on an investigation of these materials in control and tumorigenic virus-transformed cell lines. All of the ganglioside homologues larger than hematoside (G_{M3}) are

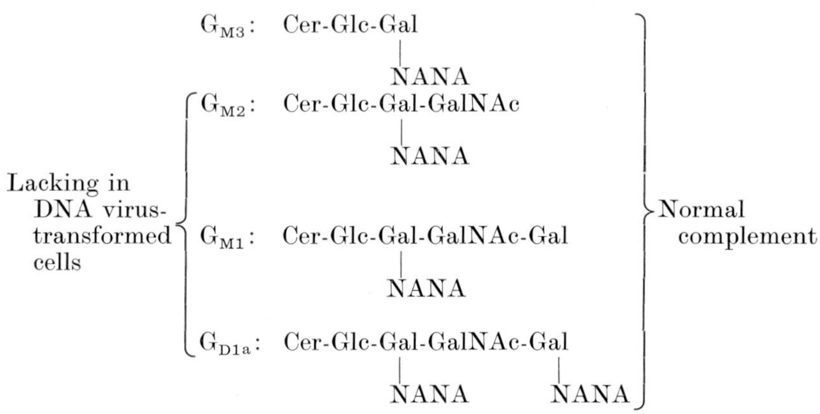

FIG. 2. Ganglioside patterns of cultured mouse cell lines.

missing in tumorigenic DNA virus transformed cells (Fig. 2) (Mora *et al.*, 1969). This phenomenon is a constant and a heritable change in the ganglioside pattern in the transformed cells. The question which had to be resolved was whether the higher gangliosides were catabolized excessively rapidly, or not synthesized after virus transformation. Investigations were begun along this line by adding labelled precursors of gangliosides to the medium in which the cells were grown. Only the short-chain (dihexosyl) gangliosides were labelled in the transformed cells. This pattern is in sharp contrast to that seen in the contact-inhibited parent cell lines (Fig. 3) (Brady & Mora, 1970).

We therefore began an investigation of ganglioside catabolism. The catabolism of the most prevalent gangliosides G_{D1a} and G_{M3} (Fig. 3) is initiated by the cleavage of the molecule of *N*-acetylneuraminic acid linked to the terminal molecule of galactose. These gangliosides were biosynthetically labelled in the sialic acid portion of the molecule (Kolodny *et al.*, 1970). Tests with these labelled materials indicated that ganglioside catabolism was not accelerated in the virally transformed cells (Cumar *et al.*, 1970). This finding made it mandatory to investigate the other possibility: an impairment of ganglioside anabolism in the transformed cells. Since all of the gangliosides whose oligosaccharide chain is longer than lactose are lacking, it was anticipated that something was wrong with the transfer of *N*-acetylgalactosamine from UDP-*N*-acetylgalactosamine to hematoside. The activity of this aminosugar transferase was found to be drastically reduced in the DNA virus-transformed cells (Cumar *et al.*, 1970). At this point it is not known whether the enzyme is not synthesized, or has an abnormal structure, or is catabolized more rapidly than normal. In any event, this alteration in the ganglioside synthesizing

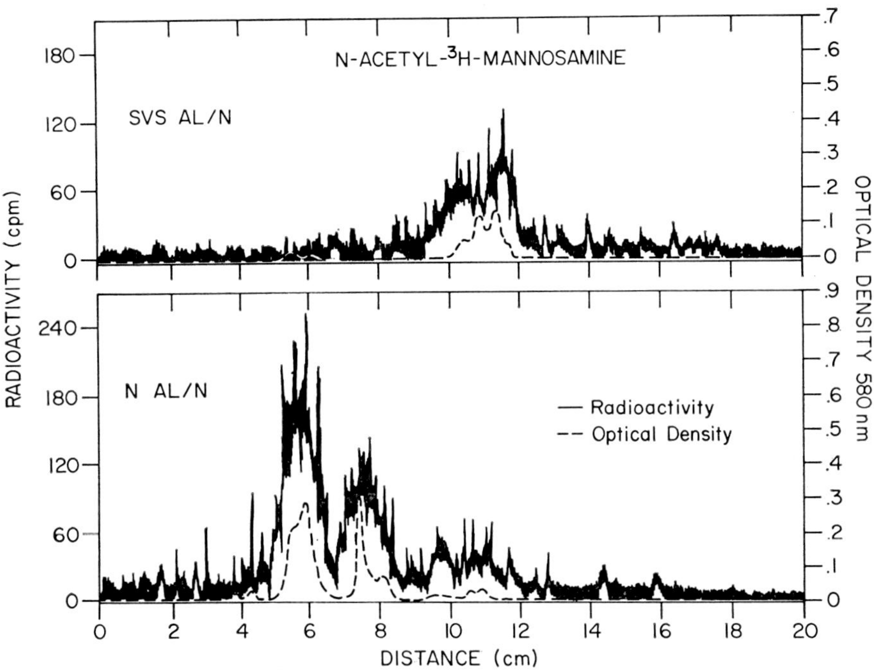

FIG. 3. Absorbance at 580 nm and radioactivity of thin-layer chromatograms of gangliosides from cultured mouse cell lines. The figures on the abscissa indicate the vertical distance from the origin (Brady & Mora, 1970).

system results in a specific phenotypic change in the chemical composition of the plasma membrane in the DNA virus-transformed cells. This attenuation in aminosugar transferase activity is not due simply to infection with the virus: the viral genome has to be inserted into the genetic apparatus of the transformed cell (Mora, Cumar & Brady, 1971).

One final experiment deserves to be mentioned here. Under certain conditions in tissue culture, a very small percentage of these cells escape from the phenotypic alteration caused by the insertion of the viral genome. These are called "flat revertant" cells. They are contact inhibited; they are less tumorigenic; the virus can be rescued from these cells, and the ganglioside pattern returns completely to normal in the Simian virus reverted cell which was available for our study. The gangliosides had partially returned to normal in a polyoma virus flat revertant; however, reversion was incomplete in the polyoma revertant cell line that we examined judging by other criteria such as morphology and cell density in culture. The aminosugar transferase in the SV 40 revertant was perfectly normal. In the polyoma incomplete revertant, this enzyme activity had partially returned to normal (Mora *et al.*, 1971). Therefore, we feel that these enzymological and phenotypic changes are genetically linked and that they are a direct consequence of transformation of cells by tumorigenic DNA viruses and the expression of the inserted viral genome.

Where do we go from here? There are many other things to do and one primary aspect of tumorigenesis which must be investigated is to try to learn what happens to cells when they are transformed with RNA viruses. Secondly, what happens during chemical carcinogenesis? So far there has been only very little experimentation in these areas regarding ganglioside metabolism. The labelling pattern of gangliosides in hepatocytes is quite different from that found in the gangliosides in a minimally deviated hepatoma cell line (Brady, Borek & Bradley, 1969). Obviously much more work is required to substantiate these findings and a satisfactory explanation must be provided for these differences. More to the point will be the unravelling of the physiologic importance of these changes in membrane glycolipids in tumorigenic cells. Hopefully, this information will provide some insight into the pathogenesis of neoplastic diseases.

REFERENCES

Austin, J., Balasubramanian, A. S., Pattabiraman, T. M., Saraswathi, S., Basu, D. K. & Bachhawat, B. K. (1963). *J. Neurochem.* **10**, 805.

Bass, N. H., Witmer, E. J. & Dreifuss, F. E. (1970). *Neurology* **20**, 52.

Brady, R. O. (1970*a*). *Ann. Rev. Med.* **21**, 317.

Brady, R. O. (1970*b*). *Clin. Chem.* **16**, 811.

Brady, R. O. (1971). *Bull. New York Acad. Med.* **47**, 173.

Brady, R. O. (1972*a*). *Seminars in Hematology*, Vol. 9, 273.

Brady, R. O. (1972*b*). In *Duncan's Diseases of Metabolism, Seventh Edition* (Ed. by Bondy, P. K.), Philadelphia: W. B. Saunders Co. (In the press.)

Brady, R. O. (1972*c*). In *Current Topics in Biochemistry: N. I. H. Lecture Series*, 1971. (Ed. by Anfinsen, C. B., Goldberger, R. F. and Schechter, A. N.), New York: Academic Press. pp. 1–48.

Brady, R. O. (1972*d*). *Angewandte Chemie.* (In press.)

Brady, R. O., Borek, C. & Bradley, R. M. (1969). *J. biol. Chem.* **244**, 6552.

Brady, R. O., Kanfer, J. & Shapiro, D. (1965). *Biochem. biophys. Res. Commun.* **18**, 221.

Brady, R. O. & King, F. M. (1972). In *Lysosomes and Storage Diseases.* (Ed. by Hers, H. G. & Van Hoof, F.), New York: Academic Press. (In press.)

Brady, R. O., Johnson, W. G. & Uhlendorf, B. W. (1971). *Am. J. Med.* **51**, 423.

Brady, R. O. & Mora, P. T. (1970). *Biochim. biophys. Acta* **218**, 308.

Brady, R. O., Uhlendorf, B. W. & Jacobson, C. B. (1971). *Science* **172**, 174.

Brady, R. O., Kanfer, J. N., Bradley, R. M. & Shapiro, D. (1966*a*). *J. clin. Invest* **45**, 1112.

Brady, R. O., Kanfer, J. N., Mock, M. B. & Fredrickson, D. S. (1966*b*). *Proc. natn. Acad. Sci. U.S.A.* **55**, 366.

Brady, R. O., Gal, A. E., Bradley, R. M., Martensson, E., Warshaw, A. L. & Laster, L. (1967). *New Eng. J. Med.* **276**, 1163.

Brady, R. O., O'Brien, J. S., Bradley, R. M. & Gal, A. E. (1970). *Biochim. biophys. Acta* **210**, 193.

Brady, R. O., Tallman, J. F., Johnson, W. G. & Quirk, J. M. (1972). In *Sphingolipids, Sphingolipidoses and Allied Disorders*. (Ed. by Volk, B. W. & Aronson, S. M.), New York: Plenum Publishing Co. pp. 277–285.

Cumar, F. A., Brady, R. O., Kolodny, E. H., McFarland, V. W. & Mora, P. T. (1970). *Proc. nat. Acad. Sci. U.S.A.* **67**, 757.

Dawson, G., Malaton, R. & Stein, A. O. (1971). *J. Pediatrics* **79**, 423.

Dawson, G. & Stein, A. O. (1970). *Science* **170**, 556.

Dawson, G. & Spranger, J. W. (1971). *New Eng. J. Med.* **285**, 122.

Durand, P., Borrone, C. & Della Cella, G. (1969). *J. Pediatrics* **75**, 665.

Epstein, C. J., Brady, R. O., Schneider, E. L., Bradley, R. M. & Shapiro, D. (1971). *Am. J. Hum. Genet.* **23**, 533.

Hakomori, S. I. (1970). *Chem. Phys. Lipids* **5**, 96.

Ho, M. W., Seck, J., Schmidt, D., Veath, L., Johnson, W. G., Brady, R. O. & O'Brien, J. S. (1972). *Am. J. Hum. Genet.* **24**, 37.

Kaback, M. M. & Howell, R. R. (1970). *New Eng. J. Med.* **282**, 1336.

Kampine, J. P., Brady, R. O., Kanfer, J. N., Feld, M. & Shapiro, D. (1967*a*). *Science* **155**, 86.

Kampine, J. P., Brady, R. O., Yankee, R. A., Kanfer, J. N. & Shapiro, D. (1967*b*). *Cancer Res.* **27**, 1312.

Kattlove, H. E., Williams, J. C., Gaynor, E., Spivack, M., Bradley, R. M. & Brady, R.O. (1969). *Blood* **33**, No. 2 pt. 2, 379.

Kint, J. A. (1970). *Science* **167**, 1268.

Kolodny, E. H., Brady, R. O. & Volk, B. W. (1969). *Biochem. biophys. Res. Commun.* **37**, 526.

Kolodny, E. H., Brady, R. O., Quirk, J. M. & Kanfer, J. N. (1970). *J. Lipid Res.* **11**, 144.

Mehl, E. & Jatzkewitz, H. (1965). *Biochem. biophys. Res. Commun.* **19**, 407.

Mora, P. T., Cumar, F. A. & Brady, R. O. (1971). *Virology* **46**, 60.

Mora, P. T., Brady, R. O., Bradley, R. M. & McFarland, V. W. (1969). *Proc. natn. Acad. Sci. U.S.A.* **63**, 1290.

O'Brien, J. S., Okada, S., Chen, A. & Fillerup, D. L. (1970). *New Eng. J. Med.* **283**, 15.

O'Brien, J. S., Okada, S., Fillerup, D. L., Veath, M. L., Adornato, B., Brenner, P. H. & Leroy, J. G. (1971). *Science* **172**, 61.

Okada, S. & O'Brien, J. S. (1968). *Science* **160**, 1002.

Okada, S. & O'Brien, J. S. (1969). *Science* **165**, 698.

Percy, A. K. & Brady, R. O. (1968). *Science* **161**, 594.

Quirk, J. M., Tallman, J. F. & Brady, R. O. (1972). *J. labeled Compounds*. (In press.)

Robinson, D. & Stirling, J. L. (1968). *Biochem. J.* **107**, 321.

Romeo, G. & Migeon, B. R. (1970). *Science* **170**, 180.

Sandhoff, K., Andreae, U. and Jatzkewitz, H. (1968). *Life Sciences* **7**, 283.

Sandhoff, K., Harzer, K., Wassle, W. & Jatzkewitz, H. (1971). *J. Neurochem.* **18**, 2469.

Schneck, L., Valenti, C., Amsterdam, D., Friedland, J., Adachi, M. & Volk, B.W. (1970). *Lancet* **1**, 582.

Sloan, H. R., Uhlendorf, B. W., Kanfer, J. N., Brady, R. O. & Fredrickson, D. S. (1969). *Biochem. biophys. Res. Commun.* **34**, 582.

Suzuki, K. & Suzuki, Y. (1970). *Proc. natn. Acad. Sci. U.S.A.* **66**, 302.

Suzuki, Y. & Suzuki, K. (1971). *Science* **171**, 73.

Tallman, J. F., Brady, R. O. & Suzuki, K. (1971). *J. Neurochem.* **18**, 1775.

Taniguichi, N. & Nanba, I. (1970). *Clin. Chim. Acta* **29**, 375.

Thannhauser, S. J. (1950). In *Lipidoses: Diseases of the Cellular Lipid Metabolism*, p. 611. (Ed. by Christian, H. A.), New York: Oxford University Press.

Trams, E. G. & Brady, R. O. (1960). *J. clin. Invest.* **39**, 1546.

Van Hoof, F. & Hers, H. G. (1968). *Lancet*, June 1, 1198.

Wolfe, L. S., Callahan, J., Fawcett, J. S., Anderman, F. & Scriver, C. R. (1970). *Neurology* **20**, 23.

METABOLISM OF MYELIN LIPIDS IN THE DEVELOPING BRAIN

By A. N. DAVISON

Department of Neurochemistry, Institute of Neurology, London, England

Electron microscopic observations indicate that the myelin sheath is built up of many membranes wrapped spirally round the central conducting axon. As many as 100 such lamellae are found around peripheral nerve fibres and up to about 40 in central nerve tracts. These radially stacked membranes have proved a very convenient source for study by low angle X-ray diffraction analysis and much useful information about the probable constitution and metabolism of the membrane has resulted. The myelin sheath consists of a repeating unit (17–18 nm in width) with two fundamental 7·5 nm unit membranes containing a bimolecular lipid layer (5·5 nm thick) sandwiched between protein. The lipid composition of the membrane in the central nervous system is remarkably constant within and even between different vertebrate species (Cuzner, Davison & Gregson, 1965) with a molar ratio of approximately 2:2:1, cholesterol:phospholipid:galactocerebroside. These molecules appear to be arranged in a tightly packed parallel order so that their polar groups interact with the proteins and their long hydrocarbon chains (sphingosine, nervonic acid, etc.) are held together by non-covalent bands (e.g. London-van der Waal's forces) in a hydrophobic environment. Stereomodels (Finean, 1965; Vandenheuvel, 1965; O'Brien, 1967) show how remarkably it is proposed that the lipid components of myelin fit together and thus suggest that the structure must have a high degree of stability.

The concept of metabolic stability of the myelin sheath is further supported by much data indicating that, in general, lipids of the adult nervous system do not undergo dynamic metabolism as is evident for most body constituents. Thus it is difficult to demonstrate active sterol synthesis in the adult brain, while cholesterol, galactocerebroside and some phospholipids have very slow turnover rates (Smith & Eng, 1965; Davison, 1970). Experiments on developing rats, mice and rabbits injected with labelled precursors (e.g. ^{14}C-acetate) or [4-^{14}C] cholesterol indicated that the labelled lipid remained in the myelin sheath for months and even years after incorporation. In addition, it has proved difficult to alter myelin lipid composition either by dietary deficiencies or modifications. It was, therefore, concluded that: (1) myelin was relatively metabolically inert, and that (2) the lipid composition of myelin was invariant. In this review data are presented to show that both these hypotheses require substantial modification.

Table 1. *Lipid analysis of myelin isolated from developing rat brain.*

The number of brains analysed are shown in parentheses (*Cuzner & Davison*, 1968)

Age (days)	Myelin lipid composition Cholesterol : phospholipid : cerebroside μmoles/whole brain			Molar lipid ratio Cholesterol = 100 : phospholipid : cerebroside		
10(2)	1·47	2·07	0·077	100:	141:	5
16(2)	3·20	4·47	0·321	100:	140:	10
21(3)	5·50	8·80	1·00	100:	160:	18
30(2)	15·40	13·90	4·50	100:	90:	29
70(1)	33·80	30·20	—	100:	89:	—
Adult (6)	34·00	32·00	11·80	100:	94:	35

Table 2. *Lipids of cat optic nerves during development.*

Whole optic nerve and adult cat myelin were analysed for lipid by quantitative thin-layer chromatography. The number of nerves analysed are shown in parentheses (*Banik* et al., 1968)

	Age in days					
	10	16	19	28	Adult	Adult optic nerve myelin
Lipid	μmoles lipid/g wet wt. whole optic nerve					
Cholesterol:	54·7 ± 5·3	60·6 ± 3·1	87·9	88·2	105·0, 102·0	70·4, 63·6
Phospholipid:	56·9 ± 4·5	62·3 ± 1·6	73·5	76·0	86·1, 77·0	51·2, 43·5
Cerebrosides:	2·2 ± 0·3	3·2 ± 0·4	4·3	4·6	35·7, 30·9	20·4, 14·5
Molar lipid ratio:	1:1·05:0·04	1:1·04:0·05	1:0·84:0·05	1:0·86:0·055	1:0·8:0·36	1:0·7:0·5
Average wet wt. of optic nerves (mg)	37·8 (4)	29·0 (5)	48·2 (1)	68·2 (1)	85·1 (2)	120·4 (2)

BIOCHEMICAL CHANGES IN MYELIN OF THE DEVELOPING BRAIN

In the adult rat brain (Cuzner & Davison, 1968) about 30% of cholesterol and phospholipid and more than half the galactocerebroside were found in myelin isolated by subcellular and gradient density centrifugation. Hence cerebroside and its 3-sulphate ester (sulphatide) are useful indices for the amount of myelin present in nervous tissue (small amounts of galactolipids are also found in the axon).

Horrocks and his colleagues (Horrocks, Meckler & Collins, 1966) first noted that crude myelin isolated from developing mouse brain was different in composition from that of adult mice. Similarly in 10–20 day-old rat brain, Cuzner & Davison (1968) found that isolated myelin contained a higher molar proportion of phospholipid to cholesterol and there was a relative deficiency in cerebroside (Table 1). Such data could be ascribed to extensive contamination of myelin fractions, for example, by microsomal fragments present in an unusually high proportion in developing brain, or the changes in lipid composition could be due to alterations in the composition of the developing myelin sheath. Evidence for the latter possibility came from observations by Banik, Blunt & Davison (1968). It was found that although myelin rings could be seen by light microscopy in 10-day-old kitten optic nerve stained with sudan black or luxol fast blue, the myelin could not be stained with osmium tetroxide until optic nerves from animals of at least 25 days of age were examined. Wolman (1957) had also observed that osmiophilia of rat brain, spinal cord and sciatic nerve myelin sheath does not coincide with the first appearance of the sudanophilic myelin rings. Electron microscopy confirmed that myelin was present in the optic nerves of 10-day-old kittens and that myelination was well established by 16 days postnatally. Chloroform-methanol (2:1, v/v) extracts were, therefore, prepared from kitten (10–28 day-old) and cat optic nerves and the washed lipid extracts analysed (Table 2). The proportion of the typical myelin-lipid cerebroside was much less in the developing optic tract than was expected from the accumulation of cholesterol and from the amount of histologically stainable myelin. Only traces of galactocerebrosides were found to be present in the 10–16-day-old kitten optic nerve and even in the 28-day-old animal the molar proportion of cerebroside to cholesterol was less than a fifth of that found in the adult cat. As development proceeds the high proportion of myelin phospholipid lessens and phosphoglyceride fatty acid chain length increases (Banik *et al.*, 1968). Thus, it appeared that the lipid composition of "early" myelin (i.e. that formed during the first period of myelinogenesis) was more similar to that of the plasma and other cytomembranes than to the unique lipid composition of myelin.

In the developing rat brain accumulation of cerebroside does not reach its maximal rate until about 25 days after birth although phospholipid and cholesterol synthesis is most rapid by about 10 days postnatally. Since myelination starts in some areas of rat brain at 10 days and appears histologically to be advanced by 15 days, the brain phospholipid and cholesterol concentrations are consistent with the morphological observations but there is less cerebroside in the brain than would be anticipated from the histologically stainable myelin in the central nervous system. As development continues (after 30 days postnatally) there is good correlation between accumulation of myelin lipids and the continuing formation of myelin. During the early phase of myelination in tissue culture of foetal dorsal root ganglia the refractile sheaths appear thin and smooth (Peterson & Murray, 1965) later, irregularities form. In rat cerebellum tissue culture Aparicio, Bradbury & Nixon (1971) demonstrated sequential incorporation of phosphoglyceride, sphingomyelin and finally cerebroside into the "early" myelin. These various observations suggested that there were indeed changes in composition of myelin during development of both central and peripheral nervous system and this hypothesis was further evaluated. The composition of crude myelin, obtained by the method of Cuzner & Davison (1968) from 15-day-old rat brain, was compared with purified myelin isolated by the technique of Norton & Poduslo (see Norton, 1971). Norton, W. T. & Davison, A. N. (unpublished) found that the crude myelin could be separated into two fractions on a caesium chloride gradient, one comparable to purified myelin and a second "myelin-like" fraction which was rich in phospholipid and contained little cerebroside. Alternatively, crude myelin isolated from a brain mitochondrial pellet may be separated on a discontinuous sucrose gradient as described by De Robertis (De Robertis *et al.*,

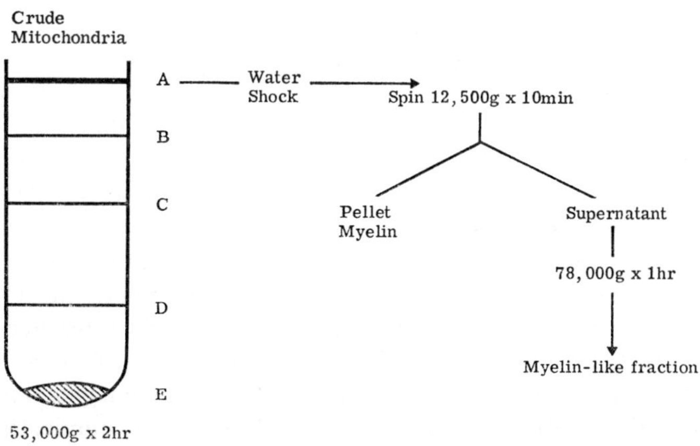

FIG. 1. Separation of the myelin-like fraction from developing rat brain (Agrawal *et al.*, 1970).

1962) and further separated into two fractions after osmotic shock (Fig. 1). In the 15-day-old rat brain about equal amounts of the two fractions are obtained (Agrawal *et al.*, 1970). Electron micrographs and chemical analyses showed one to be myelin while the second fraction consisted mainly of single membrane vesicles with a distinctly different composition from myelin. The second fraction was prepared by high speed centrifugation of the cloudy supernatant after precipitation of purified myelin by centrifugation at 12,500 g × 10 min and was designated the "myelin-like" fraction. It had a higher phospholipid to cholesterol ratio and was relatively rich in lecithin whereas only traces of cerebroside were present. Analysis of phospholipid fatty acids from the myelin and particularly the myelin-like fraction of developing brain showed that more short-chain saturated fatty acids were present than in the adult.

In order to further identify the myelin-like fraction Agrawal and his colleagues compared its protein spectrum with that of all subcellular fractions from developing brain. The proteins were first separated by disc gel electrophoresis using phenol:acetic acid:water (2:1:1, w/v/v) on polyacrylamide gel according to the technique of Takayama *et al.* (1966). The separated protein bands showed that there are similarities between those of purified myelin, the myelin-like fraction, the microsomal fraction and the nuclear fraction. The separated myelin protein bands showed a distribution similar to that described by Mehl & Halaris (1970) and other authors. Protein bands from the synaptic endings and free mitochondria were quite different as had already been shown by Cotman & Mahler (1967) in their studies on the adult rat brain. It has been claimed that basic encephalitogenic protein is a unique constituent of myelin and therefore the distribution of this protein was next examined. Electrophoresis in phenol:acetic acid:water (2:1:1, w/v/v) showed that the two bands of basic protein were present only in barely detectable amounts in the myelin-like fraction and in negligible amounts in the microsomes. This impression was confirmed by extracting the basic protein with 0·1 M HCl from subcellular fractions according to the technique of Martenson & Gaitonde (1969). Electrophoresis of the acid-extractable protein of each subcellular fraction was carried out at pH 10·6. No more than traces of the highly basic protein were present in any other fraction except pure myelin. Encephalitogenic basic protein present in the nuclear pellet was thought to be due to the presence of contaminating myelin. Using differential centrifugation procedures for the preparation of myelin, it has been shown that more than half the myelin isolated from the 12–15-day-old rat brain was accounted for by the myelin-like fraction. However, cerebroside was present in the myelin-like fraction from adult brain, suggesting it was derived from or contaminated by myelin. In addition to its behaviour on centrifugation the myelin-like fraction resembled myelin in its high 2′3′-cyclic nucleotide-3′-phosphohydrolase and leucine amino peptidase activity and in having low "marker" enzyme activity for other subcellular fractions (Banik & Davison, 1969). For example, the specific activity of succinic dehydrogenase was 11·9 μmoles substrate/mg protein/hr in the mitochondrial fraction and only 0·43 in the myelin-like fraction. While the myelin-like fraction resembled myelin in its electrophoretic protein pattern it differed in not containing the basic encephalitogenic protein. It therefore seemed unlikely that the myelin-like fraction was a form of degraded myelin. The protein patterns and the enzymatic activities (for example succinic dehydrogenase and acetylcholinesterase) were unlike those of the mitochondrial or synaptic-ending fractions. However, the appearances of the myelin-like fraction on electron microscopy and its protein pattern on disc-electrophoresis were similar to those of the microsomal fraction. Since microsomes are a highly heterogeneous collection of different membrane fractions, like smooth and rough endoplasmic reticulum as well as neuronal and glial plasma membranes, it was not surprising to find such similarities. Its low nucleic acid content makes it unlikely that the myelin-like fraction is derived from the nuclear fraction or from the rough endoplasmic reticulum. Moreover, our studies make it unlikely that it was directly related to the smooth endoplasmic reticulum, because this fraction had a lower specific activity for the total ATPase, acetylcholinesterase and 5′-mononucleotidase than that in the microsomal fraction. Dr. Louise Cuzner also has separated the microsomes from 15-day-old rat brain into subcellular fractions. All subfractions contained RNA, and the light fraction containing the highest phosphohydrolase activity also contained cerebroside and appeared on electron microscopy to be mostly myelin. It is, therefore, proposed that the myelin-like fraction is derived from the first loose whorls of single glial plasma membrane before compact mature myelin is formed (e.g. Caley & Maxwell, 1968; Peters & Vaughn, 1970). This again suggests that the first loose lamellae (de Webster, 1971) appearing in the first stages of peripheral nerve myelination, are derived from the Schwann cell plasma membrane and that compact myelin has a different composition. The initial sequence of changes seen in developing myelin fractions continues for a long period. Purified myelin isolated from the brains of rats younger than 30 days old is still relatively deficient in galactocerebroside. In addition, there are changes in phosphoglyceride fatty acid chain length and definite alterations in the ganglioside pattern. In myelin isolated from 15–425-day-old rat brain, Suzuki *et al.* (1967) showed a steady increase in the G_{M1} monosialoganglioside and a decrease in tri- and disialogangliosides. It is of interest that in certain of the neurological mutants there appears to be delayed myelination and in place of mature myelin there is accumulation of a membrane fraction similar to that of the myelin-like material found in the early stages of neural development (Greenfield, Norton & Morell, 1971; Meier & MacPike, 1970).

METABOLISM OF MYELIN LIPIDS

Changes in the composition of myelin (described above) are difficult to explain in terms of the concept of metabolic stability, which suggests that myelin composition should remain invariant in the mature nervous system; other inconsistencies were also found.

Although it had been demonstrated that labelled lipid remained for long periods in myelin after injection into developing animals it was surprising to find that radioactivity was readily incorporated into the apparently inert myelin of adult animals (see for example August, Davison & Williams, 1961). In addition, it was observed that ^{14}C-cholesterol remained for as long in mitochondrial membranes as in myelin (Cuzner *et al.*, 1966; Khan & Folch, 1967) and re-utilization of the labelled sterol was suggested as one of the possible explanations of the results. Smith & Eng (1965) also showed that different myelin lipids turned over at different rates (half-life of lecithin two months, cerebroside more than a year) and the polyphosphoinositides of myelin were demonstrated by Eichberg & Dawson (1965) to be metabolically quite active. It was next found that desmosterol (24-dehydrocholesterol) could be detected in small amounts in myelin from young animals but unexpectedly this cholesterol precursor did not remain in the myelin of mature animals. After injection of anti-cholesterolaemic drugs (e.g. Triparanol or AY-9944) into developing animals, much of the total sterol appeared in the brain as either 24-dehydro- or 7-dehydrocholesterol, whereas on discontinuing drug treatment these sterol precursors were soon replaced in myelin and subcellular fractions by cholesterol (Fumagalli *et al.*, 1969). Thus when AY-9944 was injected into 5-day-old rats followed by ^{14}C-acetate given intraperitoneally, myelin 7-dehydrocholesterol was extensively labelled when the rats were 12 days old. From then onwards the precursor disappeared and an equivalent of radioactive cholesterol was found subsequently in the myelin, indicating a possible localized conversion of 7-dehydrocholesterol to cholesterol. This suggested that contrary to earlier views, myelin sterols in the developing brain are metabolically active and that either the myelin sterols could undergo exchange or that the cholesterol precursors could be reduced to cholesterol within the developing sheath. However, the 7-dehydrocholesterol reductase responsible for the conversion was present in brain microsomes and not in purified myelin (Banik & Davison, 1971). These observations, therefore, suggest that in developing rat brain the 7-dehydrocholesterol can migrate from myelin, become reduced by microsomal reductase and the resultant cholesterol molecules are re-incorporated into myelin. Myelin labelled with ^{14}C-lipid or with dehydrocholesterol was, therefore, incubated at 37°C for 30 min with whole of rat brain homogenate or with isolated microsomes labelled with ^{14}C-protein. After incubation myelin and microsomes were re-isolated. Contamination was checked by determining ^{14}C-protein and cyclic 2′3′-nucleotide 3′-phosphohydrolase in the separated fractions. Labelled cholesterol and 7-dehydrocholesterol were found to undergo up to 10% exchange between myelin and microsomes. Phospholipid and cerebroside also underwent exchange from one membrane to another. Graham & Green (1967) have already demonstrated a 57% exchange of cholesterol from adult myelin with plasma lipoproteins after 24 hr incubation at 37°C. It is also well established that cholesterol of erythrocyte stroma undergoes ready exchange with plasma sterols. In our experiments, enhanced exchange occurred with whole-brain suspensions rather than myelin plus isolated microsomes. It has been demonstrated that exchange *in vitro* requires a heat-labile supernatant factor, identified as a lipoprotein carrier (Wirtz & Zilversmit, 1968; McMurray & Dawson, 1969; Beattie, 1969) for the exchange of phospholipid molecules between rat liver mitochondria and microsomes (Jungalwala & Dawson, 1970); hence a similar protein carrier may be important in the exchange of cerebral lipid. In an extension of this work to adult animals (Spohn, M. & Davison, A. N., unpublished work) ^{14}C-acetate was injected into 16-day-old animals and the relative specific activities of cholesterol of cerebral subcellular fractions were determined; at different times (10 min, 6 and 7 months) after injection, there was uniform specific activity of cholesterol in all fractions. In another series of experiments adult rats were given intravenous injections of [4-^{14}C]cholesterol and killed at intervals. Ten days later, the cholesterol specific activity of myelin had increased to that of the sterol in all other fractions—a finding completely unexpected if adult myelin were inert. Similar results have been obtained in adult animals after labelling with 7-dehydrocholesterol. Jungalwala & Dawson (1971) have examined incorporation of [^{32}P]phosphate, [U-^{14}C]glycerol and [2-^{14}C]ethanolamine into microsomal and a highly purified myelin fraction of adult rat brain. Initially the specific activity of microsomal phospholipids was higher than that of myelin but by 20 days after intracerebral injection the reduced specific activities were similar. Persistence of labelling occurred after 10-day-old rats had been injected, thus supporting the previous concept that two metabolic pools exist in myelin, one a rapidly-exchanging pool and the other a more persistent and slowly-turning-over part which can be significantly labelled in the developing animal (although exchange of phospholipids between a labelled brain microsomal fraction and unlabelled myelin particles was not demonstrable *in vitro*, Miller, E. K. & Dawson, R. M. C., unpublished work). This recent work indicates that the pool of exchangeable myelin lipid is larger and more accessible than had been previously assumed.

Exchange of myelin lipid also depends on the binding of lipid to myelin protein. Braun & Radin (1969) have demonstrated special interactions between sulphatide, phosphatidylserine and phosphatidylinositol with proteolipid protein which may be compared to the weaker binding of the nonionic lipids cholesterol and cerebrosides. It has also been shown that polyphosphoinositides and lecithin readily combine with the myelin basic protein (Palmer & Dawson, 1969). These myelin components are all known to undergo rapid turnover and one may, therefore, be combined in some form in the myelin sheath. Despite the ready exchange of lipid within the nervous system it appears to be difficult to modify the brain lipids by altering whole brain metabolism. Thus no change in brain lipids was noted in severely-starved rats by Dobbing (1968) and only minor changes were seen in brain fatty acids in chronic deficiency of essential polyunsaturated fatty acids (Galli, White & Paoletti, 1970). Rathbone (1965) fed 9–10-week-old rats diets containing sunflower seed oil (high in $C_{18:2}$) and coconut oil (low in $C_{18:2}$ and high in short chain fatty acids) for periods up to 34 weeks. Small changes were seen in the fatty acid composition of the brain as a result of the prolonged dietary regime. In myelin the proportion of lineoleic acid increased from 0·9% to 1·3% in the sunflower-seed fed animals and dropped to 0·4% in those given supplements of coconut oil. There were, therefore, no very marked changes in fatty composition and it would appear that endogenous fatty acid synthesis or re-utilization of pre-formed long chain molecules predominates in the adult brain.

CONCLUSION

Changes in the lipid composition of the developing nervous system have been correlated with the quantitatively important process of myelination. It is postulated that myelination begins with promulgation of the plasma membrane of the glial or Schwann cell. This extended membrane is thought to be separable from other subcellular structures during the early stages of myelination and has been designated the myelin-like fraction. The basic encephalitogenic protein and cerebroside are regarded as critical components in the formation of compact myelin; failure to form either results in accumulation of a myelin-like fraction similar to that found in developing brain. Even after the appearance of compact lamellae, isolated myelin has not the composition of that from mature brain. The continuous change in composition correlates with experimental evidence for the exchangeability of myelin lipids. Despite the flux between membrane lipids within the brain, metabolism of sterols, fatty acids and many other lipids is slow because the nervous system seems to conserve such molecules for re-utilization in membrane structures. Such a conclusion emphasizes the critical importance of normal development of the central nervous system especially with regard to synthesis of structural protein and formation of complex lipid molecules.

REFERENCES

Agrawal, H. C., Banik, N. L., Bone, A. H., Davison, A. N., Mitchell, R. F. & Spohn, M. (1970). *Biochem. J.* **120**, 635–642.

Aparicio, S. R., Bradbury, K. & Nixon, H. (1971). *Proc. R. microsc. Soc.* **6**, 7–8.

August, C., Davison, A. N. & Williams, F. M. (1961). *Biochem. J.* **81**, 8–12.

Banik, N. L., Blunt, M. J. & Davison, A. N. (1968). *J. Neurochem.* **15**, 471–475.

Banik, N. L. & Davison, A. N. (1969), *Biochem. J.* **115**, 1051–1062.

Banik, N. L. & Davison, A. N. (1971). *Biochem. J.* **122**, 751–758.

Beattie, D. A. S. (1969). *J. Membrane Biol.* **1**, 383–401.

Braun, P. E. & Radin, N. S. (1969). *Biochemistry* **8**, 4310–4318.

Caley, D. W. & Maxwell, D. S. (1968). *J. comp. Neurol.* **133**, 45–70.

Cotman, C. W. & Mahler, H. R. (1967). *Archs Biochem. Biophys.* **120**, 384–396.

Cuzner, M. L. & Davison, A. N. (1968). *Biochem. J.* **106**, 29–34.

Cuzner, M. L., Davison, A. N. & Gregson, N. A. (1965). *J. Neurochem.* **12**, 469–481.

Cuzner, M. L., Davison, A. N. & Gregson, N. A. (1966). *Biochem. J.* **101**, 618–626.

Davison, A. N. (1970). In *Myelination* (Ed. by Davison, A. N. & Peters, A., pp. 80–161), Springfield, Illinois: C. C. Thomas.

De Robertis, E., Pellegrino de Iraldi, A. P., Rodriguez de Iraldi, G. R. R. & Salganicoff, L. (1962). *J. Neurochem.* **9**, 23–25.

Dobbing, J. (1968). In *Malnutrition, Learning and Behaviour* (Ed. by Schrinshaw, N. S. & Gordon, J. E.), p. 181, Cambridge, Mass: M.I.T. Press.

Eichberg, J. & Dawson, R. M. C. (1965). *Biochem. J.* **96**, 644–650.

Finean, J. B. (1965). *Ann. N.Y. Acad. Sci.* **122**, 51–76.

Fumagalli, R., Smith, M. E., Urna, G. & Paoletti, R. (1969). *J. Neurochem.* **16**, 1329–1339.

Galli, C., White, H. B. & Paoletti, R. (1970). *J. Neurochem.* **17**, 347–355.

Graham, J. M. & Green, C. (1967). *Biochem. J.* **103**, 16C–18C.

Greenfield, S., Norton, W. T. & Morell, P. (1971). *J. Neurochem.* **18**, 2119–2128.

Horrocks, L. A., Meckler, R. J. & Collins, R. L. (1966). In *Variations in Chemical Composition of the Nervous System* (Ed. by Ansell, G. B.), p. 46, Oxford: Pergamon Press.

Jungalwala, F. B. & Dawson, R. M. C. (1970). *Biochem. J.* **117**, 481–490.

Jungalwala, F. B. & Dawson, R. M. C. (1971). *Biochem. J.* **123**, 683–693.

Khan, A. A. & Folch-Pi, J. (1967). *J. Neurochem.* **14**, 1099–1105.

Martenson, R. E. & Gaitonde, M. K. (1969). *J. Neurochem.* **16**, 333–347.

McMurray, W. C. & Dawson, R. M. C. (1969). *Biochem. J.* **112**, 91–108.

Mehl, E. & Halaris, A. (1970). *J. Neurochem.* **17**, 659–668.

Meier, H. & MacPike, A. D. (1970). *Exp. Brain Res.* **10**, 512–525.

Norton, W. T. (1971). In *Chemistry and Brain Development* (Ed. by Paoletti, R. & Davison, A. N.), pp. 327–337, New York: Plenum Press.

O'Brien, J. S. (1967). *J. theoret. Biol.* **15**, 307–324.

Palmer, F. B. & Dawson, R. M. C. (1969). *Biochem. J.* **111**, 637–646.

Peterson, E. R. & Murray, M. R. (1965). *Ann. N.Y. Acad. Sci.* **122**, 39–50.

Peters, A. & Vaughn, J. E. (1970). In *Myelination* (Ed. by Davison, A. N. & Peters, A.), pp. 3–79, Springfield, Illinois: C. C. Thomas.
Rathbone, L. (1965). *Biochem. J.* **97**, 620–628.
Smith, M. E. & Eng, L. F. (1965). *J. Am. Oil Chem. Soc.* **42**, 1013–1018.
Suzuki, K., Poduslo, S. E. & Norton, W. T. (1967). *Biochim. Biophys. Acta* **144**, 375–381.
Takayama, K., MacLennan, D. H., Tzagoloff, A. & Stoner, C. D. (1966). *Archs. Biochem. Biophys.* **114**, 223–230.
Vandenheuvel, F. A. (1965). *Ann. N.Y. Acad. Sci.* **122**, 57–85.
Webster, de H. F. (1971). *Neurosci. Res. Prog. Bull.* **9**, 470–477.
Wirtz, K. W. A. & Zilversmit, P. B. (1968). *J. biol. Chem.* **243**, 3596–3602.
Wolman, M. (1957). *Bull. Res. Coun., Israel E.* **6**, 163–167.

COMPARISON OF PROPERTIES OF THE ENZYMES INVOLVED IN METACHROMATIC LEUKODYSTROPHY AND IN TAY-SACHS DISEASE

By H. JATZKEWITZ

Max-Planck-Institut für Psychiatrie, Neurochemische Abteilung, Munich, Germany

Both metachromatic leukodystrophy and infantile amaurotic idiocy arise from an accumulation of sphingoglycolipids. In the case of metachromatic leukodystrophy (a demyelinating disease), cerebroside sulphates (= sulphatides) (Fig. 1) accumulate (Jatzkewitz, 1958, 1960; Austin, 1959). Sulphatides are constituents of brain white matter especially of myelin.

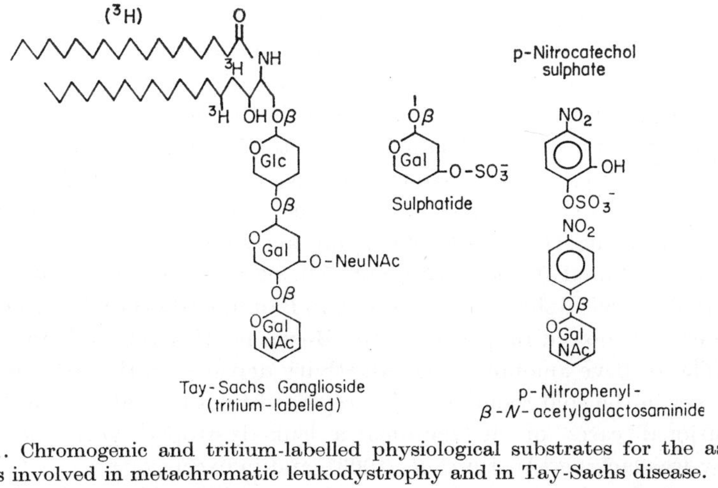

Fig. 1. Chromogenic and tritium-labelled physiological substrates for the assay of enzymes involved in metachromatic leukodystrophy and in Tay-Sachs disease.

In patients suffering from infantile amaurotic idiocy, gangliosides (= ceramide oligosaccharides + N-acetylneuraminic acid) (Klenk, 1939) and to a lesser extent their neuraminic acid-free residues are stored (Gatt & Berman, 1963; Jatzkewitz & Sandhoff, 1963). In the conventional type, called Tay-Sachs disease, this ganglioside, which contains a single N-acetylneuraminic acid, is derived from a trihexosylceramide and has the structure shown in Fig. 1 (Svennerholm, 1962, 1963). Gangliosides are constituents of the grey matter. Accumulation of the Tay-Sachs ganglioside leads to swollen and disintegrated neurons.

Both these human diseases, following a recessive mode of inheritance, are caused by a disturbance of the turnover of the accumulated sphingolipids. This is due to a genetically induced deficiency in the catabolic enzyme system which resides in the lysosomes. Therefore, these diseases are sometimes termed "lysosomal diseases".

Normally the degradation of cerebroside sulphates should be initiated by a sulphatase and that of the Tay-Sachs ganglioside by a β-N-acetylhexosaminidase. For a long time three arylsulphatases called arylsulphatases A, B, and C were known (see Dodgson & Spencer, 1957). Later on two β-N-acetylhexosaminidases, called aryl- or β-N-acetyl-hexosaminidases A and B were noted. This notation indicates that these enzymes were first recognized from their action on aryl compounds, which are artificial chormogenic substrates, and that there existed different forms with similar chromogenic substrate specificities as indicated by the letters A, B and C. These different forms can now be distinguished by their different electrophoretic mobilities.

The enzyme deficiency in these two inborn errors of metabolism enabled us to find the physiological substrates for the arylsulphatase A and the hexosaminidases A and B. For this purpose it was necessary to study the aryl enzyme patterns in controls and pathological tissues, on the one hand, and on the other hand to correlate these patterns with the accumulated physiological substances. Then, the particular aryl enzyme which was lacking in the tissues of the patients but which was present in controls should act on the accumulated compounds. In order to demonstrate this, the compounds were tritium-labelled in the sphingosine moiety as shown in Fig. 1. After such relatively simple labelling the entire activity remains in the ceramide-containing fragment after enzymic release of sulphate or of the individual sugars.

Let us now turn to metachromatic leukodystrophy. As can be seen in Fig. 2, when the solubilized arylsulphatases of normal human tissues were analysed by isoelectric focusing, three peaks of enzymic activities towards p-nitrocatechol sulphate were found at isoelectric points 5·0, 8·2 and >8·5. The acidic one was recognized as arylsulphatase A according to Baum, Dodgson & Spencer (1959), the other two showed properties of the B-enzyme (Harzer *et al.*, in preparation). The relative amount of the A-activity depends on the storage time of the tissues in the cold room and their repeated freezing and thawing). In the conventional cases of metachromatic leukodystrophy arylsulphatase A activity was lacking (Austin *et al.*, 1963), and in exceptional cases B (and C) are also absent (Austin, Armstrong & Shearer, 1965; Murphy *et al.*, 1971). The conventional cases where B is present are called variant B, while the exceptional cases where A and B (and C) are lacking are called variant O. In the conventional cases cerebroside sulphates accumulate. Thus it seems reasonable to assume that the physiological substrate for the arylsulphatase A is cerebroside sulphate; in other words arylsulphatase A is a cerebroside sulphatase. This was shown experimentally to be true (Mehl & Jatzkewitz, 1964, 1968; Jatzkewitz & Mehl, 1969). In exceptional cases an additional accumulation of glycosaminoglycans and of steroid sulphates was observed (Murphy *et al.*, 1971). These substances could be substrates for the arylsulphatases B and C.

Similarly, following isoelectric focusing of extracts from normal human tissues (Fig. 3) two peaks with β-N-acetylhexosaminidase activity were found. The acidic one (i.p. 5·0) corresponds to Robinson & Stirling's hexosaminidase A and the other (i.p. 7·3) to B. Again, in conventional cases of Tay-Sachs disease (of Jewish origin) the acidic A-enzyme is missing, while the B-activity in brain increases. It was therefore called variant B. Again, in a less frequent

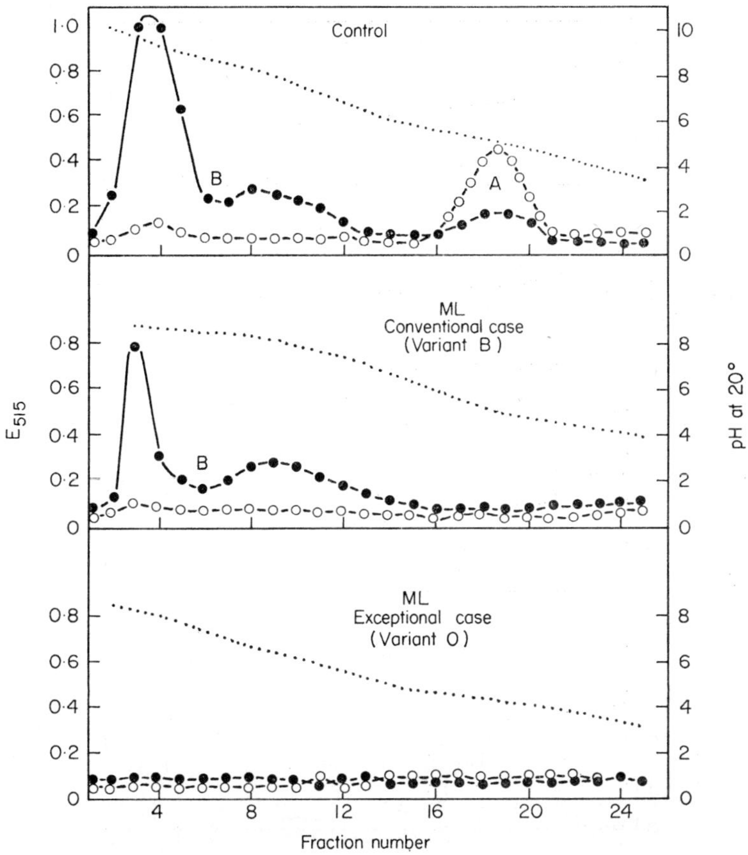

FIG. 2. Variations of arylsulphatase A patterns in the kidney tissue of cases with metachromatic leukodystrophy (ML). Method: Isoelectric focusing (.....pH). Assay of activity: Extinction at 515 nm with p-nitrocatechol sulphate as substrate: —●—●— in acetate buffer pH 6; —○—○— pH 5 (+pyrophosphate + 10% NaCl, according to Baum, Dodgson & Spencer, 1959).

variant the A and B-enzymes are both absent, and it is therefore termed variant O. But in contrast to the enzymically different variants of metachromatic leukodystrophy, one case of Tay-Sachs disease was observed where both activities were enhanced in the brain and it was named variant AB (Sandhoff, 1969; Sandhoff *et al.*, 1971).

In the variant B Tay-Sachs ganglioside and its neuraminic acid-free residue accumulate in the brain. The same was true in the variant AB and variant O.

In the variant O, however, kidney globoside was additionally deposited in the visceral organs (Fig. 4) (Sandhoff, Andreae & Jatzkewitz, 1968). Thus, apart from variant AB, which will be discussed later on, it seemed likely that the physiological substrate for *N*-acetyl-hexosaminidase A is Tay-Sachs ganglioside and that trihexosylceramide and kidney globoside are physiological substrates for *N*-acetylhexosaminidase B.

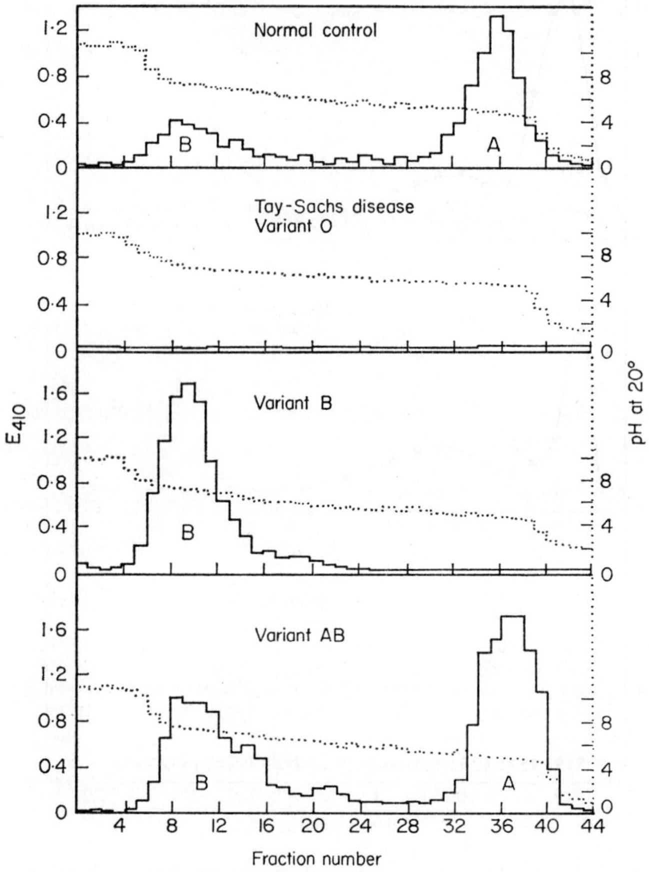

FIG. 3. Variations of β-N-acetylhexosaminidase patterns in the brain tissue of three variants of Tay-Sachs disease. Method: Isoelectric focusing (..... pH). Assay of activity: Extinction at 410 nm with *p*-nitrophenyl-β-N-acetylglucosaminide as substrate.

Let us now go back and consider the cerebroside sulphatase activity of the arylsulphatase A. Only the arylsulphatase A peak which was obtained after isoelectric focusing and on further purification exhibited a significant cerebroside sulphatase activity, when incubated with tritium-labelled sulphatide (Fig. 5) (Harzer *et al.*, in preparation). It was found that the cerebroside sulphatase activity of the arylsulphatase A was enhanced seven-to thirteen-fold by the addition of a fraction which was devoid of arylsulphatase activity and which moved more anodally in carrier-free high voltage electrophoresis

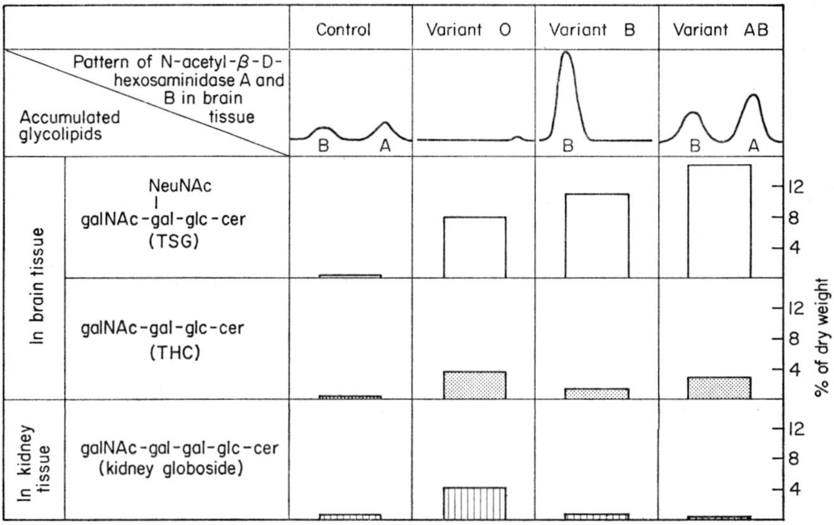

FIG. 4. Correlation between the alterations of the hexosaminidase patterns and glycolipid accumulation in three variants of Tay-Sachs disease.

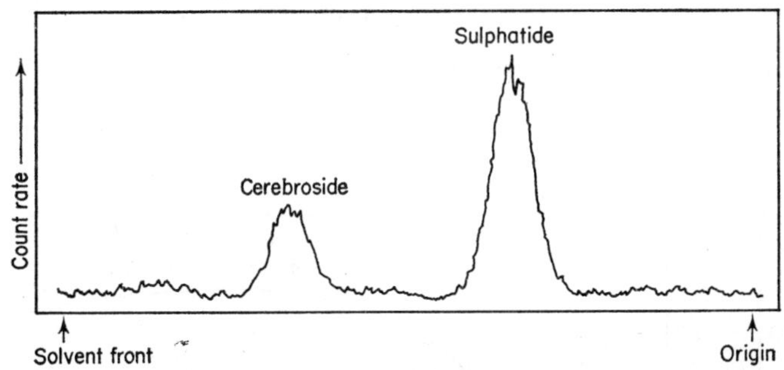

FIG. 5. Action of the isoelectrofocused arylsulphatase A peak on tritium-labelled sulphatide (scan of a thin-layer radiochromatogram). (Source of enzyme: human kidney.)

(Mehl *et al.*, 1964). The existence of this so-called "activating" or "complementary factor" could also be demonstrated by isoelectric focusing (Fig. 6). It showed an isoelectric point at a lower pH than the aryl enzyme, in agreement with the previous study (Stinshoff & Jatzkewitz, in preparation).

As can be seen in Fig. 7, the cerebroside sulphatase activity of the arylsulphatase A is dependent upon the buffer concentration in the incubation mixture. In 0·01 M sodium formate buffer the arylsulphatase A shows the highest cerebroside sulphatase activity. It decreases rapidly, and at a buffer concentration of 0·2 M (where the assays are mostly performed) higher enzyme concentrations and/or addition of the activating fraction or detergents are necessary for obtaining a significant degradation of sulphatide. This figure further shows that the arylsulphatase activity is inhibited in the presence of sulphatide in buffer concentrations where it shows the highest cerebroside sulphatase activity, and is activated in those cases where no cerebroside sulphatase activity can be detected. This is yet another evidence for the

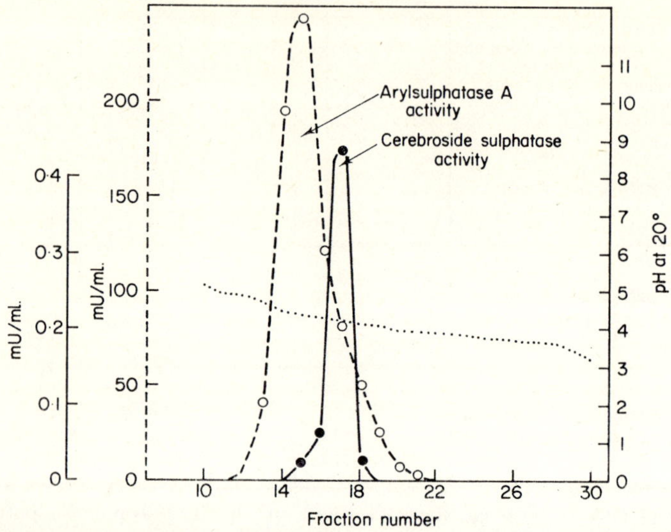

FIG. 6. Cerebroside sulphatase activity of arylsulphatase A from pig kidney after isoelectric focusing, ... pH. Assay of arylsulphatase A activity: — — —. Extinction at 515 nm with *p*-nitrocatechol sulphate as substrate, 0·2 M acetate buffer pH 5 (+pyrophosphate + 10% NaCl). Assay of cerebroside sulphatase activity: ——. Labelled sulphatide, 0·2 M acetate buffer pH 4·5.

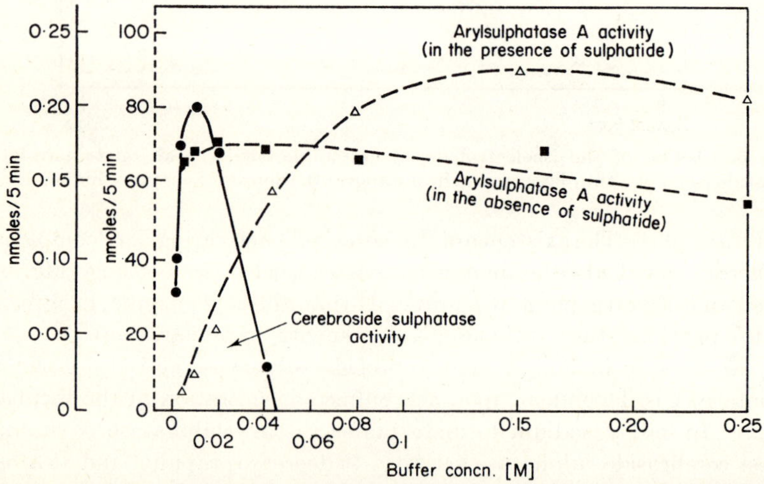

FIG. 7. Differences in the dependence of cerebroside sulphatase and arylsulphatase A activities on buffer concentration (source of purified enzyme: human kidney). Conditions: Buffer: Sodium formate, pH 5·2. The same enzyme concentration in all incubations. Assay of arylsulphatase A activity: Substrate concentration: 0·5 mM *p*-nitrocatechol sulphate in the presence of 0·1 mM sulphatide and in the absence of sulphatide, respectively. Assay of cerebroside sulphatase activity: Substrate concentration: 0·28 mM cerasine sulphate.

identity of the arylsulphatase A with the cerebroside sulphatase (Stinshoff *et al.*, in preparation).

When an isoelectrofocused highly purified arylsulphatase A preparation was stored for a longer time, and refocused, instead of the original single A-peak two peaks appeared, one showing the i.p. of the original A-enzyme and the other an i.p. of 7·2. (The i.p. of a physiological B-enzyme is, as mentioned before, about 8·2) (Fig. 8.) In contrast to the physiological B-enzyme, however, the basic form obtained by conversion of the A-form was as active towards cerebroside sulphate as the original A-form (Stinshoff *et al.*, in preparation). I shall return to this point at the end of this paper.

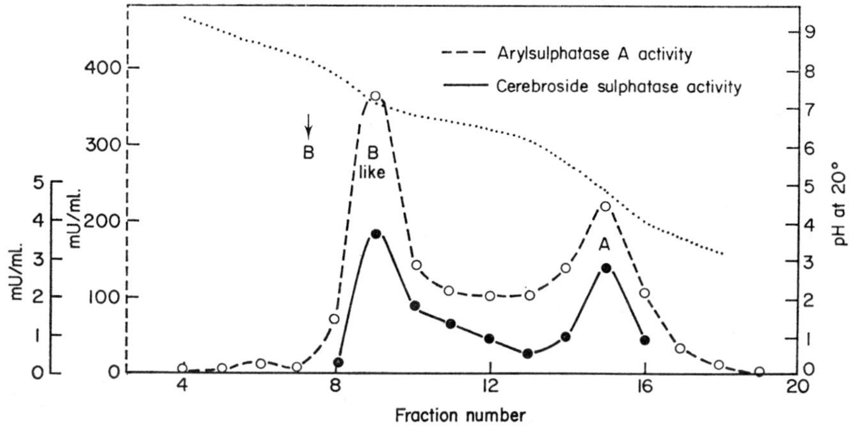

FIG. 8. Conversion of a purified arylsulphatase A into a component with an isoelectric point similar to that of a B-like enzyme but with both arylsulphatase and cerebroside sulphatase activities (Source of enzyme: human kidney). Conditions: Storage of an isoelectrofocused A-enzyme at $-10°$ for about 3 months with repeated freezing and thawing. Method: isoelectric focusing (..... pH). Assay of cerebroside sulphatase activity without detergent or activating fraction.

But first I should like to go back and consider the relationships between the different enzyme patterns and the lipid storage patterns (Fig. 4) in Tay-Sachs disease. It was mentioned that one could assume from these data that the hexosaminidase A is responsible for the degradation of the Tay-Sachs ganglioside, while the B-enzyme most probably degrades the trihexosylceramide and kidney globoside. More evidence could be obtained after the hexosaminidases were purified and their action on the tritium-labelled physiological substrates was investigated (Sandhoff & Wässle, 1971; Sandhoff *et al.*, 1971).

It could be shown that only β-*N*-acetylhexosaminidase A degraded Tay-Sachs ganglioside to a lesser extent (Sandhoff, 1970) to give a product with an R_F value of hematoside (Fig. 9). But at the same time it could also be shown that β-*N*-acetylhexosaminidase A as well as B degraded the tri-hexosylceramide and the kidney globoside in the presence of a detergent. Storage of these compounds in the AB variant can be explained by the fact that the β-*N*-acetylhexosaminidase A in this variant is essentially reduced only in its ability to split Tay-Sachs ganglioside (Table 1) (Sandhoff *et al.*, 1971).

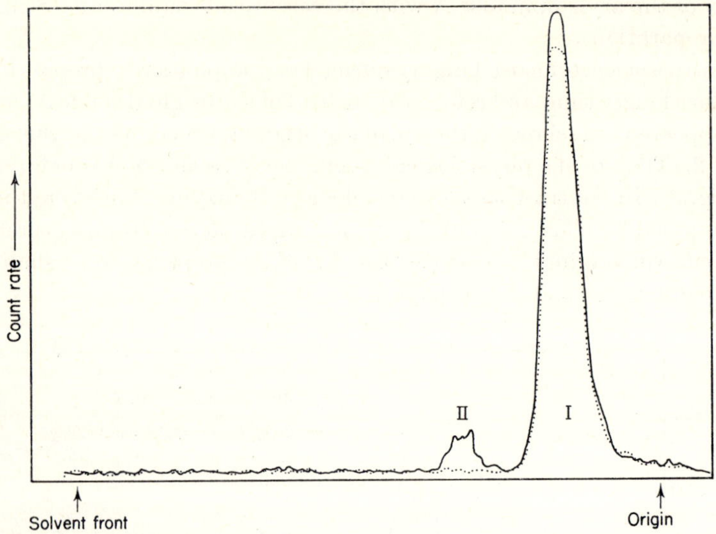

Fig. 9. Scan of a thin-layer radiochromatogram showing the result of the action of a purified *N*-acetyl-β-D-hexosaminidase A on tritium-labelled Tay-Sachs ganglioside. (Source of enzyme: human liver). Experimental conditions see text. ——, active enzyme preparation; ..... heat denatured enzyme preparation. Peak I = Tay-Sachs ganglioside; peak II = degradation product with the mobility of hematoside.

Table 1. *Degradation of Tay-Sachs ganglioside (TSG) by a crude mixture of both* N-*acetyl-β-*D-*hexosaminidases A and B in relation to the degradation of* p-*nitrophenyl*-N-*acetyl-β-*D-*glucosaminide*

	μmoles TSG split/mol *p*-nitrophenyl-N-acetyl-β-D-glucosaminide split*	
Control	0·7	(0·6–0·8)
Variant O	0	
Variant B	0	
Variant AB	0·1	(0·0–0·2)
Control + variant AB (1:1)	0·5	(0·2–0·8)

* Mean values of two or three experiments. Maximum deviations in brackets. The values are calculated from 24 hr incubations with TSG and from 30 min incubations with the chromogenic substrate.

The kinetic data of the hexosaminidases A and B towards the chromogenic and physiological substrates are very similar (Sandhoff *et al.*, 1971). The acidic form (i.p. 5·0) can be converted by heating into a component which has the same i.p. (7.3), the same thermolability and the same molecular weight as the physiological B-enzyme (Fig. 10) (Sandhoff & Markianos, unpublished work; see also Dance, Price & Robinson, 1970). Both lysosomal hexosaminidase A and arylsulphatase A are glycoproteins and contain neuraminic acid. Both acidic forms are transformed into basic forms under mild conditions, probably as a result of a loss of *N*-acetylneuraminic acid (Goldstone, Konecny & Koenig, 1971). But while the B-form in the case of the hexosaminidase A is quite similar to or identical with the B-enzyme, the B-form in the case of the aryl-sulphatase is different from the natural B-enzymes.

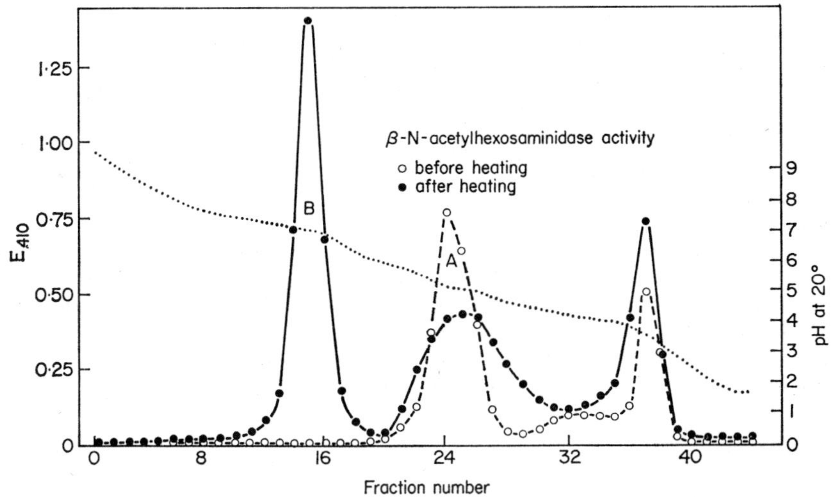

FIG. 10. Conversion of a purified β-*N*-acetylhexosaminidase A into a component with the same isoelectric point and thermolability as the B-enzyme. (Source of enzyme: human liver.) Conditions: Heating of a purified β-*N*-acetylhexosaminidase A for 30 min at 52° in 0·01 M phosphate buffer pH 5·0 with 0·1 M ammonium sulphate. Loss of total activity: 75%. Method: Isoelectric focusing (..... pH). Assay of activity: Extinction at 410 nm with *p*-nitotrophenyl-β-*N*-acetylglucosaminide as substrate.

From a genetic point of view it is difficult to understand how a possible one-point mutation can lead to a partial or total deficiency of the different forms of the same type of enzyme. One should include in these considerations the question of a deficient sialyltransferase, discussed elsewhere (Goldstone *et al.*, 1971; Sandhoff *et al.*, 1971). I put this question at the end of this paper because not far from here, in Dr. Bachhawat's laboratory, another glycoprotein, an alkaline phosphatase, was neuraminidase-treated and then resialylated with CMP-*N*-acetylneuraminic acid (Saraswathi & Bachhawat, 1970). Both forms of the enzyme obtained by this treatment showed the same activity and kinetic properties. Thus, similar experiments could be carried out with the sulphatases and hexosaminidases to get a deeper insight into the formation, the inter-relationship and genetic dependence of the lysosomal enzymes.

REFERENCES

Austin, J. H. (1959). *Proc. Soc. exp. Biol. Med.* **100**, 361.

Austin, J., Armstrong, D. & Shearer, L. (1965). *Archs Neurol.* **13**, 593.

Austin, J. H., Balasubramanian, A. S., Pattabiraman, T. N., Saraswathi, S., Basu, D. K. & Bachhawat, B. K. (1963). *J. Neurochem.* **10**, 805.

Baum, H., Dodgson, K. S. & Spencer, B. (1959). *Clin. Chim. Acta* **4**, 453.

Dance, N., Price, G. R. & Robinson, D. (1970). *Biochim. biophys. Acta* **222**, 662.

Dodgson, K. S. & Spencer, B. (1957). In *Methods of Biochemical Analysis*, Vol. IV (Ed. by Glick, D.), p. 211, New York: Interscience Publishers, Inc.
Gatt, S. & Berman, E. R. (1963). *J. Neurochem.* **10**, 43.
Goldstone, A., Konecny, P. & Koenig, H. (1971). *FEBS Lett.* **13**, 68.
Jatzkewitz, H. (1958). *Hoppe-Seyler's Z. physiol. Chem.* **311**, 279
Jatzkewitz, H. (1960). *Hoppe-Seyler's Z. physiol. Chem.* **318**, 265.
Jatzkewitz, H. & Mehl, E. (1969). *J. Neurochem.* **16**, 19.
Jatzkewitz, H. & Sandhoff, K. (1963). *Biochim. biophys. Acta* **70**, 354.
Klenk, E. (1939). *Hoppe-Seyler's Z. physiol. Chem.* **262**, 128.
Mehl, E. & Jatzkewitz, H. (1964). *Hoppe-Seyler's Z. physiol. Chem.* **339**, 260.
Mehl, E. & Jatzkewitz, H. (1968). *Biochim. biophys. Acta*, **151**, 619.
Murphy, J. V., Wolfe, H. J., Balazs, E. A. & Moser, H. W. (1971). In *Lipid Storage Diseases* (Ed. by Bernsohn, J. & Grossman, H. J.), p. 67, New York, London: Academic Press Ltd.
Sandhoff, K. (1969). *FEBS Lett.* **4**, 351.
Sandhoff, K., Andreae, U. & Jatzkewitz, H. (1968). *Pathologia eur* **3**, 278.
Sandhoff, K. & Wässle, W. (1971). *Hoppe-Seyler's Z. physiol. Chem.* **352**, 1119.
Sandhoff, K., Harzer, J., Wässle, W. & Jatzkewitz, H. (1971). *J. Neurochem.* **18**, 2469.
Saraswathi, S. & Bachhawat, B. K. (1970). *Biochim. biophys. Acta* **212**, 170.
Svennerholm, L (1962). *Biochem. biophys. Res. Commun.* **9**, 436.
Svennerholm, L. (1963). *J. Neurochem.* **10**, 613.

THE METABOLISM OF PHOSPHATIDYLCHOLINE IN BRAIN TISSUE

By G. B. ANSELL AND SHEILA SPANNER

Department of Pharmacology (Preclinical), The Medical School, Birmingham, England

PHOSPHATIDYLCHOLINE (1,2-diacyl-*sn*-glycero-3-phosphorylcholine) is a universal constituent of mammalian cell membranes and about 9% of the dry weight of brain tissue is accounted for by this phospholipid. In Table 1 the levels

Table 1. *Phosphatidycholine distribution in brain cells and subcellular fractions and its fatty acid composition*

		Major fatty acids as wt % of the total			
	% lipid P	$C_{16:0}$	$C_{18:0}$	$C_{18:1}$	$C_{20:4}$(n-6)
*Neurons	47	38·1	10·7	30·0	8·9
*Glial cells	43	45·8	10·7	27·8	6·1
*Mitochondria	46	39·2	9·5	28·9	9·8
*Synaptosomes	41	48·0	9·9	26·4	6·3
*Myelin	32	32·9	18·6	34·7	2·9
†Microsomes	34	42·7	13·7	30·7	4·7

* Choline phosphoglycerides of rabbit cortex (Hamberger & Svennerholm, 1971).
† Phosphatidylcholine of mouse brain (Sun & Horrocks, 1970).

of phosphatidylcholine in neurones, glial cells and some subcellular fractions of brain tissue are given. Small amounts of choline plasmalogen (1-alkenyl-2-acyl-*sn*-glycero-3-phosphorylcholine) (Webster, 1960; Wells & Dittmer, 1965) and lysophosphatidylcholine (1-acyl-*sn*-glycero-3-phosphorylcholine) (Cuzner & Davison, 1967; Owens, 1966) are also present. The major fatty acids present are also given in Table 1 from which it can be seen that the phosphatidylcholine of all membranes contains large amounts of $C_{16:0}$ and $C_{18:1}$. Myelin is richer in $C_{18:0}$ and poorer in $C_{20:4}$ than other components. Recent work by O'Brien & Geison (1971) has demonstrated the presence of relatively large amounts of the dipalmitoyl species in the synaptosomal fraction and the combined microsomal and soluble fractions.

BIOSYNTHESIS

The capacity of brain tissue to synthesize phosphatidylcholine by the transfer of phosphorylcholine from CDP-choline to a diglyceride acceptor was first demonstrated by McMurray *et al.* (1957). It is now clear that the enzymes

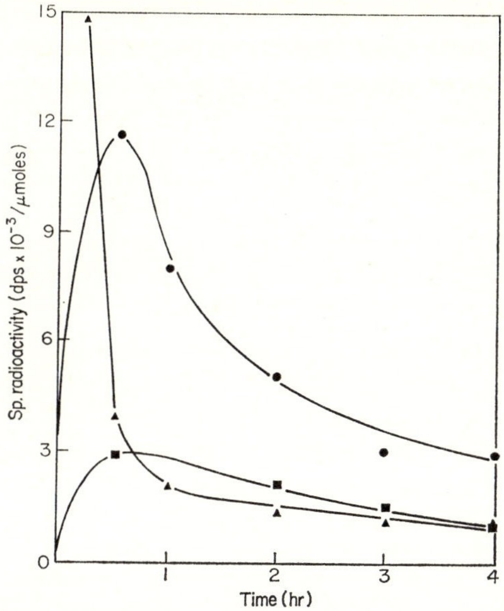

FIG. 2. Specific radioactivity of acid-soluble choline components of rat brain at various time intervals after intracerebral injection of 1 μCi (0·017 μmol) of [*Me*-^{14}C]choline. ▲, Choline; ●, phosphorylcholine; ■, CDP-choline.

shown (radioactivity was not found in significant amounts other than as free or bound choline). After 0·5 hr the largest amount was in the soluble fraction but subsequently there was a rapid transfer to the crude mitochondrial and microsomal fractions. The major lipid-containing component of the crude mitochondrial fraction is myelin, and in subsequent experiments in which this

Table 3. *The percentage distribution of radioactivity in crude subcellular fractions of whole brain after intracerebral injection of* [*Me*-^{14}C]*choline* (1 *μCi*, 0·017 *μmol*)

Time after injection (hr)	% radioactivity in the crude fractions:			
	Nuclear	Mitochondrial	Microsomal	Soluble
0·5	12	18	12	58
2	11	55	9	25
5	18	46	23	18

fraction was subdivided into its constituents significant labelling of the myelin phosphatidycholine was seen even after 0·5 hr (e.g. Table 4). Other subcellular fractions were also labelled, particularly the plasma membranes (Table 4).

The labelling of the phosphatidylcholine of myelin is further evidence that this membranous component of the brain is metabolized at a significant rate in the adult rat and confirms earlier experiments with ethanolamine (Ansell & Spanner, 1967, 1968*b*; Horrocks, 1969). Considerable support for these

On the other hand, after an intracerebral injection of labelled ethanolamine only ethanolamine lipids in the brain are labelled (Ansell & Spanner, 1967). Furthermore, the fatty acid composition of the phosphatidylcholine of the brain microsomal fraction, where methylation would presumably occur (cf. Bremer & Greenberg, 1961), is completely different from that of the phosphatidylethanolamine (Sun & Horrocks, 1970). A summary of additional evidence is given by Ansell & Spanner (1968*a*).

METABOLISM *in vivo*

The operation of the cytidine pathway can be demonstrated readily *in vivo* by labelling the free choline pool with an intracerebral injection of [*Me*-^{14}C] choline which is rapidly assimilated into the free choline pool of less than 0·2 μmol/g. Such a "pulse labelling" demonstrates a rapid transfer of free holine to a lipid-bound form (Fig. 1). The base is first phosphorylated and is

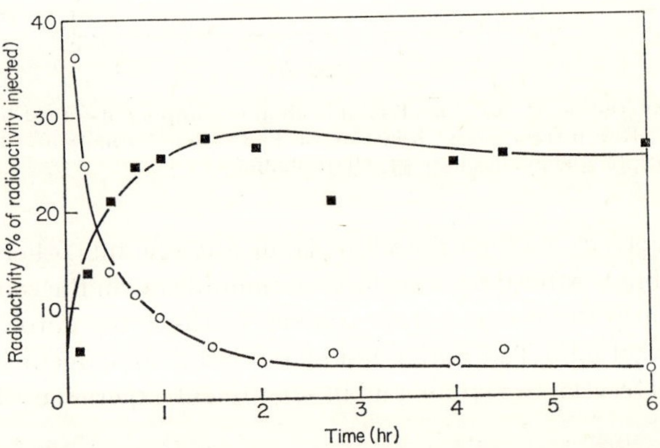

FIG. 1. Percentages of radioactivity injected as 1 μCi [*Me*-^{14}C]choline recovered in the brain of the rat at various time intervals after intracerebral injection. ○, Acid-soluble fraction; ■, lipid fraction. Each point represents the mean obtained from two or three animals. (Reproduced from Ansell & Spanner (1968*a*) by permission of the Biochemical Journal.)

then transferred to CDP-choline (Fig. 2). It has also been shown that CDP-choline behaves as the precursor of phosphatidylcholine which is the choline lipid most rapidly labelled (Ansell & Spanner, 1968*a*).

It was noted earlier that the final stage in the biosynthesis of phosphatidylcholine takes place in the materials isolated as the microsomal fraction by centrifugation procedures. It is not surprising, therefore, that Mandel & Nussbaum (1966) found that, after an intraperitoneal injection of [^{32}P]orthophosphate, the microsomal phosphatidylcholine was more rapidly labelled than that of other subcellular fractions. A more pronounced difference between the microsomal fraction and myelin was found by means of intraventricular "pulse labelling" (Jungalwala & Dawson, 1971).

In Table 3 the distribution of radioactivity in crude subcellular fractions after an intracerebral injection of [*Me*-^{14}C]choline into adult female rats is necessary to form this lipid by the cytidine pathway are all present in adult mammalian brain tissue and their subcellular distribution is reasonably well established. It will be seen from Table 2 that choline kinase (ATP:choline

Table 2. *Choline kinase activity in subcellular fractions of rat brain*

	μmol/ brain/hr	Specific activity (μmol/mg protein/hr)
Whole brain	11·5 ± 1·1 (7)	0·09
Mitochondrial fraction	2·61	0·25
Mitochondrial fraction (water-shocked)	0·80	0·08
Synaptosomal	1·90	0·09
Microsomal fraction	1·10(4)	0·22
Soluble fraction	11·30(3)	0·30

The choline kinase activity was measured by the method of McCaman (1962) at pH 10 which is the optimum. Subcellular fractions were prepared essentially by the method of Gray & Whittaker (1962). The figures in brackets refer to separate fractionations.

phosphotransferase, EC 2.7.1.32) is exclusively found in the soluble fraction of the tissue, as originally indicated by the work of McCaman & Cook (1966). At its optimum pH this enzyme has more than twenty times the activity of the cholinephosphate cytidylyltransferase (CTP:cholinephosphate cytidylyltransferase, EC 2.7.7.15) on a brain weight basis (Chojnacki & Ansell, 1967); the latter is also an enzyme of the soluble fraction which can be separated from the analogous enzyme which forms CDP-ethanolamine (Chojnacki *et al.*, 1967). The final step catalysed by the enzyme cholinephosphotransferase (CDP choline:1,2-diglyceride cholinephosphotransferase, EC 2.7.8.2) can transfer maximally 13 μmol of phosphorylcholine/g tissue/hr according to McCaman & Cook (1966). An extensive study of the diglyceride requirement for this step has not been made though those containing unsaturated fatty acids are more active (McCaman & Cook, 1966). The overall rate of uptake of free choline into choline lipids in a brain slice is relatively slow at 0·05 μmol/g tissue/hr (Pritchard, 1962) and the rate-limiting step could well be the formation of CDP-choline.

The phosphatidylcholine of brain mitochondria is rich in polyunsaturated fatty acids (Skrbic & Cumings, 1970). These could be introduced by the acylation system described by Webster & Alpern (1964) subsequent to the production of a 1-acyl-*sn*-glycero-3-phosphorylcholine by an endogenous phospholipase A_2, on the assumption that this is a mitochondrial enzyme in brain as it is in the liver (Waite & van Deenen, 1967; Nachbaur & Vignais, 1968).

There is no evidence that the phosphatidylcholine can be formed in brain by the stepwise methylation of phosphatidylethanolamine though this is a most important pathway in the liver as originally demonstrated by Bremer, Figard & Greenberg (1960). Thus, if the phosphatidylethanolamine of liver is labelled *in vivo* by an intraperitoneal injection of ^{14}C-labelled ethanolamine, the radioactivity rapidly appears in the choline moiety of phosphatidylcholine.

Table 4. *The distribution of radioactivity in components of the crude mitochondrial fraction of rat brain tissue after intracerebral injection of* [Me-^{14}C]*choline*

	Acid-soluble	Lipid*
Myelin	2	10
Fraction A	6	33
B	4	7
C	2	4
D	3	7
E	3	13
F	2	6

* 90% of this radioactivity was in phosphatidylcholine. Two rats were each injected with 1 μCi [Me-^{14}C]choline and the crude mitochondrial fraction, obtained after 0·5 hr exchange, was subjected to zonal centrifugation by the method shown in Fig. 3 of Spanner & Ansell (1971). A, plasma membranes; B, unknown; C, synaptosomal fraction; D, synaptosomal fraction; E, mitochondrial fraction; F, lysosomal fraction.

findings has come from recent extensive investigations by Jungalwala & Dawson (1971) who injected a variety of labelled phospholipid precursors by the intraventricular route and concluded that, by all criteria, the myelin lipids were in a state of metabolic flux. The seemingly extraordinary finding of Dhopeshwarkar, Subramanian & Mead (1971*b*) that 15 sec after a close arterial injection of labelled acetate, radioactivity was found in lipids characteristic of myelin, is further evidence.

If the phosphatidylcholine of all the membranes of brain is in a state of flux then there should be active phospholipases in the tissue. The presence of the A_1 variant of phospholipase A (phosphatidate acyl-hydrolase, EC 3.1.1.4) was unequivocally shown by Gatt, Barenholz & Roitman (1966) and by Gatt (1968) and the A_2 variant was demonstrated by Cooper & Webster (1970). A lysophospholipase (lysolecithin acyl-hydrolase, EC 3.1.1.5) active towards 1-acyl-*sn*-glycero-3-phosphorylcholine has been partially purified by Liebovitz & Gatt (1968) and glycero-3-phosphorylcholine diesterase (*L*-3-glycerylphosphorylcholine glycerophosphohydrolase*, EC 3.1.4.2) demonstrated many years ago by Webster, Marples & Thompson (1957). Nothing can be said with certainty about the subcellular distribution of these enzymes, the combined action of which leads to the release of fatty acids, glycerol-3-phosphate and choline from phosphatidylcholine though the work of Illingworth & Glover (1969) would suggest that they are not confined to any particular fraction. Measurement of the rate of production of glycero-3-phosphorylcholine in the brain *in vivo* has not been made and it is presumably hydrolysed by the diesterase thereby contributing to the free pool of choline.

The fate of any choline liberated within the brain from the hydrolysis of choline lipids has not been studied but we have detected very little labelled betaine after an intracerebral injection of [Me-^{14}C]choline (cf. Diamond, 1971); if there is little oxidation then the rapid phosphorylation would suggest significant re-utilization. It is known, for example, that there is a rapid uptake of free choline into isolated synaptosomes (Marchbanks, 1968) where it can be

* Old nomenclature.

acetylated to yield acetylcholine. Such a mechanism is presumably useful for the re-utilization of choline released post-synaptically by the hydrolysis of acetylcholine. The possible relationship between the acetylcholine cycle and the phosphatidylcholine cycle, in both of which choline is a starting product and end product, is intriguing but no definite evidence linking the one with the other has so far been produced (Ansell & Spanner, 1971*a*).

THE ORIGIN OF BRAIN CHOLINE

Because the brain cannot synthesize choline it must receive it from an external source via the bloodstream, though the amount required will depend on the capacity of the tissue to re-utilize the base. Groth, Bain & Pfeiffer (1958) demonstrated that intraperitoneally injected [*Me*-^{14}C] and [1,2-^{14}C]choline could be detected in the acid-soluble and lipid fractions of brain tissue. Schuberth, Sparf & Sundwall (1969) clearly showed that, if the blood was heavily labelled with [*Me*-^{3}H]choline, 0·5% of the injected dose could be detected in the acid-soluble fraction of brain tissue within 15 sec, of which 75% was in the form of free choline. Very recently Diamond (1971) has observed that intravenously injected [*Me*-^{3}H]choline can apparently enter the brain by an active transport mechanism which exhibits saturation kinetics; this choline was rapidly assimilated into a lipid-bound form in a variety of membranous constituents. The uptake of free choline into isolated brain tissue was demonstrated by Schuberth *et al.* (1966) and into isolated synaptosomes by Marchbanks (1968) among others, but the idea that a water-soluble substance containing a quaternary nitrogen can readily pass from the blood to the brain is a novel one, particularly since the level of free choline in the plasma is as low as 10^{-5} M (Bligh, 1952).

Although these observations would appear to demonstrate that the source of brain choline is the free choline of the plasma they were all made after a direct labelling of the circulating base. It is therefore pertinent to recall the experiments of Bjørnstad & Bremer (1966) who confirmed the work of Bremer & Greenberg (1961) that the synthesis of choline by the methylation of lipid-bound ethanolamine is significant only in the liver. Since this is the only way in which choline can be synthesized it has to be assumed not only that the free choline of the plasma is produced from liver phosphatidylcholine by the combined action of phospholipase A_1 and A_2 and glycero-3-phosphorylcholine diesterase, but that the liver is the primary source of choline in extra-hepatic tissues. However, Bjørnstad & Bremer (1966) found that the labelling of the choline of extra-hepatic tissues could not be diluted out by an injection of "cold" choline if the liver phosphatidylcholine was labelled from [*Me*-^{3}H]-methionine. Only if the free plasma choline was separately labelled could the labelling of extra-hepatic tissues be affected by the injection of "cold" choline. The implication was that choline is transported from the liver in a form other than as free choline.

Ansell & Spanner (1971*b*) have recently reported a series of experiments in which the phosphatidylcholine of the liver was labelled by intraperitoneal injections of [1,2-^{14}C]dimethylaminoethanol or [1,2-^{14}C]ethanolamine and the labelling of the choline compounds was followed in the blood and brain of the rat. It was found that at no time from 1 to 7 hr was there any significant labelling of the free choline in the blood or in the brain. Neither was the phosphorylcholine of the blood labelled, though after 2 hr that of the brain was labelled. To judge from the labelling of the phosphatidylcholine (and lysophosphatidylcholine) in the blood and brain, lipid-bound choline was the most likely form in which the choline was being transported, presumably as a lipoprotein complex in the plasma. It was also observed that there was negligible labelling of the free choline of the liver itself since after the injection of [1,2-^{14}C]dimethylaminoethanol, for example, most of the radioactivity in the acid-soluble fraction of the tissue other than that in the form of phosphoryldimethylaminoethanol was located in the betaine or phosphorylcholine

Table 5. *The distribution of radioactivity in the acid-soluble fraction of rat liver after intraperitoneal injection of* [*Me*-^{14}C]*choline or* [1,2-^{14}C]*dimethylaminoethanol*

Base injected	Time after injection (hr)	% radioactivity in the acid-soluble fraction as: Choline	Phosphorylcholine	Phosphoryldimethylaminoethanol	Betaine
Choline	1	0·17	28	—	70
	6	0·14	32	—	70
Dimethylaminoethanol	1	0·01	6·4	79	7·4
	2	0·01	3·0	47	4·4
	4	0·01	4·0	62	5·0
	6	0·01	4·4	80	5·0

Adult female rats were injected with 20 μCi of choline (0·4 μmol), or dimethylaminoethanol (15 μmol), the animals killed and the livers rapidly extracted with 10% trichloroacetic acid. Phosphate esters, choline and betaine were separated as described by Ansell & Spanner (1971*b*).

(Table 5). This indicated a rapid oxidation or phosphorylation of any free choline produced. After an intraperitoneal injection of [*Me*-^{14}C]choline the phosphatidylcholine of the liver was labelled and so, as might be expected, was the phosphorylcholine. Almost no labelling of the free choline was found (Table 5) but 70% of the acid-soluble radioactivity was located in betaine. These experiments indicated that unless there is a rapid *post-mortem* loss of choline, the level of free choline in the liver is kept very low *in vivo* by either phosphorylation or oxidation.

There are precedents for postulating that choline is transported in whole or in part to the brain in a lipid-bound form, apart from the long held view that lipophilic molecules pass more readily from the blood to the brain than others. Thus, Davison *et al.* (1958, 1959) showed that [4-^{14}C]cholesterol was incorporated intact into chick and rabbit brain from the blood, while Siek & Newburgh (1965) concluded that in the later stages of the development of the chick phospholipids are carried intact from the yolk to the brain by the bloodstream.

Stein & Stein (1966) demonstrated that lysophosphatidylcholine was carried intact to various organs of the rat though they did not study the brain. Hoelzl & Franck (1969) reported briefly that phosphatidylcholine, doubly labelled with ^{32}P and ^{14}C, when injected intravenously into rats, could be recovered from the brain with an unchanged isotopic ratio. Intravenously injected fatty acids can also enter the brain and be assimilated rapidly into complex lipid molecules (Dhopeshwarkar & Mead, 1970; Dhopeshwarkar *et al.* 1971*a*).

CONCLUSIONS

The incorporation of choline into phosphatidylcholine by the cytidine pathway is an energy-consuming process and it would be of great interest to know why a major component of all the membranous materials of adult rat brain, particularly the myelin, is apparently in a state of metabolic flux. It is tempting to suggest that this turnover is related to a specific function of the lipid in the tissue but evidence is completely lacking. Miller & Dawson (1972) have produced good evidence that the mitochondria and myelin themselves do not synthesize phospholipids but receive them from the endoplasmic reticulum by a transport mechanism. It remains to be seen whether the synthetic capacity of the endoplasmic reticulum of glial cells is different from that of neurones.

ACKNOWLEDGEMENT

Some of the work reported here was made possible by a generous grant from the Multiple Sclerosis Society of Great Britain. The interest of Professor P. B. Bradley is very much appreciated.

REFERENCES

Ansell, G. B. & Spanner, S. (1967). *J. Neurochem.* **14**, 873–885.

Ansell, G. B. & Spanner, S. (1968*a*). *Biochem. J.* **110**, 201–206.

Ansell, G. B. & Spanner, S. (1968*b*). *J. Neurochem.* **15**, 1371–1373.

Ansell, G. B. & Spanner, S. (1971*a*). In *Drugs and Cholinergic Mechanisms in the CNS* (Ed. by Heilbronn, E. & Winter, A.), pp. 143–162, Research Institute of National Defence, Stockholm: Almqvist & Wiksell.

Ansell, G. B. & Spanner, S. (1971*b*). *Biochem. J.* **122**, 741–750.

Bjørnstad, P. & Bremer, J. (1966). *J. Lipid Res.* **7**, 38–45.

Bligh, J. (1952). *J. Physiol.* **117**, 234–240.

Bremer, J., Figard, P. H. & Greenberg, D. M. (1960). *Biochim. biophys. Acta* **43**, 477–488.

Bremer, J. & Greenberg, D. M. (1961). *Biochim. biophys Acta* **46**, 205–216.

Chojnacki, T. & Ansell, G. B. (1967). *J. Neurochem.* **14**, 413–420.

Chojnacki, T., Radomińska-Pyrek, A. & Korzybski, T. (1967). *Acta biochim. Polon.* **14**, 383–388.

Cooper, M. F. & Webster, G. R. (1970). *J. Neurochem.* **17**, 1543–1554.

Cuzner, M. L. & Davison, A. N. (1967). *J. Chromatog.* **27**, 388–397.

Davison, A. N., Dobbing, J., Morgan, R. S. & Payling-Wright, G. (1958). *J. Neurochem.* **3**, 89–94.

Davison, A. N., Dobbing, J., Morgan, R. S. & Payling-Wright, G. (1959). *Lancet* **i**, 658–660.

Dhopeshwarkar, G. A. & Mead, J. F. (1970). *Biochim. biophys. Acta* **210**, 250–256.

Dhopeshwarkar, G. A., Subramanian, C. & Mead, J. F. (1971*a*). *Biochim. biophys. Acta* **231**, 8–14.

Dhopeshwarkar, G. A., Subramanian, C. & Mead, J. F. (1971*b*). *Biochim. biophys. Acta* **248,** 41–47.

Diamond, I. (1971). *Arch. Neurol.* **24,** 333–339.

Gatt, S. (1968). *Biochim. biophys. Acta* **159,** 304–316.

Gatt, S., Barenholz, Y. & Roitman, A. (1966). *Biochem. biophys. Res. Commun.* **24,** 169–171.

Gray, E. G. & Whittaker, V. P. (1962). *J. Anat.* (*Lond.*) **96,** 79–88.

Groth, D. P., Bain, J. A. & Pfeiffer, C. C. (1958). *J. Pharmacol. exp. Ther.* **124,** 290–295.

Hamberger, A. & Svennerholm, L. (1971). *J. Neurochem.* **18,** 1821–1829.

Hoelzl, J. & Franck, H. P. (1969). In *Proc. 2nd Int. Meet. International Society for Neurochemistry* (Ed. by Paoletti, R., Fumagalli, R. & Galli, C.), pp. 219–220. Milan: Tamburini Editore.

Horrocks, L. A. (1969). *J. Neurochem.* **16,** 13–18.

Illingworth, D. R. & Glover, J. (1969). *Biochem. J.* **115,** 16–17P.

Jungalwala, F. B. & Dawson, R. M. C. (1971). *Biochem. J.* **123,** 683–693.

Liebovitz, Z. & Gatt, S. (1968). *Biochim. biophys. Acta* **164,** 439–441.

McCaman, R. E. (1962). *J. biol. Chem.* **237,** 672–676.

McCaman, R. E. & Cook, K. (1966). *J. biol. Chem.* **241,** 3390–3394.

McMurray, W. C., Strickland, K. P., Berry, J. F. & Rossiter, R. J. (1957). *Biochem. J.* **66,** 634–644.

Mandel, P. & Nussbaum, J. L. (1966). *J. Neurochem.* **13,** 629–642.

Marchbanks, R. M. (1968). *Biochem. J.* **110,** 533–541.

Miller, E. K. & Dawson, R. M. C. (1972). *Biochem. J.* **126,** 823–835.

Nachbaur, J. & Vignais, P. M. (1968). *Biochem. biophys. Res. Commun.* **33,** 315–320.

O'Brien, J. F. & Geison, R. L. (1971). *J. Neurochem.* **18,** 1615–1623.

Owens, K. (1966). *Biochem. J.* **100,** 354–361.

Pritchard, E. T. (1962). *Can. J. Biochem. Physiol.* **40,** 353–361.

Schuberth, J., Sparf, B. & Sundwall, A. (1969). *J. Neurochem.* **16,** 695–700.

Schuberth, J., Sundwall, A., Sörbo, B. & Lindell, J. O. (1966). *J. Neurochem.* **13,** 347–352.

Siek, T. J. & Newburgh, R. W. (1965). *J. Lipid Res.* **6,** 556–564.

Skrbic, T. R. & Cumings, J. N. (1970). *J. Neurochem.* **17,** 85–90.

Spanner, S. & Ansell, G. B. (1971). In *Separations with Zonal Rotors* (Ed. by Reid, E.), pp. V–3.1–V–3.7, Guildford, Surrey: Wolfson Bioanalytical Centre.

Sun, G. Y. & Horrocks, L. A. (1970). *Lipids* **5,** 1006–1012.

Stein, Y. & Stein, O. (1966). *Biochim. biophys. Acta* **116,** 95–107.

Waite, M. & van Deenen, L. L. M. (1967). *Biochim. biophys. Acta* **137,** 498–517.

Webster, G. R. (1960). *Biochim. biophys. Acta* **44,** 109–116.

Webster, G. R. & Alpern, R. J. (1964). *Biochem. J.* **90,** 35–42.

Webster, G. R., Marples, E. A. & Thompson, R. H. S. (1957). *Biochem. J.* **65,** 374–377.

Wells, M. A. & Dittmer, J. C. (1965). *Biochemistry* **6,** 3169–3175.

THE SYNTHESIS OF NATURALLY OCCURRING PHOSPHONOLIPIDS

By ERICH BAER

Banting and Best Department, Medical Research, University of Toronto, Ontario, Canada

IT is well known by now that the isolation and identification of 2-aminoethylphosphonic acid (Table 1), from proteolipid-like fractions of ciliated protozoa of sheep rumen by Horiguchi & Kandatsu (1959) marked the beginning of a

Table 1

	Reference
ciliatin 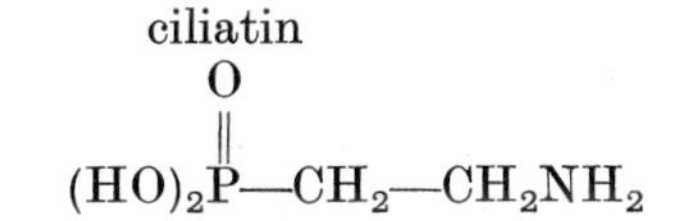 2-aminoethylphosphonic acid	Horiguchi & Kandatsu (1959)
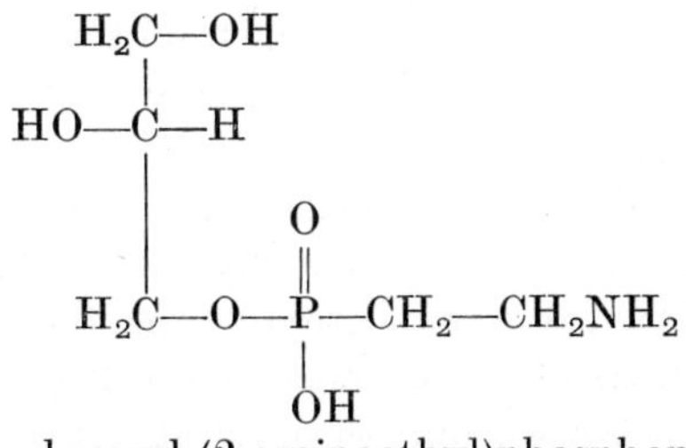 L-α-glyceryl-(2-aminoethyl)phosphonate	Kittredge, Roberts & Simonsen (1962)
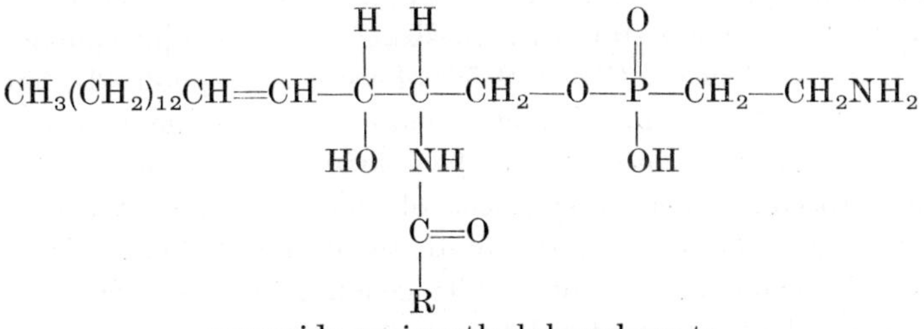 ceramide aminoethylphosphonate	Rouser *et al.* (1963)

new chapter in the chemistry of complex lipids. It provided the first indication of a possible occurrence in nature of phosphorus-containing lipids in which the nitrogenous moiety is attached to phosphorus via a carbon to phosphorus bond. The presence of ciliatin in naturally occurring lipids was confirmed three years later by Kittredge, Roberts & Simonsen (1962) who isolated this compound as well as *its glycerol ester* from ethanolic extracts of the sea anemone *Anthopleura elegantissima*. Subsequently Rouser *et al.* (1963) by a combination of paper and column chromatography succeeded in isolating from the same extracts an intact lipid containing 2-aminoethylphosphonic acid. On the basis of its hydrolysis products it was tentatively assigned the structure of a ceramide aminoethylphosphonate (Simon & Rouser, 1967).

There were reasons to believe that phosphonic acid analogues of other phospholipids also may occur in nature, awaiting their discovery. In fact, I was so firmly convinced of this possibility that I committed the rather limited resources of my research group for the last nine years almost exclusively to the synthesis of phosphonic *acid analogues* of *phospholipids* and their biological intermediates.

Their synthesis was undertaken to provide pure individual phosphonolipids of known structure and configuration so as to serve both as reference compounds for the elucidation of the structure and configuration of naturally occurring phosphonolipids, and as substrates for biochemical and biological investigations. In this undertaking, which opened up a completely new field and which presented new and exciting challenges to the ingenuity of the synthetic chemist, I was ably assisted by my following colleagues: Dr Nicola Z. Stanacev (1962–1965); Dr K. V. Jagannadha Rao (1963–1966); Dr G. Raghupati Sarma (1964–1968); Dr Bimal C. Pal (1964–1967); Dr Ranga Robinson (1965–1972); Dr Hemendra Basu (1966–1970); Dr Darius J. Nazir (1967–1969); Dr Sripada K. Pavanaram (1968–1971). Except for Dr Stanacev, who was the first to work with me in this field, all others have come from India and are graduates of Indian universities. Thus, the work reported in my paper can truly be described as a joint Indian–Canadian enterprise.

The idea of synthesizing phosphonolipids was conceived on the occasion of a visit in 1962 to the City of Hope Medical Centre at Duarte, California. There I was shown by Drs Roberts and Kittredge 2-aminoethylphosphonic acid which they had isolated from lipid extracts of the sea anemone *Anthopleura elegantissima*. When we began our synthetic work, only the natural occurrence of phosphonic acid analogues of sphingolipids was firmly established, while that of glycerophospholipids was still somewhat problematic. Nevertheless, I thought we would be more successful by synthesizing first glycerophosphonolipids, since we had experience in the synthesis of glycerophospholipids, but none in the synthesis of sphingolipids. Furthermore, the glyceride moiety is far more readily accessible for synthesis than the lipid moiety of sphingolipids.

In general our procedures for the synthesis of phosphonolipids resemble those developed in our laboratory for the synthesis of phospholipids. These procedures, in which the introduction of the nitrogenous moiety is usually the penultimate operation, had to be modified, however, since the formation of the carbon-to-phosphorus bond (Fig. 1) requires experimental conditions that would be far too drastic for the glyceride moiety to survive unharmed. We circumvented this difficulty by preparing independently the phosphonic acid derivatives of the nitrogenous moieties to be introduced. In order to prevent a

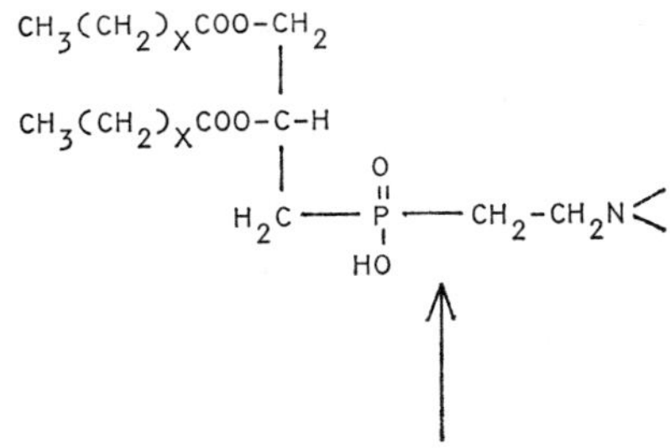

FIG. 1

possible interference of the amino group with the phosphonylation procedure, it was protected by the readily removable phthaloyl, benzyl or carbobenzoxy groups (Table 2). The phthaloyl group, which is removed by hydrazine, permits the synthesis of phosphonolipids with unsaturated lipid moieties.

Table 2

(2-phthalimidoethyl)phosphonic acid

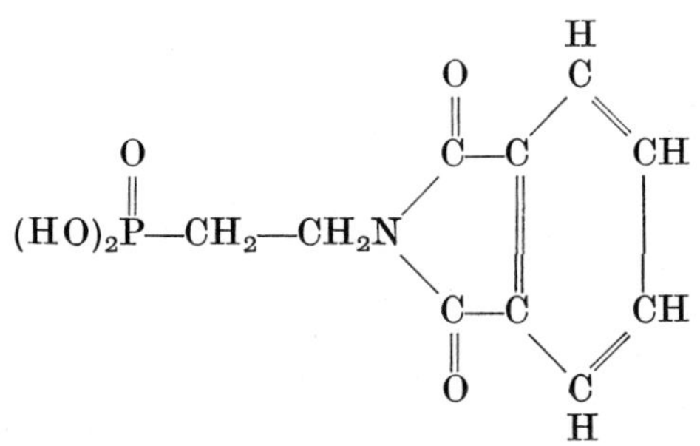

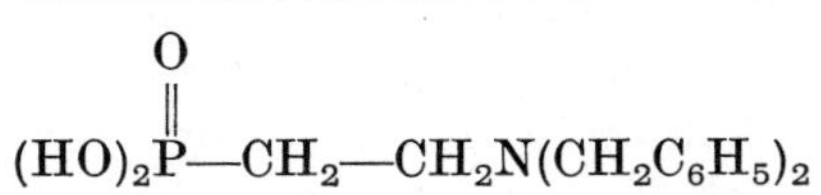

N,*N*-dibenzyl-2-aminoethylphosphonic acid

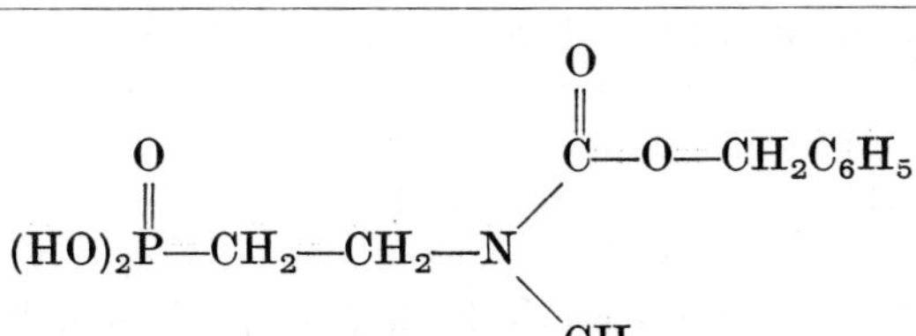

N-carbobenzoxy, *N*-methyl-(2-aminoethyl)phosphonic acid

Earlier work in our laboratory had revealed that naturally occurring glycerophospholipids, with few exceptions, have the L configuration. Expecting their phosphonic acid analogues to have the same configuration we prepared exclusively their L isomers.

Since the isolation of glyceryl (2-aminoethyl)phosphonate from the lipid extracts of the sea anemone *Anthopleura elegantissima* suggested a natural occurrence of phosphonocephalins, priority was given to their synthesis (Baer & Stanacev, 1964*a*). The phosphonocephalins (Reaction Scheme 1) were

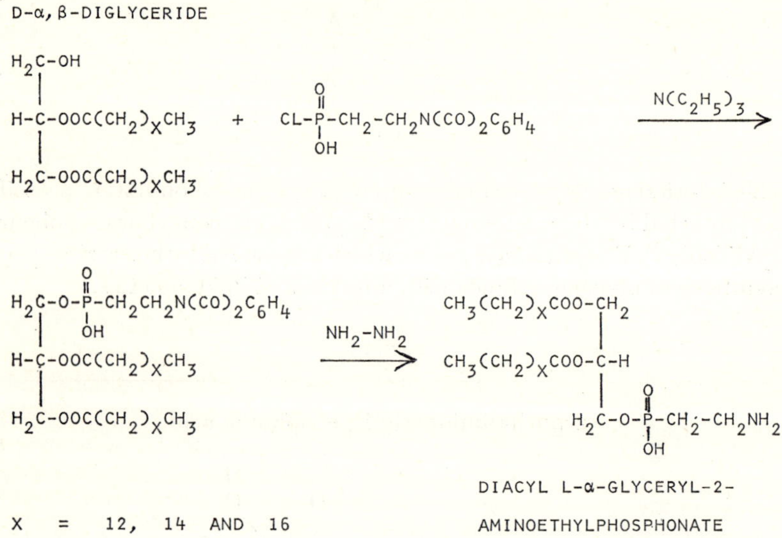

SCHEME 1. Synthesis of L-α-phosphonocephalins (Baer & Stanacev, 1963).

obtained by esterifying (2-phthalimidoethyl)phosphonic acid monochloride with the appropriate diglyceride in the presence of triethylamine, and by removing the protective phthaloyl group by hydrazinolysis in boiling ethanol. In its physical properties the dipalmitoyl L-α-phosphonocephalin resembled closely a phosphonolipid isolated by Horiguchi from *Tetrahymena*.

The widespread occurrence in nature of ether-phospholipids suggested the possibility that phosphonic acid analogues of these compounds may occur in nature also. This was confirmed by Thompson by isolating from *Tetrahymena pyroformis* a phosphonolipid containing one ether group. Anticipating the discovery of glycerophosphonolipids containing two ether groups, we have synthesized phosphonic acid analogues of both diether cephalins and diether-lecithins. The diether-phosphonocephalins (Fig. 2) were obtained by the same procedure as the ester-phosphonocephalins, by replacing the ester-diglycerides by ether-diglycerides (Baer & Stanacev, 1964*b*).

The presence of monomethylamino-, dimethylamino- and trimethylaminoethyl phosphonic acid (Table 3) in the hydrolysates of ethanolic extracts of the sea anemone *Anthopleura elegantissima* (Kittredge, 1967) and *Anthopleura xanthogrammica* (Kittredge, Isbell & Hughes, 1967) suggested a natural occurrence of *N*-methyl substituted phosphonolipids. The facts that they

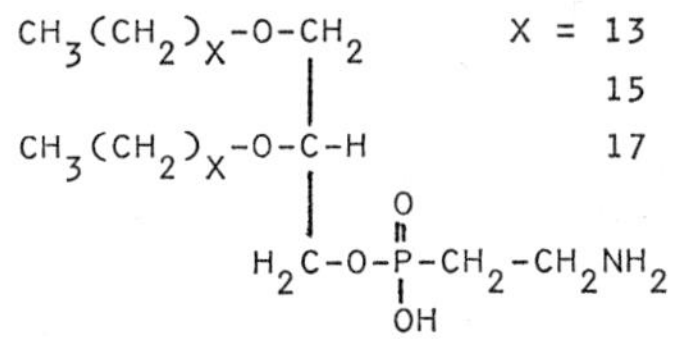

FIG. 2. Diether phosphonocephalins. Dialkyl L-α-glyceryl-(2-aminoethyl) phosphonate (Baer & Stanacev, 1965).

Table 3. Mono-, di- and trimethyl-2-aminoethylphosphonate

$$(HO)_2\overset{O}{\overset{\|}{P}}—CH_2—CH_2NH(CH_3)$$

$$(HO)_2\overset{O}{\overset{\|}{P}}—CH_2—CH_2N(CH_3)_2$$

$$(HO)_2\overset{O}{\overset{\|}{P}}—CH_2—CH_2N(CH_3)_3{}^+OH^-$$

appear to be normal cell products and that some of them may play a role as intermediates in the biosynthesis of phosphonolecithins from phosphonocephalins by progressive methylation of the amino group, prompted us to undertake the synthesis of *N*-methyl substituted glycerophosphonolipids.

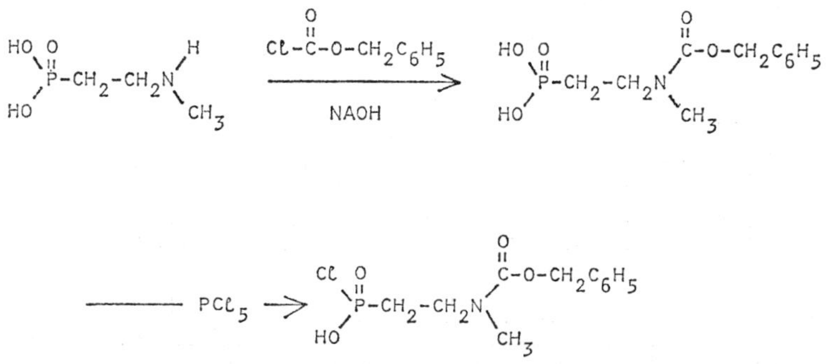

SCHEME 2. *N*-carbobenzoxy, *N*-methyl-(2-aminoethyl) phosphonic acid monochloride (Baer & Pavanaram, 1970*a*).

Reaction Schemes 2 and 3 illustrate the synthesis of L-α-phosphono(*N*-methyl)-cephalins (Baer & Pavanaram, 1970*a*). They were obtained by esterification of D-α, β-distearin or dipalmitin with (*N*-methyl-2-aminoethyl)-phosphonic acid monochloride. A possible reaction of the amino group with phosphorus pentachloride, during the course of preparation of the acid chloride of *N*-methylaminoethylphosphonic acid (Reaction Scheme 2), was prevented by protecting it with the readily removable carbobenzoxy group. Esterification

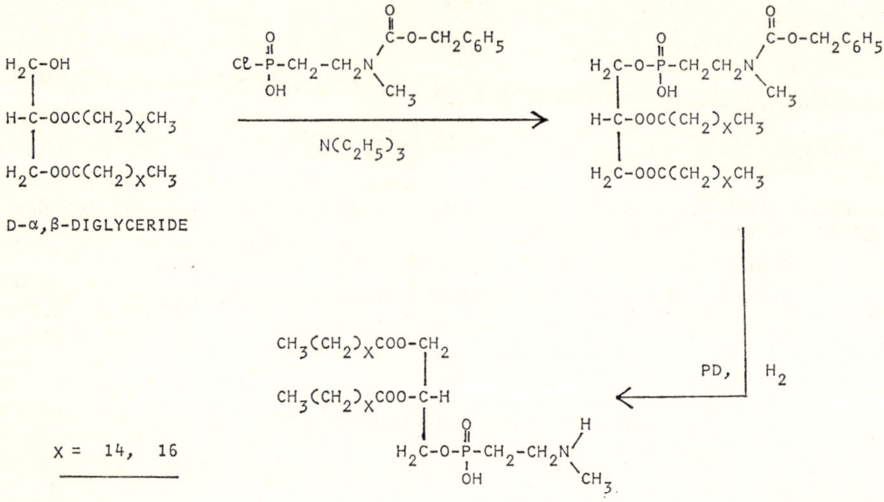

SCHEME 3. Diacyl L-α-glyceryl-(*N*-methyl-2-aminoethyl)phosphonate. Synthesis of phosphonic acid analogues of L-α-(*N*-methyl)cephalins (Baer & Pavanaram, 1970*a*).

of the diglycerides with the acid chloride (Reaction Scheme 3), and removal of the protective group by catalytic hydrogenolysis gave the corresponding L-α-phosphono(*N*-methyl)cephalins.

We have not yet accomplished the synthesis of phosphono (*N*,*N*-dimethyl)-cephalins. We succeeded, however, in obtaining their ether analogues (Baer & Rao, 1970), by phosphonylating *ether*-diglycerides with 2-bromo-ethylphosphonic acid monochloride, and by treating the reaction product with dimethylamine at 50–55° for 3 days (Reaction Scheme 4).

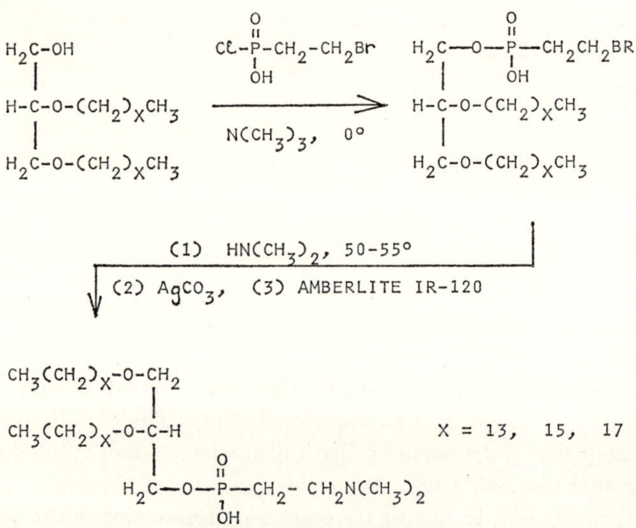

SCHEME 4. Dialkyl L-α-glyceryl-(*N*,*N*-dimethyl-2-aminoethyl)phosphonate. Phosphonic acid analogues of diether-(*N*,*N*-dimethyl)cephalins (Baer & Rao, 1970).

The formation *in situ* of the nitrogenous moiety of a phosphonolipid by treatment of the appropriate bromo-compound with an *N*-methyl substituted amine was found to be a useful reaction in the preparation of *ester*-phosphonolecithins also (Baer & Stanacev, 1965) (Reaction Scheme 5), and of *ether*-phosphonolecithins (Baer & Rao, 1967) (Reaction Scheme 6).

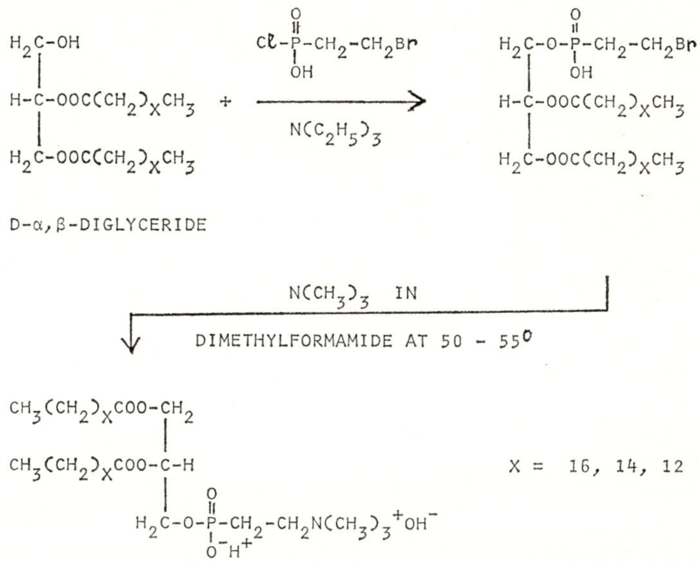

SCHEME 5. Diacyl L-α-glyceryl-(2-trimethylammoniumethyl)phosphonate. Synthesis of phosphonic acid analogues of L-α-lecithins (Baer & Stanacev, 1965).

The glycerophospholipids and their phosphonic acid analogues can be distinguished from each other by thin-layer chromatography of the intact lipids or their fatty acid-free moieties.

The existence in nature of phosphonolipids obviously aroused speculations regarding the pathways of their biosynthesis and metabolism. In order to facilitate the study of these processes, we have synthesized the fatty acid-free moieties of phosphonocephalins, phosphono(*N*-methyl) cephalins and phosphonolecithins, expecting these to be intermediates in the metabolism of phosphonolipids. The starting material, D-acetone glycerol, was the same in all three procedures. Based on previous experience, we assumed that the acid chlorides of the phosphonic acids to be esterified with the acetone glycerol would make satisfactory phosphonylating agents. This proved to be true in only one case. In the other two, the presence of chloride ions created considerable difficulties in the isolation of the final products in pure state. These difficulties were overcome, when we discovered that *meta*phosphonates, which are the anhydrides of phosphonic acids, make satisfactory phosphonylating agents. They combine directly with the hydroxyl-containing compounds forming the desired ester.

Thus, the fatty acid-free moiety of phosphonocephalins (Baer & Sarma, 1967) was obtained (Reaction Scheme 7) by condensing (2-phthalimidoethyl)-metaphosphonate with D-acetone glycerol, and by removing the protective

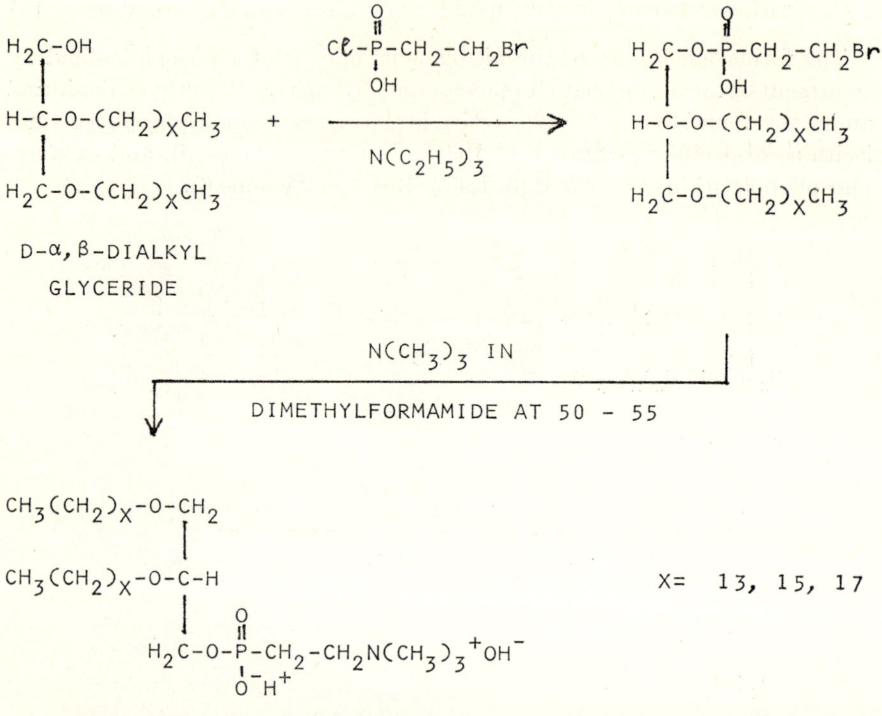

SCHEME 6. Dialkyl L-α-glyceryl-(2-trimethylammoniumethyl)phosphonate. Synthesis of phosphonic acid analogues of ether lecithins (Baer & Rao 1966, 1967).

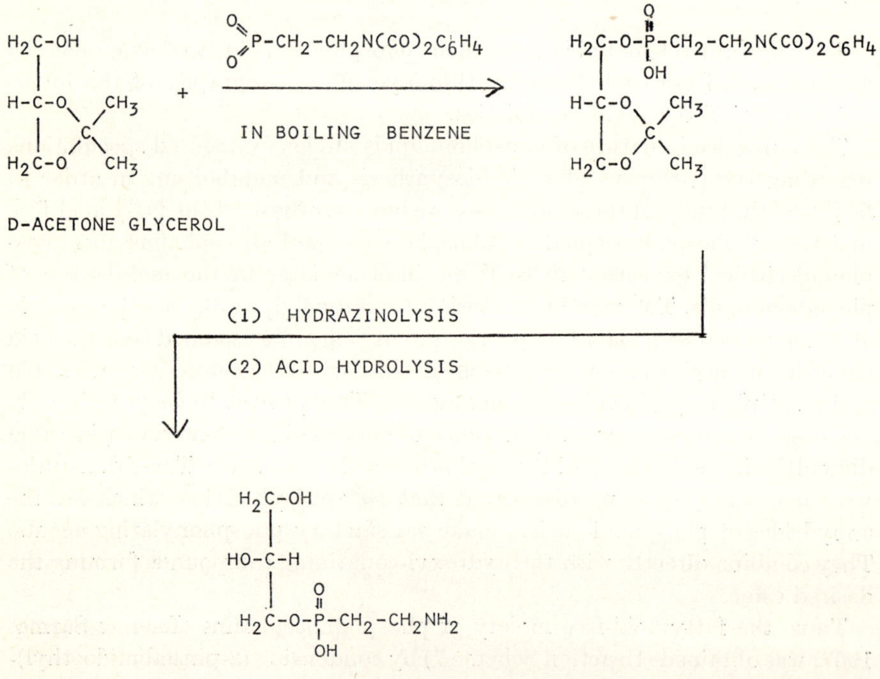

SCHEME 7. Synthesis of L-α-glyceryl-(2-aminoethyl)phosphonate (Baer & Sarma, 1967).

groups of the condensation product by hydrazinolysis and mild acid hydrolysis, respectively. By replacing in this procedure the phthalimidoethyl-metaphosphonate by (2-bromoethyl)metaphosphonate (Reaction Scheme 8), by removing the protective acetone group by acid hydrolysis, and by treating the

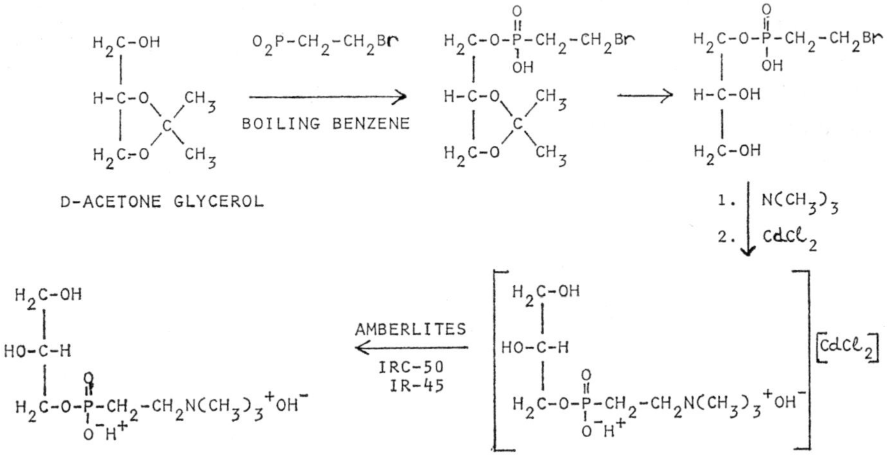

SCHEME 8. Synthesis of L-α-glyceryl-(2-trimethylammoniumethyl)phosphonate (Baer & Robinson, 1967).

bromo-compound in dimethylformamide with trimethylamine at 60° for 3 days, L-α-glyceryl-(2-trimethylammoniummethyl)phosphonate, the fatty acid-free moiety of phosphonolecithins (Baer & Robinson, 1967) was obtained. It was isolated and purified via its cadmium chloride compound. Finally, the fatty acid-free moiety of L-α-phosphono(*N*-methyl)cephalins (Reaction Scheme 9) was obtained by phosphonylating D-acetone glycerol with (*N*-carbobenzoxy-*N*-methyl-2-aminoethyl)phosphonic acid monochloride, and by

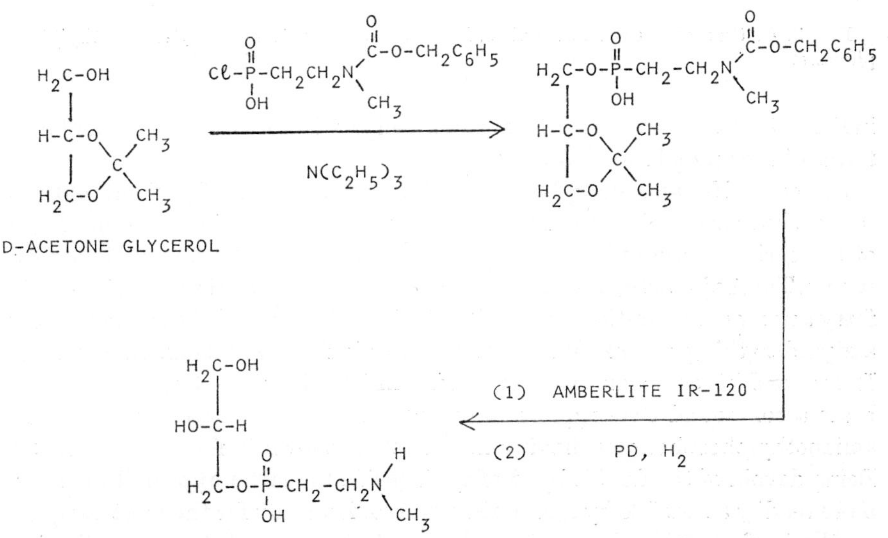

SCHEME 9. Synthesis of L-α-glyceryl-(*N*-methyl-2-aminoethyl)phosphonate (Baer & Pavanaram, 1970).

freeing the reaction product of its protective acetone and carbobenzoxy groups by hydrolysis and hydrogenolysis, respectively (Baer & Pavanaram, 1970*b*).

Since Rouser's original report on the isolation of a ceramide aminoethylphosphonate from the lipid extracts of the sea anemone, *Anthopleura elegantissima*, similar compounds have been isolated from a variety of biological sources. These include certain species of shellfish, snails and cephalopods. They were also obtained from the viscera of the oyster *Ostrea gigas*, and the spermatozoa of a fresh water mussel, *Hyriopsis schlegelii*. A recent report, however, by Alam & Bishop (1968), describing the isolation of lipid-bound cholinephosphonate from human aortas with athero-sclerotic plaques, seems to

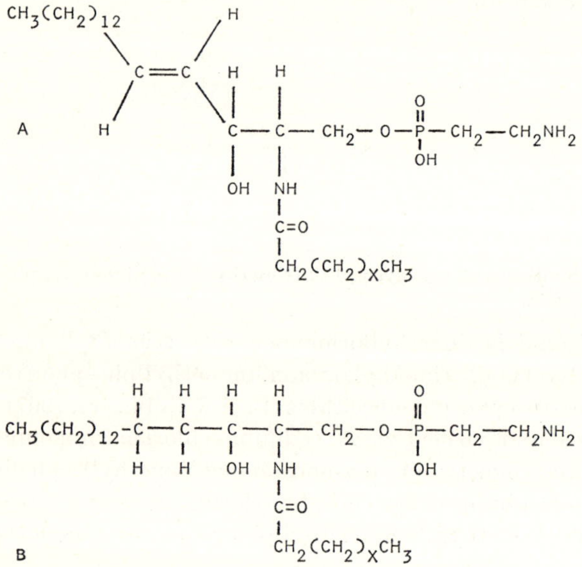

FIG. 3. A. Ceramide aminoethylphosphonate; B. Dihydroceramide aminoethylphosphonate.

indicate that the occurrence of phosphonolipids is not confined to the lower forms of the animal kingdom.

All ceramide aminoethylphosphonates isolated thus far from natural sources have been mixtures of homologous compounds, differing by the length of the carbon chain of both the sphingosine base and the fatty acid substituents. The major base, however, appeared to be *erythro*-sphingosine. The main fatty acids were palmitic and stearic acid. The chemical structure tentatively assigned to this group of compounds is shown in formula A (Fig. 3). Recently, Rouser and his colleagues have succeeded in isolating from a marine invertebrate a compound that appears to be a representative of saturated ceramide aminoethylphosphonates, having the structure shown in formula B (Fig. 3). Being devoid of the double bond of sphingosine the latter appeared to us the more readily accessible by synthesis. Even so, it was by far the most complex synthesis of any phosphonolipid we had undertaken until then, necessitating the preparation of eighteen intermediate compounds (Baer & Pal, 1968).

Eleven of these were required for the synthesis of the lipid moiety (Reaction Scheme 10) *erythro-N*-palmitoyl-3-0-benzoyl-DL-dihydrosphingosine, compound 12. The reaction scheme, describing its synthesis by the procedure of Shapiro and his colleagues, is shown primarily to explain our hesitation in undertaking the synthesis of phosphonic acid analogues of sphingolipids.

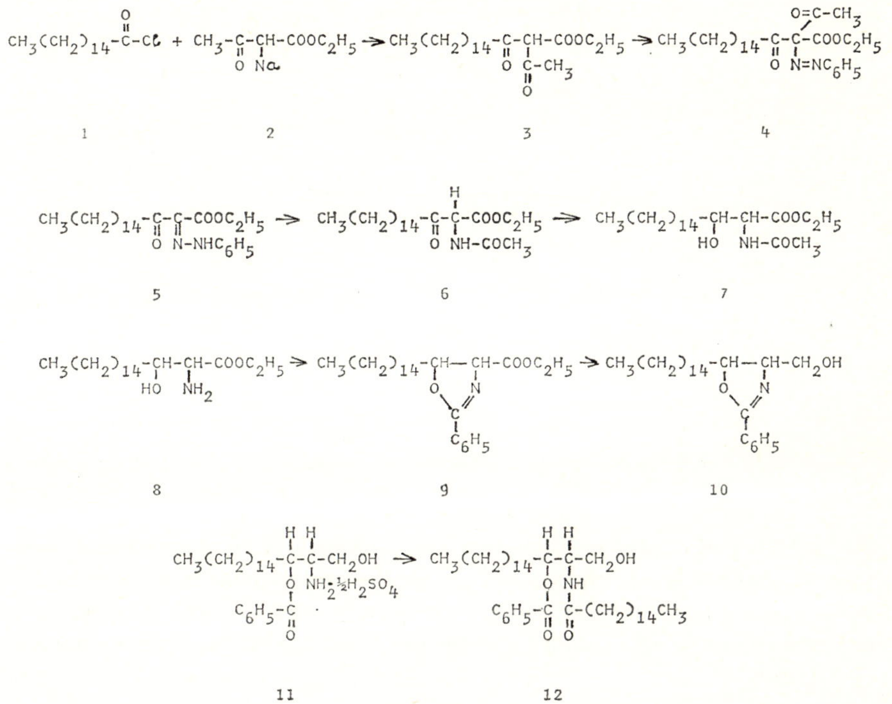

SCHEME 10.

Next, we had to prepare a derivative of 2-aminoethylphosphonic acid suitable for esterification with the lipid moiety. For this purpose we chose (dibenzyl-2-aminoethyl)phosphonic acid monochloride, compound 16 (Reaction Scheme 11).

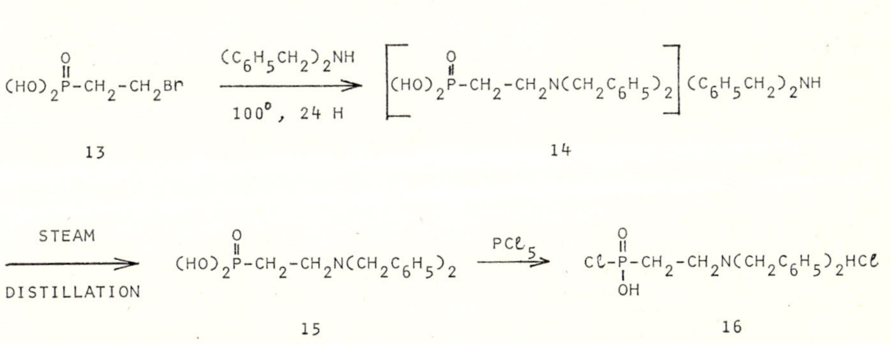

SCHEME 11. Synthesis of (2-*N*,*N*-dibenzylaminoethyl)phosphonic acid.

Treatment of the dihydroceramide derivative, compound 12, with dibenzylaminoethylphosphonic acid monochloride (Reaction Scheme 12) and removal of the protective benzyl groups of compound 17 by catalytic hydrogenolysis, and of the benzoyl group of compound 18 by saponification yielded the desired

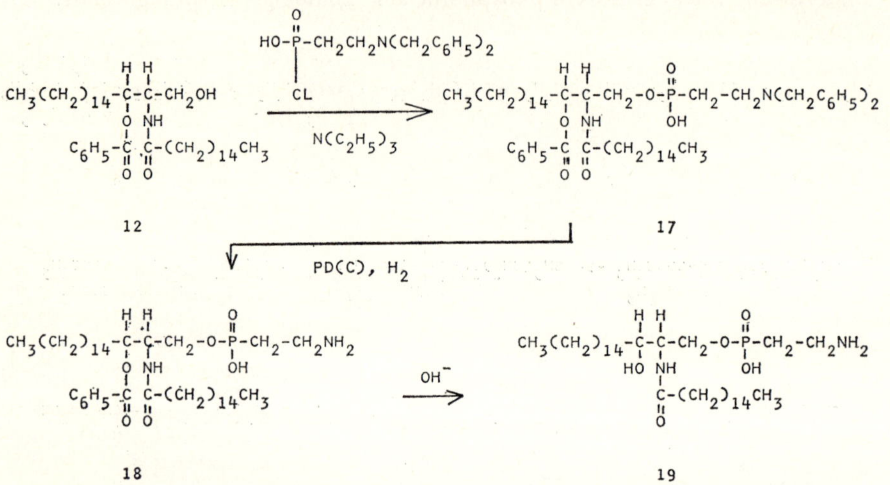

SCHEME 12. Synthesis of erythro-*N*-palmitoyl-DL-dihydrosphingosyl-1-(2-aminoethyl) phosphonate (Baer & Pal, 1968).

dihydroceramide aminoethylphosphonate, compound 19. Its infrared spectrum resembled closely those reported for the ceramide aminoethylphosphonates isolated from natural sources, excepting that the synthetic compound did now show the absorption band for the trans-double bond of the natural compounds.

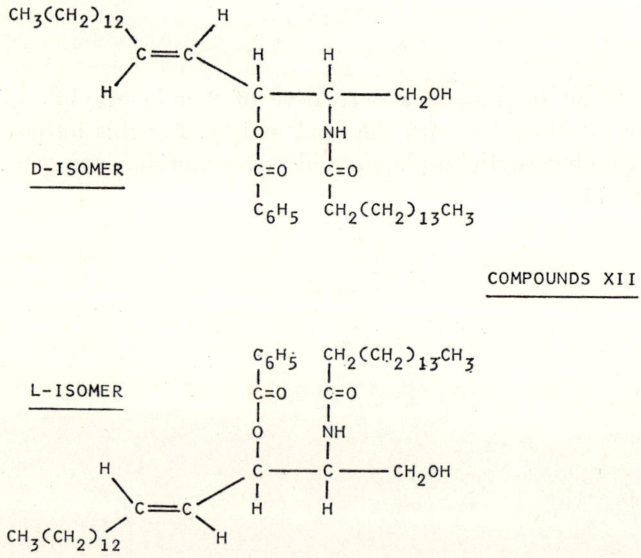

FIG. 4. Erythro-*N*-Palmitoyl-3-O-benzoyl-D-and L-sphingosine.

The synthesis of the *dihydro*ceramide aminoethylphosphonate in general confirmed the structure tentatively assigned to natural ceramide aminoethylphosphonates. However, for establishing their structure and configuration beyond doubt, it was deemed necessary to synthesize a representative ceramide aminoethylphosphonate, preferably in both of its enantiomeric forms.

Preparation of the ceramide derivative, compound XII, in both of its optically isomeric forms (Fig. 4) by the procedure of Shapiro and his colleagues proved to be as lengthy and time-consuming as that of the corresponding dihydroceramide derivate. It too required preparation of no less than eleven intermediate compounds. The final steps of the synthesis of the D-isomer of ceramide aminoethylphosphonate (Baer & Sarma, 1969) are outlined in Reaction Schemes 13 and 14. It was obtained by treating the appropriate

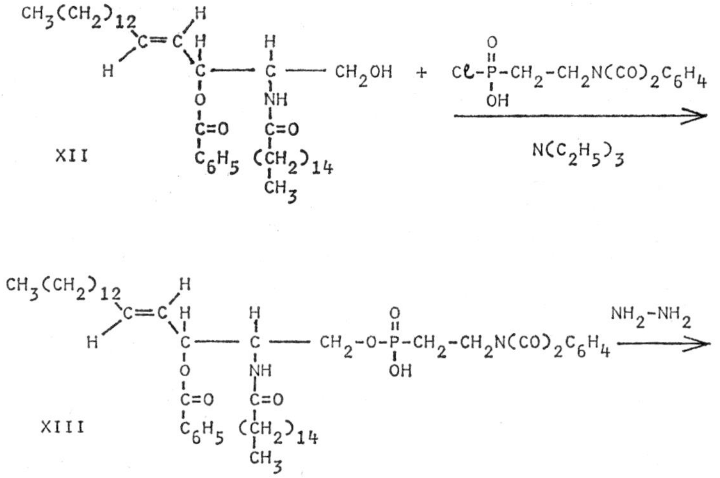

SCHEME 13. Synthesis of ceramide-(2-aminoethyl)phosphonates.

optical isomer of compound XII with (phthalimidoethyl) phosphonic acid monochloride and triethylamine (Reaction Scheme 13), and by removing from the reaction product (compound XIII) its protective phthaloyl group by hydrazinolysis (Reaction Scheme 14) and of its benzoyl group by alkaline hydrolysis. The L-isomer was obtained in the same manner by starting with the L-isomer of the ceramide derivative. On comparing at *low* concentrations the specific rotations of our synthetic ceramide aminoethylphosphonates with that of a homologous mixture of ceramide aminoethylphosphonates isolated from *Corbicula sandai*, generously donated by Professor Hori, we found that the natural compounds, as expected, are derivatives of D-sphingosine and have the structure and configuration as shown by Formula XV. Compound XV was thus the first natural ceramide aminoethylphosphonate to be obtained by synthesis. It is most likely that Formula XV represents the general structure and configuration of all natural ceramide aminoethylphosphonates containing the C_{18}-spingosine base and giving the same infrared spectra as the ceramide aminoethylphosphonates of *Corbicula sandai*.

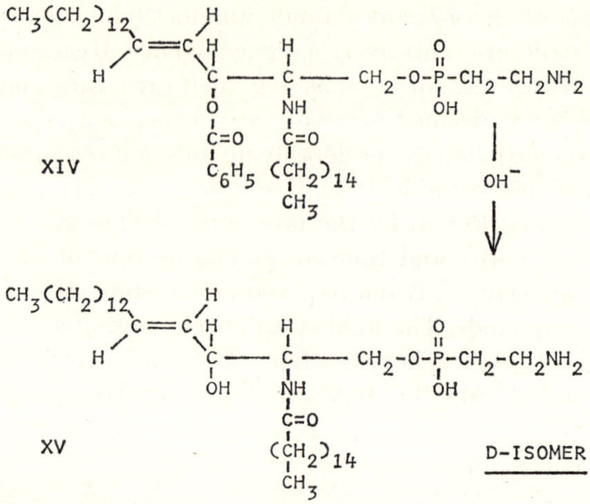

SCHEME 14. Erythro-*N*-palmitoyl-sphingosyl-1-(2-aminoethyl)phosphonate. Synthesis of ceramide-(2-aminoethyl)phosphonates (Baer & Sarma, 1969).

While we were pursuing our synthetic work, I noticed with great satisfaction the steadily growing active interest of the biochemists in natural phosphonolipids. Their work provided strong evidence for the occurrence in nature of phosphonic acid analogues of a wide variety of phospholipids, a possibility I suggested in our first paper on the synthesis of phosphonolipids (Baer & Stanacev, 1964) and which prompted us to attempt their synthesis.

It is my hope that the phosphonolipids we have synthesized will play an essential part in the elucidation of the structure and configuration of the natural compounds, as well as the pathways of their biosynthesis and metabolism.

REFERENCES

Alam, A. U. & Bishop, S. H. (1968). *Abstr. Div. biol. Chem.* 156 Meeting of the Am. chem. Soc., Atlantic City, Sept. 9–13.

Baer, E. & Pal, B. C. (1968). *Can. J. Physiol. Pharmacol.* **46**, 525.

Baer, E. & Pavanaram, S. K. (1970*a*). *Can. J. Biochem.* **48**, 979.

Baer, E. & Pavanaram, S. K. (1970*b*). *Can. J. Biochem.* **48**, 988.

Baer, E. & Rao, K. V. J. (1967). *Can. J. Biochem.* **45**, 317.

Baer, E. & Rao, K. V. J. (1970). *Can. J. Biochem.* **48**, 184.

Baer, E. & Robinson, R. (1967). *Can. J. Biochem.* **45**, 1747.

Baer, E. & Sarma, G. R. (1967). *Can. J. Biochem.* **45**, 1755.

Baer, E. & Sarma, G. R. (1969). *Can. J. Biochem.* **47**, 603.

Baer, E. & Stanacev, N. Z. (1964*a*). *J. biol. Chem.* **239**, 3209.

Baer, E. & Stanacev, N. Z. (1964*b*). *J. biol. Chem.* **240**, 44.

Baer, E. & Stanacev, N. Z. (1965). *J. biol. Chem.* **240**, 3754.

Horiguchi, M. & Kandatsu, M. (1959). *Nature* **184**, 901.

Kittredge, J. S. K. (1967). *Fedn Proc. Abstr. III*, Coll. IV–4, 453.

Kittredge, J. S. K., Isbell, A. F. & Hughes, R. R. (1967). *Biochemistry* **6**, 289.

Kittredge, J. S. K., Roberts, E. & Simonsen, D. G. (1962). *Biochemistry* **1**, 624.

Rouser, G., Kritchevsky, G., Heller, D. & Lieber, E. (1963). *J. Am. Oil Chem. Soc.* **40**, 425.

Simon, G. & Rouser, G. (1967). *Lipids* **2**, 55.

WAX ESTERS: MAJOR MARINE METABOLIC ENERGY SOURCES

BY A. A. BENSON, R. F. LEE AND J. C. NEVENZEL

Scripps Institution of Oceanography, University of California, San Diego La Jolla, California, USA

ENERGY metabolism in marine organisms is generally focused on production and utilization of lipids. There being no fresh water available, marine animals like those in arid terrestrial regions (Schmidt–Nielsen, 1964) resort to lipid metabolism for production of water. Pelagic marine algae, predominantly diatoms and dinoflagellates, accumulate oil globules within and outside of their chloroplasts. Whether these globules, largely triglycerides of long chain polyunsaturated acids, may act in part as buoyancy regulators has not yet been established. Such regulation in some marine animals has been attributed to metabolic control of liver lipid content. Thermal insulation in cetaceans, pinnepeds and birds is provided by lipid storage tissues. Focusing and conduction of high frequency sounds in cetaceans is often mediated by lipids including wax esters. The milk of pinnipeds like the California sea lion usually contains no lactose, only protein and lipids. Lipids serve many and varied purposes in marine organisms.

Marine lipids include triglycerides, wax esters, squalene and glycerol ethers. Of these, the wax esters (Fig. 1) are the simplest in structure and probably the

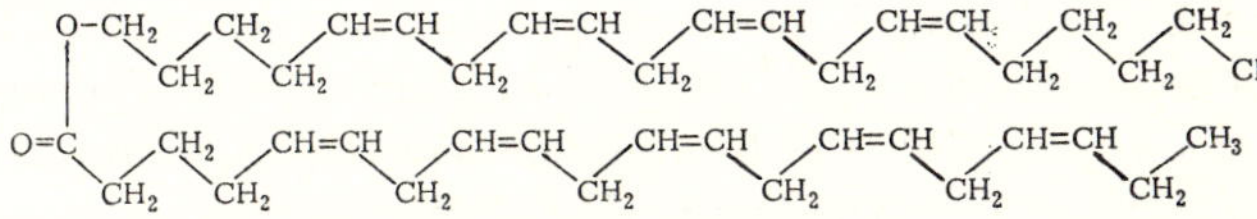

Polyunsaturated Marine Wax Ester, 40:9

FIG. 1

predominant energy storage materials in the sea. Nevenzel (1970) and Nevenzel *et al.* (1969) found that lantern fish accumulate wax esters. The vertically-migrating lantern fish of the sea utilize wax esters for buoyancy, presumably as an aid in overcoming density changes of their daily one kilometer migration. Isaacs (personal communication, J. D. Isaacs) deduced that the buoyancy regulation material of such fish should be less compressible than sea water; he found only one such biological material in the literature, beeswax. The oil of another remarkable group of vertical migrators, the copepods, was analysed by Lee, Nevenzel & Paffenhöfer (1970) and found to include much wax ester. These tiny crustacean "camels of the sea" (S. Wakil, personal communication) may migrate vertically as much as 500 metres each day. This paper reviews salient information now available on the scope of this process in the marine food chain and biochemical principles which work with organisms specialized in wax ester production and utilization has revealed.

WAX ESTERS IN MARINE ORGANISMS

The physical, physiological, and biochemical properties of wax esters appear to have directed their utilization in a great variety of marine animals. Nevenzel (1970) listed thirty species in seventeen orders and three phyla where wax esters occurred as the major lipid. Our data now includes a wider range of organisms. The most populous group of fishes in the seas, the cyclothones or bristlemouths, are rich in wax ester. Molluscs, cephalapods, chaetognaths, shrimps, decapods, anemones, corals and marine birds accumulated wax esters. Certain fishes were unique. The coelacanth (*Latimeria chalumnae*) contained 70% wax ester (Nevenzel *et al.*, 1966). The lipid of the castor oil fish (*Ruvettus pretiosus*) was 92% wax ester (Nevenzel, Rodegker & Mead, 1965). Mullets, croakers, codfishes (Moridae), gouramis and sharks contain appreciable amounts of wax ester.

A list of zooplankton and micronekton taken in a 3 m Isaacs–Kidd mid-water trawl from 2500 m is given in Table 1. The high percentage of wax ester in the lipids of the amphipods, polychete worms, arrowworms (Chaetognatha), and mysid shrimp revealed what appears to be a fundamental relationship in marine metabolism (Lee *et al.*, 1972). Every organism living below 1000 m in the sea (average depth, 5000 m) utilizes or stores wax esters. The evolution of the ability to do so began early in the development of migratory marine animals. Concurrent development of wax synthetase and lipase regulation ability provided an important dimension in ecological resilience.

Lipid metabolism in marine copepods is a most elegant example of environmental adaptation. Their world-wide function in grazing the photosynthetic algae of the seas and inland waters means that at least half of the earth's photosynthetic production is converted, for a time, to wax. The first isolation of wax ester from a copepod, the deep-sea species *Gaussia princeps* in 1967 (Lee *et al.*, this laboratory) led to the recognition of wax ester metabolism as a major metabolic activity in Nature.

WAX DISTRIBUTION IN MARINE COPEPODS

A survey of lipids of zooplankton from a wide range of marine environments (Lee, Hirota & Barnett, 1971) has revealed some general principles related to wax esters storage and utilization. Copepod genera belonging to the families Calanidae, Euchaetidae, Lucicutiidae, Heterorhabdidae and Augaptilidae accumulated greater than 20% of their lipid as wax esters. These genera generally occur in deep or near-surface cold waters. The families Candaciidae and Pontellidae whose members are found near the surface of tropical or temperate waters accumulated less than 10% wax ester. Other families such as the Eucalanadidae have species inhabiting wide ranges of temperature and depth.

Table 1. *Zooplankton and micronekton from 2500 metres*

Taxonomic group	Lipid*	Wax ester**	Triglyceride**
Alciopidae sp. (Polychaeta)	55	76	11
Gammaridae sp. (Amphipoda)	57	73	13
Hyperiidae sp. (Amphipoda)	7	21	17
Gnathophausia sp. (Mysidacea) Shrimp	42	69	12
Gennadas sp. (Decapoda) shrimp	21	10	35
Chaetognatha sp. 1	38	34	12
Chaetognatha sp. 2	40	71	11
Pteropoda sp. Pelagic mollusc	5	0	8
Oegopsidae sp. (Cephalopoda) squid	23	27	6
Argyropelecus sp. (Sternoptychidae) Hatchet fish	—	22	7
Stomias atriventer (Stomiatidae)	—	9	32
Cyclothone sp. 1 (Gonostomatidae) Bristlemouth	22	53	16
Cyclothone sp. 2	13	18	22
Myctophidae† Lantern fish	50·6	86·7††	1·7

* Percent of dry weight. ** Percent of total lipid. † *Lampanyctus ritteri.* †† Muscle tissue.

Adaptations to the vicissitudes of marine environment include storage of wax ester where the food supply is not constant or predictable. Those in warm tropical seas where the supply of phytoplankton is reasonably constant throughout the year accumulate very little wax ester. Species from Arctic and Antarctic waters, where photosynthetic production is vigorous but brief, store the greatest amounts of wax. Deep cold water species, too, are adapted to the low probabilities of finding food in copious quantities. When they do, they are able to store the excess energy supply as wax ester. Lee, Nevenzel & Paffenhöffer (1971) observed that wax ester storage was a function of dietary levels.

Unsaturation and low temperatures

In general, organisms adapted to low temperatures store more highly unsaturated lipids than those living at higher temperatures. Copepod wax ester, however, is more saturated in animals from the cold deep waters. Genera such as *Bathycalanus* and *Paraeuchaeta* of the near 0° waters below 250 m were rich in lipid, up to 61% dry wt., and in wax ester, up to 82% of the lipid.

Bathycalanus bradyi and *Lucicutia bicornuta* collected at 1600 to 2500 m had 59% and 15% lipid containing 77% and 50% wax ester respectively. These bathypelagic species produced wax esters containing saturated and mono-unsaturated alcohols (Table 2). Their fatty acids, too, were more saturated

Table 2. *Wax esters of bathypelagic copepods**

Fatty alcohol	*Calanus helgolandicus* (Laboratory-cultured)	*Gaussia princeps*	*Gaetanus brevicornis*
14:0	1·5	7·6	6·7
16:0	13·0	54·9	44·5
18:1	3·8	11·6	11·3
20:1	18·4	4·9	5·1
20:4	10·9	—	—
22:1	—	11·5	5·2
Fatty acids			
14:0	11·8	0·3	0·4
16:0	29·0	0·8	0·5
16:1	13·2	8·8	12·7
18:1	12·1	70·5	63·7
20:4	7·6	0·2	—
20:5	2·0	5·8	11·4

* Lee *et al.*, 1972.

Table 3. *Wax esters of marine copepods*

Carbon number	*Calanus helgolandicus*		*Megacalanus longicornis**	*Paraeuchaeta rubra**	*Metridia princeps**
28	0·2**	0·2†			
29	0·3	0·1			
30	21·1	2·8	1·6	3·2	3·1
31	—	—	0·6	1·5	0·6
32	32·3	8·3	8·6	14·9	7·5
33	—	—	3·6	5·4	3·4
34	11·6	16·9	40·5	28·9	42·5
35	—	—	6·6	4·8	5·6
36	18·2	23·2	18·2	14·8	13·0
37	—	—	2·8	2·1	3·0
38	12·6	17·9	7·2	7·5	5·7
39	—	—	0·9	0·9	—
40	3·7	9·5	4·8	9·9	8·8
41		—	0·2	0·7	—
42		8·9	4·3	4·9	6·8
43		—	—	—	—
44		7·1	tr	0·6	—
Total lipid	36·6††	12·4			

* 1900 metres. ** Laboratory grown animals fed Skeletonema. † Wild type. †† Percent lipid of dry weight.

than those of *C. helgolandicus* grown in the laboratory on diatoms. The explanation of the predominance of more saturated fatty alcohols and acids in the deep water animals is clear. Their carnivorous diet lacks the polyunsaturated lipids of the photosynthetic algae and diatoms near the sea surface. As *de novo* synthesis of fatty acids proceeds in the deeper members of the marine food chain the fraction directly related to the polyunsaturated lipids derived from marine algae becomes negligible. The total chain lengths (Table 3), too, vary with diet and environment of the animals. The unsaturated 20:4, 20:5, 22:3, 22:6 and 24:5 chains of wax esters in *C. helgolandicus* grown in the laboratory on diatoms and dinoflagellate algae are probably directly formed from the algal triglyceride esters. During starvation, some of these components are rapidly depleted. Others, particularly the 22:6 acid occurring in the ethanolamine lysophospholipid of *Calanus plumchrus* (Table 4), are specific

Table 4. *Wax esters of* Calanus plumchrus *copepodid V from Strait of Georgia, British Columbia, Canada, May* 1971

Fatty chain	Wax acid	Wax alcohol	Tri-glyceride	PE	PC	Sphingo-lipids	PE Lyso-lipid
14:0	22·1*	3·0	9·3	2·6	2·0	t	t
14:1	t		t	t	t	—	t
15:0	0·2	0·1	t	0·1	t	—	t
15:1	t		t	t	t	—	t
16:0	16·9	11·4	7·1	28·0	6·1	16·7	5·4
16:1	8·3	8·8	1·5	0·1	6·3	0·7	0·5
16:2	3·1		1·6		0·2		
16:3	8·8		0·7				
16:4			0·9				
17:0	t		t	0·2	t		
17:1	0·2	1·4	t	0·1	t		
18:0	t	0·3	0·2	2·8	2·7	11·8	2·9
18:1	0·2	4·3	38·7	5·0	6·2	5·0	7·1
18:2	5·9	0·7	2·6	0·2	2·0		0·1
18:3	3·3	t	0·9	0·5	0·1		
18:4	6·2		6·7	0·7	0·9		
20:0	t			1·0	1·3		0·1
20:1	3·2	37·8	0·5	0·4	0·2		0·4
20:2	1·5		0·8				
20:4	4·1		0·8	0·1	4·4		0·4
20:5	8·4		14·9	12·0	22·8		7·0
21:0		0·1					
22:0	0·8		0·6	1·5	0·1		
21:1		0·2					
22:1		31·9	0·8	1·2		12·8	
22:2	0·9		0·2				
22:4	0·6						
22:5	1·0		0·1		0·2		0·7
22:6	4·3		10·7	43·1	44·3	53·6	78·3
24:1			1·3				
24:4							

* Percent of total.

components of essential membranes in the animal. They are not depleted by starvation as are the storage lipids of the animal (Lee, Nevenzel & Paffenhöffer, 1971).

Wax ester accumulation in developmental stages of copepods

Examination of wax ester content in several species of copepods (Tables 5 and 6) revealed dramatic changes in the ratios of wax ester and triglyceride content. The developmental stages requiring resilience in the event of environmental adversity contained largest amounts of wax ester (Lee *et al.*, 1972).

Table 5. *Lipid transformations during the life cycle of* Calanus helgolandicus

Stage	% Lipid	% Wax ester	% Triglyceride
Eggs	—	absent	60
Naupliar I	—	absent	15
Copepodid I	—	absent	12
Copepodid III	—	1	5
Copepodid III (late)	—	10	2
Copepodid IV	—	16	6
Copepodid IV (late)	—	22	5
Copepodid V	—	50	3
Adult	28	41	12

Lipid % dry wt. and % wax and triglyceride of total lipid. 800 eggs were analysed for lipids. 200 eggs were used for Naupliar I. Nauplii were reared on mixture of *Lauderia borealis* and *Gymnodinium*.

Table 6. *Lipid transformations during the life cycle of Euchaeta japonica**

Stage	Lipid**	Wax ester†	Triglyceride†
Eggs	64·4	58	19
Eggs (late stage)	59·1	50	17
Naupliar 2	43·8	61	17
Naupliar 3	30·8	56	5
Naupliar 4	25·0	20	3
Naupliar 5	21·2	15	1
Naupliar 6	17·1	12	0
Copepodid I	14·2	9	0
Copepodid I (field-collected)	23·6	29	0
Copepodid II	11·6	12	0
Copepodid IV (field-collected)	31·2	40	3
Copepodid V (field-collected)	50·1	81	2
Adult (field-collected)	41·3	54	18

* Lee *et al.*, 1972. ** Percent of dry weight. † Percent of total lipid.

The near-surface copepod, *Calanus helgolandicus*, contained 50% wax ester prior to final molting, mating and egg production. As the adult stage accomplishes its biological purposes the lipid content decreases. The eggs of *C. helgolandicus* with 60% triglyceride, are produced by an animal containing largely wax ester. Eggs of calanoid copepods usually are released by the female to float to the surface where the early naupliar stages can graze on small algae. Density, then, is an important factor in determining the type of lipid stored. The eggs of the deep water carnivorous copepod *Euchaeta japonica*, on the other hand, contained 58% wax ester. They are carried by the female. The naupliar stages do not feed and must use the stored wax ester in the yolk sac for their energy requirements. On development to the copepodid stages the animals are carnivorous and again begin to accumulate wax ester in large amounts (Table 6).

Oil storage in copepods

Marine planktonic biologists had for years observed oil storage in many species. This is not unusual in marine algae and animals. Its identification as nearly pure wax ester, however, places the storage bodies in a different

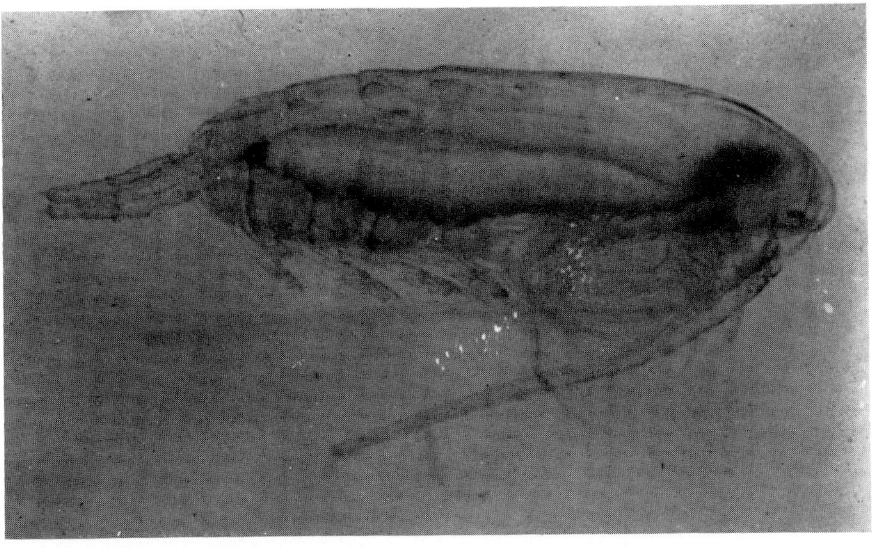

FIG. 2. Marine copepod, *Calanus plumchrus* copepodid Stage V, Strait of Georgia, British Columbia, June, 1971. Note dark mass of undigested algae in gut and the long sac of colourless liquid wax ester.

category. Isolation of oil from individual animals, 1 mm long, by M. Zalokar revealed that the oil sac of *C. helgolandicus* contained pure wax ester and that the animal's triglyceride component must be distributed elsewhere. Calanoid copepods of temperate waters often have a well organized oil sac such as that in *Calanus plumchrus* shown in Fig. 2. The colourless oil is sharply separated from globules of intensely coloured carotenoid solutions adjacent to the sac, in antennae, legs and thoracic regions.

Copepods of the benthos and polar seas living in near freezing waters have adapted to the problem of lower viscosity of their wax ester by storing their wax in small globules distributed throughout the body rather than in a single large oil sac. Biodegradation and transport of their oil therefore is possible at lower temperatures. These species are usually richest in wax ester content and most resistant to starvation. Presence of membrane around the individual globules has not yet been ascertained.

Role of wax esters in starvation

The natural environments of many copepod species offer great variations in food supply. Arctic copepods, for example, have only two months for grazing on rich algal populations. In this time they accumulate metabolic fuel for molting, mating, egg production and other life processes during the dark months of the year. Likewise, the copepods of the deep sea must be prepared for uncertain periods between appearance of food sources. These groups accumulate massive reserves of wax ester and can withstand long periods of starvation. They are among the most remarkable of species in this respect.

During starvation the copepod draws upon two sources of energy, its triglycerides and wax esters, at controllable and differing rates. The control point appears to be the relative activity of two enzymes, was ester lipase and triglyceride lipase. Being such similar lipid esterases one might anticipate overlap in substrate specificity. There seems to be none. The wax lipase activity of wild copepod preparations was zero while their triglyceride lipase activity was measurable. In starvation experiments, Lee, Hirota & Barnett (1971), Lee, Nevenzel & Paffenhöffer (1970, 1971) found most copepod species readily utilized the triglyceride component of their stored lipid and more slowly metabolized their wax ester. Activation of wax lipase by the stress of starvation provides the control limiting the depletion of the animal's reserves.

Biodegradation of wax esters in fishes

Being the major metabolic fuel in the sea, the site and efficiency of wax ester digestion and metabolism in predator fishes are of obvious interest. Fishes living on copepods which are 70% wax (dry wt basis) must be adapted to remarkably difficult solubilization, hydrolysis and transport problems. It was anticipated that the hydrophobic wax esters should form wax chylomicrons upon passage through the gut. This seems not to be the case. Wax lipase activity is exceptionally high in the digestive system of fishes. Neither wax ester nor fatty alcohol have been observed in serum from fishes living on copepods.

We have examined gut contents of small chum salmon fry. They varied greatly but the fishes were usually monotonic in their preferences for zooplankters. Those preferring copepods predominated and were gorged. The chum fry (4 cm) devour so much copepod that they become saturated with wax ester and their floating fecal pellets are rich in oil. In spite of the fact that their rich diet was only incompletely hydrolysed no wax ester passes into the circulating blood.

The site of wax lipase production in fishes is the pyloric caecum, a mass of tubular visceral appendages leading from the junction of the stomach and the small intestine. Their production of lipase was known but the high activity of wax lipase had not been recognized.

Wax ester and triglyceride lipase activities in fishes and copepods

The pyloric caecum of fishes includes the diffuse hepatopancreas tissue presumed to be the source of digestive lipases. From the appearance of oil globules in the lumen of pyloric caeca of fishes following wax injestion we presume that digestion of copepods progresses in this organ.

Assay of wax lipase activity

A comparison of pyloric caecum lipases was undertaken by using ^{14}C-triolein and polyunsaturated C_{18}-alkyl oleate-^{3}H, a synthetic wax ester (Table 7). In every experiment the capability for triglyceride hydrolysis of the

Table 7. *Relative lipase activities for wax ester and triglyceride in pyloric caecum and copepod extracts*

Lipase source	Wax lipase	Triglyceride lipase	Cholesterol ester lipase	Ratio TG/WE
Chum salmon fry*	5·6	66·8	—	12
Jack Mackerel**	0·913‡	—		
Anchovy†	20·4§	25·6		1·2
Pacific sardine	active			
Calanus plumchrus	0·4	16·9, 40·2	15·8	100
Hog pancreatic lipase	15·5	228, 15·2		15

* *Onchyrinchus keta.* ** *Trachurus symmetricus.* † *Engraulis mordax.* †† nmoles fatty acid released per gram wet tissue in 1 hr at 16°. § nmoles fatty acid released per mg protein in 1 hr at 37°.

enzyme preparation exceeded that for wax ester by 10- to 15-fold. Experiments with small chum salmon (*Onchyrinchus keta*) revealed appreciable but unexpectedly low activities of wax lipase. Preparations from pyloric caeca of anchovy and jack mackerel were more promising. The synthetic wax ester was hydrolysed at rates usually lower than those for ^{14}C-triolein. Cholesteryl oleate was hydrolysed as readily as triglyceride by preparations from both sources.

The clear intermediate layer obtained by centrifugation at 2° at 27,000 $\times$ g of homogenates prepared in 0·01 M, pH 7·0 phosphate buffer contains lipase active on wax ester, triglyceride, and cholesteryl ester. Optimal rates were obtained at pH 7·0. Activity assay involved measurement of ^{3}H or ^{14}C in the fatty acid released. Porcine pancreatic lipase, in a single experiment, released 3·5 times as much fatty acid from triglyceride as from wax ester.

Absence of wax lipase activity in copepodid stage V of C. plumchrus

Homogenates of *C. plumchrus* in buffer at pH 7·0 failed to hydrolyse synthetic wax ester whereas they hydrolysed cholesteryl oleate-^{14}C and triolein-^{14}C almost equally well. The rates for triglyceride were at least 10 fold higher than those determined by assay of activities of ^{3}H-oleate produced from the synthetic wax ester.

Copepods which accumulate wax ester can do so because their wax lipase activity is near zero until stressed by starvation. Wax lipase activity in *C. plumchrus* was measured with homogenates of the animals and compared with triglyceride lipase activity. The difference was striking. It is clear that the copepodid Stage V is practically devoid of wax lipase activity. During moulting, mating, egg production in the adult the wax reserve is completely depleted. Certainly wax lipase is active during this period in the animal's life.

In Table 7 one sees that the anchovy, like the herring and the sardine, is particularly well adapted to wax digestion. The major diet of these fishes is copepod. The chum salmon, on the other hand, which arrives from its river spawning grounds at the time that the copepod population nears its maximum oil content, is not nearly so well adapted to a wax diet. Its triglyceride lipase activity was ten times that of its wax lipase activity. The subsequent oceanic life of the salmon, unlike those of the sardine, herring and anchovy involves a carnivorous diet not restricted to copepods.

It had been anticipated that copepod wax lipase activity would be low and that salmon and anchovy pyloric caecal wax lipase would be very high. Only the first assumption proved to be correct. While anchovy wax ester lipase activity was appreciable it did not exceed that of the triglyceride lipase.

Experiments to determine fatty acid transferase capability in these lipases, especially with cholesterol as acceptor alcohol have not been done.

FATE OF WAX ESTERS IN FISHES

Digestion of wax esters in fishes must proceed rather directly in the pyloric caecum. Appearance of labelled cholesterol ester in serum lipoproteins following feeding of ^{14}C-labelled wax ester indicated action of a fatty acyl transferase acting to transfer fatty ester from wax to cholesterol. It has been shown (D. L. Puppione, unpublished) that adult salmon serum contains very high levels of cholesterol ester, up to seventy times the level normal for human high density serum lipoproteins (HDL).

No fatty alcohol or wax-ester has been found in the serum of salmon or anchovies. It must be deduced that the fatty alcohol moiety is oxidized in the lumen of the gut and converted to plasmalogen or fatty acid ester. The occurrence of plasmalogen in pyloric caecum tissue has been revealed by TLC of acid-treated phospholipids but actual turnover of fatty alcohol in this reservoir remains to be demonstrated. Identification of an extremely unsaturated (22:6) lysophosphatidyl ethanolamine lipid in *C. plumchrus* by Patton & Lee (1972) could be an indication of plasmalogen involvement in formation of the 20:1 and 22:1 alcohols of copepod wax esters.

WAX ESTER BIOSYNTHESIS

The early studies of Nevenzel & Kayama (1968) demonstrated that wax esters could be synthesized from ^{14}C-labelled acetates *in vivo*. In recent experiments Sargent, Lee & Nevenzel (1971) succeeded in producing wax ester from labelled fatty acid, by using cell-free copepod preparations. The soluble enzyme system included the fatty acid reductase to produce the necessary fatty alcohol. Addition of exogenous oleyl alcohol, however, increased the yield of wax ester two to three-fold. The experimental results revealed the expected requirement for ATP and CoA. Triglyceride production was only one tenth that of wax ester. Synthesis of wax ester by a soluble enzyme system implies that wax ester biosynthesis in the copepod could be occurring in the cytoplasm.

A remarkable aspect of wax ester synthetase in the copepod is the specificity of the fatty acid reductase. *Calanus plumchrus* contains wax esters with 20:1 and 22:1 alcohols in major quantities. The dietary lipids of the animal are the varied and highly unsaturated triglycerides of marine diatoms. Since no free alcohol is observed in copepods or predator fishes, one can anticipate that the site of specificity lies with the fatty acyl reductase or its substrate rather than with the wax synthetase. Related copepod species in deep water synthesize 16:0 alcohols in major amounts.

Ultrastructural examination of the oil sac of *C. plumchrus* reveals that the sac is actually a multiple layer of cells. Since the sac contains nearly pure wax ester there was reason to suspect that the sac and its cell membranes contained enzyme systems which produced was ester (J. Kalley & T. Bisalputra, personal communication).

"FOSSIL" WAX FROM COPEPODS

Wax esters remaining after biological degradation provide information on microbial selectivities for the variety of wax esters in copepods. In our recent research expedition in British Columbia the waxes of the predominant native copepod, *Calanus plumchrus*, were compared with submarine deposits of wax which had baffled previous investigators. In a fjord known as Bute Inlet there occasionally accumulates copious amounts of wax, apparently rising from underwater deposits. Nevenzel, Lee & Benson (1971) analysed "Bute Inlet Wax" and found it much more saturated than that of its possible local source, *Calanus plumchrus* wax. In Table 8 we reproduce the analyses of the submarine deposit and that of the copepod stage accumulating most wax. Predominance of C_{34} and C_{36} wax esters is apparent. The alcohol and acid moieties, however, showed less similarity (Table 9).

It is clear that the polyunsaturated acids of *C. plumchrus* waxes have been degraded by bacterial action in the sediments or bottom waters to which they had settled after a massive kill of copepods by unrecorded environmental, factors. Unlike laboratory-reared copepods and certain other species. *C. plumchrus*, synthesizes no polyunsaturated fatty alcohol in spite of the dietary availability of the polyunsaturated acids of the diatoms upon which these animals graze. The alcohols of the copepod are similar to those of Bute Inlet wax but the acid

Table 8. *Comparison of Bute Inlet wax with that of local* Calanus plumchrus

Carbon number	Bute Inlet wax 1960	Bute Inlet wax 1971	*Calanus plumchrus* Copepodid V
30	6·5	7·8	8·2
32	5·7	5·5	11·9
34	34·2	35·5	22·2
36	31·2	31·1	28·2
38	10·5	8·5	14·0
40	4·9	2·8	6·2
42	3·4	4·1	5·3

Table 9. *Wax esters of Bute Inlet and* C. plumchrus

	Alcohols		Fatty acids	
Component	Bute Inlet	*C. plumchrus*	Bute Inlet	*C. plumchrus*
14:0	0·9	3·0	53·2	22·1
16:0	3·2	11·4	17·5	16·4
16:1	8·0	8·8	6·1	8·3
18:1	3·8	4·3	2·9	0·2
20:1	45·3	37·8	8·8	3·2
22:1	37·4	31·9	7·3	—
C_{16}–C_{22} polyunsat.	—	0·7	—	4·1

moieties differ greatly. It is then seen that the 53% of 14:0 acid in Bute Inlet wax must be derived from the 22% of 14:0 acid in the copepod as a result of bacterial destruction of the 44% of polyunsaturated acid components. The disparity in C_{14} to C_{16} ratios of the marine and animal waxes must be attributed to bacterial oxidative selectivity and preference for the 16:0 acid moiety.

These bacterial degradative specificities are no less remarkable than those involved in biosynthesis of the fatty alcohol of *C. plumchrus* wax. The predominant 20:1 and 22:1 alcohols may even be related to the relative absence of those chains in the fatty acid complement of *C. plumchrus* wax.

ACKNOWLEDGEMENTS

This work was supported by the U.S. National Science Foundation (Grant GB 24834) and by the Foundation for Ocean Research, San Diego Helpful collaboration of Elizabeth Baker and Professor Stuart Patton is appreciated These studies were facilitated by the Department of Fisheries and Forestry of Canada and the Fisheries Research Board of Canada.

REFERENCES

Lee, R. F., Hirota, J. & Barnett, A. M. (1971). *Deep Sea Res.* (In press.)

Lee, R. F., Nevenzel, J. C. & Paffenhöffer, G.-A. (1970). *Science* **167**, 1508.

Lee, R. F., Nevenzel, J. C. & Paffenhöffer, G.-A. (1971). *Mar. Biol.* **9**, 99.

Lee, R. F., Nevenzel, J. C., Paffenhöffer, G.-A. & Benson, A. A. (1970). *J. Lipid Res.* **11**, 237.

Lee, R. F., Hirota, J., Nevenzel, J. C., Sauerheber, R., Lewis, A. & Benson, A. A. (1972). *CALCOFI* (California Cooperative Fisheries Institute) *Reports* **16**. (In press.)

Nevenzel, J. C. (1970). *Lipids* **5**, 308.

Nevenzel, J. C. & Kayama, M. (1968). *Fedn Proc. Fedn Am. Socs exp. Biol.* **26**, 647.

Nevenzel, J. C., Lee, R. F. & Benson, A. A. (1971). *Am. chem. Soc. Abstracts*, p. 34.

Nevenzel, J. C., Rodegker, W. & Mead, J. F. (1965). *Biochem.* **4**, 1589.

Nevenzel, J. C., Rodegker, W., Mead, J. F. & Gordon, M. S. (1966). *Science* **152**, 1753.

Nevenzel, J. C., Rodegker, W., Robinson, J. S. & Kayama, M. (1969). *Comp. Biochem. Physiol.* **31**, 25.

Patton, S. & Lee, R. F. (1972). (In press.)

Sargent, J. R., Lee, R. F. & Nevenzel, J. C. (1971). (In press.)

Schmidt-Nielson, K. (1964). In *Handbook of Physiology. Sec. 4, Adaptation to the Environment* (Ed. by Dill, D. B.), pp. 493–507. Am. Physiol. Soc. Washington, D.C.

PHOSPHOLIPASE A AND ITS ZYMOGEN FROM PORCINE PANCREAS

By P. P. M. BONSEN, W. A. PIETERSON, J. J. VOLWERK and G. H. de HAAS

Biochemistry Laboratory, State University of Utrecht, Vondellaan, Utrecht, The Netherlands

INTRODUCTION

Porcine pancreatic phospholipase A_2 (E.C. 3.1.1.4) catalyses the specific hydrolysis of fatty acid ester bonds at the 2-position of 1,2-diacyl-*sn*-phosphoglycerides (van Deenen & de Haas, 1963). The enzyme is secreted by the pancreas as an enzymatically inactive zymogen, which by limited tryptic proteolysis is converted to the active enzyme. The zymogen and active enzyme are single chain proteins with a molecular weight of around 14,000 and are characterized by a high content of intra-molecular disulphide bridges. In the activation process the *N*-terminal heptapeptide of the zymogen chain is split off. The complete amino acid sequence and the position of the six disulphide bonds has been reported (de Haas *et al.*, 1970*a*,*b*).

The enzyme has an absolute requirement for calcium ions in the assay procedure used, namely, an aqueous emulsion of egg-yolk in the presence of sodium deoxycholate. This system, however, is too complicated for an investigation of substrate binding and the role of metal ions in the catalysis. On the other hand, bile salts are absolutely required in the hydrolysis of long-chain lecithins. Aqueous dispersions of such lecithins consist of large lamellar structures, called liposomes, which are not attacked by the enzyme in the absence of bile salts. The bile salts break the tightly packed lipid bilayer structure down to small mixed micelles and rapid hydrolysis by phospholipase A then occurs. Apparently the action of the enzyme is controlled by specific properties of the lipid–water interface. For further characterization of the enzyme with respect to its physicochemical requirements for organized lipid surfaces, the function of metal ions and amino acid residues involved in the catalytic breakdown, use has been made of shorter-chain lecithin analogues. These compounds are water soluble and produce clear micellar solutions which are rapidly hydrolysed by the enzyme in the absence of detergents. By varying the fatty acyl chain length of the lecithin, different lipid–water interfaces can be offered to the enzyme and the influence on activity can be studied. Moreover, by introducing minor chemical modifications in these short-chain lecithins, it is possible to prepare water-soluble substrate analogues which are not hydrolysed by the enzyme and whose inhibitory properties can be investigated by kinetic studies.

MATERIALS AND METHODS

Porcine pancreatic phospholipase A was obtained by activation of the pure zymogen with trypsin as described previously (de Haas *et al.*, 1968*a*). 3-*sn*-Phosphatidylcholines (L-α-lecithins) with two identical acyl chains varying in chain length between pentanoic acid (C_5) and dodecanoic acid (C_{12}) were prepared as described by Cubero Robles and his associates; (Cubero Robles & de Iongh, 1967; Cubero Robles & van den Berg, 1969). 1-*sn*-Phosphatidylcholines (D-α-lecithins) and "lecithins" containing an acylamide bond instead of the 2-acyl ester function were prepared as described by Bonsen *et al.* (1972). *p*-Bromophenacylbromide (analytical grade) was obtained from Fluka, A. G. Methylacetimidate. HCl was prepared according to Ludwig & Hunter (1967). Enzymic hydrolysis of the synthetic lecithins was followed by continuous titration of the liberated fatty acids at constant pH (between pH 5 and 10) at 40° with a Radiometer pH-stat equipment. The assay system was standardized to 4 ml total volume and contained lecithin, $CaCl_2$ and NaCl in varying amounts. Depending upon the pH, 0·5 mM Tris or acetate buffer was used. Lyophilized enzyme, dissolved in distilled water, was added in aliquots of 10–100 μl. Titration with 0·04 M NaOH was carried out under nitrogen. In the pH range used, non-enzymatic hydrolysis of the substrate was negligible. Initial rates of hydrolysis were measured under conditions where a good proportionality between enzyme concentration and activity existed. Analysis of the reaction products showed the formation of lysolecithin and fatty acid only. At pH values below 5·0, the pH-stat technique cannot be used to measure enzyme activity and the hydroxamate method as modified by Augustyn & Elliott (1969) was applied. Specific activity of the enzyme is expressed as the number of μmoles of fatty acid released/min/mg of protein (dry weight). Details on the kinetic analysis of the enzyme by the monolayer technique were described recently (Zografi, Verger & de Haas, 1971).

RESULTS AND DISCUSSION

Activity of phospholipase A with short-chain lecithins as substrates

The activity of phospholipase A was examined with short-chain lecithins. These lecithins (diheptanoyllecithin and lower homologues) give optically clear solutions in water. Micelle formation occurs at low concentrations of dihexanoyllecithin and diheptanoyllecithin (13·8 mM and 2 mM, respectively). Up to about 0·6 mM dioctanoyllecithin also gives a clear micellar solution, but at higher concentrations a phase separation occurs (coacervate) and no stable dispersion can be obtained. Dinonanoyllecithin behaves like the naturally occurring long-chain lecithins; it is insoluble in water and produces a stable aqueous dispersion of the liposomal type.

Figure 1 shows the influence of a gradually changing lipid–water interface on enzyme activity. Large differences in enzyme activity are found for these chemically very related substrates. It is difficult to explain these dramatic differences in the rates of enzymic breakdown on the basis of the small differences in chemical composition of the substrates. Apparently the differences in architecture of the micellar structures of these lecithins have very significant bearing on either binding of the enzyme (K_m) or on subsequent reaction steps during catalysis (V_{max}). Assuming a classical acylation–deacylation mechanism for this special hydrolase, we have to conclude that at least the rate-limiting step must be strongly influenced by one (or more) of the parameters of the interface. Support for the idea that the relatively low activity of

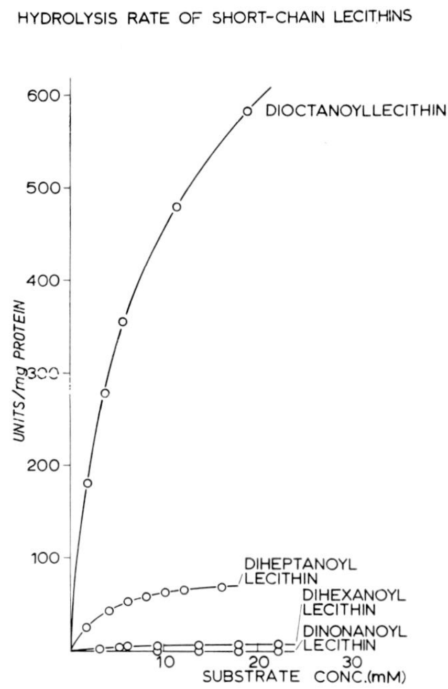

FIG. 1. Michaelis curves for the hydrolysis of dinonanoyl-, dioctanoyl-, diheptanoyl- and dihexanoyllecithin by phospholipase A. Assay conditions: pH 7·0, 0·1 M NaCl, 0·5 mM Tris and 1 mM $CaCl_2$. Substrate concentrations are expressed in mmoles/l of substrate present as micelles or liposomes.

the enzyme against the dihexanoyl- and diheptanoyllecithin compared with the dioctanoyl homologue is caused mainly by physico-chemical differences of the interface can be derived from the finding that high ionic strength of the medium greatly enhances the reaction velocities. Addition of large amounts of NaCl (up to 4 M) to aqueous solutions of such short-chain lecithins markedly increases the activity of the enzyme against these substrates (Table 1).

Ultra-sedimentation and light-scattering studies on aqueous solutions of these substrates in the presence or absence of increased NaCl concentrations did not reveal significant differences in the micellar structures with respect to the micellar size and form or number of molecules per micelle. However, such increased salt concentrations will undoubtedly lead to subtle changes in the fine architecture of the lipid–water interface. The water structure surrounding the micellar surface and charge–charge interactions involving the polar head-groups of the lecithins will change and such changes are believed to induce alterations in orientation or conformation of the substrate molecules at the

Table 1. *Maximal velocities (μmoles of fatty acid released/min/mg of protein) of phospholipase A with short chain lecithins as substrate at different ionic strengths*

Substrate	V_{max}	
	0·1 M NaCl	3·9 M NaCl
Dihexanoyllecithin	10	400
Diheptanoyllecithin	45	800
Dioctanoyllecithin	820	5000

Assay conditions: pH 7·0, 1 mM $CaCl_2$, 0·5 mM Tris.

interface which might be responsible for the observed enhanced enzymic activity. A technique for inducing changes in orientations of molecules in an interface is spreading the compound as a monomolecular layer at the air–water interface of a Langmuir trough. By changing the surface pressure π, that is to say the compression state of the film, one can change continuously the surface area per molecule and its orientation at the air–water interface. Such surface technique is useful for kinetic studies of a soluble enzyme if (*a*) the substrate is insoluble in water and gives a stable film at the air–water interface, (*b*) the products of the enzymatic reaction are freely water soluble and diffuse away into the subphase. During the enzymatic reaction substrate disappears, and by diminishing the surface available to the remaining substrate molecules, one can keep the surface pressure π (= surface density) constant. The enzymatic reaction then follows pseudo first-order kinetics and its rate constant k can be determined. Figure 2 shows how the rate constant of the enzymatic reaction depends upon the surface pressure π of the film, in this case dioctanoyllecithin.

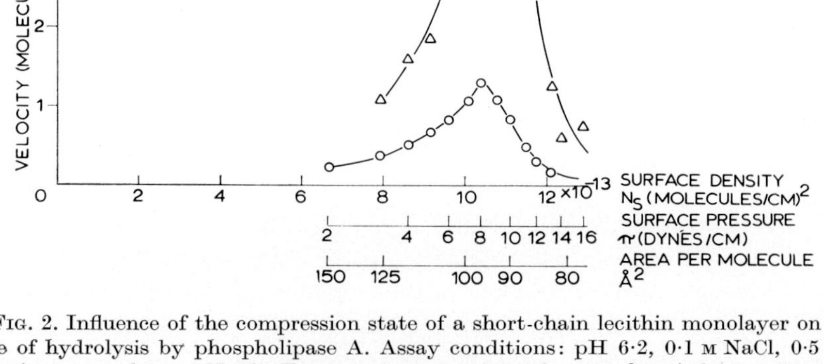

FIG. 2. Influence of the compression state of a short-chain lecithin monolayer on the rate of hydrolysis by phospholipase A. Assay conditions: pH 6·2, 0·1 M NaCl, 0·5 mM Tris-Acetate and 1 mM $NaCl_2$. Enzyme concentration 5·7 × 10^{-6} mg/ml (○—○) without stirring; (△—△) with stirring of the subphase.

There is a pronounced optimum in the hydrolysis of the lecithin film at 8 dynes/cm. Surface films of substrates like didecanoyl-(C_{10}) and didodecanoyllecithin (C_{12}), which are not hydrolysed by the enzyme in bulk, because of their liposomal structures, are hydrolysed at the same rate as the dioctanoyllecithin and again maximal hydrolysis is observed at about 8 dynes surface pressure. Our tentative conclusion is that at a surface pressure of 8 dynes/cm the lecithin molecule reaches an orientation or conformation which is optimal for enzyme binding or catalysis. Work is in progress for the determination by the monolayer technique of the Michaelis constants, V_{max} and K_m in order to find out the parameter which is influenced by the orientation of the substrate molecule.

Influence of metal ions on the hydrolysis of short-chain lecithins by phospholipase A

Under physiological conditions, that is to say in the assay systems containing long-chain phospholipids in the presence of bile salts, the pancreatic enzyme is completely inactive if Ca^{2+} is omitted. Figure 3 shows that the enzyme has

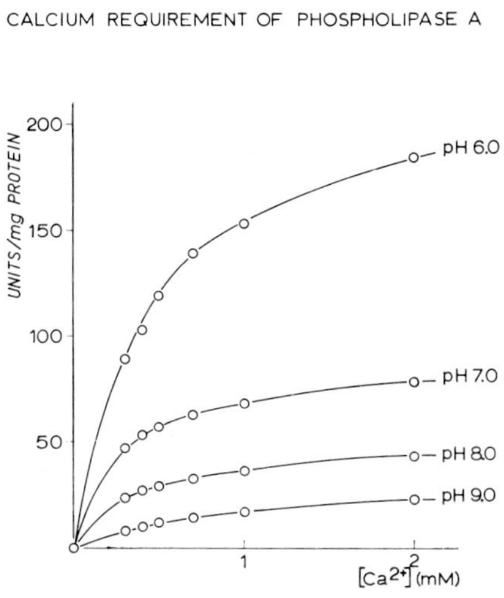

FIG. 3. Effect of pH on the requirement for Ca^{2+} in the hydrolysis of diheptanoyllecithin with phospholipase A. Assay conditions: 0·5 mM Tris, 0·1 M NaCl, 14 mM Diheptanoyllecithin and varying $CaCl_2$ concentrations.

an absolute requirement for Ca^{2+} in the assays with water soluble short-chain lecithins also.

At all pH values studied, at least 2 mM Ca^{2+} is required for maximal velocity, suggesting a role for this metal ion in the fixation of the enzyme to its substrate, in the reaction steps of the catalytic process itself, or in inducing the active conformation of the enzyme. The requirement of the enzyme for Ca^{2+} is highly specific—the related ions Ba^{2+} and Sr^{2+} turned out to be pure competitive inhibitors for the enzyme, whereas Mg^{2+} was fully inert. Direct binding studies by equilibrium dialysis between the enzyme and Ca^{2+} showed that phospholipase A is able to bind one metal ion per enzyme molecule and this binding gives rise to a conformational change in the protein which can be detected by ultraviolet difference spectroscopy at 242 nm. In agreement with the kinetic results, Ba^{2+} and Sr^{2+} are also bound to the enzyme and they also show the optical change at 242 nm. Mg^{2+}, which is kinetically inert, does not produce any change in absorption at this wavelength. Neither the kinetics nor the features of the ultraviolet difference spectrum of the enzyme with Ca^{2+} are influenced by the initial presence of Mg^{2+} which indicates that the calcium-binding site is unable to bind Mg^{2+}.

In order to investigate more closely the pathway leading to the formation of the active enzyme–calcium–substrate complex, initial reaction rates of the enzyme with a short-chain lecithin as substrate were measured as a function of varying substrate and calcium concentrations. Both double reciprocal plots of enzyme activity *versus* Ca^{2+} concentration and *versus* substrate concentration showed straight lines with common intercepts on the X axis. In this special case (random mechanism) the dissociation constants $K_{Ca}{}^{2+}$ of the enzyme–metal complex and K_S of the enzyme–substrate complex can be calculated. Moreover, it indicates that the active enzyme–calcium–lecithin complex can be formed by two mutually independent pathways as shown below:

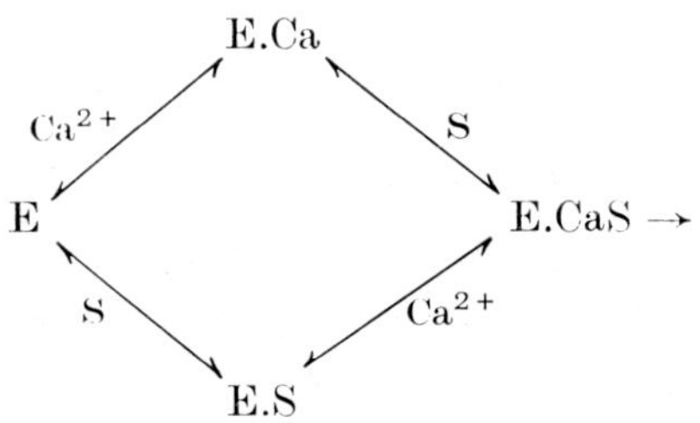

From the double reciprocal plots of velocity against Ca^{2+} and substrate concentration one can calculate the velocity of the enzyme at infinite Ca^{2+} and substrate concentrations (V_{max}). Figure 4 shows how the maximal velocity depends on pH.

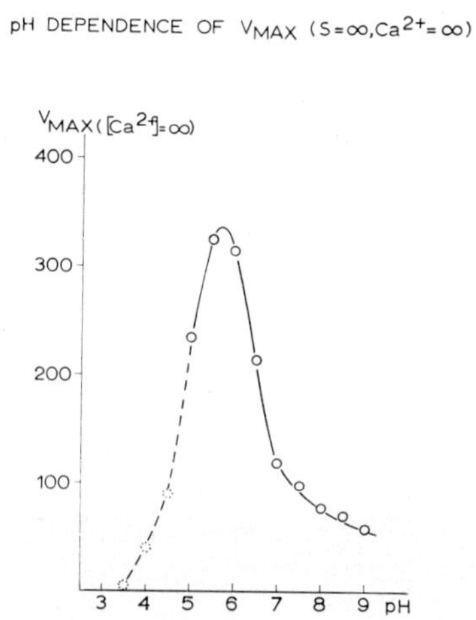

FIG. 4. pH optimum of pancreatic phospholipase A hydrolysis of diheptanoyllecithin. Maximal velocities were obtained from Lineweaver–Burk plots at infinite substrate and Ca^{2+} concentrations. The values below pH 5 were obtained by the hydroxamate method at a fixed high Ca^{2+} concentration.

With short-chain lecithins as substrate and in absence of detergents the enzyme has a pronounced pH optimum between pH 5·5 and pH 6·0. The same pH optimum of about pH 6·0 was also obtained by the monolayer technique with dioctanoyllecithin as substrate (Zografi *et al.*, 1971). From this result one might speculate that the decomposition of the enzyme–calcium–substrate complex is controlled by at least two active-site amino acids with pK values around 4·8 and 6·5.

An indication that carboxyl groups might be involved in the catalytic centre of phospholipase A was obtained by inactivation studies of the enzyme with

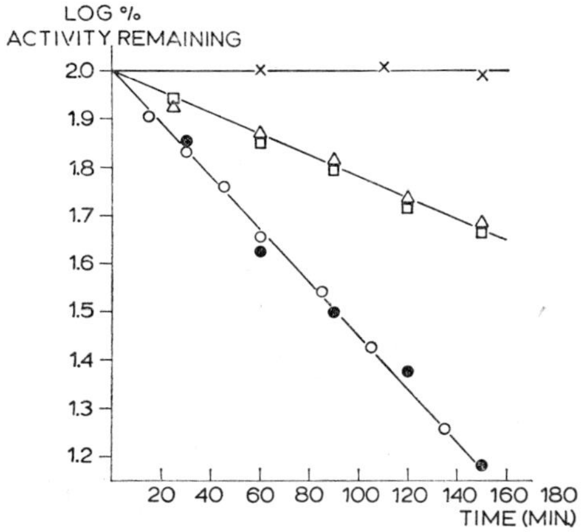

FIG. 5. Loss of phospholipase A activity as a function of time during incubation with *p*-bromophenacylbromide at 30°. Phospholipase (1 mg/ml) was treated with 50 fold molar excess of *p*-bromophenacylbromide in a solution containing 0·1 M CaCl–0·1 M cacodylate-HCl (pH 6·0). Phospholipase A activities (X—X) in the absence of inhibitor; (□—□) in the presence of inhibitor and 5 mM of Ca^{2+}; (△—△) in the presence of inhibitor and 5 mM of Ba^{2+}; (○—○) in the presence of inhibitor and absence of metal ions; (●—●) in the presence of inhibitor and 50 mM of Mg^{2+}.

p-bromophenacylbromide. This reagent has been used to demonstrate the presence of a carboxyl group close to the active centre of pepsin (Erlanger *et al.*, 1967).

As shown in Fig. 5, incubation of phospholipase A with a small excess of *p*-bromophenacylbromide (50 moles per mole of enzyme) caused a rapid first-order decrease in activity which was dependent on pH. After incubation for 6 hr at pH 6·0 no enzyme activity was left and chemical analysis of the modified enzyme showed an incorporation of 1·0 to 1·2 Moles of reagent per mole of enzyme. Increasing protection against inactivation by the reagent was observed in the presence of increasing concentrations of Ca^{2+} or Ba^{2+}, whereas Mg^{2+} gave no protection at all.

The general electrophilic character of the *p*-bromophenacylbromide makes it impossible to draw conclusions about the type of amino acid residues modified by the reagent. Although efficient protection by Ca^{2+} (and Ba^{2+}) against inactivation points to a probable involvement of carboxyl group(s), the study of the inactivation reaction as a function of the pH (see Fig. 6) seems to indicate that the pK of the modified amino acid is close to 6·2. Therefore, other amino acids such as histidine cannot be precluded *a priori*.

For the identification of the amino acid(s) which are modified by the reagent, work is in progress to prepare ^{14}C-labelled *p*-bromophenacylbromide and to isolate the radioactive peptide after proteolysis. However, the fact that one residue seems to be incorporated into the enzyme (indicating that possibly non-functional residues are also partially modified) makes it necessary to apply differential labelling techniques, and competitive inhibitors may also be required. From previous work we know the minimal substrate requirements of the enzyme: one fatty acid ester bond has to be present in a position adjacent to the alcohol–phosphate ester function and the acyl chain should occupy a

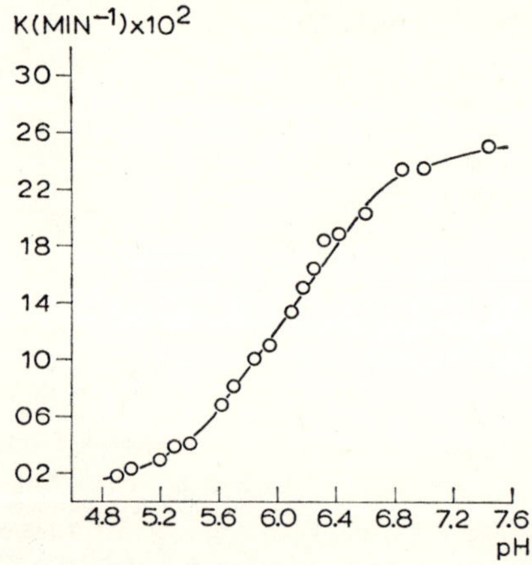

FIG. 6. Dependence of the pseudo first-order rate constant of inactivation (k) of phospholipase A by *p*-bromophenacylbromide on pH. A cacodylate-HCl buffer 0·1 M was used. For incubation conditions, see Fig. 5.

well-defined stereochemical orientation in space (de Haas *et al.*, 1968*b*). Therefore, modification of this part of the lecithin structure, which is supposed to be important either for binding with phospholipase A or in the catalytic reactions, could possibly give phospholipid analogues with inhibitory properties. Indeed, two classes of water-soluble pure competitive inhibitors were obtained: short-chain lecithins of the opposite stereochemical configuration, the 2,3-diacyl-1-*sn*-phosphatidylcholine and short-chain "lecithins" in which the susceptible ester bond had been replaced by an acylamido linkage.

As shown in Fig. 7, the D-isomer of dioctanoyllecithin behaves as a pure competitive inhibitor for the enzyme in assays with L-dioctanoyl-lecithin as substrate. In spite of their rather weak inhibition constants, these stereoisomeric D-lecithins are very useful inhibitors because of the fully identical affinity of stereoisomeric lecithins to the enzyme ($K_I = K_S$). Apparently the stereochemical specificity of the enzyme with respect to configuration of its substrate is an expression of catalytic specificity and not of binding specificity.

Scheme 1 shows the chemical synthesis of the other class of competitive inhibitors, the so-called "Amide-lecithins". The "amide-lecithins" are up till now the most potent competitive inhibitors we have prepared. Their affinity constants to the enzyme are 10–20 times higher than those of the corresponding lecithin substrates with the same acyl chain length.

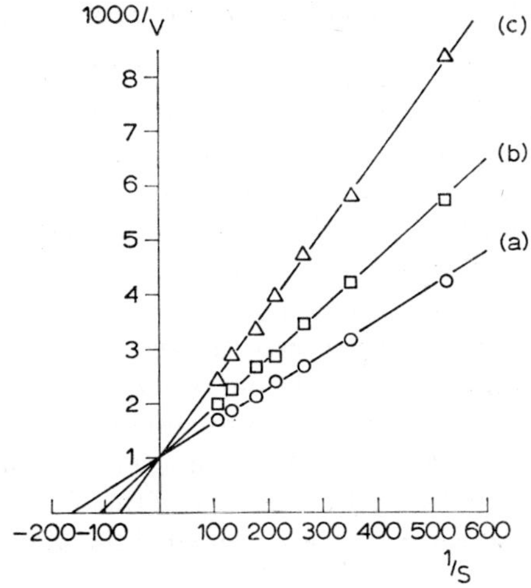

FIG. 7. Lineweaver–Burk plots for the hydrolysis of L-α dioctanoyllecithin by phospholipase A in 0·1 M NaCl (curve a). Curve b: inhibition with 1·9 mM D-α-dioctanoyllecithin. Curve c: inhibition with 4·75 mM D-α-dioctanoyllecithin. $K_I = K_S = 6$ mM. Assay conditions: pH 7·0, 40°, 0·1 M NaCl and 1 mM $CaCl_2$.

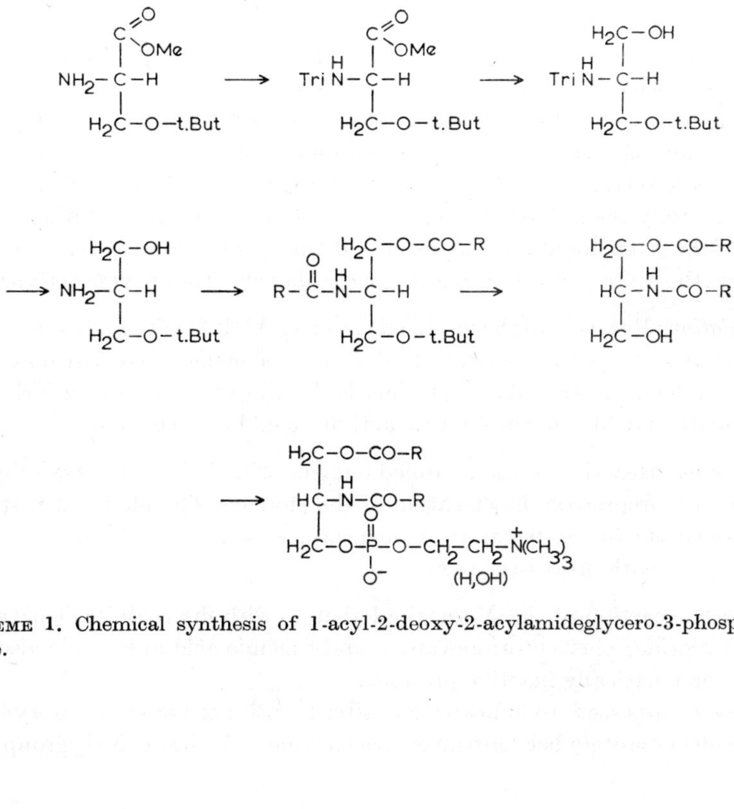

SCHEME 1. Chemical synthesis of 1-acyl-2-deoxy-2-acylamideglycero-3-phosphorylcholine.

Role of the N*-terminal residue in enzymatic activity*

Limited proteolysis of the zymogen of phospholipase A by trypsin splits off the *N*-terminal heptapeptide of the precursor and liberates the α-NH_2 group of *N*-terminal amino acid alanine of the active enzyme.

The important function of this α-NH_2 group in protonated form for enzyme activity could be demonstrated by chemical modification reactions. However, in order to study specifically the α-NH_2 group of the active enzyme, the reactive ε-NH_2 functions of the lysine residues had to be protected first. Therefore, the inactive zymogen was treated with methyl-acetimidate-HCl which converts, under mild conditions, the zymogen into a derivative in which all lysine groups are transformed into amidino-lysine residues (Ludwig & Hunter, 1967). The obtained modified zymogen can still be converted into an active phospholipase A by tryptic cleavage of the *N*-terminal heptapeptide pyroglu-glu-gly-ile-ser-ser-arg. Table 2 compares the kinetic constants of the

Table 2. *Kinetic constants of the native phospholipase A* (P_A) *and of its* ε-NH_2 *amidinated derivative (AMPA). Substrate: Diheptanoyllecithin*

Enzyme	K_S (mM)	V_{max}	$K_{Ca^{2+}}$ (mM)	pH optimum
P_A	3·9	100	4·0	6·0
AMPA	6·7	71	6·0	6·0

native phospholipase A (P_A) with the modified enzyme (AMPA) having all its lysine groups converted into amidinolysine residues.

It can be concluded from Table 2 that the specific chemical transformation of lysine into the much less reactive amidinolysine groups has not seriously changed the tertiary structure of the enzyme. With this modified enzyme we can now study the influence on enzyme activity of specific modification reactions of the *N*-terminal alanine residue. It turned out that any reaction which removes the proton from this α-NH_3^+ group abolishes the enzyme activity, e.g.:

Maleylation: Reaction of the modified enzyme (AMPA) with maleic acid anhydride shows a rapid incorporation of 1 mole of maleic acid and no enzyme activity is left. Under acidic conditions deblocking of the alanine α-NH_2 group takes place and 75% of the enzyme activity could be recovered.

Acetylation: Reaction of the modified enzyme with ^{14}C-acetic acid anhydride results in a progressive inactivation of the enzyme. The identical first-order rate constants found by activity measurements and by ^{14}C-incorporation studies that both processes are related.

Finally, reaction of the *N*-terminal alanine with fluorodinitrobenzene and dansylchloride, phenylisothiocyanate and succinic acid anhydride also gives rise to enzymatically inactive proteins.

These results seem to indicate that after tryptic activation of the zymogen, the resulting protein becomes an active enzyme if the free α-NH_2 group of the

N-terminal alanine becomes protonated. By analogy with the results obtained with chymotrypsin, one could speculate that creation of the active centre of the enzyme is related to a conformational change in the protein caused by a shift of the protonated *N*-terminal NH_3^+ group from an exposed position into the interior of the molecule. Such a shift of the *N*-terminal ala-residue during tryptic activation of the zymogen seems to be supported by the finding that during the conversion of the inactive zymogen into the active phospholipase A molecule there is a dramatic decrease in tryptophan fluorescence. Tryptic activation of the zymogen as well as the decrease in fluorescence are, under conditions of the experiment, both first-order processes showing exactly identical rate constants. As the enzyme contains only one tryptophan residue in a position very close to the *N*-terminal alanine residue (position 3), fluorescence is a very sensitive tool for study of changes in the environment of the *N*-terminal amino acid. X-ray studies are in progress and it is hoped that the various speculations on the mechanism of activation will soon find some support from the tertiary structure of the enzyme and its precursor.

SUMMARY

The kinetic behaviour of pancreatic phospholipase A_2 (E.C. 3.1.1.4) has been investigated with water-soluble short-chain lecithins as substrate. The physico-chemical state of the substrate molecules forming an organized lipid–water interface greatly influences the activity of the enzymes. The enzyme has an absolute requirement for Ca^{2+} and a random mechanism, in which the metal ion and substrate combine independently of each other with the enzyme, is proposed. Two classes of water-soluble short-chain lecithins were found to be pure competitive inhibitors of the enzyme: lecithins of the stereochemically opposite configuration (D-α-lecithins) and "lecithins" in which the 2-acyl ester bond is replaced by an acylamido linkage. *p*-Bromophenacylbromide was found to inhibit the enzyme irreversibly, probably by reacting with one or more functional amino acid residues. The important role in activity of the *N*-terminal alanine and the mechanism of activation of the zymogen are discussed.

ACKNOWLEDGEMENT

The authors are much indebted to Mrs G. J. Burbach-Westerhuis and Mrs M. C. M. H. de Waal-Bouman for their skilful technical assistance. Thanks are due to Professor G. Zografi (Ann Arbor, U.S.A.) and Dr R. Verger (Marseille, France) for the information on monolayer experiments and for many stimulating discussions. Collaboration with Professor M. Lazdunski and Dr J. P. Abita in experiments on the activation kinetics of the zymogen is greatly appreciated. Thanks are due to Mr D. Ottevanger for his contributions to the inhibition studies of the enzyme. The authors thank Drs R. J. M. Tausk (Van't Hoff Laboratory, State University, Utrecht) for giving information on the ultracentrifugation runs and light scattering experiments prior to publication.

REFERENCES

Augustyn, J. & Elliott, W. B. (1969). *Anal. Biochem.* **31**, 246.

Bonsen, P. P. M., Burbach-Westerhuis, G. J., Haas, G. H. de & Deenen, L. L. M. van (1972). *Chem. Phys. Lipids* **8**, 199.

Cubero Robles, E. & Berg, D. van den (1969). *Biochim. biophys. Acta* **187,** 520.

Cubero Robles, E. & Iongh, H. de (1967). *Rec. Trav. Chim.* **86,** 762.

Deenen, L. L. M. van & Haas, G. H. de (1963). *Biochim. biophys. Acta* **70,** 538.

Erlanger, B. F., Vratsanos, S. M., Wassermann, N. & Cooper, A. G. (1967). *Biochem. biophys. Res. Comm.* **28,** 203.

Haas, G. H. de, Postema, N. M., Nieuwenhuizen, W. & Deenen, L. L. M. van (1968*a*). *Biochim. biophys. Acta* **159,** 118.

Haas, G. H. de, Postema, N. M., Nieuwenhuizen, W. & Deenen, L. L. M. van (1968*b*). *Biochim. biophys. Acta* **159,** 103.

Haas, G. H. de, Slotboom, A. J., Bonsen, P. P. M., Nieuwenhuizen, W., Deenen, L. L. M. van, Maroux, S., Puigserver, A. & Desnuelle, P. (1970*a*). *Biochim. biophys. Acta* **221,** 31.

Haas, G. H. de, Slotboom, A. J., Bonsen, P. P. M., Nieuwenhuizen, W., Deenen, L. L. M. van, Maroux, S., Dlouha, V. & Desnuelle, P. (1970*b*). *Biochim. biophys. Acta* **221,** 54.

Ludwig, M. L. & Hunter, M. J. (1967). In *Methods in Enzymology,* Vol. XI (Ed. by S. P. Colowick and N. O. Kaplan), p. 595.

Zografi, G., Verger, R. & Haas, G. H. de (1971). *Chem. Phys. Lipids* **7,** 185.

III

METABOLISM OF ISOPRENOID COMPOUNDS

POLYPRENOLS, THEIR NATURE, OCCURRENCE AND BIOLOGICAL ROLES

By R. A. MORTON

The University of Liverpool, Liverpool, England

GERANIOL (1) and farnesol (2) as well as their isomerides linalool (3) and nerolidol (4) have long been known but solanesol $C_{45}H_{73}OH$ (7) was discovered as late as 1956 (Rowland, Latimer & Giles, 1956). It is present in green tobacco leaves at 0·05% and in flue-cured tobacco at 4%, so that little destruction can occur on curing. The interesting feature of the constitution is that solanesol is a regular polyprenol:

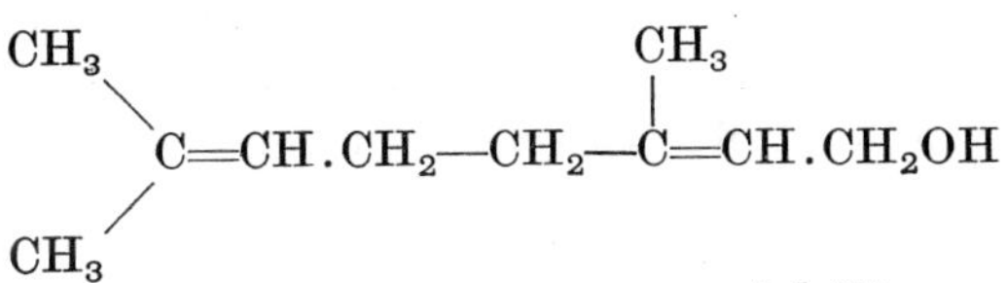

geraniol (1)

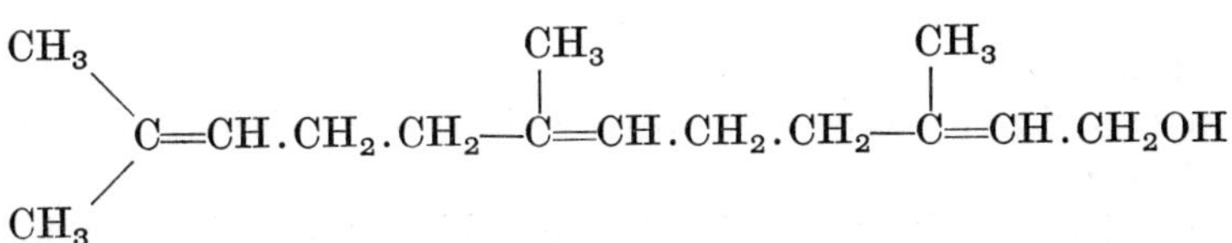

farnesol (2)

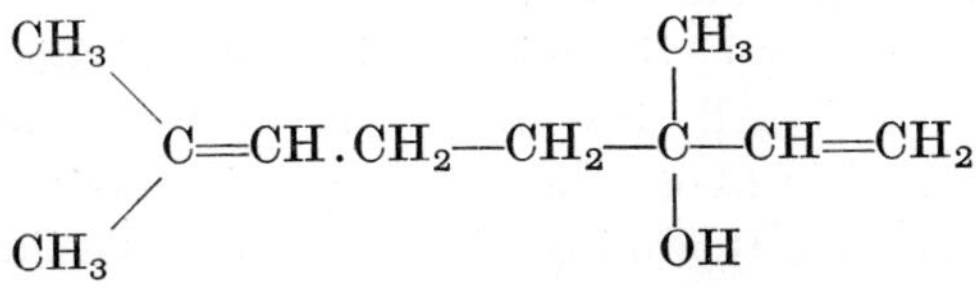

linalool (3)

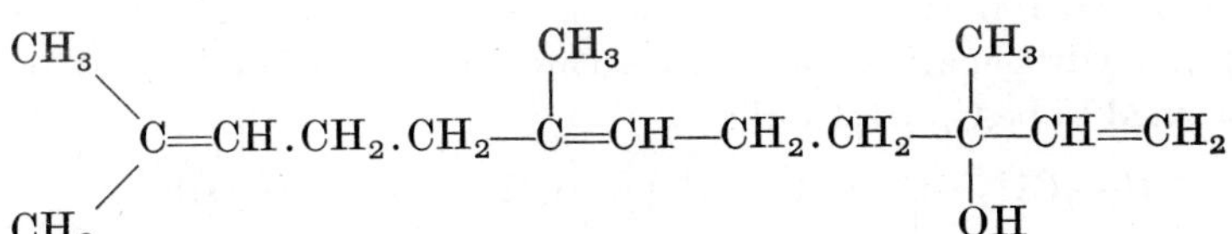

nerolidol (4)

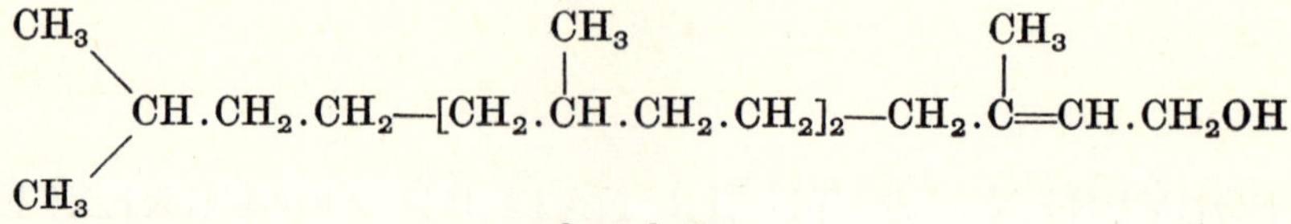

phytol (5)

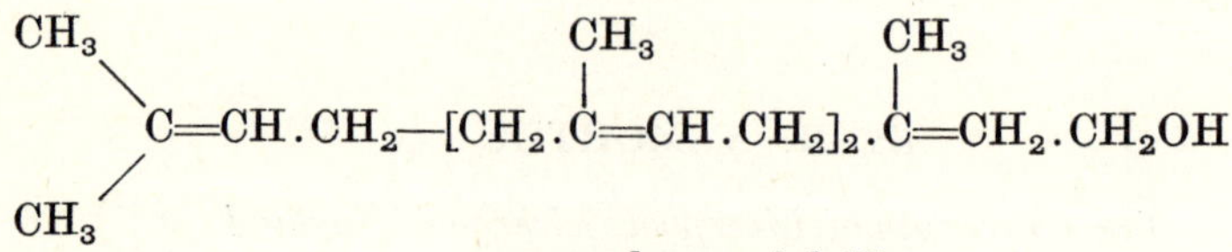

geranyl geraniol (6)

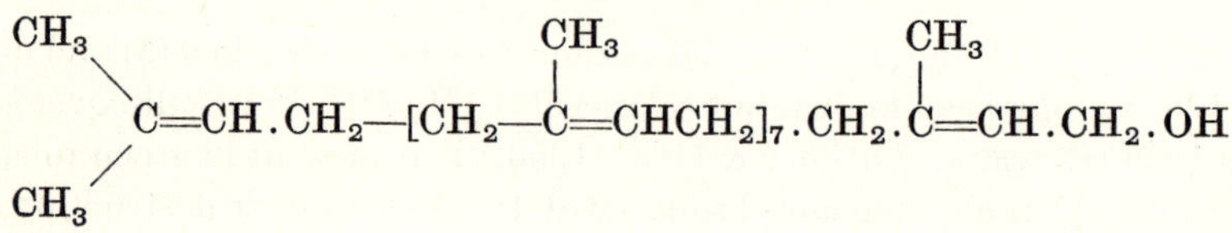

solanesol (7)

It was first regarded as a decaprenol $C_{50}H_{81}OH$ but it is in fact the all-*trans* nonaprenol (Erickson, Shunk, Trenner, Arison & Folkers, 1959; Rüegg, Gloor, Langemann, Kofler, von Planta, Ryser & Isler, 1960). Squalene $C_{30}H_{50}$ (8) has six isoprene units but the presence of a central grouping:

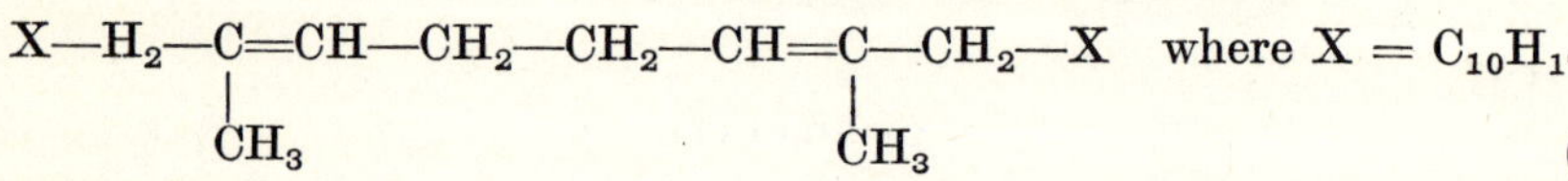

(8)

indicates the irregular attachment which makes squalene a symmetrical molecule. Phytoene, a biosynthetic precursor of carotenoids, is similarly a symmetrical molecule.

The nonaprenyl grouping occurs in plastoquinone which is itself a precursor of solanachromene.

At Liverpool while examining tissues like liver, kidney and spleen for vitamin A, chromatographic fractions of unsaponifiable matter were found to contain sterols and a diluent transparent to ultraviolet light which might have contained hydrocarbons and/or unsaturated alcohols. A large amount of kidney tissue was worked up at Liverpool in connexion with the isolation of ubichromenol and a substantial fraction, virtually free from ultraviolet absorption and made up of unsaturated alcohols, was obtained. The material was acetylated and the acetate purified (Pennock, Hemming & Morton, 1960). The infrared absorption resembled that of solanesol except that a band at 10 μm characteristic of the allylic group (=C—CH_2OH) was missing, whereas a new band (C—OH) appeared at 9·4 μm. The molecular weight, assuming homogeneity, indicated 100 carbon atoms and the substance was given the name dolichol (9). Hydrogenation indicated about 19 double bonds, and so the formula seemed to be $C_{100}H_{163}OH$:

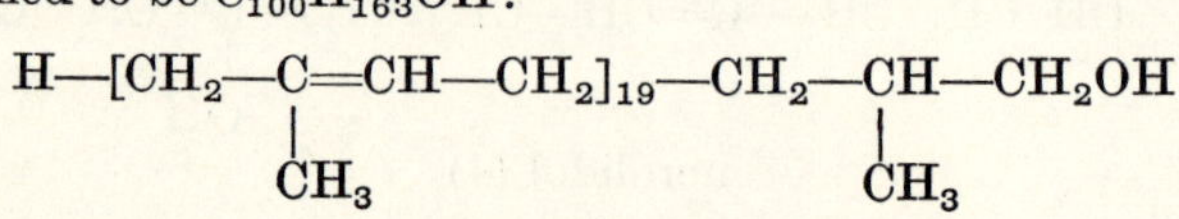

(9)

This fitted the mass spectra, cryoscopic determinations, intensity of absorption of coloured esters and a molecular weight of the ^{14}C-acetate fixed by isotope dilution.

It is interesting that as early as 1926, Channon & Marrian had obtained from liver a fraction having some resemblances to squalene. Channon, Devine & Loach (1934) returned to the topic and obtained analyses indicating $C_{45}H_{76}$ or $C_{50}H_{84}$. Dimter (1941) isolated a similar fraction and recorded a very high yield (290 mg/kg) from pig liver; he called the product hepene. It now appears that the hydrocarbon of Channon & Marrian and Dimter's "hepene" were mainly dolichol. Elementary analysis at a time when oxygen was estimated by difference could scarcely have sufficed particularly on polyprenol fractions of ill-defined purity.

Nuclear magnetic resonance spectra pointed to a structure containing 15 or 16 internal isoprene units in the *cis* configuration (Burgos *et al.*, 1963). When obtained from pig liver, dolichol was optically active (dioxan, 25°, —1·8° at 360 nm and —0·4° at 700 nm). One sample of pig liver yielded 110 mg of dolichol/kg of wet tissue. The substance, largely esterified, was fairly evenly distributed between the different cellular fractions at about 1 mg/g protein (Burgos & Morton, 1962). In later experiments rather more than half the pig liver dolichol occurred as ester. Most of the free dolichol (*ca* 77%) occurred in the mitochondrial fraction and about half the combined dolichol was in the "cell debris plus nuclei" fraction (Butterworth & Hemming, 1968). Improved methods of separation involving reversed phase chromatography on silica gel thin-layers resulted in the resolution of pig liver dolichol into at least six components: C_{85}, C_{90}, C_{95}, C_{100}, C_{105} and C_{110} (Dunphy *et al.*, 1967).

Continuous reversed phase chromatography and gas liquid chromatography (Lindgren, 1965; Wellburn & Hemming, 1966*a*) confirmed the heterogeneity. R_M values ($R_M = \log R_F - 1$) for fully hydrogenated acetates (perhydropolyprenyl acetates) plotted against chain-length were linear and for the dolichols from pig tissues (liver, kidney, pancreas and spleen) the results showed the predominant members to be C_{90}, C_{95}, C_{100} and C_{105} with smaller amounts of C_{110} and C_{115}. Gough & Hemming (1970*b*) obtained similar results for rat liver dolichols.

Yeast also contains dolichols C_{70}, C_{75}, C_{80}, C_{85} and C_{90}, and the last two alcohols are the same as the first two in the mammalian series (Dunphy *et al.*, 1966; Wellburn & Hemming, 1966*b*). It seems likely that there is a single series and that chain-lengthening consists of an increment of one *cis* isoprene unit. *Aspergillus fumigatus* mycelium yielded a mixture of seven polyprenols (C_{90}–C_{120}) with an average molecular weight corresponding with C_{110} (Packter & Glover, 1960, 1964; Butterworth, Burgos & Hemming, 1966; Stone, Butterworth & Hemming, 1967; Burgos *et al.*, 1964).

The plant "Lords and Ladies" (*Arum maculata*) is characterized by a spadix with properties of considerable biochemical interest, among them the fact that the spadix non-saponifiable fraction contains polyprenols very closely resembling those of pig tissues and human kidney. The C_{95} and C_{100} isoprenologues are the most plentiful (Hemming, Morton & Pennock, 1963). This work led Hemming and Pennock and their research students to undertake systematic searches for polyprenyl alcohols. The first raw material was the leaves of horse chestnut (*Aesculus hippocastanum*) chosen because they were known to be a good source of plastoquinone with its C_{45} side-chain and might have been expected to yield solanesol. A small amount of solanesol was indeed found but several other polyprenols were more plentiful (Wellburn *et al.*, 1967). These were designated castaprenols, written castaprenol-*n* where *n* denoted the number of isoprene (C_5H_8) units. The newer methods of separation were used and molecular weights were later obtained by mass spectroscopy. The constitutions were further studied by nuclear magnetic resonance.

The castaprenols include castaprenol-10 (C_{50}), castaprenols -11, -12 and -13 (C_{55}, C_{60} and C_{65}). As an example the mass spectrum of castaprenol-12 ($C_{60}H_{97}OH$) indicated a molecular weight of 834. The molecular ion lost water (M-18 = 8 16), then the ω terminal residue was lost (M-18–69 = 747) followed by losses of C_5H_8 units until 747 $- 9 \times 68 =$ 135 was reached (i.e. 153-18 ($C_{10}H_{16}OH—H_2O = C_{10}H_{15}$). This fragment was very prominent. The strongest low peaks were at m/e 41, 55, 69 and 81, the 69 peak corresponding with a charged fragment from the ω end of the molecule. Each castaprenol has three *trans* residues excluding the ω terminal isoprene group while the residue at the hydroxyl end of the chain is in the *cis* configuration.

A decorative rubber plant (*Ficus elasticus*) contained ficaprenols-10, -11, -12 and -13, very closely akin to the castaprenols (Stone *et al.*, 1967). A similar series of alcohols was obtained from the plant *Aglaonema robelinii* (Dunphy *et al.*, 1966). The distinction between *cis* and *trans* isoprenoid units is made largely on nuclear magnetic resonance measurements.

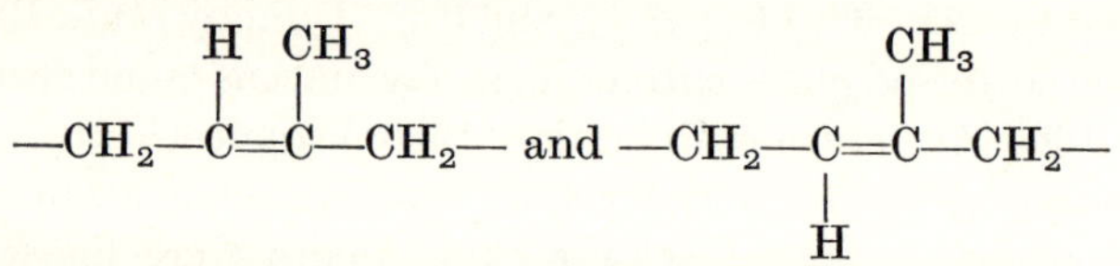

As the isomerism may be biochemically meaningful it will be useful to recall the evidence as set out in a thorough study by Feeney & Hemming (1967).

Table 1 applies to NMR spectra (at 100 Mc/sec) of polyprenols.

A more detailed analysis of the spectra in the region 8·2 to 8·5τ is based on the work of Bates & Gale (1960), Bates *et al.* (1962), Bates, Gale & Gruner (1963) who used a 60 Mc/sec instrument while Feeney & Hemming (1967) used the 100 Mc/sec apparatus.

In the case of 60 Mc/sec spectra there is incomplete separation of peaks and therefore it is difficult to estimate individual peak areas. Chen (1962) noted the superior resolution of *cis* and *trans* methyl groups in polyisoprenes when benzene was used as solvent but Feeney & Hemming (1967) found it worth while to carry out measurements in both benzene and carbon tetrachloride. In carbon tetrachloride the *cis* OH-terminal residue (c) has a peak at 8·28τ separated from the other peaks but the *trans* "OH terminal" peak would be hidden under the *cis* internal residue peak 8·34τ. In benzene the *cis* "OH-terminal" peak (8·355τ) is hidden under the *trans* internal residue peak (8·38τ) while the *trans* "OH-terminal" (8·51τ) residue is clear of the other peaks. Benzene thus allows a *trans* "OH-terminal" residue to be identified and carbon

Table 1 *NMR spectra of polyprenols*

Chemical shifts (τ)(± 0·01) In benzene	In CCl_4	Assignments
8·5 → 8·2	8·5 → 8·2	R—C(*CH*$_3$)=CH—R
7·95w, 7·87∼, 7·83, 7·80∼	8·00, 7·97	—*CH*$_2$—C(—)=
6·06, 5·90 (d)	6·10, 6·02 (d)	=CH—*CH*$_2$—OH
4·66, 4·59, 4·52 (t)	4·74, 4·66, 4·58 (t)	—C(—)=*CH*—CH_2OH

w = weak, ∼ = shoulder, d = doublet, t = triplet. The resonating proton is italicized.

Table 2. *NMR of protons of methyl groups of polyprenols in the region* 8·2–8·5 τ

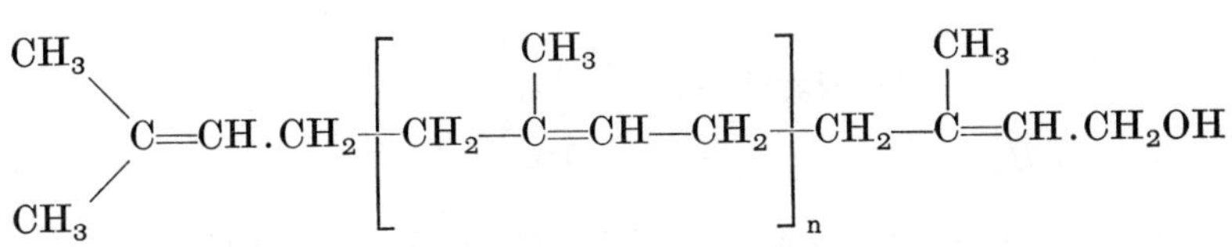

a Residue containing the methyl group	*b* In benzene *cis* (τ)	*b* In benzene *trans* (τ)	*c* In carbon tetrachloride *cis* (τ)	*c* In carbon tetrachloride *trans* (τ)
a (ω-terminal)	8·32	8·43	8·34	8·42
b (internal)	8·25	8·38	8·34	8·42
c (OH-terminal)	8·355	8·51	8·28	8·34

tetrachloride allows a *cis* "OH-terminal" residue to be recognized. With a *cis* "OH terminal" group on benzene there would be no 8·28τ peak and the relative intensities of the 8·38τ and 8·355τ peaks increase by an amount equivalent to three protons. The *cis* "OH-terminal" residue of nerol gave a peak at 8·37τ while the *trans* "OH-terminal" (8·51τ) was given a peak by solanesol and by phytol. [In solanesol the methyl groups are one *cis* and one *trans* ω-terminal, seven *trans* internal and one *trans* "OH-terminal" and four distinct peaks are shown in benzene]. For citronellol (10), examined

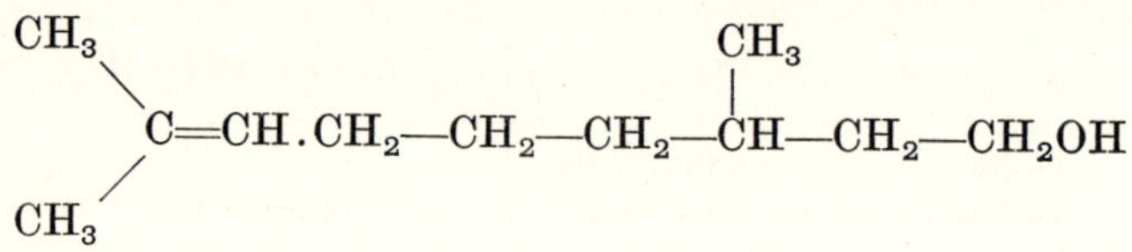

(10)

in benzene the ω-residue peaks occur at 8·32 and 8·43τ with intensities corresponding to three protons but are absent from the phytol spectrum.

In unsaturated polyprenols the doublet at 5·99 and 6·06τ is due to the α-methylene $=CH—CH_2OH$ but on hydrogenation to $—C—CH_2—CH_2OH$ a triplet appears (6·46, 6·53, 6·60τ). Citronellol and perhydropolyprenols show this triplet near 6·5τ. The natural dihydropolyprenols like dolichols and the hexahydropolyprenols of *Aspergillus* (see below) lack the 6·0τ doublet but show the 6·5τ triplet. The saturated isoprene residues also display in benzene

9·19 and 9·09τ
9·12 and 9·02τ $\quad —CH_2—CH(CH_3)—CH_2—CH_2OH$

caused by the above with

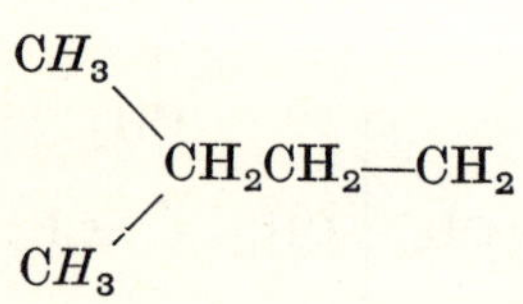

and 8·8–8·5τ for $\quad CH_2—CH(CH_3)—CH_2—CH_2—C$

The ethylenic proton in citronellol gives rise to a triplet 4·71, 4·78, 4·84τ in benzene, the splitting being caused by the protons of the adjacent $—CH_2$-group. In phytol the triplet occurs at 4·51, 4·57 and 4·63τ in benzene.

Feeney & Hemming (1967) describe the quantitative determination of the number of *cis* and *trans* isoprene residues based on the relative areas under well-resolved peaks. All leaf polyprenols studied showed 8·6τ(±0·4 S.D.) protons on methyl groups in *trans* internal isoprene residues. Each of these alcohols contains a *cis* "OH-terminal" isoprene residue. It is possible but not certain that the three *trans* internal residues are adjacent to the ω-terminal residue with all *trans* geranylgeranyl-pyrophosphate. Perhaps the best method of determining chain length involves the perhydropolyprenyl acetates already mentioned. Here of course *cis-trans*-differences are abolished but the linear relationship in gas–liquid chromatography between log retention time and the number of isoprene units enables fractions to be monitored.

Lindgren (1964) isolated from birch wood (*Betula verrucosa*) an alcohol fraction at first regarded as $C_{40}H_{65}OH$ but later found to be a mixture of C_{30}, C_{35}, C_{40} and C_{45} polyprenols designated betulaprenols. Hemming's group studied both the leaves and the wood and obtained betulaprenol-10 with small amounts of betulaprenols-6 to -9 (all unesterified and resembling the castaprenols). The silver birch wood contained esterified betulaprenols-6 to -9 (confirming Lindgren's results) with very small amounts of higher isoprenologues (Wellburn & Hemming, 1966*b*; Lindgren, 1965; Gough & Hemming, 1967, 1970*a*).

The leaves of the rubber tree (*Hevea brasiliensis*) contain solanesol and heveaprenols-10, -11, -12 and -13, all with *cis* moieties predominating over *trans*. These alcohols have a primary allylic hydroxyl and in the C_{55} compound seven isoprene units, including the nearest to the hydroxyl are *cis*, leaving three units (Dunphy *et al.*, 1966, 1967).

The polyprenols of leaves are allylic alcohols predominantly *cis* with three internal *trans* isoprene residues. This has been shown for castaprenols-11 and -12 which have the *cis* configuration at the hydroxyl end and only three *trans* internal residues. This is also true for ficaprenols-10, -11 and -12 and for heveaprenols -11 and -12. Betulaprenol-7 seems to have two *trans* units and betulaprenol-11 three *trans* units (Hemming, 1967). Solanesol-9 (all *trans*) is thus a very unusual compound.

If dolichol-20 were fully unsaturated the formula would be $C_{100}H_{161}OH$. Actually the isoprene unit bearing the hydroxyl group is saturated making the formula $C_{100}H_{163}OH$. The manganese dioxide oxidation discovered in connexion with the preparation of retinal from retinol (Ball, Goodwin & Morton, 1948) is effective with allylic alcohols but with dolichols oxidation is quite slow and the solution must be heated under reflux. The entire dolichol series is an extensive one and all the alcohols from C_{70} to C_{115} (dolichols-14 to -23) are poly *cis* with two or three internal *trans* moieties. The principal biosynthetic step is the addition of a *cis* unit.

When leaves age on plants the polyprenol concentration rises but the proportions of the individual isoprenologues do not change and the alcohols do not form fatty acid esters. Birch wood is anomalous in the sense that wood from other trees has not so far yielded polyprenols (Wellburn & Hemming, 1966*b*).

Three species of lactobacilli contain a polyprenol designated bactoprenol by Thorne & Kodicek (1962, 1966). This is mainly a C_{55} alcohol with one double bond filled in so that it resembles a dolichol-11. Mevalonic acid acts as a biosynthetic precursor. *Lactobacillus plantarum* yields a C_{55} polyprenol with two internal *trans*, one α-*cis* and seven internal *cis*-isoprene units per molecule (Gough *et al.*, 1970).

The polyprenol from *Aspergillus fumigatus* Fresenius was thought to be partially saturated (Burgos *et al.*, 1964). Further study (Stone *et al.*, 1967) showed that at least seven alcohols were present with 18, 19, 20, 21, 22, 23 and 24 isoprene units respectively. Each of the alcohols is a hexahydropolyprenol in which the "hydroxy-terminal" residue is saturated as are the ω-terminal

Table 3 *Polyprenols*

DOLICHOLS
$$H\left[CH_2{-}\overset{\displaystyle CH_3}{\overset{|}{C}}{=}CH{-}CH_2\right]_{n-1}CH_2{-}\overset{\displaystyle CH_3}{\overset{|}{C}H}{-}CH_2{-}CH_2OH$$

Pig	$n =$ 17–22 (100 mg/kg liver)
Rat	16–21 (*ca.* 10 mg/kg liver)
Man	18–20 mainly
Yeast	14–18

	isoprene residue at hydroxyl end	internal residues	ω residue
LEAF POLYPRENOLS			
Decaprenol	*cis*	3 *trans* 5 *cis*	*trans*
Undecaprenol	*cis*	3 *trans* 6 *cis*	*trans*
Dodecaprenol	*cis*	3 *trans* 7 *cis*	*trans*
SILVER BIRCH WOOD POLYPRENOLS			
Betulaprenol-7	*cis*	2 *trans* 3 *cis*	*trans*
Betulaprenol-8	*cis*	2 *trans* 4 *cis*	*trans*
LACTOBACILLUS PLANTARUM POLYPRENOL			
Undecaprenol	*cis*	2 *trans* 7 *cis*	*trans*

ASPERGILLUS FUMIGATUS FRESENIUS POLYPRENOLS

Hexahydropolyprenols $n = 18$–24

a, ψ and ω residues saturated

and the adjacent isoprene residues. The major components are C_{105} (38%), C_{100} (30%) and C_{110} (20%). There are two *trans* internal isoprene residues. In this respect the alcohols resemble the dolichols. Stone *et al.* (1967) discussing nomenclature, preferred to restrict the term dolichol to the 2-3-dihydropolyprenols with mainly *cis* internal residues and two internal *trans* residues. To call the *Aspergillus* alcohols tetrahydrodolichols is less satisfactory than $a\psi\omega$-hexahydropolyprenols and it might be appropriate to use the prefix *Aspergillus*. Stone & Hemming (1968) examined the total lipid from the mycelium and found that more than half the hexahydropolyprenol mixture was esterified, predominantly with saturated acids. Most of the free polyprenol occurred in the mitochondrial fraction but the ester mixture was distributed more evenly over the cell fractions. The mycelium was also rich in ergosterol, free and esterified, but a study of the distribution indicated no association between the acids of sterol esters (93% unsaturated, 77·8% diunsaturated) and prenyl esters (47·5% unsaturated, 26·8% mono-unsaturated). Stone & Hemming (1967) studied the stereochemistry of hexahydroprenol biosynthesis and suggested addition of *cis*-isoprene units to all-*trans* geranylgeranyl pyorphosphate or a hydrogenated derivative thereof.

Wellburn & Hemming (1966*a*) examined the leaves of eleven angiosperms and found castaprenol-like material in all of them at variable concentrations between 0·1 and 3·0 mg/kg wet weight for young leaves. *Chlorella pyrenoidosa* contained a mixture of higher alcohols more like dolichols than castaprenols but ten other cryptograms contained no detectable amounts of polyprenols.

Four plants were examined for seasonal changes. *Laurus vulgare* showed a doubling of the polyprenol content of old leaves compared with young leaves (0·5 to 1·0 mg/kg). *Ficus elasticus* showed an enormous rise from very young to young, old and very old leaves (2·5 to 27·1, 150 and 1500 mg/kg). In the case of horse chestnut leaves the variations in chlorophyll, castaprenols, solanesol and squalene were followed at monthly intervals after terminal bud unfolding

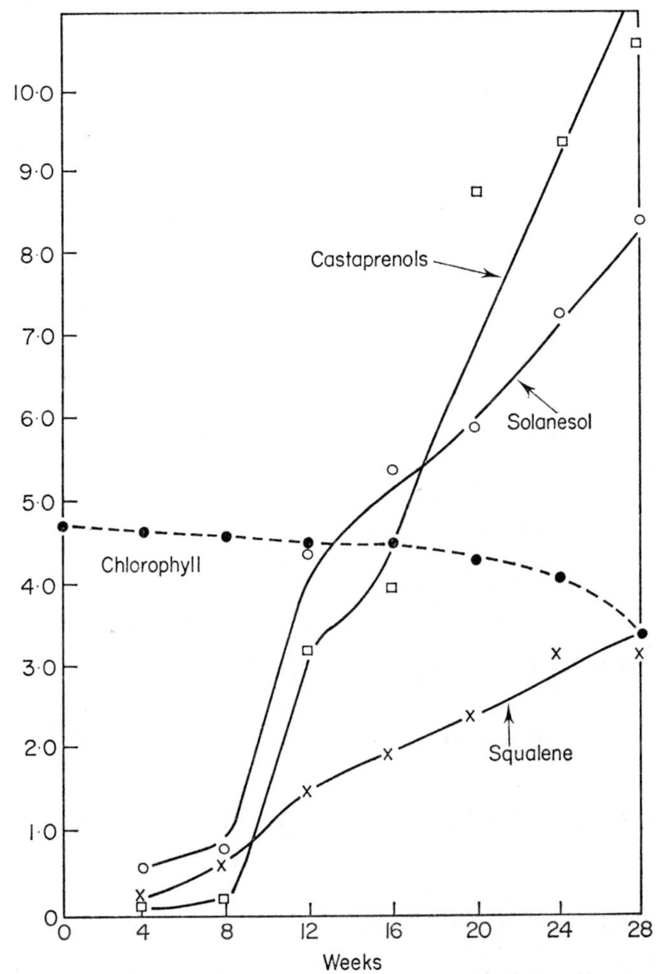

Fig. 1. Variations of isoprenoid constituents of horse chestnut leaves with time in weeks after terminal bud unfolding. Chlorophyll μmoles/10 g tissue, castaprenols μg/g tissue, solanesol and squalene mμmoles/g tissue. (Based on Wellburn & Hemming, 1966*a*.)

Over six months the chlorophyll content was fairly constant from 4·7 μmoles/g of tissue in April to 4·1 μmoles/g in September. The castaprenols, solanesol and squalene all rose quite strikingly (Fig. 1). The plastoquinone content of horse chestnut leaves was fairly stable from 8 weeks to 24 weeks (average 153 μmoles/kg tissue). In horse chestnut leaves it can be said that from week 8 to week 28 the molar ratio chlorophyll to total plastoquinones is approximately steady and the chlorophyll/PQ-9 molar ratio approaches constancy. There is some tendency for the total tocopherol content to increase more rapidly than for the chlorophyll content to fall but the chlorophyll/total tocopherol molar ratio does not vary greatly except in the very young and very old leaves. On the other hand the chlorophyll/castaprenol ratio falls from about 3000 in young leaves to 25–30 in old leaves. In the 28 week leaves the molar ratios in chloroplasts were chlorophyll/castoprenols *ca* 34/1, chlorophyll/plastoquinone *ca* 50/1. In both cases approxinately 80% of the polyprenol or plastoquinone occurred in osmiophilic globules rather than lamellae. The globules contained no chlorophyll.

There is now no doubt that plastoquinone is a necessary component of the photosynthetic electron transfer chain or that only a small proportion of the chlorophyll is concerned in the primary light absorption process. Much of the polyprenol and of the plastoquinone in the fully mature leaf by its occurrence in osmiophilic globules is not participating in photosynthesis. The absolute amounts and variations with time present a *prima facie* case that plastoquinones and total tocopherols share with chlorophyll a direct role in the light reactions of photosynthesis whereas the roles to be attributed to squalene, solanesol and the castaprenols are almost certain to be indirect.

Wright *et al.* (1967) studied the biosynthesis of the *Salmonella* O-antigen. This is a component of the lipopolysaccharide of the cell surface and it has a mannosyl–rhamnosyl–galactose repeating unit. The sequence is assembled on a membrane-bound lipid designated antigen carrier lipid (ACL) (see Wright, Dankert & Robbins, 1965). The biosynthesis proceeds as follows:

$$\text{UDP-Gal} + \text{ACL-P} \rightleftharpoons \text{Gal-P-P-ACL} + \text{UMP}$$

$$\text{Gal-P-P-ACL} + \text{TDP-Rh} \rightharpoonup \text{Rh-Gal-P-P-ACL} + [\text{TDP}]$$

$$\text{Rh-Gal-P-P-ACL} = \text{GDP Man} \rightharpoonup \text{Man -Rh-Gal-P-P-ACL} + [\text{GDP}]$$

$$n(\text{Man-Rh-Gal-P-P-ACL}) \rightharpoonup (\text{Man-Rh-Gal})_n\text{-P-P-ACL} + (n-1)\text{Pi} + (n-1)\text{ACL-P}$$

$$(\text{Man-Rh-Gal})_n\text{-P-P-ACL} + \text{“core”} \rightharpoonup (\text{Man-Rh-Gal})_n\text{-“core”} + \text{Pi} + \text{ACL-P}$$

(Gal = galactose; Rh = Rhamnose; Man—mannose; UDP = uridediphosphate; TDP = thymidinediphosphate; GDP = guanosinediphosphate)

The disaccharidepyrophosphate-ACL was prepared (Dankert *et al.*, 1966) using *Salmonella newington* as a source of enzyme. Mild alkaline hydrolysis produced disaccharide cyclic phosphate and ACL-phosphate. Mild acid hydrolysis of either the disaccharide pyrophosphate ACL or of ACL-phosphate split off a weakly polar substance extractable by light petroleum. Thin layer chromatography gave rise to four or five components regarded mainly as artefacts of the acid hydrolysis. The antigen carrier lipid (ACL) showed 17–19% C-methyl content (20% is required for a polyisoprenoid compound). Infrared absorption carried out on ACL-phosphate in carbon tetrachloride showed a strong peak at 1150–1090 cm^{-1} (P—O—C?) and a band at 1375 cm^{-1} ($C—CH_3$) with 3020 cm^{-1} (ethylenic linkages). Broadly, the infrared spectrum resembled that of phytol. Mass spectroscopy showed a peak at m/e 748 for the product obtained by acid hydrolysis together with M-18-69, M-18-69-*n*(68) to 135 with 69 as the strongest line ($C_{55}H_{89}OH = 766$, minus $H_2O = 748$).

The disaccharide–pyrophosphate–ACL was reduced catalytically (Pt catalyst and borohydride) to give the perhydro ACL which on chromatography seemed homogeneous but by mass spectroscopy yielded an alcohol $C_{55}H_{109}OH$ and an olefine $C_{55}H_{110}$, the latter predominant. The succession of peaks shows a difference of 70 mass units. Pairs of peaks 14 mass units apart indicate that the reduced molecule lacks symmetry.

The evidence suggests that ACL is a C_{55} polyprenol with a terminal allyl–alcohol group to form an allyl pyrophosphate. The infrared spectrum and the C-methyl determinations are consistent with this view.

Biosynthesis of the peptidoglycan of bacterial cell walls has attracted considerable attention, (Anderson *et al.*, 1965; Anderson & Strominger, 1965; Anderson *et al.*, 1966). Particulate enzyme preparations from *Staphylococcus aureus* and *Micrococcus lysodeikticus* utilize UDP–muramyl–pentapeptide and UDP–acetylglucosamine to form a linear peptidoglycan made up of alternate residues of acetylglucosamine, and acetylmuramyl–pentapeptide. Both of the D-alanine moieties present in the uridine nucleotide precursor are retained. The other reaction products are UMP, inorganic phosphate and UDP. Phospholipid intermediates were postulated and regarded as complex glycerophosphatides related to the cardiolipins (Anderson & Strominger, 1965).

The structure of a lipid intermediate in cell wall peptidoglycan synthesis became clearer as a result of the work of Higashi, Strominger & Sweeley (1967) on *M. lysodeikticus*. The cells were treated with lysozyme and a particulate preparation was used as a source of enzyme. The lipid —P—P— disaccharide-pentapeptide was generated after incubation with uridine nucleotide substrates. The enzyme preparation could not however use the intermediates for peptidoglycan synthesis. ^{14}C-Lysine was used as a marker in the biosynthesis and ingenious steps in purification were introduced. The isolated lipid intermediate was shown to be biologically effective. Analysis gave the following results (molar ratios to glutamic acid):

Glutamic acid 1·0; alanine 1·91; lysine 0·86; glycine 0·2;
MurNAc. 0·98; GlcNAc. 0·84 and organic phosphate 2·14.

The two moles of organic phosphate/mole of disaccharide pentapeptide were very readily split off by acid but no glycerol and no fatty acids were detected. The lipid intermediate was readily hydrolysed by mild alkaline treatment. Mass spectrometry showed a weak molecular ion at m/e = 766 and a stronger ion at m/e 748 (M − 18). The fragmentation followed the general pattern observed for solanesol (Rowland, Latimer & Giles, 1956; Erickson *et al.*, 1959). No evidence was found for a saturated isoprene unit. Higashi *et al.* (1967) did not exclude the presence of up to 5% of lower homologues (C_{45}, C_{50}) of the principal lipid, undecaprenol ($C_{55}H_{89}OH$).

Scher, Lennarz & Sweeley (1968) studied the biosynthesis of mannosyl lipids concerned in the enzymic production of a homopolysaccharide of mannose. Using *M. lysodeikticus* it was already known that GDP-mannose could transfer the sugar to three lipids and a mannan. Two of the lipids were mannosyl diglyciderides and the third was shown to be a mannosyl–phosphoryl polyprenol (MPP). The new product, after chromatographic purification, released mannose on acid hydrolysis (0·01 N HCl, 100°, 10 min) and milder treatment set free mannose-1-phosphate. The hydrolysate extracted with ether yielded a lipid which on mass spectrometry gave m/s 766 (weak), 748 (stronger) and a sequence m/e 748–69n) down to m/e 135. The mass spectrum agreed with that of castaprenol-11 (Wellburn *et al.*, 1967) and ficaprenol-11 (Stone *et al.*, 1967) and with the polyprenols concerned in the biosynthesis of the *Salmonella* O-antigen and the bacterial cell wall peptidoglycans already discussed. Scher *et al.* (1968) had evidence of the presence of a small proportion of dodecaprenol ($C_{60}H_{97}OH$ M-18, 816) and there was also a little decaprenol (C_{50}). The undecaprenol however accounted for 91% of the total polyprenol. NMR measurements indicated 11 double bonds, two of the internal double bonds being *trans* and eight *cis*. The castaprenol-11 and ficaprenol-11 had three internal *trans* residues and seven *cis* isoprene residues.

Behrens & Leloir (1970) found that a liver microsomal fraction catalysed the transfer of glucose from UDPG to a lipid acceptor, apparently identical with dolichol monophosphate prepared chemically.

$$\text{UDPG} + \text{acceptor lipid} \rightharpoonup \text{G-acceptor lipid} + \text{UDP}$$

$$\text{G-acceptor lipid} + \text{protein} \rightharpoonup \text{acceptor lipid} + \text{G-protein.}$$

Dolichol was prepared from pig liver following the Liverpool work (Burgos *et al.*, 1963). It was phosphorylated using trichloracetonitrile as condensing agent (Cramer & Böhm, 1959; Popják *et al.*, 1962). The acceptor lipid was prepared from pig liver and enzymic glucose transfer to it was demonstrated using UDPG labelled in the glucose moiety. Purification of the acceptor lipid was not easy, but dolichol monophosphate proved fully effective and behaved in the same way as the acceptor lipid on chromatography. It must be concluded that the acceptor lipid is dolichol monophosphate (DMP) but it remains to be seen whether one dolichol is more effective than another. Transfer from the glucosylated compound (DMPG) opens up a large field. Little is known yet about the glycoproteins formed. Caccum, Jackson & Eylar (1969) used cell fractions from liver in a study of the biosynthesis of glycoproteins containing mannose. Again a lipid intermediate was found and it proved very similar to the mannosyl-1-phosphoryl polyprenol already noted as having been obtained from *M. lysodeikticus*.

Douglas & Baddiley (1968), Archibald, Baddiley & Button (1968) and Brooks & Baddiley (1969) studied the role of lipid intermediates in the biosynthesis of cell wall teichoic acids in *Staphylococcus lactis* strains. The amounts of lipid were small but a strong case was presented that a polyprenol might be concerned. In *S. lactis* 13 it seemed possible that the same lipid could function in the biosynthesis of two wall polymers, namely peptidoglycan and teichoic acid. [The peptidoglycan contains *N*-acetylmuramic acid, *N*-acetylglucosamine, alanine, lysine, glutamic acid and glycine while the repeating unit of the wall teichoic acid consists of glycerol-1-phosphate attached to the 3-position of the 6-0-alanyl-*N*-acetyl-glucosamine-1-phosphate]. Competition for the

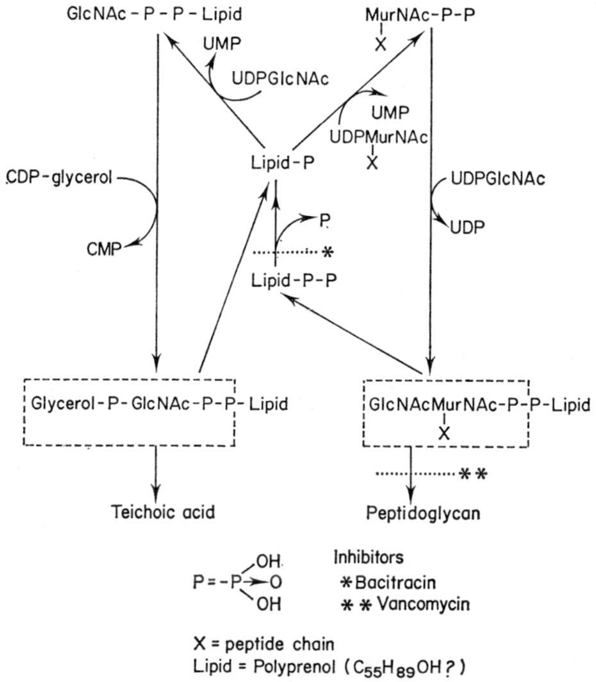

FIG. 2. Biosynthesis of teichoic acid and peptidoglycan through a shared membrane lipid in *S. lactis* 13. (After Watkinson, Hussey & Baddiley, 1971.)

lipid is obviously possible. Figure 2 illustrates this. Watkinson, Hussey & Baddiley (1971) make the point that the membrane lipid of choice seems always to be the undecaprenol although it remains to be seen whether in all cases the same specific isomer (castaprenol-11) is the one involved.

Alam *et al.* (1970) accepted the evidence that prenolphosphate sugars are concerned in biosynthesis in plant cell walls and of mammalian glycoproteins (Villemez & Clark, 1969; Kauss, 1969; Caccum *et al.*, 1969; Behrens & Leloir, 1970). They studied the responses to different prenolphosphates of mannosyl transferase activities from GDP-mannose in mammalian, plant and fungal systems. With pig liver endoplasmic reticulum and GDP-U-^{14}C mannose 10% of the ^{14}C was recovered as lipid when no prenol phosphate was added but at a concentration of 178 μM the monophosphates of dolichols resulted in a four-fold increase. The betulaprenol monophosphates were equally effective and solanesyl phosphate produced a three-fold increase and the ficaprenol phosphates a two-fold increase. Farnesyl phosphate and cetyl phosphate had no effect. The dolichyl and betulaprenyl phosphates had a maximal effect in 5 min. Two labelled lipids were formed. With mung-bean seedlings a fraction transferring mannose to cell wall polysaccharides also responded to dolichols and betulaprenols selectively and not to the other prenols tried.

This work with its interesting implications shows that much remains to be done. DeLuca, Schumacher & Wolf (1970), DeLuca & Wolf (1968) and DeLuca, Rosso & Wolf (1970) obtained a subcellular fraction was thought to be concerned in the biosynthesis of a mannolipid. When the fraction was prepared from "mildly vitamin A-deficient rats" (presumably the liver reserves were not far from zero) it was incubated with 15-^{14}C-retinol and GDP-mannose-^{14}C. The mannolipid produced contained ^{14}C from both the retinol and the mannose, that from the labelled retinol being present in a polar metabolite of the vitamin. The metabolite resembled retinoic acid in polarity but the two were not identical. The metabolite was not retinoyl glucuronide.

Vitamin A is a tetraprenol ($C_{20}H_{29}OH$) of a special character. It is necessary for the maintenance of normality in mucus-secreting structures and in the light of the work described here it would be reasonable to imagine a role for retinol in the biosynthesis of specific glycopeptides. It is important however to note that DeLuca, Schumacher & Wolf, 1970 attribute the catalytic role to a polar metabolite which has so far not been identified. Whether a carrier role for vitamin A affords a clue to the general systemic mode of action remains to be seen.

REFERENCES

Alam, S. S., Barr, R. M., Richards, J. B. & Hemming, F. W. (1970). *Biochem. J. Proc.* **121**, 19P.

Anderson, J. S. & Strominger, J. L. (1965). *Biochem. biophys. Res. Commun.* **21**, 516.

Anderson, J. S., Matsuhashi, M., Haskin, M. A. & Strominger, J. L. (1965). *Proc. natn Acad. Sci. U.S.A.* **4**, 881.

Anderson, J. S., Meadow, P. M., Haskin, M. A. & Strominger, J. L. (1966). *Archs Biochem. Biophys.* **116**, 487.

Archibald, A. R., Baddiley, J. & Button, D. (1968). *Biochem. J.* **110**, 543.

Ball, S., Goodwin, T. W. & Morton, R. A. (1948). *Biochem. J.*, **42**, 516.

Bates, R. B. & Gale, D. M. (1960). *J. Am. chem. Soc.* **82**, 5749.

Bates, R. B., Gale, D. M. & Gruner, B. J. (1963). *J. Org. Chem.* **28**, 1086.

Bates, R. B., Carnighan, R. H., Rakutis, R. & Schauble, J. H. (1962). *Chem. Ind.* 1020.

Behrens, N. & Leloir, L. F. (1970). *Proc. natn Acad. Sci. U.S.A.* **66**, 153.

Brooks, D. & Baddiley, J. (1969). *Biochem. J.* **115**, 307.

Burgos, J. & Morton, R. A. (1962). *Biochem. J.* **82**, 454.

Burgos, J., Hemming, F. W., Pennock, J. F. & Morton, R. A. (1963). *Biochem. J.* **188**, 470.

Burgos, J., Butterworth, P. H. W., Hemming, F. W. & Morton, R. A. (1964). *Biochem. J.* **91**, 22P.

Butterworth, P. H. W., Burgos, J. & Hemming, F. W. (1966). *Archs Biochem. Biophys.* **114**, 398.

Butterworth, P. H. W. & Hemming, F. W. (1968). *Archs Biochem. Biophys.* **128**, 503.

Caccum, J. F., Jackson, J. J. & Eylar, E. H. (1969). *Biochem. biophys. Res. Commun.* **35**, 505.

Channon, H. J., Devine, J. & Loach, J. V. (1934). *Biochem. J.* **28**, 2071.

Channon, H. J. & Marrian, G. F. (1926). *Biochem. J.* **20**, 409.

Chen, H. Y. (1962). *Anal. Chem.* **34**, 1793.

Cramer, P. & Böhm, H. J. (1959). *Angew. Chem.* **71**, 775.

Dankert, M., Wright, A., Kelley, W. S. & Robbins, P. W. (1966). *Archs Biochem. Biophys.* **116,** 425.

DeLuca, L., Rosso, G. & Wolf, G. (1970). *Biochem. biophys. Res. Commun.* **41,** 615; (1971). *Fedn Proc. Fedn Am. Socs exp. Biol.* **30,** 1134.

DeLuca, L., Schumacher, M. & Wolf, G. (1970). *J. biol. Chem.* **245,** 4551.

DeLuca, L. & Wolf, G. (1968). *Archs Biochem. Biophys.* **123,** 1.

Dimter, A. (1941). *Hoppe-Seyl. Z. Physiol. Chem.* **271,** 293.

Douglas, L. J. & Baddiley, J. (1968). *FEBS Lett.* **1,** 114.

Dunphy, P. J., Kerr, J. D., Pennock, J. F. & Whittle, K. J. (1966). *Chem. Ind.* 1549.

Dunphy, P. J., Kerr, J. D., Pennock, J. F., Whittle, K. J. & Feeney, J. (1967). *Biochim. biophys. Acta* **136,** 136.

Erickson, R. E., Shunk, C. H., Trenner, N. R., Arison, B. H. & Folkers, K. (1959). *J. Am. chem. Soc.* **81,** 4999.

Feeney, J. & Hemming, F. W. (1967). *Anal. Biochem.* **20,** 1.

Gough, D. P. & Hemming, F. W. (1967). *Biochem. J.* **105,** 10C.

Gough, D. P. & Hemming, F. W. (1970*a*). *Biochem. J.* **117,** 309.

Gough, D. P. & Hemming, F. W. (1970*b*). *Biochem. J.* **118,** 163.

Gough, D. P., Kirby, A. L., Richards, J. B. & Hemming, F. W. (1970). *Biochem. J.* **118,** 167.

Hemming, F. W. (1967). In *Terpenoids in Plants* (Ed. by Pridham, J. B.), p. 223. London and New York: Academic Press.

Hemming, F. W., Morton, R. A. & Pennock, J. F. (1963). *Proc. R. Soc.* **B158,** 291.

Higashi, Y., Strominger, J. L. & Sweeley, C. C. (1967). *Proc. natn Acad. Sci. U.S.A.* **57,** 1878.

Kauss, H. (1969). *FEBS Lett.* **5,** 81.

Lindgren, B. O. (1964). *Acta chem. Scand.* **18,** 836.

Lindgren, B. O. (1965). *Acta chem. Scand.* **19,** 1317.

Packter, N. M. & Glover, J. (1960). *Nature, Lond.* **187,** 413.

Packter, N. M. & Glover, J. (1964). *Abstr. 6th Int. Congr. Biochem.* **7,** 113.

Pennock, J. F., Hemming, F. W. & Morton, R. A. (1960). *Nature* **186,** 470; (1960). *Biochem. J.* **74,** 38P.

Popják, G., Cornforth, J. W., Cornforth, R. H., Ryhage, R. & Goodman, D. S. (1962). *J. biol. Chem.* **237,** 56.

Rowland, R. L., Latimer, P. H. & Giles, J. A. (1956). *J. Am. chem. Soc.* **78,** 4680.

Rüegg, R., Gloor, U., Langemann, A., Kofler, M., von Planta, C., Ryser, G. & Isler, O. (1960). *Helv. chim. Acta* **43,** 1745.

Scher, J., Lennarz, W. J. & Sweeley, C. (1968). *Proc. natn Acad. Sci. U.S.A.* **59,** 1313.

Stone, K. J., Butterworth, P. H. W. & Hemming, F. W. (1967). *Biochem. J.* **102,** 443.

Stone, K. J. & Hemming, F. W. (1965). *Biochem. J.* **96,** 14C.

Stone, K. J. & Hemming, F. W. (1967). *Biochem. J.* **104,** 43.

Stone, K. J. & Hemming, F. W. (1968). *Biochem. J.* **109,** 877.

Stone, K. J., Wellburn, A. R., Hemming, F. W. & Pennock, J. F. (1967). *Biochem. J.* **102,** 325.

Thorne, K. J. L. & Kodicek, E. (1962). *Biochim. biophys. Acta* **59,** 273, 280, 295.

Thorne, K. J. L. & Kodicek, E. (1966). *Biochem. J.* **99,** 123.

Villemez, C. L. & Clark, A. F. (1969). *Biochem. biophys. Res. Commun.* **36,** 57.

Watkinson, R. J., Hussey, H. & Baddiley, J. (1971). *Nature New Biology* **229,** 57–59.

Wellburn, A. R. & Hemming, F. W. (1965). *J. Chromatography* **23,** 51.

Wellburn, A. R. & Hemming, F. W. (1966*a*). *Phytochem.* **5,** 969.

Wellburn, A. R. & Hemming, F. W. (1966*b*). *Nature, Lond.* **212,** 1364.

Wellburn, A. R., Stevenson, J., Hemming, F. W. & Morton, R. A. (1967). *Biochem. J.* **102,** 313.

Wright, A., Dankert, M. & Robbins, P. W. (1965). *Proc. natn Acad. Sci. U.S.A.* **54,** 235.

Wright, A., Dankert, M., Fennessey, P. & Robbins, P. W. (1967). *Proc. natn Acad. Sci. U.S.A.* **57,** 1798.

CHEMISTRY AND METABOLISM OF β-APOCAROTENALS AND THEIR EPOXIDES

By HARMEET SINGH, A. KRISHNA MALLIA AND H. R. CAMA

Department of Biochemistry, Indian Institute of Science, Bangalore, India

Ever since Moore first demonstrated (1930) the provitamin A nature of β-carotene in the animal body, it has been known to be a precursor of vitamin A. Since then considerable interest has been shown in the possible mechanism of conversion of carotene to vitamin A, but it was only recently that the mechanism has become clearer. Though Moore first postulated that the liver is the site of conversion in mammals, it was later shown that intestine can accomplish the conversion much more effectively (Glover, Goodwin & Morton, 1947, 1948; Mattson, Mehl & Deuel, 1947; Thompson, Ganguly & Kon, 1947, 1949). The conversion of β-carotene to retinyl esters by intestinal sections in the presence of conjugated bile acids was first demonstrated by Olson (1964), using ^{14}C-labelled β-carotene. It can also be effected by the isolated perfused rat liver (Zachman & Olson, 1965); the kidney tissue is considerably less active in this respect (Olson & Hayaishi, 1965).

Metabolic studies on β-carotene labelled with tritium as well as with ^{14}C at 15, 15′-positions (Crain, Lotspeich & Krause, 1963) showed that both labels are retained in retinyl esters. After feeding labelled β-carotene to rats, Huang & Goodman (1965) showed that 90% of the radioactivity is absorbed from the intestine and appears in the lymph as retinyl esters. Similar observations were reported in human subjects also (Goodman *et al.*, 1966). These observations supported the concept of central cleavage of β-carotene, which was further substantiated by the work on α-carotene carried out by Mcanally & Szymanski (1966), who could detect both retinyl esters and α-retinyl esters in the liver of vitamin A-deficient rats fed α-carotene.

Using enzyme preparations from rat intestinal mucosa Goodman and his associates (1965), Goodman, Huang & Shiratori (1966), Goodman *et al.* (1967), showed that the β-carotene molecule is cleaved centrally at the 15,15′-double bond. This cleavage enzyme from intestine and liver requires molecular oxygen (Goodman & Huang, 1965; Olson & Hayaishi, 1965). Similar results were obtained with enzyme preparations from hog intestinal mucosa (Fidge, Smith & Goodman, 1969) and with partially purified enzyme from rabbit intestinal mucosa (Olson, 1969). The enzyme from rabbit mucosa cleaved β-apo-8′-carotenal and β-apo-10′-carotenal at a faster rate as compared to β-carotene. But the ring substituted carotenoids like 5,6-monoepoxy-β-carotene and 5,6:5′,6′-diepoxy-β-carotene were not cleaved by this enzyme (Lakshmanan, Pope & Olson, 1968).

By using the rat-growth assay procedure Subbarayan, Lakshmanan & Cama (1966) observed that 5,6-monoepoxy-β-carotene is biologically 21% as active as all-*trans*-β-carotene, whereas 5,6:5′,6′-diepoxy-β-carotene is completely inactive. Only retinyl esters were detected in the liver after oral administration of 5,6-monoepoxy-β-carotene to vitamin A-deficient rats. Work of Kishore, Subbarayan & Cama (1970) with 5,6-monoepoxy-α-carotene showed that it gives rise to only α-retinyl esters. *In vivo* studies on cryptoxanthin (Jacob John *et al.*, 1970) and on citroxanthin (Jacob John & Cama, 1972 also showed that only retinyl esters are formed from cryptoxanthin or citroxanthin. In these *in vivo* studies administration of various ring substituted carotenoids to vitamin A-deficient rats did not give rise to ring substituted retinal or retinol. Studies on the absorption of epoxides of retinal, namely, 5,6-epoxyretinal and 5,8-epoxyretinal (Lakshmanan, Jungalwala & Cama, 1965) and of 4-hydroxyretinal (Jacob John *et al.*, 1970) showed that these compounds are readily reduced to their corresponding retinols, absorbed, esterified and stored in the liver as epoxides of retinyl and 4-hydroxyretinyl esters, respectively. Thus, it is clear that these ring substituted retinals are absorbed *per se* and also adequately stored in the liver.

Although it has been clearly established in *in vivo* (Huang & Goodman, 1965; Goodman *et al.*, 1966*a*; Olson, 1961) as well as in *in vitro* studies (Goodman & Huang, 1965; Goodman *et al.*, 1967; Fidge *et al.*, 1969; Olson, 1969) that β-carotene is cleaved at the central 15,15′-double bond, *in vivo* work on ring substituted carotenoids, namely, 5,6-monoepoxy-β-carotene, 5,6:5′,6′-diepoxy-β-carotene, cryptoxanthin, citroxanthin and 5,6-monoepoxy-α-carotene did not substantiate the theory of central cleavage mechanism. Moreover, the inability of the enzyme from rabbit intestinal mucosa (Lakshmanan *et al.*, 1968) to cleave 5,6-monoepoxy-β-carotene and 5,6:5′,6′-diepoxy-β-carotene prompted us to study the mode of cleavage of ring substituted carotenoids. With this end in view various β-apo-carotenals and their epoxides were prepared and their metabolism was studied in detail.

OBSERVATIONS AND DISCUSSION

Various β-apo-carotenals were prepared according to the method of Hasani & Parrish (1967) which is briefly described as follows: β-Carotene (250 mg) dissolved in 100 ml benzene was oxidized with 30% hydrogen peroxide (15 ml) in the presence of potassium permanganate (1·3 g dissolved in 100 ml water). After stirring for 3½ hr the reaction was stopped by filtration, extracted with diethyl ether, washed free of potassium permanganate, dried over anhydrous sodium sulphate and reduced in volume *in vacuo*. Various β-apo-carotenals namely, β-apo-8′-carotenal, β-apo-10′-carotenal and β-apo-12′-carotenal and unreacted β-carotene were separated over a 3% (v/w) water-deactivated alumina column. The β-apo-carotenals were further purified by rechromatography over 3% (v/w) water-deactivated alumina and crystallized thrice from light petroleum. The yield of crystalline β-apo-8′-carotenal, β-apo-10′-carotenal and β-apo-12′-carotenal was 4·5%, 1·5–2·0% and 2·0% and their absorption maxima in light petroleum were at 455 nm, 437 nm and 410 nm, respectively. The R_f values of the three β-apocarotenals on thin-layer chromatography using 6% (v/v) acetone in light petroleum as solvent system were 0·562, 0·500 and 0·687, respectively. Their structures were confirmed by their ultraviolet, visible, infrared absorption spectra and by co-chromatography with authentic samples.

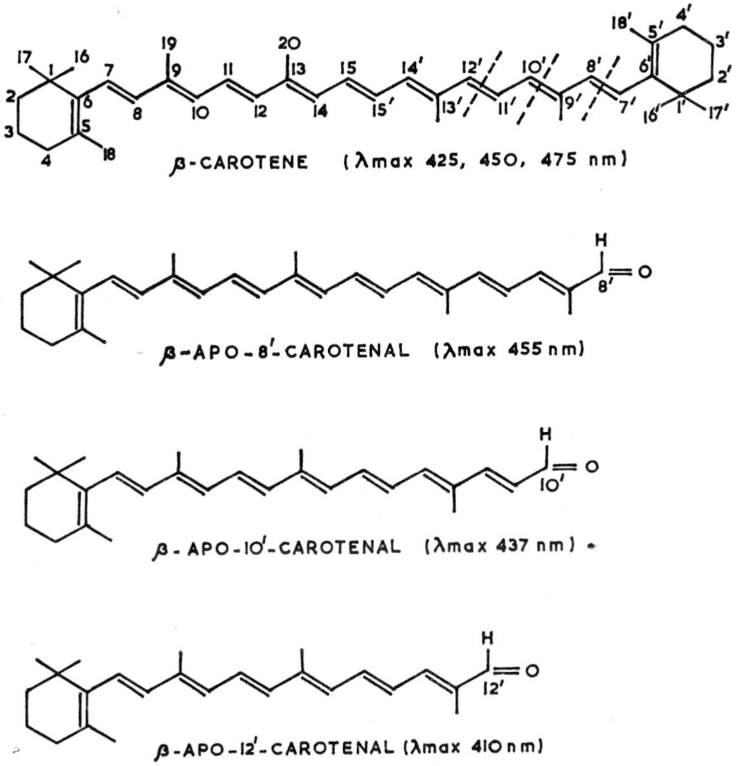

FIG. 1. Structures of β-carotene and various β-apocarotenals.

The absorption spectra of various β-apocarotenals in different solvents were recorded with a Unicam SP-700 recording spectrophotometer. The absorption maxima and the $E_1^{1\%}$ cm values for the apocarotenals in different solvents are given in Table 1. The $E_1^{1\%}$ cm values for the apocarotenals were highest in light petroleum and their absorption maxima showed bathochromic shift of 10–20 nm in polar solvents.

The absorption spectrum of β-apo-8′-carotenal in light petroleum showed a maximum at 455 nm and a shoulder at 475 nm which was transformed to a broad-banded spectrum in ethanol and chloroform and showed a bathochromic shift of 10–15 nm. The shoulder at 475 nm observed in light petroleum completely disappeared, probably due to hydrogen bond formation between the carbonyl group of the apocarotenal and the proton of the solvent. The absorption spectra in different solvents are shown in Fig. 2.

Similar changes in absorption spectra in different solvents were also noticed in the case of β-apo-10′-carotenal and β-apo-12′-carotenal.

Table 1. *Spectroscopic properties of β-apocarotenals in different solvents*

Solvent	8′-Carotenal λ_{max}(nm)	8′-Carotenal $E_1^{1\%}$ cm	10′-Carotenal λ_{max}(nm)	10′-Carotenal $E_1^{1\%}$ cm	12′-Carotenal λ_{max}(nm)	12′-Carotenal $E_1^{1\%}$ cm
Light petroleum (b.p. 40–60°)	455	2652	437	2406	410	2198
Cyclohexane	457	2354	440	2338	415	2053
Carbon tetrachloride	465	2238	450	1855	420	1971
Chloroform	470	2084	460	1958	430	1811
Ethanol	465	2116	450	2060	425	1872
$SbCl_3$-colour test	825	2717	770	3922	710	2918

The apo-carotenals gave blue colour with the antimony trichloride reagent. The colour was stable for 1–2 min. The $E_1^{1\%}$ cm values are shown in Table 1. It is evident that an increase of one double bond caused a bathochromic shift of the chromogen by 55–60 nm.

Infrared absorption spectra of the apo-carotenals were taken in nujol and they showed a sharp carbonyl peak at 1650 cm^{-1}. The presence of the carbonyl group was further confirmed by reduction with lithium aluminium hydride to their corresponding alcohols which showed in light petroleum a hypsochromic shift of 25–30 nm in the visible spectrum of the apo-carotenals.

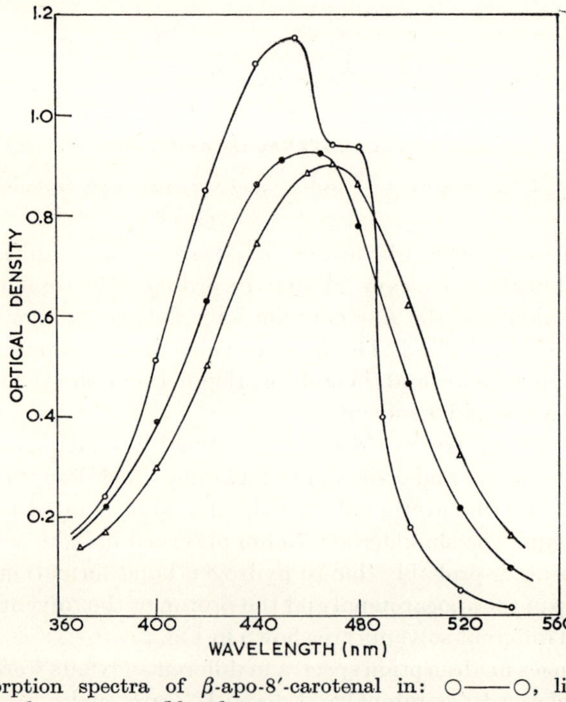

FIG. 2. Absorption spectra of β-apo-8′-carotenal in: ○——○, light petroleum; ●——●, ethanol; △——△, chloroform.

Nuclear magnetic resonance spectra of the apo-carotenals were recorded in a HA-100 D nuclear magnetic resonance spectrometer. From the nuclear magnetic resonance spectrum of β-apo-8′-carotenal (Fig. 3) all the 40 protons could be accounted for. The aldehyde proton at the end of the side chain showed a sharp signal at 0·65τ. It is difficult to assign signals to different olefinic protons in the structure due to the overlapping of various signals and spin–spin coupling of various protons. Methyl group protons showed signals between 7·90–9·00τ. Gem-dimethyl group protons showed one 6 protons signal at the highest field, i.e. at 9·00τ. The 3 protons signal at 8·3τ may be assigned to C_{18} methyl group protons. Similarly 3 protons signal at 8·15τ was assigned to C_{19} methyl protons and 9 protons signal at 8·05τ to C_{20}, C'_{20}, C'_{19} methyl protons. Two short signals at 8·5τ and 8·75τ were assigned to C_2 and C_3 methylene protons in the β-ionone ring, respectively.

The nuclear magnetic resonance spectra of the other two apo-carotenals were similar to those of β-apo-8′-carotenal except for the absence of one methyl proton signal due to C'_{19} methyl protons and some minor differences in the number of olefinic proton signals.

When tested with vitamin A-deficient rats the three β-apo-carotenals were found to be biologically active. A single dose (by oral or subcutaneous means) of 1 mg/rat of various β-apocarotenals could support the growth of vitamin A-deficient rats for over 4 weeks. In order to study the metabolism, each β-apocarotenal was administered orally to vitamin A-deficient rats for 7 days (1 mg/rat/day). They were killed 24 hr after the last dose, after which tissues like stomach, intestine, liver, lungs, kidneys and spleen were analysed for vitamin A and various β-apocarotenals. The results are shown in Table 2.

Table 2. *Recovery of vitamin A and various β-apocarotenals from the liver of vitamin-A deficient rats after oral administration of various β-apocarotenals for 7 days (1 mg/rat/day)*

Compound fed	Retinyl esters* (μg)	β-Apocarotenal* (μg)
β-Apo-8′-carotenal	107·5	10·6
β-Apo-10′-carotenal	91·2	2·0
β-Apo-12′-carotenal	101·2	8·4

* Mean of value obtained from a group of 4 rats.

β-Apocarotenals and retinyl esters could be detected only in liver. There was no significant difference in the storage of retinyl esters when different β-apocarotenals were administered orally to vitamin A-deficient rats (Table 2). Traces of apocarotenals were detected in the stomach also. Thus it is evident that β-apocarotenals are not reduced to their respective β-apocarotenols as was observed after oral administration of retinal. Retinal has been shown to be reduced to retinol and then absorbed from the small intestine whereas absorption and storage of retinal as such was reported to be negligible. In contrast

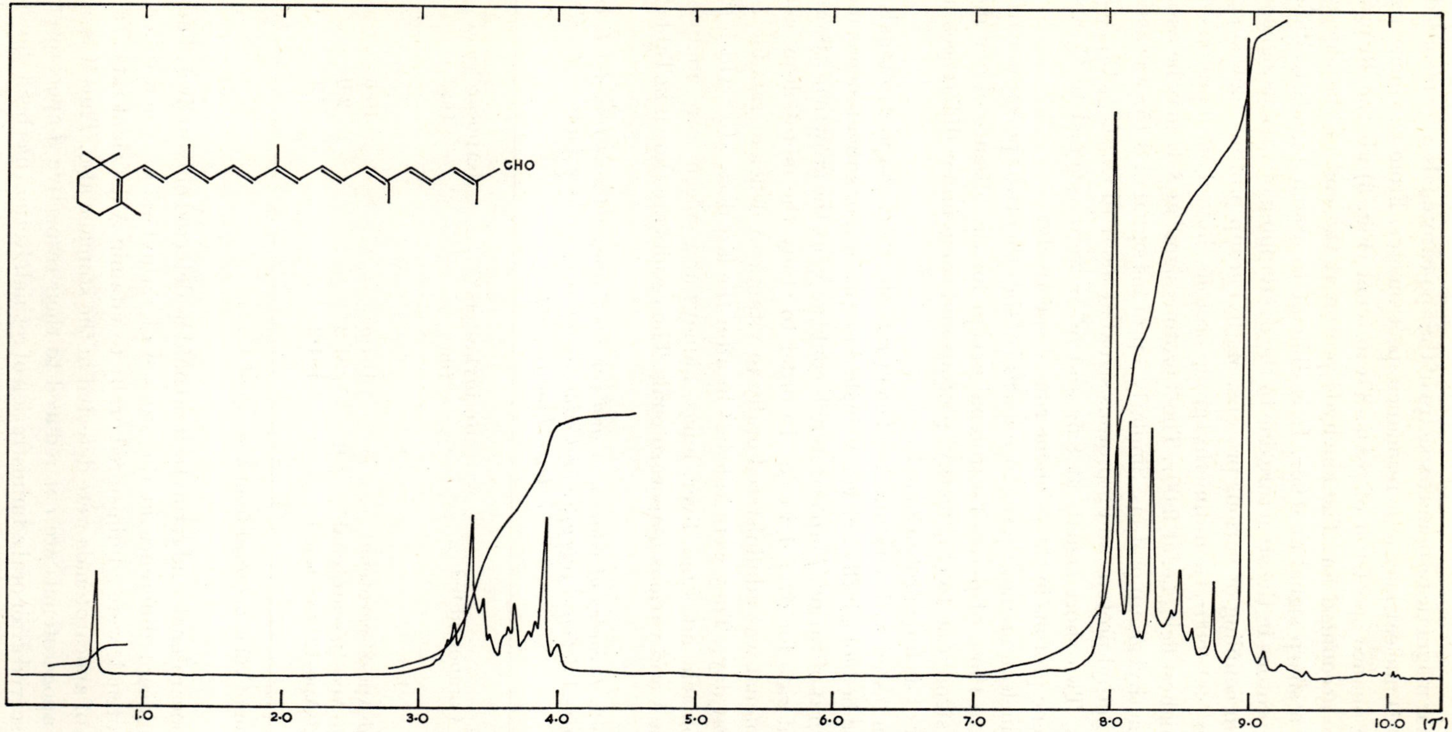

FIG. 3. Nuclear magnetic resonance spectrum of β-apo-8′-carotenal.

evidence presented here and our unpublished observations suggest that β-apocarotenals are absorbed as such and are stored in the liver thereby indicating that either they are not reduced to apocarotenols or that the enzyme which converts retinal to retinol may not be effective against the β-apocarotenals. Efforts to detect β-apocarotenal or retinyl esters in other tissues were not successful.

Various β-apocarotenals were also administered intra-peritoneally to vitamin A-deficient rats (1 mg/rat/day) for 7 days. They were killed 24 hr after the last dose and tissues like stomach intestine, liver, lungs, kidneys and spleen were analysed. The recovery of retinyl esters and of β-apocarotenals from the tissues of such is given in Table 3.

While retinyl esters could be detected only in the liver, β-apocarotenals were detected in the liver as well as in the intestine. Retinyl estes and β-apocarotenals were not detected in tissues other than liver and intestine. The retinyl esters content of the liver of rats given β-apo-10′-carotenal was higher; however, the amount of β-apo-10′-carotenal in intestine and liver was much lower as compared to other β-apocarotenals (Table 3).

Table 3. *Recovery of vitamin A and β-apocarotenals from the intestine and liver of vitamin A-deficient rats after intraperitoneal administration of various β-apocarotenals for 7 days (1 mg/rat/day)*

Compound administered	Retinyl esters* (μg)		β-Apocarotenal* (μg)	
	Intestine	Liver	Intestine	Liver
β-Apo-8′-carotenal	—	81·0	92·0	66·6
β-Apo-10′-carotenal	—	101·2	36·3	18·9
β-Apo-12′-carotenal	—	56·8	58·9	92·5

* Mean of values obtained from a group of 4 rats.

By rat-growth assay procedure the biopotency of the three β-apocarotenals was determined, with *all-trans* β-carotene as standard. The biopotencies of β-apo-8′-carotenal, β-apo-10′-carotenal and β-apo-12′-carotenal were 31·3%, 75·9% and 44·2%, respectively of that of *all-trans* β-carotene. It is interesting to note that β-apo-10′-carotenal exhibited the highest activity. These observations suggested that the cleavage may very well take place centrally at the 15,15′-double bond, because β-apocarotenals were biologically less active than *all-trans* β-carotene. *In vivo* and *in vitro* experiments have clearly shown that one molecule of β-carotene can give rise to 2 molecules of vitamin A; however, only one molecule of vitamin A can be expected from one molecule of β-apocarotenals. The differences observed in the biopotencies of various β-apocarotenals may be attributed to the differences in the *in vivo* specificity of the cleavage enzyme towards these apocarotenals.

The above results were further supported by the findings of Lakshmanan *et al.* (1968) that the β-carotene cleavage enzyme, isolated from rabbit intestinal mucosa, which specifically splits β-carotene at the 15,15′-double bond giving two molecules of retinal, could also cleave various β-apocarotenals. This suggests that the same enzyme may cleave β-carotene and β-apocarotenals *in vivo* to form retinal and that its specificity towards different substrates may vary from species to species. Although *in vivo* studies have shown that β-carotene is biologically more active than the β-apocarotenals, the *in vitro* activity of the carotene cleavage enzyme has been shown to be much higher against β-apocarotenals compared to β-carotene (Lakshmanan *et al.*, 1968).

The rabbit enzyme does not cleave 5,6-monoepoxy-β-carotene, but *in vivo* studies (Subbarayan *et al.*, 1966) have shown that 5,6-monoepoxy-β-carotene is converted to vitamin A. In order to elucidate the mode of cleavage of ring substituted carotenoids the epoxides of β-apo-8′-carotenal and β-apo-12′-carotenal were prepared and their metabolism was studied in detail.

The 5,6-epoxide of β-apo-8′-carotenal was prepared by treatment of β-apo-8′-carotenal with monoperphthalic acid in the ratio of 1:2. β-Apo-8′-carotenal (100 mg) was dissolved in peroxide-free dry diethyl ether (100 ml) and freshly prepared monoperphthalic acid (200 mg) was added, after which the reaction was carried out at room temperature for 72 hr in an atmosphere of nitrogen. The reaction was stopped by adding dilute sodium bicarbonate (5%) solution. The products were extracted with diethyl ether, washed with distilled water, dried over anhydrous sodium sulphate, reduced in volume *in vacuo* and chromatographed over 5% (v/w) water-deactivated alumina. The β-apo-8′-carotenal and 5,6-epoxide of β-apo-8′-carotenal were separated by gradient elution from the alumina column. Further purification of the 5,6-epoxide was effected by rechromatography over 5% (v/w) water-deactivated alumina. Yield of the purified compound was 60 mg.

The 5,6-epoxide was crystallized thrice from light petroleum (m.p. 154°) and was characterized by its absorption spectrum in the visible region, when it showed a triple-banded spectrum with absorption maxima at 430 nm, 450 nm and 477 nm. On treatment with ethanolic hydrogen chloride (0·5N) it was converted to its furanoid form (5,8-epoxy-β-apo-8′-carotenal). The conversion to the furanoid form was accompanied by a hypsochromic shift of 20 nm in the absorption spectrum. The furanoid compound was purified by chromatography over 5% (v/w) water-deactivated alumina and crystallized thrice from light petroleum (m.p. 160°). It showed a triple-banded spectrum with absorption maxima at 405 nm, 430 nm and 455 nm in light petroleum. The structures of both epoxides were further confirmed by infrared absorption spectra, nuclear magnetic resonance spectra and elemental analyses.

With the antimony trichloride reagent both 5,6- and 5,8-epoxides of β-apo-8′-carotenal gave stable blue chromogens with absorption maxima at 600 nm and 645 nm, respectively. Spectroscopic properties of the two epoxides are given in Table 4.

The $E_{1}^{1\%}$ cm values of both epoxides were highest in light petroleum (Table 4) and both showed a triple banded spectrum in light petroleum and cyclohexane. A bathochromic shift of 10–15 nm was observed in chloroform and carbon

Table 4. *Spectroscopic properties of epoxides of β-apo-8′-carotenal in different solvents*

Solvent	5,6-epoxide		5,8-epoxide.	
	λ_{max}(nm)	$E_{1}^{1\%}$ cm	λ_{max}(nm)	$E_{1}^{1\%}$ cm
Light petroleum (b.p. 40–60°)	430, 450, 477	1902, 2710, 2251	405, 430, 455	1855, 2561, 2212
Cyclohexane	430, 454, 484	1844, 2510, 2083	410, 435, 457	1819, 2461, 2121
Carbon tetrachloride	—	—	420, 440, 465	1666, 2152, 1788
Chloroform	450	1961	445	1907
Ethanol	457	2102	445	2020
$SbCl_3$-colour reaction	600	1866	645	1721

tetrachloride in the case of 5,8-epoxide. The triple-banded spectra of both epoxides in light petroleum changed to broad single-banded spectra in chloroform and ethanol, probably due to hydrogen bond formation between the carbonyl group of the epoxides and the proton of the solvent. The absorption spectra of the 5,6-epoxide in different solvents are shown in Fig. 4.

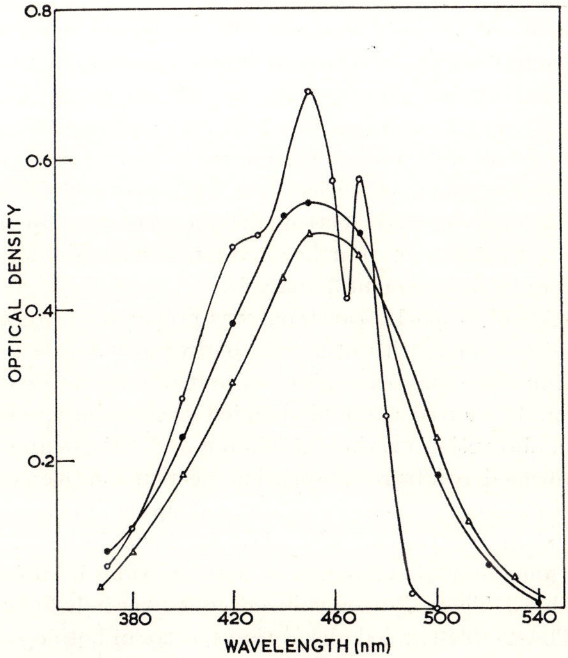

FIG. 4. Absorption spectra of 5,6-epoxy-β-apo-8′-carotenal in: ○——○, light petroleum; ●——●, ethanol; △——△, chloroform.

The nuclear magnetic resonance spectrum of the 5,6-epoxy-β-apo-8′-carotenal (Fig. 5) accounted for all the 40 protons and it was similar to that of β-apo-8′-carotenal, as described earlier in this paper. However, the following minor differences were observed: the C_{18} methyl protons signal was shifted from 8·3τ to 9·1τ, i.e. beyond the gem-dimethyl group signal; the 9 protons signal at 8·05τ due to C_{20}, C'_{20} and C'_{19} methyl group protons was split into two signals of 6 protons and 3 protons; the 3 protons signal at 8·05τ is assigned to C_{20} methyl protons and 6 protons signal at 8·00τ to C'_{20} and C'_{19} methyl protons; the olefinic protons at C_7 and C_8 got separated from the rest of the olefinic proton signals and showed signals at 4·25τ and 4·10τ, respectively.

Though the nuclear magnetic resonance spectrum of the 5,8-epoxy-8′-carotenal was similar to that of the 5,6-epoxycompound, a few differences were observed. The 6 protons signal at 9·0τ due to gem-dimethyl group was split into two equal signals of 3 protons on furanization and appeared at 8·85τ and 8·90τ. The C_{18} methyl protons signal was shifted from 9·1τ to 8·64τ on furanization. The C_7 and C_8 olefinic protons spin coupled with each other and appeared as quartet which was further shifted towards higher field and these signals appeared between 4·8τ and 5·10τ.

The 5,6- and 5,8-epoxides of β-apo-8′-carotenal were orally administered to vitamin A-deficient rats at the level of 1 mg/rat/day for 7 days. They were killed 24 hr after the last dose, after which tissues such as liver, intestine, stomach, kidneys, lungs and spleen were analysed. No epoxides of β-apo-8′-carotenal, of retinal or of vitamin A could be detected in any of the tissues. Moreover, these epoxides of β-apo-8′-carotenal did not support growth of the vitamin A-deficient rats at this dosage level.

The 5,6-epoxides of β-apo-12′-carotenal was prepared in the same way as described earlier and was characterized by the hyposochromic shift of 7 nm in light petroleum in the visible spectrum of β-apo-12′-carotenal. It was separated from unreacted β-apo-12′-carotenal over 3% (v/w) water-deactivated alumina column, further purified by rechromatography over 3% (v/w) water deactivated alumina and crystallized thrice from light petroleum. The pure compound showed a double-banded spectrum in light petroleum having absorption maxima at 403 nm and 423 nm. Further characterization was done by infrared absorption spectrum and elemental analysis.

The 5,8-epoxide of β-apo-12′-carotenal was prepared by treatment of 5,6-epoxy-β-apo-12′-carotenal with ethanolic hydrogen chloride (0·5N). It was purified by chromatography over alumina and was crystallized thrice from light petroleum. It showed a double-banded absorption spectrum in light petroleum with absorption maxima at 380 nm and 400 nm and its structure was further confirmed by infrared absorption spectrum and elemental analysis.

Both epoxides of β-apo-12′-carotenal formed stable blue chromogens with the antimony trichloride reagent with absorption maximum at 550 nm. The $E_1^{1\%}$ cm values and absorption maxima of both epoxides in different solvents are given in Table 5. Their $E_1^{1\%}$ cm values were higher in light petroleum and cyclohexane. The double-banded absorption spectra of both epoxides in light petroleum, cyclohexane and carbon tetrachloride were changed to single broad-banded absorption spectra in chloroform and ethanol, probably due to

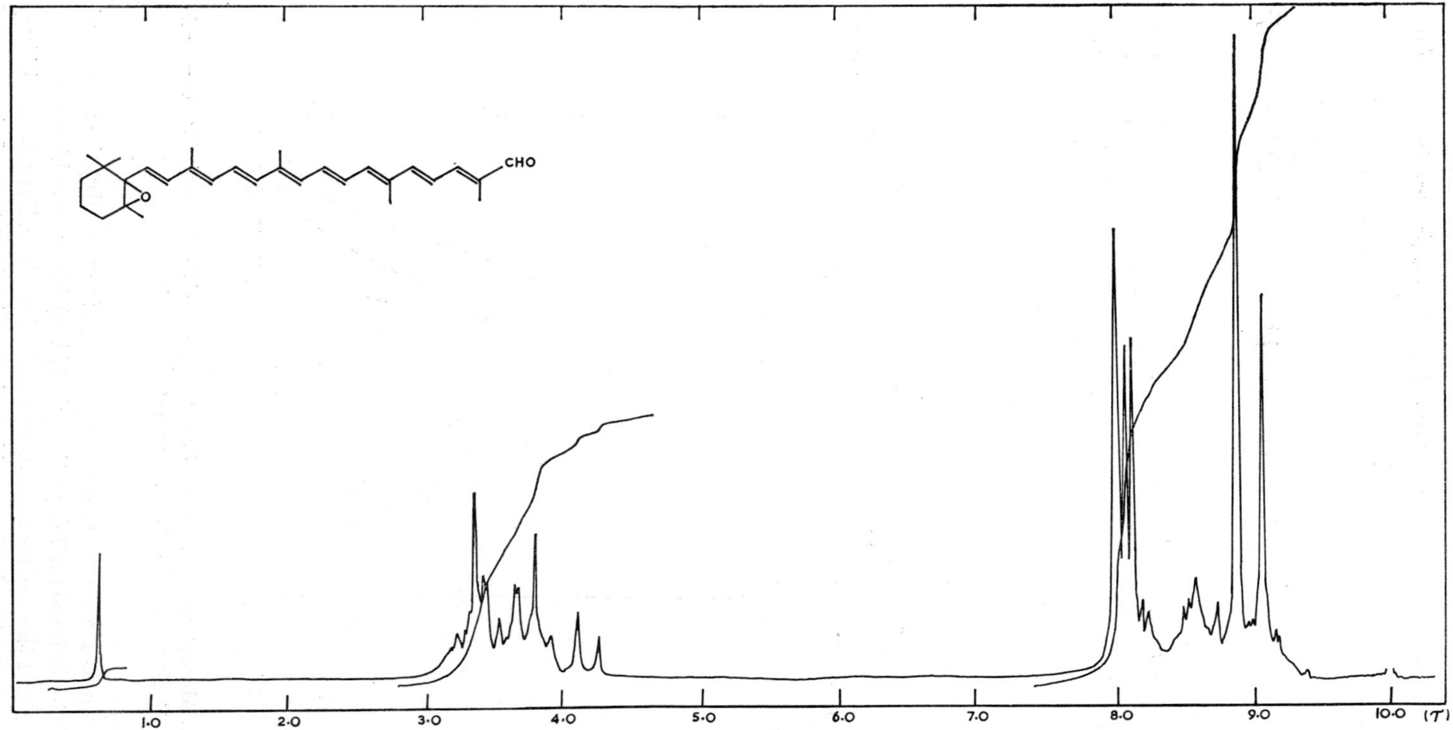

FIG. 5. Nuclear magnetic resonance spectrum of 5,6-epoxy-β-apo-8′-carotenal.

Table 5. *Spectroscopic properties of epoxides of β-apo-12′-carotenal in different solvents*

Solvent	5,6-epoxide		5,8-epoxide	
	λ_{max}(nm)	$E_1^{1\%}$ cm	λ_{max}(nm)	$E_1^{1\%}$ cm
Light petroleum (b.p. 40–60°)	403, 423	2168, 1790	380, 400	2273, 2019
Cyclohexane	407, 428	2294, 1903	380, 400	2090, 1784
Carbon tetrachloride	415, 437	1955, 1648	385, 405	1732, 1540
Chloroform	420	1680	395	1523
Ethanol	415	1719	400	1682
$SbCl_3$-colour reaction	555	1396	555	2045

hydrogen bond formation between the carbonyl group of the epoxides and the proton of the solvent. The absorption spectra of 5,6-epoxide of β-apo-12′-carotenal in different solvents are shown in Fig. 6.

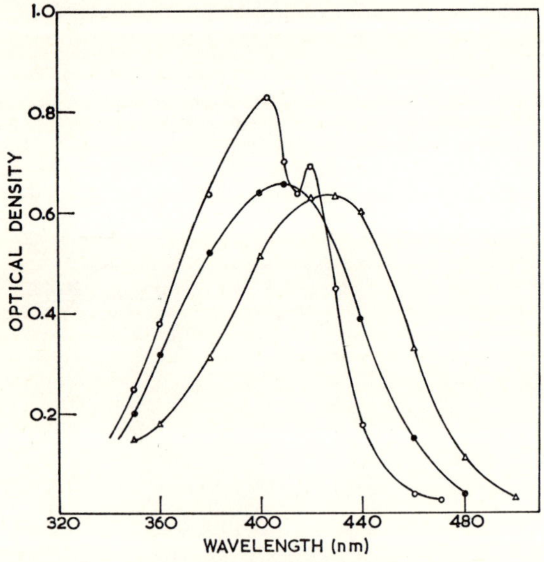

Fig. 6. Absorption spectra at 5,6-epoxy-β-apo-12′-carotenal in: ○——○, light petroleum; ●——●, ethanol; △——△, chloroform.

Vitamin A-deficient rats were separately given (orally) the two epoxides of β-apo-12′-carotenal for 7 days (1 mg/rat/day). They were killed 24 hr after the last dose and the tissues such as stomach, intestine, liver, lungs, kidneys and spleen were analysed. No epoxides of vitamin A or β-apo-12′-carotenal were detectable in any of the tissues analysed.

CONCLUDING REMARKS

The observations on 5,6-epoxyretinal (Krishna Mallia *et al.*, 1970); 5,6:5′,6′-diepoxy-β-carotene (Subbarayan *et al.*, 1966); epoxides of β-apo-8′-carotenyl acetate (Harmeet Singh & Cama, 1972) and our present studies on epoxides of β-apo-8′-carotenal and β-apo-12′-carotenal conclusively show that those compounds which cannot be converted *in vivo* to retinol are biologically inactive. The present observations also support our earlier suggestion (Krishna Mallia *et al.*, 1970) that the double bond in the β-ionone ring is a pre-requisite for vitamin A activity and that modifications of the tertiary–tertiary double bond in the β-iodone ring lead to a loss of vitamin A activity.

Again, these studies on the epoxides of β-apo-8′-carotenal and β-apo-12′-carotenal suggest that an intact β-ionone ring is absolutely essential for the cleavage to take place at the 15,15′-double bond. Such conclusions were further supported by our previous *in vivo* work on epoxides of β-apo′-8-carotenyl acetate (Harmeet Singh & Cama, 1972) and on 5,6-monoepoxy and 5,6:5′,6′-diepoxy-β-carotene (Subbarayan *et al.*, 1966), 5,6-monoepoxy-α-carotene (Kishore *et al.*, 1970), cryptoxanthin (Jacob John *et al.*, 1970) and citroxanthin (Jacob John & Cama, 1972). Earlier studies reported from our laboratory on epoxides of retinal (Lakshmanan *et al.*, 1965) and 4-hydroxyretinal (Jacob John *et al.*, 1970) have indicated that, when administered orally to vitamin A-deficient rats they are readily reduced in the intestine, absorbed, esterified and stored in the liver as esters of epoxides of retinol or of 4-hydroxyretinol. Therefore, if in the case of the ring substituted carotenoids the cleavage were to take place at the 15,15′-double bond, in all the present and previous work ring substituted vitamin A compounds should have been detected in the liver. Further, the *in vitro* work of Lakshmanan *et al.* (1968) on 5,6-monoepoxy and 5,6:5′,6′-diepoxy-β-carotene, with a partially purified carotene cleavage enzyme from rabbit intestinal mucosa has shown that the enzyme does not act upon substituted carotenoids.

Thus these findings from our laboratory suggest that in the case of ring substituted carotenoids the cleavage is not central but excentric. It is possible that the mono-ring substituted carotenoids, e.g. 5,6-monoepoxy-β-carotene, 5,6-monoepoxy-α-carotene, citroxanthin and cryptoxanthin may be cleaved from the substituted ring of the molecule by an enzyme other than the 15:15′-dioxygenase. The β-apocarotenals thus formed may be further cleaved by the usual carotene cleave enzyme or by an β-oxidation type enzyme. It may very well be that an enzyme other than the usual β-carotene cleavage enzyme which acts upon the ring substituted carotenoids may be present in the intestinal mucosa of animals.

ACKNOWLEDGEMENTS

We are grateful to Messrs Jacob John, G. S. Kishore and C. Venkateshan for help in the preparation of this manuscript and we thank the Council of Scientific and Industrial Research (India) for financial assistance.

REFERENCES

Crain, F. D., Lotspeich, F. J. & Krause, R. F. (1963). *Fedn Proc. Fedn Am. Soc. exp. Biol.* **22**, 434.
Fidge, N. H., Smith, F. R. & Goodman, DeW. S. (1969). *Biochem. J.* **114**, 689.
Glover, J., Goodwin, T. W. & Morton, R. A. (1947). *Biochem. J.* **41**, 45.
Glover, J., Goodwin, T. W. & Morton, R. A. (1948). *Biochem. J.* **43**, 512.
Goodman, DeW. S. & Huang, H. S. (1965). *Science* **149**, 879.
Goodman, DeW. S., Huang, H. S. & Shiratori, T. (1966). *J. biol. Chem.* **241**, 1929.
Goodman, DeW. S., Bloomstrand, R., Werner, B., Huang, H. S. & Shiratori, T. (1966). *J. clin. Invest.* **45**, 1615.
Goodman, DeW. S., Huang, H. S., Kanai M. & Shiratori, T. (1967). *J. biol. Chem.* **242**, 3543.
Hasani, S. M. Al. & Parrish, D. B. (1967). *J. Agri. Fd Chem.* **15, 943.**
Huang, H. S. & Goodman, DeW. S. (1965). *J. biol. Chem.* **240**, 2839.
Jacob John, J. & Cama, H. R. (1972). *Int. J. Vit. Nutr. Res.* **42**, 278.
Jacob John, J., Kishore, G. S., Subbarayan, C. & Cama, H. R. (1970). *Indian J. Biochem.* **7**, 222.
Kishore, G. S., Subbarayan, C. & Cama, H. R. (1970). *Indian J. Biochem.* **7**, 97.
Krishna Mallia, A., Jacob John, J., K. V., Lakshmanan, M. R., Jungalwala, F. B. & Cama, H. R. (1970). *Indian J. Biochem.* **7**, 102.
Lakshmanan, M. R., Jungalwala, F. B. & Cama, H. R. (1965). *Biochem. J.* **95**, 27.
Lakshmanan, M. R., Pope, J. L. & Olson, J. A. (1968). *Biochem. biophys. Res. Commun.* **33**, 347.
Mattson, F. H., Mehl, J. W. & Deuel, H. J. (1947). *Arch. Biochem.* **15**, 65.
Mcanally, J. S. & Szymanski, C. D. (1966). *Nature, Lond.* **210**, 1366.
Moore, T. (1930). *Biochem. J.* **24**, 692.
Olson, J. A. (1961). *J. biol. Chem.* **236**, 349.
Olson, J. A. (1964). *J. Lipid. Res.* **5**, 402.
Olson, J. A. (1969). *Am. J. clin. Nutr.* **22**, 953.
Olson, J. A. & Hayaishi, O. (1965). *Proc. natn Acad. Sci. U.S.A.* **54**, 1364.
Singh, H. & Carma, H. R. (1972). *Int. J. Vit. Nutr. Res.* **42.**
Subbarayan, C., Lakshmanan, M. R. & Cama, H. R. (1966). *Biochem. J.* **99**, 308.
Thompson, S. Y., Ganguly, J. & Kon, S. K. (1947). *Brit. J. Nutr.* **1**, 5.
Thompson, S. Y., Ganguly, J. & Kon, S. K. (1949). *Brit. J. Nutr.* **3**, 50.
Zachman, R. D. & Olson, J. A. (1965). *J. biol. Chem.* **240**, 541.

RECENT DEVELOPMENTS IN THE BIOSYNTHESIS OF CAROTENOIDS

By T. W. GOODWIN

Department of Biochemistry, The University, Liverpool, England

INTRODUCTION

THE overall picture of carotenoid biosynthesis has recently been fully reviewed (Goodwin, 1971*a*) and will not be considered further here. Only recent developments in the author's laboratory concerned with the use of stereospecifically labelled mevalonic acids and a strongly carotenogenic *Flavobacterium* sp. will be discussed in detail.

Mevalonic acid (1), the well-known specific precursor of terpenoids, is characterized by three prochiral centres at C-2, C-4 and C-5. A prochiral

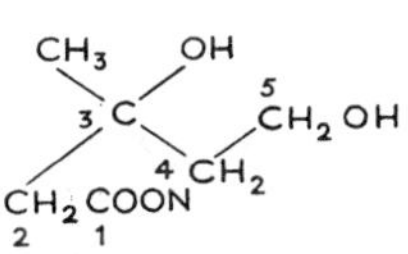

(Mevalonic acid)

(1)

centre can be considered in the following way: "If a chiral assembly is obtained when a point ligand in a finite achiral assembly of point ligands is replaced by a new point ligand, the original assembly is prochiral" (Hanson, 1966). In other words replacement of one of the hydrogens (point ligands) at C-2, or C-4 or C-5 by another substituent results in the appearance of chiral centres, in these cases asymmetric carbon atoms. (For details of the subtilties, both scientific and nomenclatural, of chirality and prochirality the reader is referred to Bentley's (1970) two volume monograph.)

The conversion of mevalonic acid into carotenoids involves, at some stage in the biosynthetic process, the removal of one hydrogen from C-2, C-4 and C-5 of each participating mevalonic acid molecule. As enzymes are stereospecific reagents it was highly likely that their removal would be stereospecific and that only one hydrogen, the pro-*R* or the pro-*S*-hydrogen, would be removed. (The terms pro-*R* and pro-*S* denote the absolute configurations which would ensue when a point ligand at a prochiral centre is replaced by another point ligand.) The only way to test this was to synthesize various species of mevalonic acid in which one hydrogen at either C-2, C-4 or C-5 was stereospecifically replaced with either deuterium or tritium. This was achieved by Cornforth & Popják and their colleagues (see for example, Cornforth & Cornforth, 1970; Cornforth & Ross, 1970) and set the stage for developments in the study of terpenoid biosynthesis at a level of sophistication which had not previously even been contemplated. The utilization of these compounds in defining the biosynthesis of squalene in stereospecific detail has been reviewed by Popják & Cornforth (1966); and their great value in studying triterpene biosynthesis and carotenoid biosynthesis has also been fully discussed (Goodwin, 1971*a*,*b*; Goad & Goodwin, 1972).

DESATURATION OF PHYTOENE

The first C_{40} carotenoid precursor is phytoene (2) formed from the tail to tail condensation of two molecules of geranylgeranyl pyrophosphate.

Desaturation of phytoene to lycopene (3) results in the formation of four additional double bonds which, in every case, involves C-2 and C-5 of mevalonic

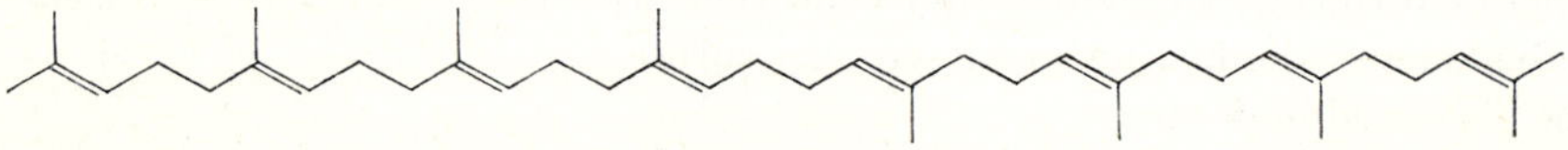

(Phytoene)
(2)

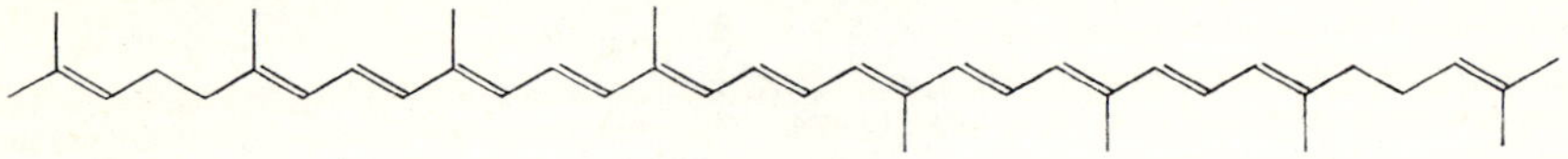

(Lycopene)
(3)

acid. Experiments with [5*R*-5-$^3H_1$]mevalonic acid and [5-$^3H_2$]mevalonic acid clearly revealed some time ago that there was stereospecific removal of a 5-*pro-R* hydrogen during the formation of each of the double bonds (Goodwin, 1971*a*). On the other hand, experiments with [2*R*-2-$^3H_1$]mevalonic acid and [2*S*-2-$^3H_1$]mevalonic acid gave in many systems equivocal data which were the result of scrambling of the label; the indications were, however, that the 2 *pro-S* hydrogen was lost (J. R. Vose and G. Britton, unpublished observations). The explanation for this appears to be that the prenyl transferase, which condenses dimethyl allyl pyrophosphate with isopentenyl pyrophosphate to form the C_{10} compound geranyl pyrophosphate as the first step in chain elongation in terpene biosynthesis (equation 1), is sluggish in higher plants and most protista. This allows the enzyme, isopentenyl pyrophosphate-dimethyl allyl pyrophosphate isomerase, which catalyses the only freely reversible reaction in the biosynthetic sequence, to scramble the label. Fig. 1 shows how [4-$^3H_1$]isopentenyl pyrophosphate arising from [2*R*-2-$^3H_1$]mevalonic acid can yield [4-$^3H_1$]isopentenyl pyrophosphate with a configuration derived from [2*S*-2-$^3H_1$]mevalonic acid; the reverse is of course true and the formation of unlabelled isopentenyl pyrophosphate is also a possibility. This

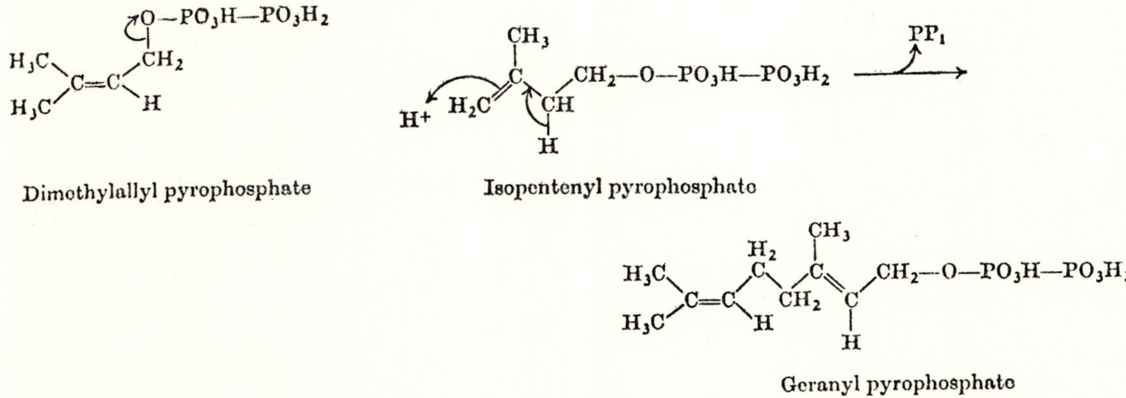
Dimethylallyl pyrophosphate
Isopentenyl pyrophosphate
Geranyl pyrophosphate

Equation 1

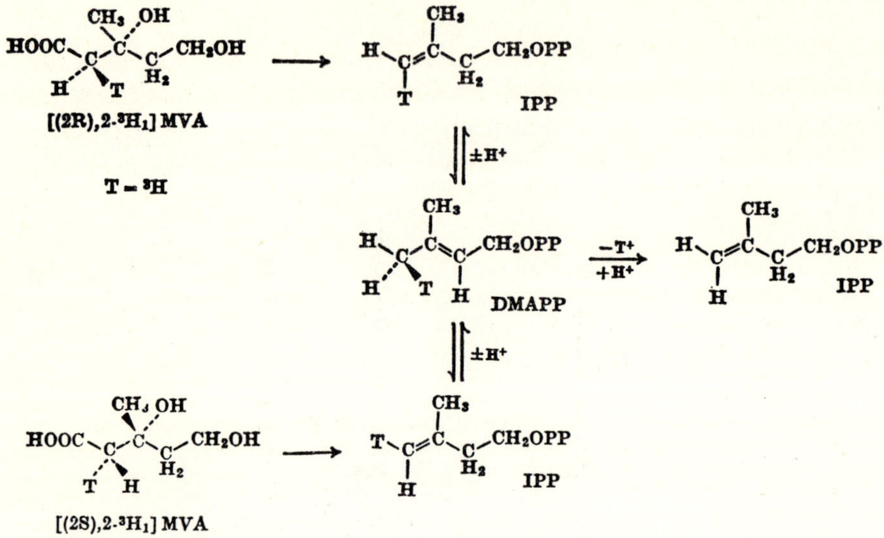

FIG. 1. How the reversible action of isopentenyl pyrophosphate dimethylallyl pyrophosphate isomerase can randomize the label from [$2R$-2-$^3H_1$]- and [$2S$-2-$^3H_1$/mevalonic acid.

difficulty was almost completely overcome by using a strongly carotenogenic strain of *Flavobacterium* kindly provided by Roche Products Ltd. This organism under normal cultural conditions synthesizes zeaxanthin (4), but in the presence of nicotine (see later) it accumulates mainly lycopene with smaller

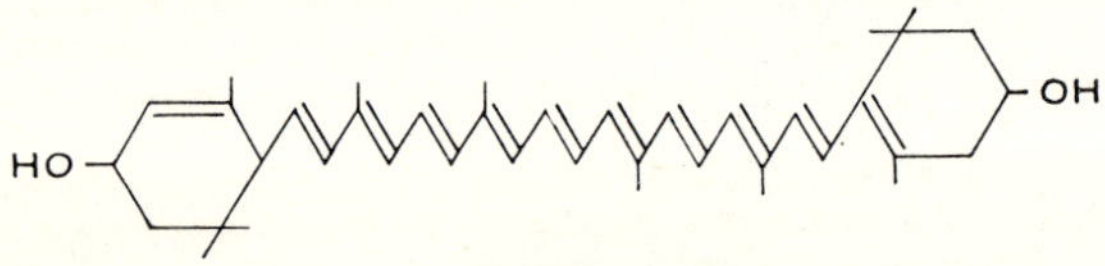

(zeaxanthin)
(4)

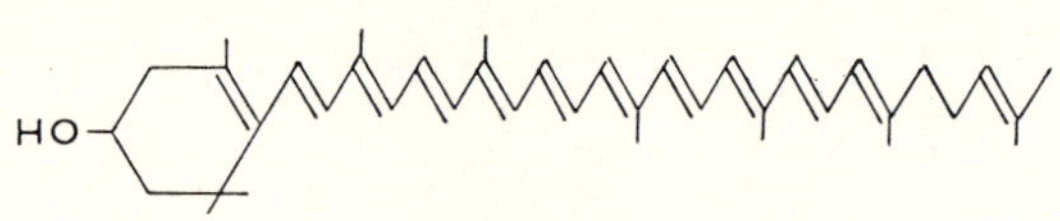

(rubixanthin)
(5)

amounts of phytoene, rubixanthin (5) and zeaxanthin. Cell-free homogenates of this organism very effectively incorporate labelled mevalonic acid into carotenoids and significant activity is found even after 30 min incubation. Short incubations allowed the randomization, just referred to, to be reduced to a minimum, and the presence of nicotine allowed lycopene to accumulate. The results given in Table 1 clearly demonstrate that the *pro*-2*S* hydrogen is lost during the formation of the four double bonds in converting phytoene into lycopene and the results with the other fully unsaturated carotenoids, rubixanthin and zeaxanthin, confirmed this (J. C. B. McDermott, G. Britton and T. W. Goodwin, unpublished observations). It will be seen that when [2-^{14}C, 2*S*-^{3}H$_1$]mevalonic acid was the substrate the ^{3}H : ^{14}C ratio dropped substantially in lycopene compared with phytoene and, on an atomic basis, this worked

Table 1. *Incorporation of* [2-^{14}*C*,2*R*-2-^{3}*H*$_1$] *and* [2-^{14}*C*-2*S*-2-^{3}*H*$_1$]*mevalonate into carotenoids by cell-free extracts from a* Flavobacter *sp.* (*Incubation at* 20° *for* 60 *min with nicotine* (2 mM) *and substrate*)

Compound	Activity ^{3}H	(dpm) ^{14}C	Activity ratio ^{3}H/^{14}C	Atomic ratio ^{14}C : ^{3}H
A. [2-^{14}C,2*R*-2^{3}H$_1$]mevalonic acid				
Mevalonic acid	45,306	7580	5·78	—
Phytoene	286	50	5·72	8:8·00
Lycopene	10,920	1976	5·52	8:7·70
Rubixanthin	3,270	607	5·39	8:7·51
Zeaxanthin	7,703	1469	5·24	8:7·30
B. [2-^{14}C,2*S*-2-^{3}H$_1$]mevalonic acid				
Mevalonic acid	41,254	6508	6·32	—
Phytoene	11,622	1876	6·39	8:8·00
Lycopene	6,412	1760	3·64	8:4·70
Rubixanthin	815	225	3·62	8:4·68
Zeaxanthin	2,565	700	3·66	8:4·73

out to just less than 4 hydrogens per mole. The slightly low result was undoubtedly due to some randomization of label. With [2-^{14}C,2*R*-^{3}H$_1$]mevalonic acid on the other hand, essentially no label was lost on transforming phytoene into lycopene, but again there was a slight indication of randomization. Thus the overall stereochemistry of the desaturation process can be summarized in Fig. 2. It should be noted from Fig. 2 that in the dimerization of geranylgeranyl pyrophosphate to form phytoene there is stereospecific loss of two *pro-S*-hydrogens from carbon atoms originating from C-5 of mevalonic acid. This was demonstrated with a non-aqueous chloroplast preparation which effectively incorporates labelled mevalonic acid into both geranylgeraniol and phytoene (Buggy, Britton & Goodwin, 1969).

CYCLIZATION OF LYCOPENE

A direct demonstration of the cyclization of lycopene to β-carotene (6) has been reported in chloroplasts (Hill & Rogers, 1969; Hill, Calderwood & Rogers, 1971) and in partly purified preparations from green tissues of higher plants (Kushwaha *et al.*, 1969). Howes & Batra (1970) showed that if a *Mycobacterium* species, which normally synthesizes β-carotene, is grown in the presence of nicotine, then lycopene accumulates; if the cells are washed free of

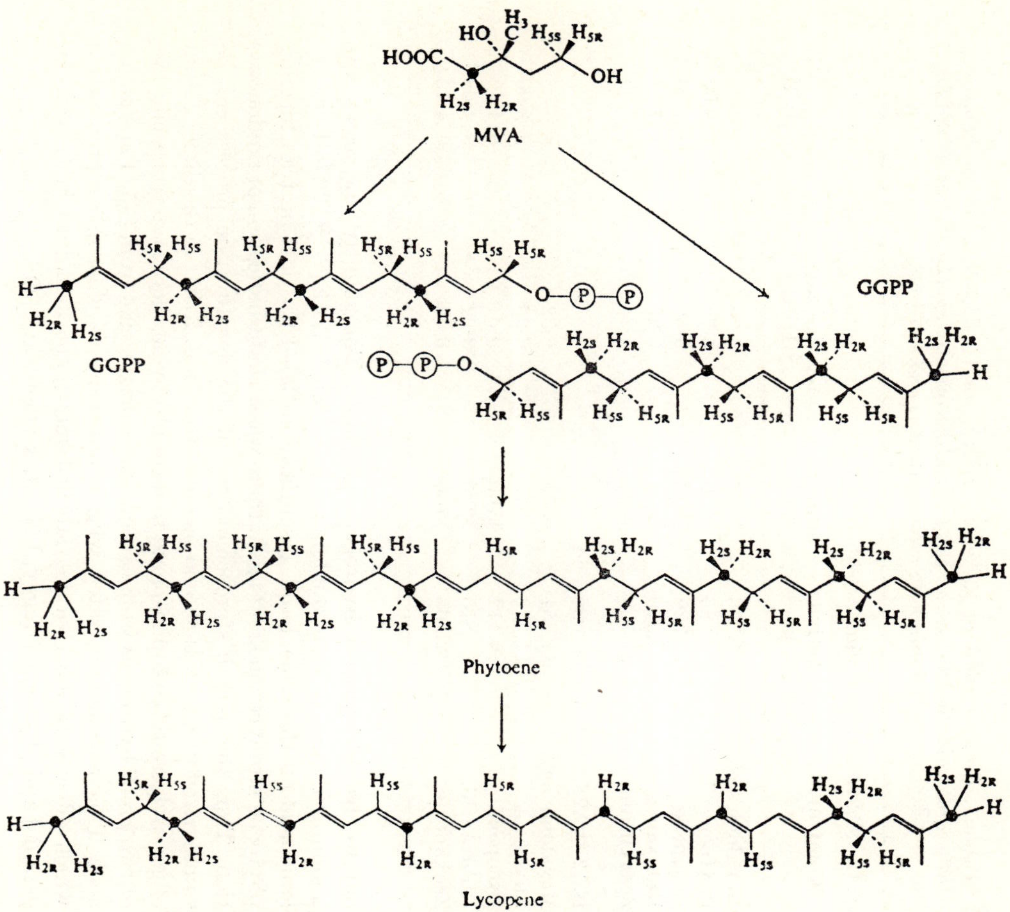

FIG. 2. The overall sterochemistry involving C-2 and C-5 of mevalonic in its conversion into lycopene.

inhibitor and resuspended, then the lycopene is rapidly converted into β-carotene. We have confirmed this effect in *Phycomyces blakesleeanus* and in our *Flavobact.* sp. Furthermore, we have shown that if nicotine-grown cells of *Flavobact.* sp. are washed and resuspended anaerobically then lycopene is

(β-carotene)
(6)

Table 2. *The anaerobic conversion of lycopene into β-carotene by* Flavobact *sp.* (Cells grown in the presence of nicotine (7·5 mM) then washed free from inhibitor and resuspended in buffer anaerobically for 24 hr)

Compound	Pigment composition (% of total pigment present)	
	Before resuspension	After resuspension
Lycopene	93	5·6
β-Carotene	0	81·0
Rubixanthin	3·6	4·4
Cryptoxanthin	0	5·1
Zeaxanthin	0·8	3·7
Unknown	2·6	0·2

rapidly converted into β-carotene (Table 2), thus demonstrating unequivocally that the cyclization is an anaerobic process and also confirming our previously proposed mechanism for cyclization which involved proton attack at the terminal double bond (Fig. 3) (Williams, Britton & Goodwin, 1967).

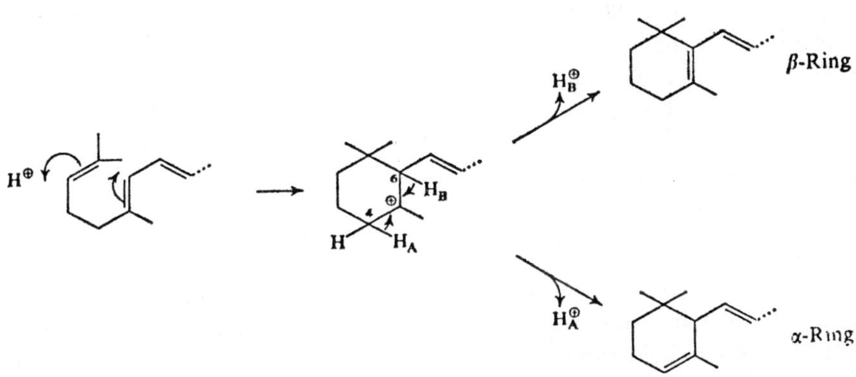

FIG. 3. The probable mechanism for cyclization of an acyclic carotenoid to yield α- and β-rings.

If, on the other hand, the nicotine-grown cells are washed and resuspended aerobically then most of the lycopene runs through to zeaxanthin with apparently β-carotene as an intermediate (Table 3).

Table 3. *The aerobic conversion of lycopene into zeaxanthin by* Flavobact *sp.* (Conditions are given in Table 2 except that the resuspension was carried out aerobically)

Compound	Pigment composition (% of total pigment present)	
	Before resuspension	After resuspension
Lycopene	76·1	13·4
β-Carotene	0	23·6
Rubixanthin	18·7	9·2
Cryptoxanthin	0	3·5
Zeaxanthin	4·5	50·3
Unknown	0·7	0

Further confirmation of the conversion of lycopene into β-carotene was obtained from experiments with cultures grown in the presence of nicotine and [2-^{14}C]mevalonic acid. It will be seen from Table 4 that the control cultures grown for 24 hr in the presence of 7·5 mM nicotine and [2-^{14}C]mevalonic acid then washed and resuspended in new medium with nicotine for 5 hr, produced lycopene with a specific activity of 442 d.p.m./μg. Resuspension for the final 24 hr in the presence of nicotine and under anaerobic conditions resulted in some further synthesis of lycopene (s.a. 522 d.p.m./μg). Similar resuspension but in the absence of nicotine resulted in the disappearance of lycopene and the appearance of β-carotene with a specific activity of 602 d.p.m./μg. Within the limits of this type of experiment, this demonstrated a direct conversion of lycopene into β-carotene and clearly eliminated the breakdown of lycopene and resynthesis of small molecules into β-carotene. The inaccuracy in the experiments arises from the difficulty in removing from the cells all unmetabolized [^{14}C]mevalonic acid. Table 4 shows that a considerable synthesis of lycopene from [2-^{14}C]mevalonic acid had occurred during the second 24 hr reincubation in the presence of nicotine. In a similar experiment in which the synthesis was allowed to run through to zeaxanthin by incubating the washed cells aerobically, the specific activities of lycopene, β-carotene and zeaxanthin were 28·0, 43·0 and 28·8 respectively.

HYDROXYLATION OF CAROTENES

The hydroxylation of β-carotene (6) to form lutein (7) involves carbons 3 and 3′ which originally arose from C-5 of mevalonic acid. By the use of [2-^{14}C-5R-5^{3}H$_1$]mevalonic acid and [2-^{14}C-5-^{3}H$_2$]mevalonic acid it was possible to show that in maize seedlings two hydrogens were lost during the conversion of α-carotene into lutein (7), and that in the formation of cryptoxanthin (8) from

Table 4. *Incorporation of* [2-^{14}*C*]*mevalonic acid into lycopene and β-carotene in* Flavobacter *sp.*

(Cells grown aerobically in the presence of [2-^{14}C]mevalonic acid and nicotine (7·5 mM) for 24 hr, then washed and reincubated in medium plus nicotine but minus mevalonic acid for a further 5 hr; these are control cells. The portions of these cells were then washed and resuspended, anaerobically for 24 hr either in medium plus nicotine or medium minus nicotine)

Reincubation	Lycopene				β-Carotene			
Conditions	Amount μg	% of total pig-ment	total counts (d.p.m.)	sp.ac. (d.p.m./μg)	Amount μg	% of total pig-ment	total counts (d.p.m.)	sp.ac. (d.p.m./μg)
Control	466	79·5	20,623	44·2	0	0	0	0
Anaerobic + Nicotine	550	80·2	28,692	52·2	20	3·0	16	—
Anaerobic − Nicotine	56	8·1	2,027	—	528	76·0	32,041	602

β-carotene by *Physalis alkekengi* fruit only one was lost. These were the *pro-R*-hydrogens at C-3 and C-3′ (Goodwin, Britton & Walton, 1968; Walton, Britton & Goodwin, 1969). Table 5 shows that with [2-^{14}C-5-^{3}H$_2$]mevalonate only two hydrogens were lost, thus ruling out a keto intermediate which would require the loss of four hydrogens. Similar results were obtained with zeaxanthin in *Flavobact.* sp. (Table 7), and only one hydrogen was lost in the case of cryptoxanthin (Table 6).

Table 5. *Incorporation of* (3*RS*)[2-^{14}C,(5*R*)5-^{3}*H*$_1$]*MVA into carotenoids of maize seedlings*

Compound	^{14}C radioactivity	^{3}H radioactivity	^{3}H^{14}C radioactivity ratio	^{3}H/^{14}C atomic ratio
	d.p.m.	d.p.m.		
α-Carotene	85	485	5·70	3·90:8
β-Carotene	3,780	22,126	5·85	4·00:8
Lutein	3,197	9,130	2·86	1·95:8
Lutein diacetate	28	83	2·92	2·00:8

Incorporation of (3*RS*)[2-^{14}*C*,5^{3}*H*$_2$]*MVA into carotenoids of maize seedlings*

Compound	^{14}C radioactivity d.p.m.	^{3}H radioactivity d.p.m.	^{3}H^{14}C radioactivity ratio	^{3}H^{14}C atomic ratio
α-Carotene	212	1,737	8·21	9·76:8
β-Carotene	4,841	41,175	8·51	10·00:8
Lutein	917	6,304	6·88	8·09:8
Lutein diacetate	1,859	12,535	6·74	7·93:8

Table 6. *Incorporation of* (3*RS*)[2-^{14}*C*,(5*R*)5-^{3}*H*$_1$]*MVA and* (3*RS*)[2-^{14}*C*,5-^{3}*H*$_2$] *MVA into carotenoids of* P. alkekengi *fruit*

Compound	^{14}C radio-activity d.p.m.	^{3}H radio-activity d.p.m.	^{3}H^{14}C radio-activity ratio	^{3}H^{14}C atomic ratio
[2-^{14}C,(5*R*)5-^{3}H$_1$]MVA				
Phytoene	5,466	37,918	6·94	8·02:8
β-Carotene	5,442	18,779	3·45	4·00:8
β-Cryptoxanthin	855	2,314	2·70	3·13:8
β-Cryptoxanthin acetate	944	2,435	2·58	2·98:8
[2-^{14}C,5-^{3}H$_2$]MVA				
β-Carotene	2,965	21,105	7·12	10·00:8
β-Cryptoxanthin	4,120	27,321	6·63	9·32:8
β-Cryptoxanthin acetate	1,754	11,435	6·52	9·16:8

Table 7. *The* ^{14}*C*:^{3}*H atomic ratio in products derived from zeaxanthin formed by* Flavobact. *sp. in the presence of* [2-^{14}C, 5*R*-5-^{3}*H*-]*mevalonic acid*

Compound	Radioactivity	(d.p.m.)	Ratio	Atomic ratio
	^{14}C	^{3}H	^{3}H:^{14}C	^{14}C:^{3}H
β-Carotene	2,254	15,837	7·03	8:4
Zeaxanthin	365	1,541	4·22	8:2·4
Zeaxanthin diacetate	952	3,850	4·04	8:2·3
β-Citraurin	370	1,773	4·79	6:2·0
3-Hydroxyretinol	461	1,870	3·05	4:1·1
Rhodoxanthin	256	1,026	4·00	8:2·2
Recovered zeaxanthin	35	132	3·76	8:2·1

Experiments with [2-^{14}C-5*R*-5^{3}H$_1$]mevalonic acid as just stated indicated that there was stereospecific removal of the pro-*R* hydrogen at C-3 and C-3′ as OH enters. If one assumes that a mixed function oxidase is involved and that no inversion takes place when OH enters at C-3 and C-3′, then these biochemical results lead to the absolute configuration (9) of zeaxanthin around C-3 which is the same as that arrived at from chemical investigations (de Ville *et al.*, 1969).

Degradation of zeaxanthin according to the scheme indicated in Fig. 4, followed by radio-assay of the products β-citraurin, 3-hydroxyretinol, and rhodoxanthin confirmed that with [2-^{14}C,5*R*-5^{3}H$_1$]mevalonic acid as substrate no tritiums are present at C-3 and C-3′ and that those two which remain are located at the centre of the molecule attached to C-15 and C-15′. The results given in Table 7 were obtained with our *Flavobact.* spp. (J. C. B. McDermott, G. Britton and T. W. Goodwin, unpublished observations).

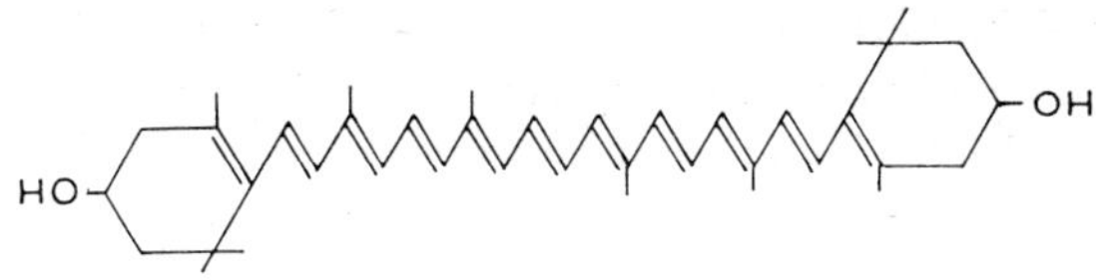

(Lutein)
(7)

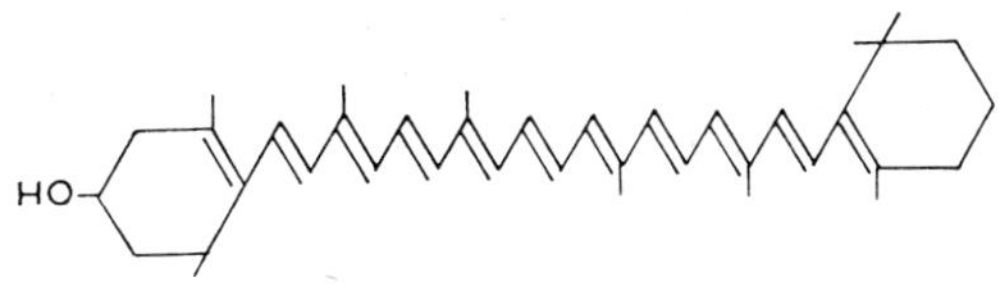

(Cryptoxanthin)
(8)

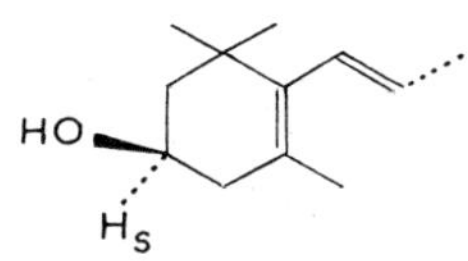

(Zeaxanthin)
(9)

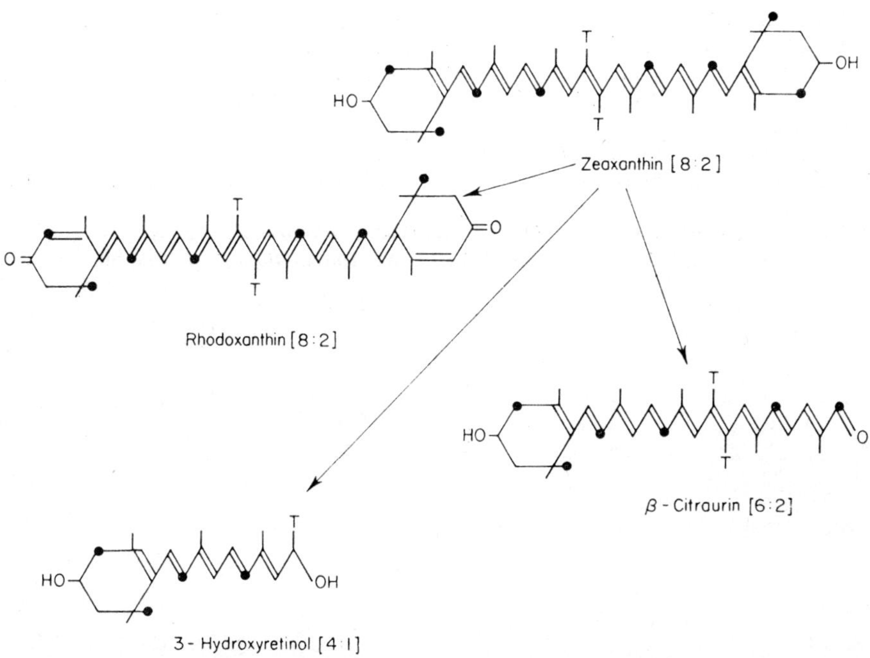

FIG. 4. The degradation of zeaxanthin to rhodoxanthin, β-citraurin and 3-hydroxyretinol. (*T* represents the expected location of tritium and ● of ^{14}C when [2-^{14}C,5 *R*-5-$^{3}H_1$] mevalonate is the substrate. The figures in square brackets represent the expected $^{14}C{}^{3}H$ atomic ratios; see also Table 6.)

The results in Tables 5, 6 and 7 appear to rule out the idea that xanthophylls (e.g. zeaxanthin) are converted in any significant degree to carotenes (e.g. β-carotene) (Costes, 1963). In our experiments the alleged precursor contains two fewer tritium atoms than the alleged final product.

ACKNOWLEDGEMENTS

The work reported here has been generously supported by the Science Research Council and by Roche Products Ltd. The author is grateful to Dr G. Britton for the important part he played in these investigations.

REFERENCES

Bentley, R. (1970). *Molecular Asymmetry in Biology*. 2 Vols. New York: Academic Press.

Buggy, M. J., Britton, G. & Goodwin, T. W. (1969). *Biochem. J.* **114**, 641.

Cornforth, J. W. & Cornforth, R. H. (1970). In *Natural Substances formed Biologically from Mevalonic Acid*. Biochem. Soc. Symp. **29** (Ed. by Goodwin, T. W.), p. 5. London: Academic Press.

Cornforth, J. W. & Ross, F. P. (1970). *J. chem. Soc. D.* 1395.

Costes, C. (1963). *Ann. Physiol. Vegetale* **5**, 115.

Goad, L. J. & Goodwin, T. W. (1972). In *Progress in Phytochemistry* (Ed. by Reinhardt, L.). London: Wiley.

Goodwin, T. W. (1971*a*). In *Carotenoids* (Ed. by Isler, O.), p. 577. Basel: Birkhauser.

Goodwin, T. W. (1971*b*). *Biochem. J.* **123**, 293.

Goodwin, T. W., Britton, G. & Walton, T. J. (1968). *Pl. Physiol.* **43**, 5–46.

Hanson, K. R. (1966). *J. Am. chem. Soc.* **88**, 2731.

Hill, H. M., Calderwood, S. K. & Rogers, L. J. (1971). *Phytochemistry* **10**, 2051.

Hill, H. M. & Rogers, L. J. (1969). *Biochem. J.* **113**, 31P.

Howes, C. D. & Batra, P. P. (1970). *Biochim. biophys. Acta* **222**, 174.

Kushwaha, S. C., Subbarayan, C., Beeler, D. A. & Porter, J. W. (1969). *J. biol. Chem.* **244**, 3635.

Popják, G. & Cornforth, J. W. (1966). *Biochem. J.* **101**, 553.

de Ville, T. E., Hursthouse, M. B., Russell, S. W. & Weedon, B. C. L. (1969). *J. chem. Soc. D.* 1311.

Walton, T. J., Britton, G. & Goodwin, T. W. (1969). *Biochem. J.* **112**, 383.

Williams, R. J. H., Britton, G. & Goodwin, T. W. (1967). *Biochem. J.* **105**, 99.

BIOGENETIC INTERRELATIONSHIP OF UBIQUINONE AND CHOLESTEROL

By T. RAMASARMA

Department of Biochemistry,
Indian Institute of Science,
Bangalore, India

BIOGENESIS of cholesterol and ubiquinone involves a common sequence of reactions. The isoprene groups constitutive of these molecules are derived from acetyl CoA-mevalonate pathway via isopentenyl pyrophosphate. This view was substantiated by the findings that both radioactive acetate and mevalonate were readily incorporated into these compounds in the animals. The ratio of cholesterol/ubiquinone concentrations in the liver is about 20:1, while the ratio of their synthesis as measured by mevalonate incorporation is greater than 50:1. Thus, quantitatively, ubiquinone may be considered to be a minor product in isoprene synthesis. Also, in transfer RNA some of the adenosine bases were found to be substituted at the N^6 position with isopentenyl groups biogenetically derived from mevalonate. The skeleton of the pathway is shown below:

Acetyl CoA
↓
Mevalonate
↓
Isopentenyl pyrophosphate
↙ ↓ ↘
Cholesterol t-RNA Ubiquinone

Most of the tissues of the rat are capable of synthesis of both cholesterol and ubiquinone. A number of inhibitors of biosynthesis of cholesterol are known and they affect different steps in the isoprene pathway, some of which are also common for ubiquinone. It was of interest to study their effects on the concentration and synthesis of ubiquinone in the liver.

Both cholesterol and ubiquinone are lipid in nature and are absorbed by the animal without being degraded, although polymeric in character, and accumulate in large concentration in the liver. This fortuitously afforded an opportunity to test the possibility of end-product inhibition in the intact animals. The results indicated that ubiquinone may play a central role in the natural regulatory mechanism of cholesteral biogenesis and also in the effects obtained under conditions such as treatment with α-*p*-chlorophenoxyisobutyrate (CPIB) and excess throxine. While it appears that the main regulation of isoprene synthesis is at the level of β-hydroxy β-methyl glutaryl CoA (HMG CoA) reductase, circadian rhythm exhibited by this enzyme was not accompanied by corresponding changes in the concentration of either cholesterol or ubiquinone.

Our findings on the biogenetic interrelationship are summarized in this communication.

BIOSYNTHESIS OF CHOLESTEROL AND UBIQUINONE IN RAT TISSUES

In view of the ubiquitous occurrence of ubiquinone in rat tissues, it was of interest to investigate whether individual tissues have the capacity to synthesize the compounds in relation to cholesterol, or whether partial requirement arises in any particular tissue. Gloor & Wiss (1958) showed that (2-^{14}C)-mevalonate was incorporated into the side chain of ubiquinone in rat liver. Joshi, Jayaraman & Ramasarma (1963) extended these studies and found incorporation of (2-^{14}C)mevalonate into cholesterol and ubiquinone in many tissues with widely differing specific radioactivities (Table 1). Except in the liver, the specific radioactivity of ubiquinone in other tissues was higher than cholesterol although its quantity was small. Other studies also revealed that the tissues were capable of independent synthesis of both cholesterol and ubiquinone. In contrast to an apparent lack of exchanges of ubiquinone between tissues, cholesterol of hepatic origin was shown to be the source for the serum and some steroidogenic organs (Friedman, Byers & Michaelis, 1951; Krum, Morris & Bennet, 1964).

Table 1. *Biosynthesis of ubiquinone and cholesterol in rat tissues*

(2-^{14}C)Mevalonate (5 μCi/rat) was orally administered before killing. The procedures of extraction, separation and purification of ubiquinone and cholesterol to constant specific radioactivity are, as given in Joshi *et al.* (1963). Mean values are given for independent determinations of 8 rats

Tissues	μmoles/tissue		c.p.m./μmole	
	Ubiquinone	Cholesterol	Ubiquinone	Cholesterol
Liver	1·22	15·4	250	1620
Kidney	0·11	6·7	2700	430
Intestine	0·02	27·7	3350	350
Spleen	0·01	6·4	1000	230
Heart	0·07	1·7	120	80

EFFECT OF INHIBITION OF CHOLESTEROL SYNTHESIS ON UBIQUINONE

Several conditions and treatments are known to result in changes in the synthesis and concentration of cholesterol. Previous studies identified the steps at which these agents act (Fig. 1). Thus, α-phenyl-butyrate was found to affect the step of acetate activation (Steinberg & Frederickson, 1955) and an additional step after mevalonate (Inamdar, 1968). Supplementing of cholesterol (Bucher *et al.*, 1959; Siperstein & Fagan, 1966), cholic acid (Beher, Baker & Penny, 1963), thyroxine (Scaife & Migicovsky, 1957); and CPIB (Avoy, Swyryd & Gould, 1965), as well as starvation (Regen *et al.*, 1966) are known to

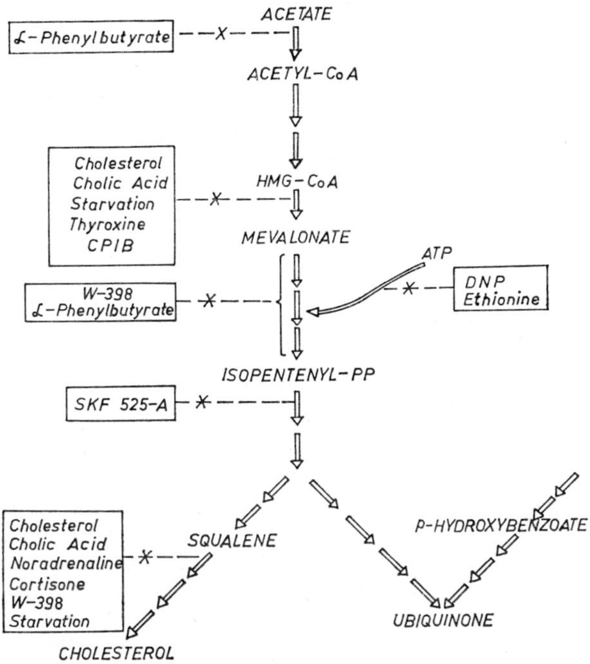

FIG. 1. Sites of inhibition of cholesterol biogenesis.

decrease biogenesis of hepatic cholesterol in the rat and the site of action in all these cases was shown to be between acetate and mevalonate, and in most cases HMG CoA reductase.

Some step between isopentenyl pyrophosphate and squalene was found to be inhibited on treatment of rats with β-diethyl-aminoethylphenyl-propylacetate hydrochloride (SKF-525A) (Holmes & Benz, 1960). Benzyl-*N*-benzylcarbethoxy hydroxamate (W-398) was considered to inhibit at a site prior to mevalonate (Douglas, 1964) but it was shown in this laboratory that this drug inhibited at two sites, one after mevalonate and another after squalene (Inamdar, 1968). Cholesterol feeding (Krishnaiah, Joshi & Ramasarma, 1967), cholic acid supplementing (Krishnaiah, 1968), noradrenaline and cortisone treatments (Inamdar, 1968) and starvation (Scaife & Migicovsky, 1957) are known to enormously increase incorporation of (2-^{14}C)mevalonate into the hydrocarbon fraction and a corresponding decrease in the sterols fraction indicating inhibition at the step of utilization of squalene. In addition to these conditions, cold exposure, vitamin A deficiency and protein deficiency, which were implicated in ubiquinone metabolism, were also included in the programme of study. The effects of these treatments on concentration, synthesis and catabolism of ubiquinone were studied. While most of the conditions listed above decreased cholesterol biogenesis, the effects on ubiquinone were variable —some increased it, some decreased it and some others had no effect.

The experimental conditions under which synthesis of ubiquinone was enhanced are listed in Table 2. Supplementing of thyroxine, adrenaline and

Table 2. *Conditions wherein concentration and synthesis of ubiquinone in the liver increased*

Experimental condition	Concentration	Biogenesis from (2-^{14}C)mevalonate	
		Intermediate stage	Late stage
Thyroxine treatment (30 mg/kg diet)	Increased (1,2)	Increased (3)	Decreased (3)
Iodinated casein feeding (Thyrotoxicosis; 0·3% in diet	Increased (4)	Increased (3)	Decreased (3)
Thyroxine + Thiouracil (30 mg/kg + 0·3% in diet)	Increased (5)	Increased (3)	—
CPIB feeding (0·5% in diet)	Increased (6)	Increased (7)	Decreased (7)
Adrenaline treatment	Increased (8)	Increased (8)	—
Vitamin A deficiency	Increased (9)	Increased (10)	Decreased (10)
Cold exposure (0–5°C)	Increased (11)	Increased (12)	Increased (12)

The experimental conditions are given in the references shown by numbers in parentheses: (1) Edwin *et al.* (1960); (2) Beyer, Noble & Hirschfeld (1962*b*); (3) Inamdar & Ramasarma (1969); (4) Aiyar *et al.* (1959); (5) Brubacher & Wiss (1961). (6) Krishnaiah, Inamdar & Ramasarma (1967); (7) Krishnaiah & Ramasarma (1970*a*); (8) Inamdar (1968); (9) Heaton, Lowe & Morton (1957); (10) Lakshmanan *et al.* (1968); (11) Beyer, Noble & Hirschfeld (1962*a*); (12) Aithal, Joshi & Ramasarma (1968).

CPIB increased hepatic ubiquinone concentration through an increase in its synthesis at an intermediate stage. The large accumulation observed following thyroxine and CPIB treatments, in vitamin A deficiency and on cold exposure was due to lowered catabolism as well as increased synthesis. It is paradoxical that CPIB, a plasma cholesterol depressing drug, known to inhibit mevalonate formation, produced a marked increase in the concentration of ubiquinone. This was made possible due to increased synthesis occurring at an intermediate stage before the above inhibition became manifest and the inhibition itself might have been caused by increased ubiquinone concentration, as discussed later. None of these agents, however, was able to activate ubiquinone synthesis in tissue slices when added *in vitro*.

Conditions wherein inhibition of ubiquinone synthesis *in vivo* was obtained are listed in Table 3. Under these conditions hepatic concentration of ubiquinone also decreased. These, therefore, appear to act in a similar fashion as on cholesterol and may be affecting the common steps in the pathway. An exception is protein deficiency. There was no effect either on concentration or on

Table 3. *Conditions wherein concentration and synthesis of ubiquinone in the liver decreased*

Experimental condition	Concentration	Biogenesis from (2-^{14}C)mevalonate
SKF 525A treatment (1% in diet)	Decreased (1)	Decreased (1)
W-398 treatment (1% in diet)	Decreased (1)	Decreased (1)
α-phenylbutyrate treatment (1% in diet)	Decreased (1)	Decreased (1)
2,4-dinitrophenol treatment (0·3% in diet)	Decreased (1) Increased per unit weight (2)	Decreased (1)
Ethionine treatment (0·5% in diet)	Decreased (1)	Decreased (1)
Protein deficiency	Decreased (3)	Decreased (4)
Starvation	Decreased (5) Increased per unit weight (2)	Decreased (5)

Experimental conditions are given in the references shown by numbers in parentheses: (1) Inamdar (1968); (2) Beyer, Noble & Hirschfeld (1962*b*); (3) Joshi, Jayaraman & Ramasarma (1963); (4) Joshi, Jayaraman & Ramasarma (1965); (5) Inamdar & Ramasarma (1971).

synthesis of cholesterol, when rats were given a protein-deficient diet or, for that matter, other diets deficient in carbohydrates for fats (G. Subba Rao & T. Ramasarma, unpublished observations). Complete deprivation of diet caused a rapid decrease in the biogenesis of cholesterol which suggested that caloric intake in any form can maintain cholesterol biogenesis. The decrease in ubiquinone under protein deficiency, therefore, appears to be due to non-availability of phenyl-amino acids for the synthesis of the ring precursor, and not due to any effect on isoprene synthesis.

Inhibitors of cholesterol synthesis which inhibit at steps after branching in the pathway should have no effect on ubiquinone. This was indeed found to be true after supplementation with cholesterol, cholic acid, cortisone and noradrenaline which are all inhibitory at the squalene stage (Table 4). Exceptions are starvation and W-398 treatment which decrease ubiquinone concentration because they also have sites of inhibition in the common pathway. Cholesterol and cholic acid supplements do not show any change in ubiquinone concentration despite the inhibition of mevalonate formation because the residual mevalonate appears to be sufficient for the small amount of ubiquinone synthesized. Decreases were, however, noted after long periods of supplementing.

Absorption of ubiquinone in the rat

Diets deficient in ubiquinone had not produced any change in the concentration of ubiquinone in rat tissues, probably because of its synthesis in all tissues (Lawson *et al.*, 1961). However, it can be absorbed when given exogenously

Table 4. *Conditions wherein concentration and synthesis of ubiquinone in the liver remain unchanged*

Experimental condition	Concentration	Biogenesis from (2-^{14}C)mevalonate
Cholesterol feeding (1% in diet)	No change (1) (decreased after long period)	Apparent increase (1)
Cholic acid feeding (0·5% in diet)	No change (2) (decreased after long period)	Apparent increase (2)
Cortisone treatment (2 mg/day/rat)	No change (3)	No change (3)
Noradrenaline treatment (0·1 mg/day/rat)	No change (3)	No change (3)

The experimental conditions are given in the references shown by numbers in parentheses: (1) Krishnaiah *et al.* (1967); (2) Krishnaiah (1968); (3) Inamdar (1968).

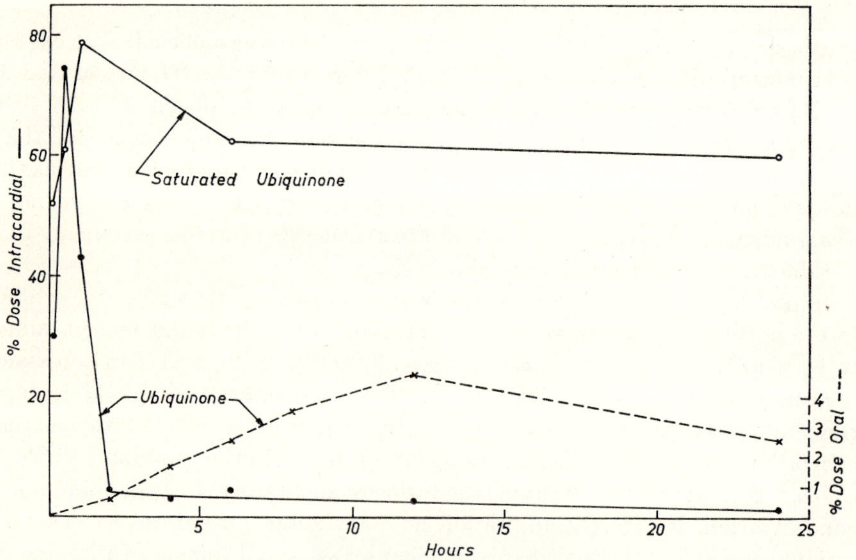

FIG. 2. Absorption and retention of exogenous ubiquinone after administration of radioactive ubiquinone (0·25 μCi]rat). Groups of rats were killed at various time intervals, livers were removed and radioactivity in the nonsaponifiable lipids was determined. The mean percentage values of the dose retained in the liver are given. Ubiquinone was labelled with ^{14}C in the first two carbons in the side chain attached to the ring (specific radioactivity, 10 μCi/mg). In one sample the double bonds in the side chain were saturated by catalytic hydrogenation. Routes of administration: intracardial (——); oral (-----). The data are taken from Jayaraman & Ramasarma (1963) and Joshi *et al.* (1964).

(Lawson *et al.*, 1961; Rudney, 1961). Given intracardially, the compound was found to be removed from the blood rapidly and was detected in the liver and to a small extent in the spleen without any change in its form. Once the compound was deposited in the liver, it was not mobilized by any other tissue even after long periods (Jayaraman & Ramasarma, 1963). On administration by oral route, ubiquinone was exclusively absorbed by the liver. Its concentration in the liver decreased after long intervals of time but it did not appear in any other tissue. These observations were interpreted to indicate total lack of internal transport of ubiquinone in rat tissues (Joshi *et al.*, 1964). The fate of administered ubiquinone is in marked contrast to that of other related lipid compounds like vitamin K (Taylor *et al.*, 1956), α-tocopherol (Weber *et al.*, 1964) and ubichromenol (Joshi *et al.*, 1963). It was also found that the double bonds in the side chain are necessary for both absorption and catabolism of ubiquinone, since saturated ubiquinone was not absorbed at all and once deposited in the liver by intracardial route, was metabolized extremely slowly (Fig. 2).

When exogenous ubiquinone was supplied in small daily doses, it accumulated in the liver and reached a maximum level after several days. When

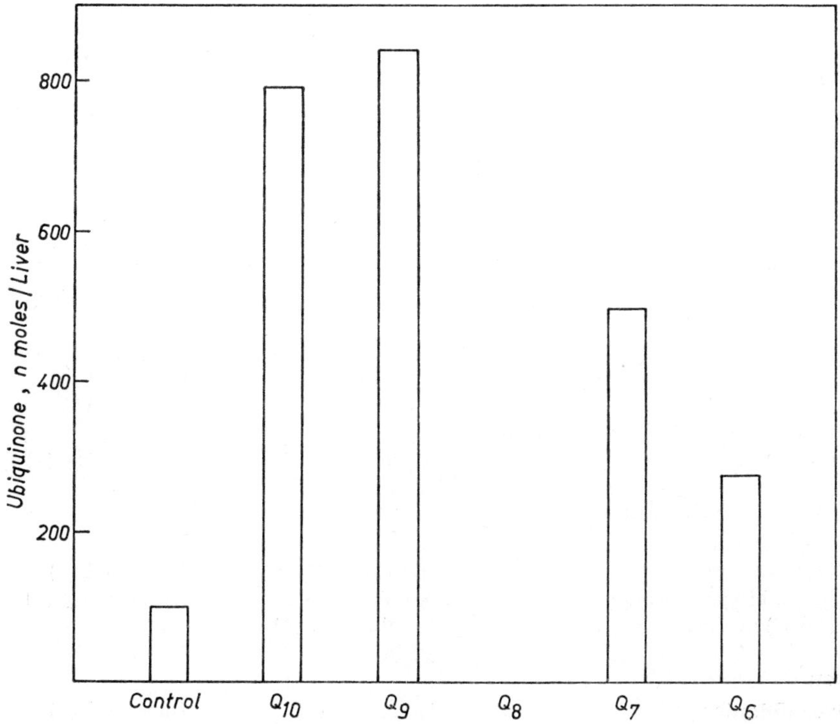

FIG. 3. Concentrations of ubiquinone in the livers of rats fed different homologues of ubiquinone. The homologues of ubiquinone, Q_{10}, Q_9, Q_7 and Q_6 (Q_8 was not tested) were fed at 1·5 mg/day/rat for 40 days. The rats were then killed and the concentrations of ubiquinone in the livers were measured. It may be noted that on molar basis higher amounts of the lower homologues were fed. The amounts retained were only small percentages of the dose and represent the steady higher levels compared to the control. Liver sterols remained unchanged under these conditions (Data from Krishnaiah, 1968).

different homologues like Q_{10}, Q_9, Q_7 and Q_6 were given maximum accumulation was obtained with the natural, major homologue Q_9, with relatively lower levels for other homologues (Fig. 3). Absorption of exogenous ubiquinone and its accumulation in the liver prompted the study of regulation of isoprene synthesis by ubiquinone on parallel lines as with dietary supply of cholesterol.

Regulation of cholesterol biogenesis by CPIB and ubiquinone

Administration of CPIB produced twin effects: an inhibition of hepatic biogenesis of cholesterol at a step prior to mevalonate (Avoy *et al.*, 1965) and a decrease in serum cholesterol (Thorp & Waring, 1962). Similar effects were also observed on feeding very low amounts of ubiquinone (Krishnaiah & Ramasarma, 1970*b*). A comparative study showed that remarkably identical effects were produced with ubiquinone and CPIB feedings (Table 5). Since ubiquinone

Table 5. *Comparison of the effects of feeding CPIB and ubiquinone*

	Ubiquinone fed	CPIB fed
NORMAL ADULT RATS		
Serum cholesterol	Decreased (1)	Decreased (1)
Liver cholesterol	No change (2)	No change (3)
Liver ubiquinone	Increased (2) (2- to 4-fold)	Increased (3) (2- to 4-fold)
Hepatic biogenesis of sterols from (1-^{14}C) acetate	Decreased (2) (50% or under)	Decreased (3) (50% or under)
Hepatic biogenesis of sterols from (2-^{14}C) mevalonate	No change (2)	No change (3)
Hepatic biogenesis of ubiquinone from (2-^{14}C) mevalonate	Decreased (2)	Increased initially followed by decrease (3)
YOUNG RATS LESS THAN 2 MTHS OLD		
Serum cholesterol	No decrease (4)	No decrease (5)
Liver ubiquinone	Increased (4) (2-fold)	Increased (4) (2-fold)
Hepatic biogenesis of sterols from (1-^{14}C) acetate	No decrease (4)	No decrease (4)
COLD EXPOSED ADULT RATS (0–5°, 5 days)		
Liver ubiquinone	Increased (4)	Increased (4)
Hepatic biogenesis of sterols from (1-^{14}C) acetate	No decrease; enhanced (4)	No decrease (4)
MICE		
Liver ubiquinone	Increased (6)	Increased (6)
Serum cholesterol	No decrease (6)	No decrease (6)

Groups of rats were fed either 0·5% CPIB in the diet or 1·5 mg/day/rat of ubiquinone-9 orally to 10 days. The experimental conditions are given in the reference shown by numbers in parentheses: (1) Krishnaiah *et al.* (1967); (2) Krishnaiah & Ramasarma (1970*b*); (3) Krishnaiah & Ramasarma (1970*a*); (4) Krishnaiah (1968); (5) Harris, Machintoch & Quackenbush (1966); (6) Krishnaiah & Jayaraman (unpublished data).

accumulated in the liver in both cases, it was conjectured that ubiquinone might be acting as the regulatory molecule in the biogenesis of hepatic cholesterol (Ramasarma, Krishnaiah, Inamdar & Aithal, 1969). It must also be pointed out here that some of these effects are small and variable and that the conditions required for the manifestation of the effects are yet to be perfected. At this stage evidence is insufficient to state that ubiquinone is functioning through a mechanism of end-product regulation of a common enzyme in the pathway. It is possible that both CPIB and ubiquinone might be acting via another unknown intermediate as represented below:

CPIB ⟶ Ubiquinone ⟶ Hepatic cholesterol ⟶ Serum cholesterol

Ubiquinone
↓
CPIB ⟶ ? ⟶ Hepatic cholesterol ⟶ Serum cholesterol

CIRCADIAN RHYTHM OF BIOGENESIS OF CHOLESTEROL

It is now recognized that cholesterol biogenesis is subject to circadian rhythm in the rat. While the increase in the activity of HMG CoA reductase, the rate-limiting enzyme in cholesterol biogenesis, was about 10-fold (Shapiro & Rodwell, 1969) at midnight compared to noon, the increase in the incorporation of radioactive acetate into cholesterol in liver slices *in vitro* (Hamprecht, Nüssler & Lynen, 1969) and *in vivo* was only about 3-fold (Fig. 4). The increase in the overall synthetic rate was less than the increase in the enzyme activity thereby implying that this enzyme may not be the rate-limiting step at midnight. Also, the data in Fig. 4 indicate that concentrations of sterols in the liver and blood-serum as well as of ubiquinone in the liver (data not shown) do not respond to rhythmic changes in the hepatic cholesterol biogenesis and HMG CoA reductase. Lack of corresponding changes in the products of isoprene synthesis make the physiological significance of the circadian rhythm obscure.

Differential response of the products to alterations in biogenesis

In the foregoing study a number of conditions or supplements were chosen which showed alterations in biogenesis of cholesterol from radioactive acetate and not from mevalonate. The effects appear to be on the formation of mevalonate and possibly on HMG CoA reductase, the rate-limiting step in the pathway. An examination of the effects on the concentrations of the end-products of the pathway showed the following remarkable variability (Table 6):

(i) liver concentration of cholesterol is usually least affected;

(ii) serum cholesterol is maintained at a constant level in starvation and at midnight compared to noon whereas the two conditions gave opposite effects of profound degree on biogenesis;

(iii) liver ubiquinone concentration remains unaltered even when the isoprene synthesis is depressed to a very low level as well as when it increased several fold;

(iv) paradoxically, in three cases—thyroxine, CPIB and ubiquinone supplements—the concentration of ubiquinone increased several fold despite a possible decrease in the formation of mevalonate.

These studies have provided two examples where an alteration of a rate-limiting step in the common pathway showed opposite responses in the two products, on the one hand, and a lack of response in the concentration of any of the products of synthesis, on the other.

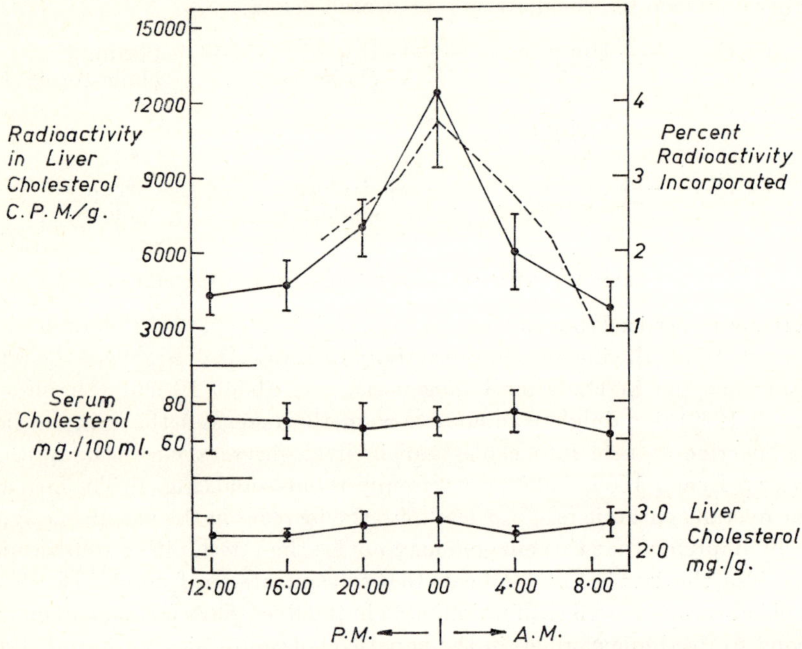

FIG. 4. Rhythmic changes in biogenesis of cholesterol *in vivo*. Male albino rats weighing 150–180 g, provided with the pellet diet (Hindustan Lever rat diet) and water *ad libitum*, were housed in a room with a light period from about 6.00 a.m. to 6.00 p.m. (1-^{14}C) Acetate (specific radioactivity 49 mCi/mmole) was injected intraperitoneally (25 μCi/rat) to the animals 30 min before killing. The concentrations of sterols in the non-saponifiable lipids of the serum and the liver were estimated and after chromatography on deactivated alumina column the radioactivity in the sterol fraction was determined. At noon and midnight, 20 rats were used per group for estimation of the concentrations of sterols in the liver and the serum of which nine rats were used for the studies on the incorporation of (1-^{14}C)acetate. At 4.00 a.m. and 9.00 a.m. four rats in each group were used. (The dotted lines in the figures represent the data from Back *et al.* (1969) on the percent incorporation of (1-^{14}C)acetate into sterols in liver slices, *in vitro*).

SUMMARY

1. Ubiquinone and cholesterol are synthesized in all rat tissues examined via the common acetate-mevalonate pathway. 2. The inhibitors of cholesterol biogenesis showed variable effects on the biogenesis and concentration of ubiquinone in the liver; some showed paradoxical accumulation of ubiquinone owing to increased synthesis at the intermediate stage. 3. Like cholesterol, ubiquinone was also absorbed into the liver and showed significant inhibition

Table 6. *A comparative study on the effects on concentration and synthesis of cholesterol and ubiquinone in rat liver conditions of altered isoprene biosynthesis*

Condition or supplement	(1-^{14}C)Acetate⟶ cholesterol	HMG CoA reductase	Concentration of cholesterol		Concentration of ubiquinone
			Liver	Serum	
Cholesterol	Decreased (over 90%) (1)	Decreased (2)	Increased (absorption)	Increased (absorption)	No change (5)
Cholic acid	Decreased (over 90%) (3)	Decreased (4)	No change	—	No change (5)
Starvation	Decreased (1) (over 90%)	Decreased (6)	Decreased (17)	No change	Decreased (7) (about 50%)
Thyroxine	Decreased (8)	—	No change (9)	Decreased	Increased (9) (2- to 4-fold)
CPIB	Decreased (10) (50% or under)	Decreased (11)	No change (10)	Decreased (12)	Increased (13) (2- to 4-fold)
Ubiquinone	Decreased (14) (50% or under)		No change (14)	Decreased (14)	Increased (14) (2- to 4-fold)
Midnight *vs* Noon	Increased (15) (3-fold)	Increased (10-fold) (16)	No change (17)	No change (17)	No change (17)

Experimental conditions are given in the references shown by numbers in parentheses: (1) Bucher *et al.* (1959); (2) White & Rudney (1971); (3) Beher *et al.* (1963); (4) Fimognari & Rodwell (1965); (5) Krishnaiah (1968); (6) Regen *et al.* (1966); (7) Inamdar & Ramasarma (1971); (8) Scaife & Migicovsky (1957); (9) Inamdar & Ramasarma (1969); (10) Avoy *et al.* (1965); (11) White (1971); (12) Thorp & Waring (1962); (13) Krishnaiah *et al.* (1967); (14) Krishnaiah & Ramasarma (1970*b*); (15) Back, Hamprecht & Lynen (1969); (16) Shapiro & Rodwell (1969); (17) Subba Rao & Ramasarma (1972).

of a step before mevalonate; this caused a decrease in serum cholesterol and resembled in many respects the effects obtained with the antihypercholesterolemic drug, α,*p*-chlorophenoxyisobutyrate. 4. The circadian rhythmic changes obtained in biogenesis of cholesterol in the liver did not show corresponding changes in concentrations of the products namely cholesterol in the liver and the serum, and ubiquinone in the liver. 5. Alterations in the regulatory step in the common pathway before mevalonate showed divergent effects on cholesterol and ubiquinone.

ACKNOWLEDGEMENTS

It is a pleasure to acknowledge participation in the various phases of this work by my colleagues: Drs J. Jayaraman, V. C. Joshi, C. K. Ramakrishna Kurup, K. V. Krishnaiah, A. R. Inamdar, H. N. Aithal and Mr G. Subba Rao. Financial assistance was received from the following agencies: University Grants Commission (India), Council of Scientific and Industrial Research (India), Indian Council of Medical Research, the Ford Foundation, the Department of Atomic Energy (India) and the Indian Institute of Science.

REFERENCES

Aithal, H. N., Joshi, V. C. & Ramasarma, T. (1968). *Biochim. biophys. Acta* **162**, 66.

Aiyar, A. S., Sulebole, G. A., Rege, D. V. & Sreenivasan, A. (1959). *Nature, Lond.* **184**, 1867.

Avoy, D. R., Swyryd, E. A. & Gould, R. G. (1965). *J. Lipid. Res.* **6**, 555.

Back, P., Hamprecht, B. & Lynen, F. (1969). *Archs Biochem. Biophys.* **133**, 11.

Beher, W. T., Baker, G. D. & Penny, D. G. (1963). *Archs Biochems. Biophys.* **133,** 11.

Beyer, R. E., Noble, W. M. & Hirschfeld, T. J. (1962*a*). *Can. J. Biochem. Physiol.* **40,** 511.

Beyer, R. E., Noble, W. M. & Hirschfeld, T. J. (1962*b*). *Biochim. biophys. Acta* **57,** 376.

Brubacher, G. & Wiss, O. (1961). *Helv. Physiol. Acta* **19,** C57.

Bucher, N. L. R., McGarrahan, K., Gould, E. & Loud, A. V. (1959). *J. biol. Chem.* **234,** 262.

Douglas, J. F. (1964). *Proc. Soc. exp. Biol. Med.* **117,** 190.

Edwin, B. E., Green J., Diplock, A. T. & Bunyan, J. (1960). *Nature, Lond.* **186,** 725.

Fimognari, G. M. & Rodwell, V. W. (1965). *Science* **147,** 1038.

Fletcher, K. & Myant, N. B. (1961). *J. Physiol.* **517,** 542.

Friedman, M., Byers, S. O. & Michaelis, F. (1951). *Am. J. Physiol.* **164,** 789.

Gloor, U. & Wiss, O. (1955). *Experientia* **24,** 410.

Hamprecht, B., Nüssler, C. & Lynen, F. (1969). *FEBS Lett.* **4,** 117.

Harris, R. A., Machintoch, J. E. & Quackenbush, F. W. (1966). *J. Nutr.* **90,** 40.

Heaton, F. W., Lowe, J. S. & Morton, R. A. (1957). *Biochem. J.* **67,** 208.

Holmes, W. L. & Benz, J. D. (1960). *J. biol. Chem.* **235,** 3118.

Inamdar, A. R. (1968). *Metabolism of Ubiquinone in Relation to Nutritional Status*, Ph.D. Thesis, Indian Institute of Science, Bangalore.

Inamdar, A. R. & Ramasarma, T. (1969). *Biochem. J.* **111,** 479.

Inamdar, A. R. & Ramasarma, T. (1971). *Indian J. Biochem. Biophys.* **8,** 271.

Jayaraman, J. & Ramasarma, T. (1963). *Archs Biochem. Biophys.* **103,** 258.

Joshi, V. C., Jayaraman, J. & Ramasarma, T. (1963). *Indian J. exp. Biol.* **1,** 113.

Joshi, V. C., Jayaraman, J. & Ramasarma, T. (1965). *Biochem. biophys. Res. Commun.* **18,** 108.

Joshi, V. C., Jayaraman, J., Ramakrishna Karup, C. K. & Ramasarma, T. (1964). *Indian J. Biochem.* **1,** 7.

Krishnaiah, K. V. (1968). *Regulation of Metabolism of Isoprenoid Compounds*, Ph.D. Thesis, Indian Institute of Science, Bangalore.

Krishnaiah, K. V., Inamdar, A. R. & Ramasarma, T. (1967). *Biochem. biophys. Res. Commun.* **27,** 474.

Krishnaiah, K. V., Joshi, V. C. & Ramasarma, T. (1967). *Archs Biochem. Biophys.* **121,** 47.

Krishnaiah, K. V. & Ramasarma, T. (1970*a*). *Biochem. J.* **116,** 321.

Krishnaiah, K. V. & Ramasarma, T. (1970*b*). *Biochim. biophys. Acta* **202,** 332.

Krum, A. A., Morris, M. D. & Bennet, L. L. (1964). *Endocrinol.* **74,** 543.

Lawson, D. E. M., Threlfall, D. R., Glover, J. & Morton, R. A. (1961). *Biochem. J.* **79,** 201.

Lakshmanan, M. R., Cama, H. R., Joshi, V. C. & Ramasarma, T. (1968). *Biochim. biophys. Acta* **170,** 392.

Ramasarma, T., Krishnaiah, K. V., Inamdar, A. R. & Aithal, H. N. (1969). In *Drugs Affecting Lipid Metabolism* (Ed. by Holmes, W. L., Carlson, L. A. & Paoletti, R.), p. 395. Plenum Press.

Regen, D., Riepertinger, C., Hamprecht, B. & Lynen, F. (1966). *Biochem. Z.* **346,** 78.

Rudney, H. (1961). *J. biol. Chem.* **236,** PC39.

Scaife, J. F. & Migicovsky, B. B. (1957). *Can. J. Biochem. Physiol.* **35,** 615.

Shapiro, D. J. & Rodwell, V. W. (1969). *Biochem. biophys. Res. Commun.* **37,** 867.

Siperstein, M. D. & Fagan, V. M. (1966). *J. biol. Chem.* **241,** 602.

Steinberg, D. & Frederickson, D. S. (1955). *Proc. Soc. exp. Biol. Med.* **90,** 232.

Subba Rao, G. & Ramasarma, T. (1972). *Environmental Physiol* (In press.)

Taylor, J. D., Miller, G. J., Jacques, L. B. & Spinks, J. W. T. (1956). *Can. J. Biochem. Physiol.* **34,** 1143.

Thorp, J. M. & Waring, W. S. (1962). *Nature, Lond.* **194,** 948.

Weber, F., Gloor, U., Wursch, J. & Wiss, O. (1964). *Biochem. biophys. Res. Commun.* **14,** 189.

White, L. W. (1971). *J. Pharmacol. exp. Ther.* **178,** 361.

White, L. W. & Rudney, H. (1970). *Biochemistry* **9,** 2725

CELL-FREE BIOSYNTHESIS OF POLYPRENYL DERIVATIVES OF *p*-HYDROXYBENZOIC ACID IN MAMMALIAN AND BACTERIAL SYSTEMS

By HARRY RUDNEY

Department of Biological Chemistry, College of Medicine, University of Cincinnati, Cincinnati, Ohio, U.S.A.

THE prenylation of *p*-hydroxybenzoate is the first step in its conversion to ubiquinone. This was postulated by Folkers, Rudney and collaborators (Olsen *et al.*, 1965; Olsen *et al.*, 1966; and Friis *et al.*, 1966) as part of the overall biosynthetic pathway for ubiquinone in *Rhodospirillum rubrum*. This postulate followed from the initial isolation of 2-decaprenylphenol and 6-methoxy-2-decaprenylphenol, the first of a series of compounds, derived from *p*-hydroxybenzoate which were shown to be plausible precursors of ubiquinone synthesis in *R. rubrum*. The pathway is outlined in Fig. 1.

The cell-free synthesis of all the intermediates in the sequence has not yet been achieved, but the pathway is receiving support by the finding of polyprenylphenol derivatives in a variety of organisms including mammals.

Previous reports (Raman *et al.*, 1969; Winrow & Rudney, 1969; Rudney, 1970) documented studies which showed that polyprenylation of PHB could occur in mammalian and bacterial cell-free systems. This reaction did not take place readily unless the system was supplemented with extracts from *Micrococcus lysodeikticus* and isopentenyl pyrophosphate. This supplemented system could form 3-polyprenyl-4-hydroxy-benzoate (PPHB) and 2-polyprenylphenol (PPP) in *R. rubrum* and *E. coli* extracts. In rat tissue homogenates, e.g. liver, kidney, brain, only PPHB could be formed. In kidney homogenates the major activity resided in the mitochondrial fraction. In this report from our laboratory some subsequent studies in collaboration with Dr. Kazutaka Momose on PPHB synthesis in rat liver will be described.

As stated earlier, no cell-free synthesis of PPHB could be observed unless IPP and MLE were present. We are now able to observe a synthesis of PPHB without the use of the MLE. With 7-^{14}C-PHB as a substrate one can easily separate out the formation of PPHB from other products because this precursor is the only one that retains the carboxyl group. An important factor was a change to a more effective extraction procedure which seemed to be able to protect the compounds against deterioration in the cell-free system. Using rat

Abbreviations: PHB—*p*-hydroxybenzoic acid; MLE—*Micrococcus lysodeikticus* extract; PPP—2-polyprenylphenol; PPHB—3-polyprenyl-4-hydroxybenzoate; UQ—ubiquinone; MVA—mevalonate; IPP—isopentenyl pyrophosphate.

liver and kidney minces and 7-^{14}C-PHB and mevalonic acid, we could now observe the appearance of a compound during TLC with all the characteristics of the PPHB previously observed by Winrow & Rudney (1969) in the MLE and IPP supplemented systems. It was labelled with tritiated mevalonic acid and the tritium and ^{14}C label coincided in the same area in the chromatogram.

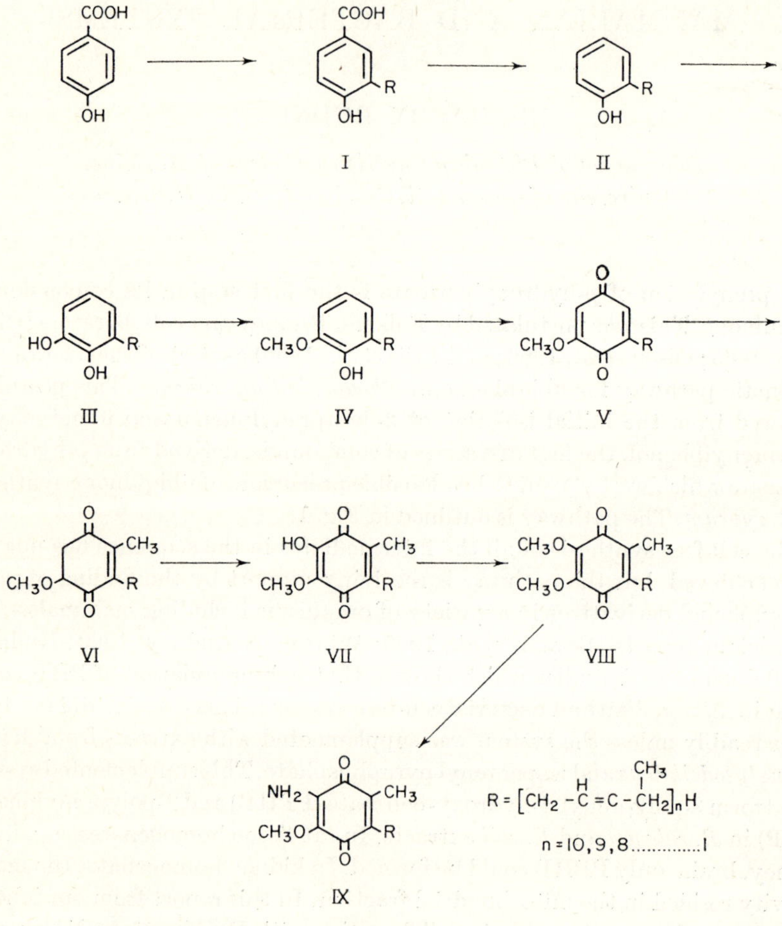

FIG. 1. Postulated pathway for the synthesis of ubiquinones from *p*-hydroxybenzoate in *R. rubrum* (Friis *et al.*, 1966).

The PPHB had an R_f of 0·35 in a TLC system of silica gel using 15% acetone in petroleum ether as the developing solvent. Since these were mince preparations, ubiquinone formation was also observed.

It was noted that the radioactivity in PPHB preceded that in UQ. Pulse label experiments were done to determine whether PPHB had the characteristics of a precursor of ubiquinone. This experiment is shown in Fig. 2. Definitive evidence of a precursor product relationship between PPHB and UQ was obtained from the shape of the curves for the incorporation of U-^{14}C-PHB into PPHB and UQ. When a pulse of non-radioactive PHB is added after 60 min to completely dilute all the radioactivity originally present in PHB it can be seen that the incorporation of radioactivity into PPHB decays rapidly within 30 min, and that ubiquinone continues to rise and remain constant apparently followed by a decay. These curves are typical of those observed between compounds with a precursor product relationship.

We then turned our attention to cell-free preparations. Full experimental details will be presented elsewhere (Momose & Rudney, 1972), but the procedure consists of careful preparation of mitochondria according to the method of Parsons & Williams (1967). The two-speed centrifugation portion of this procedure was repeated three times or more to minimize contamination by the microsomal fraction. The incubation mixture consisted of mitochondria (25–50 mg protein) suspended in 0·1 M potassium phosphate buffer in 0·02%

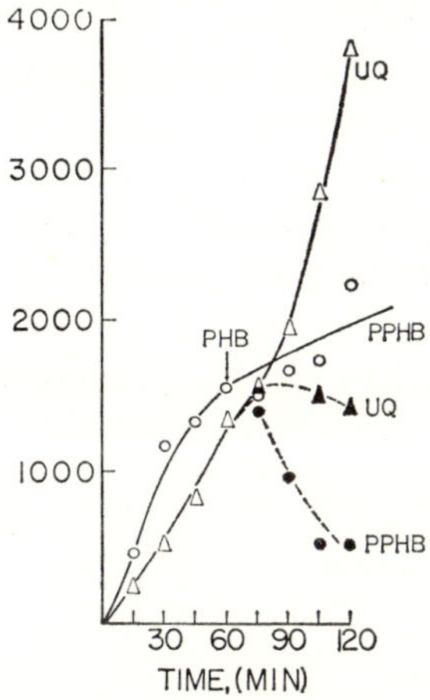

FIG. 2. Time course of the incorporation of radioactivity from U-^{14}C-PHB into PPHB and ubiquinone in rat liver mince. Rat liver mince was incubated with U-^{14}C-PHB and mevalonic acid: (——) control, (– – – –) after addition of non-radioactive PHB.

bovine plasma albumin, 10 μmoles magnesium chloride, 1 μmole manganese sulphate, 10 μmoles potassium fluoride, 1 μmole trilithium isopentenyl pyrophosphate, and 1 nanomole radioactive *p*-hydroxybenzoate in a total volume of 1·5 ml. The incubation was terminated by the addition of 3·75 volumes of chloroform-methanol (1:2), and allowed to stand overnight at room temperature. Lipids were extracted by the method of Galliard *et al.* (1965) and subsequent column and thin layer chromatography was carried out as described by Rudney (1970). Reverse phase chromatography of PPHB is outlined in the text.

When minces were incubated with PHB and MVA and the distribution of the resultant PPHB was investigated, it was found that PPHB was localized almost exclusively in the mitochondria. The purification of the mitochondria resulted in an increased concentration of PPHB, these results are shown in Table 1. Further investigation of rat liver homogenates showed that ATP and mevalonate were required for the incorporation of PHB into PPHB. Fractionation of the homogenates and recombination of the fractions clearly

Table 1. *Distribution of PPHB in cellular fractions in rat liver mince*

Cell fraction	PPHB dpm/mg protein
Homogenate	3·0
Cytosol	0·3
Crude mitochondria	2·9
Purified mitochondria	12·3

Rat liver mince was incubated with 7-^{14}C-PHB and mevalonate for 1 hr in Krebs Ringer bicarbonate buffer. The minces were washed and homogenized and the homogenate fractionated by centrifugation.

showed that mitochondria and the supernatant fraction obtained by centrifugation of the homogenate at 48,000 × g were required for the formation of PPHB. These results are shown in Table 2. It is noteworthy that the addition of microsomal fractions resulted in decreased activity in the formation of PPHB. Presumably, microsomes reduce the level of PPHB by converting a

Table 2. *Activity of rat liver subcellular fractions for PPHB synthesis*

Fraction	Radioactivity in PPHB (dpm)
Mitochondria	39
Mitochondria + F-10	148
Mitochondria + F-48	1058
Mitochondria + F-10 + F-48	632
F-10 + F-48	16

Rat liver (7 g) was homogenized in calcium free Krebs Ringer bicarbonate solution. The homogenate was fractionated by centrifugation as follows: 120–10,000 × g = mitochondria, 10,000–48,000 × g = F-10, supernatant fraction from 48,000 × g centrifugation = F-48. Each fraction incubated with 10 nanomoles 7-^{14}C-PHB, 3 mM ATP and 1 mM mevalonate in 10 ml Krebs Ringer bicarbonate buffer saturated with 95% O_2 and 5% CO_2.

portion of the polyprenyl pyrophosphate side-chains to squalene and cholesterol. The 48,000 × g fraction was necessary to convert mevalonate to isopentenyl pyrophosphate. This is borne out by the results shown in Table 3. It can be seen that isopentenyl pyrophosphate will substitute for mevalonate and ATP; however, most noteworthy was the finding that mitochondria alone without the supernatant fraction in the presence of IPP could form polyprenyl-*p*-hydroxybenzoate. This result provided the first indication that mitochondria contains a series of polyprenyl transferases capable of building up a polyprenyl pyrophosphate chain. Although the mitochondria were prepared by suspension in sucrose medium for this experiment, subsequent experiments on different buffering systems indicated that maximal synthesis was obtained when the mitochondria were suspended in isotonic phosphate buffer. Furthermore recent experiments have shown that the same results represented by experiment 2, Table 3, have been obtained with an inner membrane preparation of mitochondria.

Table 3. *Formation of PPHB in mitochondria*

Mitochondria	F-48	Substrates for Polyprenyl side chains	PPHB dpm
EXPT. 1			
16·8 mg mitochondria	Present	Mevalonate, ATP	6980
16·8 mg mitochondria	Absent	Mevalonate, ATP	143
16·8 mg mitochondria	Present	Isopentenyl pyrophosphate	4990
16·8 mg mitochondria	Absent	Isopentenyl pyrophosphate	1460
EXPT. 2			
15·3 mg mitochondria	Present	Mevalonate, ATP	5630
15·3 mg mitochondria	Absent	Mevalonate, ATP	49
15·3 mg mitochondria	Present	Isopentenyl pyrophosphate	5380
15·3 mg mitochondria	Absent	Isopentenyl pyrophosphate	3980

The mitochondria were prepared in 0·28 M sucrose according to the procedure of Parson & Williams (1967). In Expt. 1 the amount of F-48 present was 142 mg of protein and in Expt. 2, 40 mg of protein. Each flask contained 5 nanomoles 7-^{14}C-PHB, 55 μCi/μmole, 3 mM ATP, 1·0 mM isopentenyl pyrophosphate, 1·0 mM mevalonate. Incubation in air 1 hr at 37°C.

It became of interest to determine the location of the PHB:polyprenyl transferase in the mitochondria. Mitochondria were separated into the inner and outer membrane fractions according to Parsons & Williams (1967). The cross contamination of these fractions was determined by assaying for marker enzymes, i.e. succinate-neotetrazolium reductase for the inner membranes (Okui *et al.*, 1963), and mono-amine oxidase for the outer membranes (Tabor *et al.*, 1954), and was found to be minimal. The results in Table 4 clearly show that the major PHB:polyprenyl transferase activity is in the inner membrane fraction.

If mitochondria truly contained the apparatus for synthesizing polyprenyl pyrophosphate, then it would be expected that they would contain two key enzymes required for the synthesis of polyprenyl chains, i.e. isopentenyl pyrophosphate isomerase and diphenyl transferase. The combined activity of these enzymes was first determined according to the method of Shah *et al.* (1965) and indeed it was found that the activities in mitochondria and supernatant were comparable. As we applied this procedure we became aware of wide variabilities in the results. Further investigation indicated that we were overestimating the total activity because of interference due to phosphatases; subsequently the procedure of Holloway & Popják (1968), as modified by Dr

Table 4. *Localization of PHB:polyprenyl transferase in cellular and mitochondrial subfractions of rat liver*

Fraction	PPHB formation dpm/mg protein
Homogenate	31
Mitochondria	157
Cytosol	2
Mitochondrial inner membrane	153
Mitochondrial outer membrane	10

Homogenate was prepared as previously described from rat liver. The mitochondria and inner and outer mitochondrial membrane fractions were incubated with the F-48 fraction and ATP, MVA and 7-^{14}C-PHB as previously described.

T. Nishino, in our laboratory, was used with satisfactory reproducibility. In this modification the light petroleum ether extracts are washed with water. This eliminates dimethyllyl alcohol and isopentenol and leaves only the longer prenols. A comparison of the activities of these enzymes in mitochondria, microsomes and the 48,000 × g fraction is shown in Fig. 3. It will be seen that

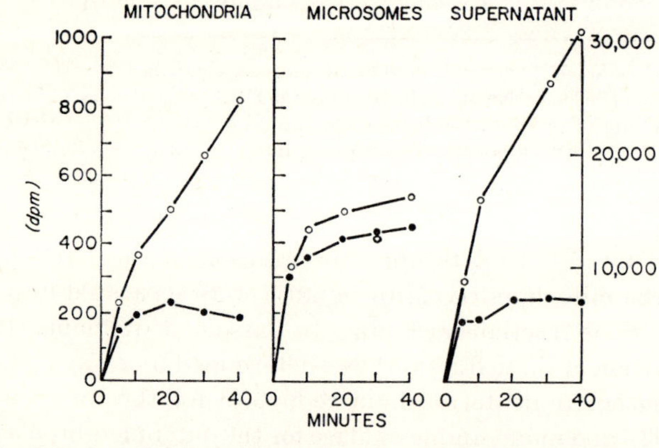

FIG. 3. Combined activities of isopentenyl pyrophosphate isomerase and diprenyl transferase in cellular fractions of rat liver. (○) control values, (●) activity in the presence of 1·0 mM iodoacetamide. The substrate was 1-^{14}C-IPP. The preparations contained the following concentrations of protein, mitochondria 2·39 mg/ml, microsomes 3·64 mg/ml, 48,000 × g supernatant fraction 4·60 mg/ml.

the mitochondria do indeed exhibit activity while, as expected there is minimal activity in microsomes. The 48,000 × g fraction, however, contains about 30-fold greater activity than the mitochondria. The activity observed in the mitochondria could not be due to contamination by entrapment of the supernatant fraction in the pellet since the mitochondria were washed four times with a 25-fold dilution of the pellet volume at each washing. Of further interest was the different pattern observed when each fraction was treated with iodoacetamide, a known inhibitor of the IPP isomerase and diprenyl transferase.

The mitochondrial and 48,000 × g fractions showed similar behaviour, but relatively little inhibition was seen with the microsomal fraction. This evidence indicated that contamination by microsomal enzymes cannot be significant. In addition, as noted previously, the synthesis of PPHB occurred in an inner membrane mitochondrial preparation when IPP and PHB were the sole substrates. The chances of the 48,000 × g supernatant fraction contaminating this preparation are extremely remote.

Finally, in view of the recent observation of Dialameh *et al.* (1970) on the stimulatory effect of coenzyme A on the conversion of PHB to 6-methoxy-2-nonaprenylphenol in rat liver, it seemed of interest to try the effect of coenzyme A in our mitochondrial system. Our results indicated that the effects were complex. In the optimal system based on isotonic phosphate buffer we could not observe any effect of CoASH. However, when the mitochondria were suspended in sucrose buffer and EDTA the overall synthesis was greatly inhibited, but a stimulation by CoASH could be observed. However, the effect was in the direction of making a higher proportion of shorter chain prenylogues of PPHB.

We examined in greater detail the nature of PPHB with reverse phase chromatography. This very sensitive technique enables one to detect the presence of a mixture of prenylogues more effectively than any other method. A reverse phase chromatography system was developed for polyprenyl-*p*-hydroxybenzoate, i.e. acetone–water–acetic acid mixture (75:25:1). Other reverse phase systems were acetone–water–acetic acid (90:10:1) for polyprenylphenols and acetic acid–water (95:5) for ubiquinones. The chromatography was carried out on Whatman paper impregnated with 5% paraffin oil in chloroform.

The distribution of quinones in rat and guinea pig liver is known. The rat liver nonaprenyl ubiquinone is generally 80–90% of the total concentration of ubiquinone, while the remainder consists of decaprenyl ubiquinone. Guinea pig liver contains only decaprenyl ubiquinone (Olson & Dialameh, 1960; Wiss *et al.*, 1961). When PPHB obtained from incubation of IPP and PHB in rat and guinea pig mitochondria was subjected to reverse phase chromatography, it was found that PPHB from rat gave two peaks whose proportions were in the same ratio observed for decaprenyl and nonaprenyl ubiquinones, while in guinea pig only one peak was found corresponding to the single moiety of decaprenyl ubiquinone, Fig. 4. When a PPHB mixture obtained by incubation of U-^{14}C-PHB, and IPP with rat liver mitochondria was purified by thin layer chromatography and added to a preparation from *R. rubrum* which can decarboxylate PPHB to polyprenylphenol, it was observed (Fig. 5B) that the product thus obtained also consisted of a mixture of nonaprenyl and decaprenylphenols in the proportions expected from the ratios seen in Fig. 4B. Polyprenylphenols obtained from *R. rubrum* have a preponderance of the decaprenyl over the nonaprenyl form (Fig. 5A). The identity of the polyprenylphenols was established by co-chromatography with known samples. The foregoing results provide definitive evidence that rat liver PPHB consists of a mixture of 3-decaprenyl-4-hydroxybenzoate and 3-nonaprenyl-4-hydroxybenzoate with the latter predominating. Guinea pig liver mitochondria by analogy form the 3-decaprenyl-4-hydroxybenzoate exclusively.

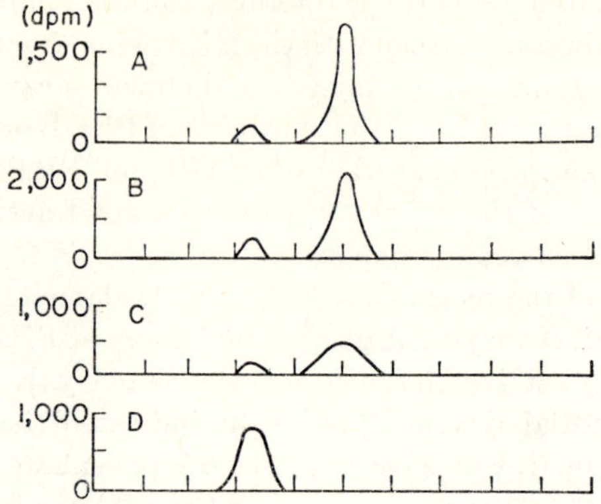

FIG. 4. Reverse phase chromatography of PPHB: (A) rat liver mince incubated with 7-^{14}C-PHB and mevalonic acid; (B) rat liver mitochondria incubated with 7-^{14}C and IPP; (C) rat liver mitochondria incubated with PHB and 1-^{14}C-IPP; (D) guinea pig liver mitochondria incubated with 7-^{14}C-PHB and IPP.

In other experiments with geranyl–geranyl pyrophosphate or farnesyl pyrophosphate as the side chain substrate, we were able to obtain shorter chain derivatives of polyprenyl-*p*-hydroxybenzoate. These results showed clearly that the PHB:polyprenyl transferase has a broad specificity with regard to chain length of the polyprenyl moiety.

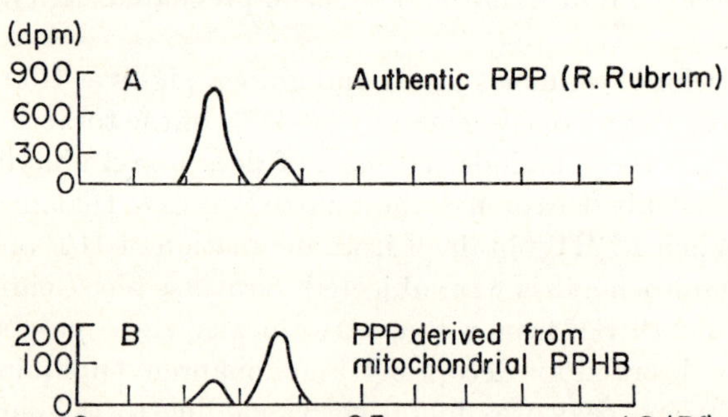

FIG. 5. Reverse phase chromatography of polyprenylphenols derived from conversion of PPHB from rat liver mitochondria: (A) authentic polyprenylphenols from *R. rubrum*; (B) polyprenylphenols from incubation of an extract of *R. rubrum* with PPHB from rat liver mitochondria.

DISCUSSION

In none of the foregoing experiments were we able to detect further metabolism of the PPHB fractions which indicates that possibly the decarboxylase system in our mitochondrial preparations is extremely sensitive and is inactivated in the process of preparing mitochondria. Another possibility is that decarboxylation may take place after hydroxylation and methylation of position 5 of 3-polyprenyl-4-hydroxybenzoate. This remains to be investigated further. In bacterial systems, as in *R. rubrum* and *E. coli* we were readily able to demonstrate decarboxylase activity and the formation of polyprenylphenols. Although we could not show a direct stimulating effect of coenzyme A on the formation of PPHB in our optimal system, stimulation by CoASH in a partially inhibited system obtained in a medium of sucrose and EDTA indicates that CoASH may act indirectly. The fact that the stimulation also leads to increased formation of short chain prenylogs of PPHB is difficult to explain and further work is required to determine this point. We think that the inability of Whistance *et al.* (1971) to observe the formation of polyprenyl derivatives of *p*-hydroxybenzoate in animal preparations may be due to problems of isolation which were a factor in our previous work.

The presence of the PHB:polyprenyl transferase in the inner membrane portion of the mitochondria is worthy of note and agrees with the observation of Sottocasa (1968) that ubiquinone is located in the inner membrane. In addition, these observations show that this enzyme has a broad specificity and can operate with a wide range of polyprenyl pyrophosphate side-chains. Thus, the previous observations of the presence of both nonaprenyl and decaprenyl ubiquinone in rat liver are readily explained now on the basis of the fact that the mitochondria are able to built up both polyprenyl side-chains, and that the PHB:polyprenyl transferase has little specificity. In the case of guinea pig, apparently more specificity is conferred upon the polyprenyl side-chain synthesizing system and so one ends up primarily with the decaprenyl derivative of ubiquinone.

A question of important biological significance arises from the demonstration of the presence of enzymes in mitochondria necessary for the synthesis of polyprenyl pyrophosphate chains. IPP can be used as a substrate, but mitochondria require the cytosol for the formation of IPP from mevalonic acid. Are these enzymes in mitochondria reflective of contamination from other parts of the cell, especially the 48,000 $\times$ g supernatant fraction? This possibility appears extremely remote in view of the 4-fold washing of the mitochondrial pellet before assay. Further evidence that the enzymes in the 48,000 $\times$ g fraction may play a minor role is our recent observation that inner membrane fractions of mitochondria incubated solely with IPP and PHB readily form PPHB.

The possibility that mitochondria might contain an endogenous polyprenyl synthesizing system is in agreement with the concept that mitochondria in eukaryotes arose from a symbiotic incorporation of an aerobic precaryote. It also agrees with the observations of Goodwin & Mercer (1963) on compartmentation of isoprenoid synthesis in plants. Of great significance, also, is the observation of Hemming and collaborators (1970) that free prenols of the dolichol type are found primarily in the mitochondria. It has also been shown that mevalonate can label these dolichols, and it is reasonable to suppose that the synthesis of dolichols is located in the mitochondria. These results are also of great interest in view of the recent finding of Sottocasa *et al.* (1971) that glycoproteins occur in the inter membrane space of the rat liver mitochondria and the observation of Behrens & Leloir (1970) that dolichol phosphates are involved in glycoprotein synthesis in animal systems.

STUDIES ON CELL-FREE SYSTEMS IN *R. rubrum*

Previous studies by Winrow & Rudney (1969), Raman *et al.* (1969) and Rudney (1970) showed that supplementation of cell-free *R. rubrum* sonicates with IPP and *Micrococcus lysodeikticus* extracts similar to those prepared by Allen *et al.* (1967) were necessary for the synthesis of measurable amounts of PPHB and PPP. This resulted from a deficiency in the *R. rubrum* system providing polyprenyl pyrophosphate the expected side-chain precursor.

Both PHB: polyprenyl transferase and PPHB decarboxylase activities were localized in the membranous portion of the *R. rubrum* sonicate. As noted earlier (Rudney, 1970), a special circumstance was the inability to observe the formation of compounds beyond PPHB and PPP, and it was considered that in these preparations the hydroxylation and methylation reactions required to proceed to intermediates outlined in Fig. 1 were inactivated.

Dr E. Markianos and Mr David Justis, in our laboratory, have reinvestigated the properties of the transferase and decarboxylase in the *R. rubrum* preparation, and they have also studied the role that the *Micrococcus lysodeikticus* extract plays in modifying the nature of the product obtained.

Conditions for the incubation of the MLE with IPP have been described elsewhere (Raman *et al.*, 1969; Rudney, 1970). After incubation for 2 hr the mixture is treated with four volumes of acetone and the resultant precipitate is extracted with 60% ethanol, and the extract evaporated to remove the ethanol. The materials remaining after the 60% ethanol extraction are taken up in 0·1 M tricine buffer, pH 8·0. This treatment results in a buffer-soluble and a buffer-insoluble fraction. This procedure provides three crude fractions with remarkably different effects in the type of product obtained when they are incubated with *R. rubrum* sonicates and PHB-U-^{14}C. This is shown in Table 5.

Table 5. *Effect of fractionation of MLE + IPP incubation mixtures on PPHB and PPP formation in* R. rubrum *sonicates*

Fraction	PPHB dpm	PPP dpm
Original incubation mixture	3,600	13,300
60% Ethanol extract	0	2,440
Buffer soluble	2,950	2,930
Buffer insoluble	12,850	7,600
Buffer insoluble reextracted	7,500	0

Each fraction was incubated with a sonicate from *R. rubrum* and U-^{14}C-PHB for 2 hr at 30°C.

The results indicate that the 60% ethanol extract is capable of providing side-chain material which is converted to PPHB and PPP, while the insoluble portion, especially when reextracted with buffer, provides polyprenyl pyrophosphate which is able to act as a substrate for the PHB-polyprenyl transferase to form PPHB only, i.e. subsequent decarboxylation does not occur.

The question then arises, why cannot the buffer-insoluble material be decarboxylated? One possibility is decreased specificity for the decarboxylase as a result of alteration of the side-chain configuration by the inclusion of *cis*-double bonds in the polyprenyl chain. A second possibility might be that the length of the side-chain is a factor.

The possibility of *cis*-double bonds insertion into the prenyl units was generally ruled out in the following manner. Incubation of mevalonate with a yeast extract whose preparation was described by Tchen (1962) provides a system for the conversion of mevalonate to IPP. This IPP can be used as a substrate by the *Micrococcus* extract to make polyprenyl pyrophosphate chains. When mevalonate labelled with tritium in the 3R–4R position was used, tritium was incorporated into PPHB and PPP. On the other hand when 3R–4R labelled mevalonate was used, negligible tritium retention could be observed. Thus, it would appear that the side-chains are entirely in the *trans*-form as expected for molecules which are precursors of ubiquinone. The role of the number of isoprenoid units in side-chains of PPHB is being investigated.

We have recently obtain evidence showing that the polyprenyl pyrophosphate chains are bound to a protein and this binding appears to be requisite for activity in the PHB:polyprenyl transferase enzyme system. When the MLE and IPP were incubated and fractionated as previously outlined, it was found that a short treatment of each fraction with trypsin led to complete inactivation of PPHB formation when incubated further with *R. rubrum* sonicates and PHB. This is shown in Table 6, experiment 1, with the buffer insoluble fraction.

Table 6. *Effect of trypsin digestion on the buffer insoluble fraction*

	PPHB dpm	PPP dpm
EXPT. 1		
(a) Untreated	6,040	0
(b) Trypsin (60 min) followed by trypsin inhibitor	1,060	0
(c) Trypsin + trypsin inhibitor	5,610	0
EXPT. 2		
(a) Untreated	35,390	146
(b) Trypsin (60 min)	1,404	0
(c) Trypsin (60 min) + fresh MLE extract	24,208	20,685

Buffer insoluble fraction prepared as previously described from *Micrococcus lysodeikticus* incubation with IPP. In Expt. 1(*c*) the trypsin inhibitor and trypsin were added at zero time. In Expt. 2(*a*) and 2(*b*) trypsin was incubated with the extract for 1 hr and its activity was terminated with trypsin inhibitor. In Expt. 2(*b*) termination of trypsin action was followed by incubation with fresh MLE. PPHB and PPP formation were measured after incubation with *R. rubrum* sonicates.

The same effect was also observed with all the fractions studied. If to a fraction inactivated by trypsin we added a fresh MLE extract, then a large conversion of PPHB to PPP was observed. This is shown in Table 6, experiment 2.

These data are interpreted to mean that hydrolysis of a protein carrier for polyprenyl pyrophosphate chains by trypsin leads to decreased activity in enzymatic reactions. Addition of a fresh extract of MLE enables transfer of the polyprenyl pyrophosphate chains to a new carrier enabling transfer of the polyprenyl chains in the PHB:polyprenyl transferase system of *R. rubrum*. In addition, these polyprenyl chains are transferred to another protein carrier which enables the formation of a PPHB which now has the capability of being decarboxylated. The MLE preparation does not contain any decarboxylase activity since no formation of PPP occurs unless the decarboxylase in *R. rubrum* is present. Further investigation of these aspects of the interaction of *R. rubrum* and *Micrococcus* systems is continuing.

In *R. rubrum* we have been able to separate the decarboxylase activity from the PHB:polyprenyl transferase activity by treatment of a sonicate with ammonium sulphate and subsequent separation on a Whatman DE-52 cellulose ion exchange column. The zero to 20% saturated ammonium sulphate cut was found to produce only PPHB regardless of which fraction of MLE plus IPP incubate was added. A 20 to 50% saturated ammonium sulphate cut contained both transferase and decarboxylase activity.

Purification of the PHB:polyprenyl transferase is proceeding and a major finding is that Mg^{++} is required but in addition the enzyme is extremely sensitive to salt concentration. This is shown in Table 7. A complicating factor

Table 7. *Effect of salt concentration on activity of PHB:polyprenyl transferase*

	Specific activity of PPHB dpm/mg protein	Turbidity
Dialysed enzyme	10·6	—
+0·05 M $MgCl_2$	33·7	++++
+0·005 M $MgCl_2$	88·5	+
+0·05 M NaCl	91·2	++

Enzyme is the 0·20% ammonium sulphate fraction of a *R. rubrum* sonicate dialysed for 15 hr against 0·015 M tricine buffer, pH 8·0. MLE and IPP were incubated as usual to generate polyprenylpyrophosphates, then dialysed against 0·015 M tricine buffer, pH 8·0, for 17 hr. The dialysed MLE and IPP preparation and U-^{14}C-PHB were incubated with the dialysed *R. rubrum* fraction as previously described.

is the development of turbidity indicative of protein solubility changes in relation to the salt effects and this undoubtedly affects the activity. Thus Mg^{++} in the presence of high and low ionic strength affects the reaction. In addition in the absence of Mg^{++}, NaCl has an activating effect. These observations are reminiscent of those obtained by Sanderman & Strominger (1971) in connection with the stimulating effect of NaCl on the ficaprenol phosphokinase reaction, and the remarkable effect of salt concentration on the activity and association and dissociation of the oxidosqualene cyclase (Yamamoto & Bloch, 1970). Studies are in progress to determine whether the salt effects are on the PHB:polyprenyl transferase itself or on the polyprenyl pyrophosphate protein complex.

ACKNOWLEDGEMENTS

Recent work from the author's laboratory described herein was supported by a grant from the National Institute for Arthritis and Metabolic Diseases, Grant No. AM-12463.

REFERENCES

Allen, C. M., Alworth, W., Macrae, A. & Bloch, K. (1967). *J. biol. Chem.* **242,** 1895

Behrens, N. H. & Leloir, L. F. (1970). *Proc. natn Acad. Sci., U.S.A.* **66,** 153.

Dialameh, G. H., Nowicki, H. G., Yekundi, K. G. & Olson, R. E. (1970). *Biochem. biophys. Res. Commun.* **40,** 1063.

Friis, P., Daves, G. D. Jr. & Folkers, K. (1966). *J. Am. chem. Soc.* **88,** 4754.

Galliard, T., Mitchell, R. H. & Hawthorne, J. N. (1965). *Biochim. Biophys. Acta* **106,** 551.

Goodwin, T. W. & Mercer, E. I. (1963). In *The Control of Lipid Metabolism,* Biochem. Society Symposium No. 24 (Ed. by Grant, J. K.), p. 37, New York: Academic Press.

Hemming, F. W. (1970). In *Natural Substances Formed Biologically from Mevalonic Acid,* Biochem. Society Symposium No. 29 (Ed. by Goodwin, T. W.), p. 105. London: Academic Press.

Holloway, P. W. & Popják, G. (1968). *Biochem. J.* **106,** 835.

Momose, K. & Rudney, H. (1972). *J. biol. Chem.* **247,** 3930.

Okui, S., Suzuki, Y. & Momose, K. (1963). *J. Biochem. Tokyo* **54,** 471.

Olson, R. E. & Dialameh, G. H. (1960). *Biochem. biophys. Res. Commun.* **2,** 198.

Olsen, R. K., Smith, J. L., Daves, G. D., Jr., Moore, H. W., Folkers, K., Parson, W. W. & Rudney, H. (1965). *J. Am. chem. Soc.* **87,** 2298.

Olsen, R. K., Daves, G. D., Jr., Moore, H. W., Folkers, K. & Rudney, H. (1966). *J. Am. chem. Soc.* **88,** 2346.

Parsons, D. F. & Williams, G. R. (1967). In *Methods in Enzymology,* Vol. X (Ed. by Estabrook, R. W.), p. 443. New York: Academic Press.

Raman, T. S., Rudney, H. & Buzzelli, N. K. (1969). *Archs. Biochem. Biophys.* **130,** 164.

Rudney, H. (1970). In *Natural Substances Formed Biologically from Mevalonic Acid,* Biochemical Society Symposium, No. 29 (Ed. by Goodwin, T. W.), p. 89. London: Academic Press.

Sandermann, H., Jr. & Storminger, J. L. (1971). *Proc. natn Acad. Sci., U.S.A.* **68,** 2441.

Shah, D. H., Cleland, W. W. & Porter, J. W. (1965). *J. biol. Chem.* **240,** 1946.

Sottocasa, G. L. (1968). *Biochem. J.* **105,** 1P.

Sottocasa, G. L., Sandri, G., Panfili, E. & DeBernard, B. (1971). *FEBS Lett.* **17,** 100.

Tabor, C. W., Tabor, H. & Rosenthal, S. M. (1954). *J. biol. Chem.* **208,** 645.

Tchen, T. T. (1962). In *Methods in Enzymology,* Vol. V (Ed. by Colowick, S. P. & Kaplan, N. O.), p. 489. New York: Academic Press.

Whistance, G. R., Field, F. E. & Threlfall, D. R. (1971). *Eur. J. Biochem.* **18,** 46.

Winrow, M. J. & Rudney, H. (1969). *Biochem. biophys. Res. Commun.* **37,** 833.

Wiss, O., Gloor, U. & Weber, F. (1961). In *Ciba Foundation Symposium on Quinones in Electron Transport* (Ed. by Wolstenholme, G. E.), p. 264. Boston: Little Brown and Co.

Yamamoto, S. & Bloch, K. (1970). In *Natural Substances Formed Biologically from Mevalonic Acid,* Biochemical Society Symposium No. 29 (Ed. by Goodwin, T. W.), p. 35. London: Academic Press.

REGULATION OF VITAMIN D METABOLISM: A NEW DIMENSION IN CALCIUM HOMEOSTASIS

By H. F. DeLUCA

Department of Biochemistry,
College of Agricultural and Life Sciences,
University of Wisconsin,
Madison, Wisconsin, USA

INTRODUCTION

In 1919 Sir Edward Mellanby (Mellanby, 1919) demonstrated conclusively that the disease rickets could be produced in experimental animals by dietary manipulation. He further demonstrated that this disease could be prevented or corrected by the administration of cod liver oil. He concluded that cod liver oil contained an accessory food substance, which he considered at that time to be fat soluble vitamin A, which was responsible for the cure and prevention of rickets. Thus it is understandable that we have regarded vitamin D as a true dietary requirement rather than a hormonal substance which it truely represents. McCollum *et al.* (1922) recognized that the substance described by Mellanby was a material distinct from his fat soluble vitamin A, which he then relegated the name of vitamin D. These important discoveries laid to waste in an intellectual sense the fact first pioneered by Huldshinsky (1919) that the disease rickets can be cured by exposure of the patient or animal to sunlight or artificially induced ultraviolet light. This process which was later more thoroughly defined by Steenbock & Black (1924) must be regarded as the first reaction which places vitamin D in the realm of a hormonal substance. Throughout the fifty years of its scientific life, the concept of vitamin D as a vitamin has led scientists and physicians to regard vitamin D as an important substance in calcium metabolism, and one whose function is catalytic in nature but not regulated. By this it was considered that vitamin D supplied in adequate amounts would catalyse the appropriate reactions leading to increased intestinal absorption of calcium and calcification of bone but that the amount of its active material fed into these reactions was not in any way regulated. It will be the purpose of this presentation to demonstrate conclusively (1) that vitamin D is metabolized to metabolically active forms, and (2) that the biosynthesis of these active forms is very tightly feed-back regulated by the product of its function, namely circulating plasma calcium concentration. The negative feed-back regulation of functional forms of vitamin D must now be considered to add a new dimension of control to the strict maintenance of plasma calcium concentration, one of the truely remarkable regulatory systems in higher animals.

ISOLATION AND IDENTIFICATION OF 25-HYDROXYCHOLECALCIFEROL

Following the discovery of vitamin D_3 by Waddell (1934) and its identification and chemical synthesis by Windaus, Schenck & von Werder (1936), it has been assumed that the unchanged vitamin functions in its biochemical systems to produce the changes in calcium metabolism by bringing about calcification of bone and cure of rickets (Kodicek, 1956; Haussler & Norman, 1967). This idea was further fortified by the early findings of Kodicek with radioactive vitamin D_2 in which he could demonstrate no metabolites of vitamin D which possess biological activity. Synthesis of radioactive vitamin D_3 of high specific activity (Neville & DeLuca, 1966) and the development of chromatographic methods capable of separating vitamin D and its metabolites in our laboratory presented for the first time the possibility of examining the important question of whether vitamin D must first be converted to metabolically active forms before it can function. Unequivocal proof for the existence of biologically active metabolites was consequently obtained (Lund & DeLuca, 1966). In these experiments, radioactive vitamin D_3 was injected intravenously and prior to the appearance of increased intestinal calcium absorption the animals were killed, their tissues removed, extracted with methanol-chloroform and chromatographed on heat activated silicic acid columns. Of considerable interest is the fact that a major metabolite fraction more polar than vitamin D_3 appears before the vitamin can bring about its well-known functions, namely increased calcium absorption and increased bone mineral mobilization. This metabolite fraction was found to cure rickets in both rats and chicks when fed orally, to increase intestinal calcium absorption and to increase bone mineral mobilization in vitamin D-deficient animals (Morii *et al.*, 1967). Most important, however, this fraction acted more rapidly than the parent vitamin in inducing calcium transport. The major biologically active component from this fraction was isolated in pure form in 1968 and identified as 25-hydroxycholecalciferol (25-HCC) (Fig. 1) (Blunt, DeLuca & Schnoes, 1968*a*,*b*). Soon

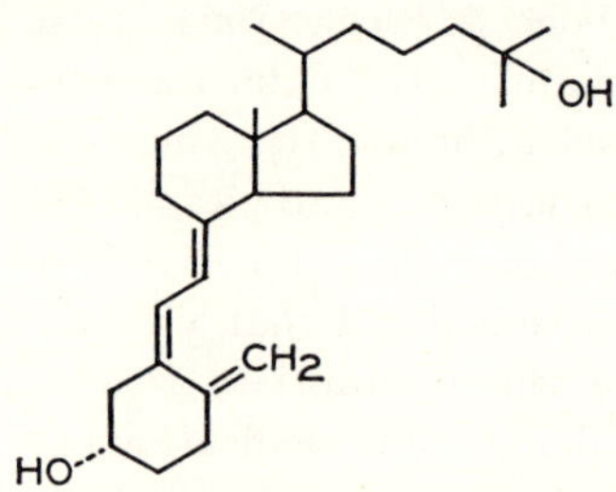

FIG. 1. 25-Hydroxycholecalciferol (25-HCC).

after this the 25-hydroxy derivative of vitamin D_2 was isolated and identified (Suda *et al.*, 1969). The 25-HCC markedly shortened the lag in response time of the intestinal calcium transport system and bone mineral mobilization (Fig. 2) (Blunt, Tanaka & DeLuca, 1968). Furthermore this metabolite, but not vitamin D_3, when added to isolated cultures of embryonic bone induces the release of calcium from bone (Trummel *et al.*, 1969) and when injected into isolated vascularly perfused intestine (Olsen & DeLuca, 1969) stimulates intestinal calcium absorption whereas massive amounts of vitamin D_3 do not. These results provided clear evidence that hydroxylation of vitamin D in the 25 position is essential for the vitamins' function *in vivo*.

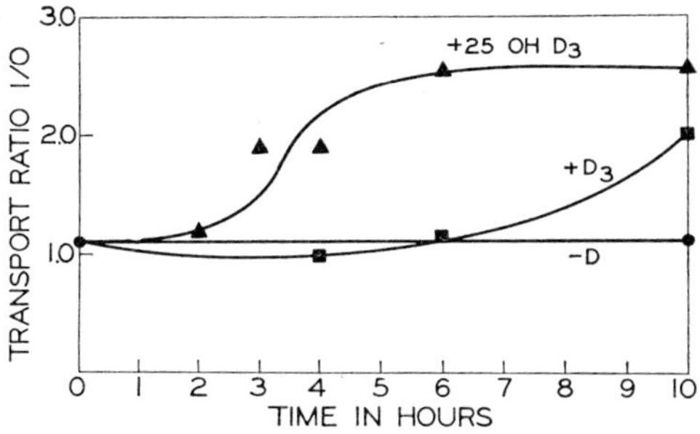

FIG. 2. Response of intestinal calcium transport of vitamin D-deficient rats to either vitamin D_3 or 25-HCC (25-OH-D_3). Deficient rats were given solvent (0·02 ml ethanol) (●), 650 pmoles 25-HCC (▲), or 650 pmoles vitamin D_3 (■) intrajugularly and at various times thereafter they were killed and intestinal calcium transport was determined using the everted sac method described by Blunt, Tanaka and DeLuca (1968).

VITAMIN D—25-HYDROXYLASE

Because the targets of vitamin D action do not respond directly to vitamin D_3 but respond to 25-HCC, it is clear that these end-organs do not possess the ability to hydroxylate vitamin D in the 25 position. In experiments not described here, work by Ponchon & DeLuca (1969*a*), Ponchon, Kennan & DeLuca (1969) and Horsting & DeLuca (1969) demonstrated conclusively that the liver is the major, if not sole, site of 25-hydroxylation of vitamin D. In fact, in hepatectomized animals, no metabolism of vitamin D- can be observed whatsoever during the 4 hr period of survival of the animals (Ponchon *et al.*, 1969). Homogenates of the livers of vitamin D-deficient rats hydroxylate vitamin D in the 25 position in a medium fortified with molecular oxygen and dithiothreitol or glutathione (Horsting & DeLuca, 1969, unpublished results). This system is not inhibited by carbon monoxide–oxygen mixtures, cyanide, azide and diphenylparaphenylenediamine. These results clearly demonstrate that this hydroxylation reaction is a unique one in that it does not involve cytochrome P-450, the electron transport chain, or the lipid peroxidation system. Of major importance is the fact that this reaction is very strongly regulated *in vivo*. For example, homogenates of liver taken from animals 2 or 3 hr after vitamin D_3 administration do not carry out the hydroxylation. Furthermore, 25-hydroxylation *in vivo* is also markedly reduced. It, therefore, seems evident that following the initial burst of 25-hydroxylation, this reaction is slowed very substantially. The exact nature of this regulation is not yet completely understood. However, the results shown in Fig. 3 clearly demonstrates that the product 25-HCC is a potent inhibitor of this reaction. It is possible that this product inhibition plus some other mechanism turns off the 25-hydroxylation reaction when sufficient 25-HCC is made for the plasma. Of major interest is the fact that 25-hydroxydihydrotachysterol, a close analog of 25-HCC, will not inhibit this reaction *in vitro* (Fig. 3). It is known that dihydrotachysterol is very effective in raising plasma calcium concentration when given in high doses. It is, in fact, markedly more effective than vitamin D itself at these dose levels. However, at low, physiological doses vitamin D is more effective than dihydrotachysterol[8] in all of the func-

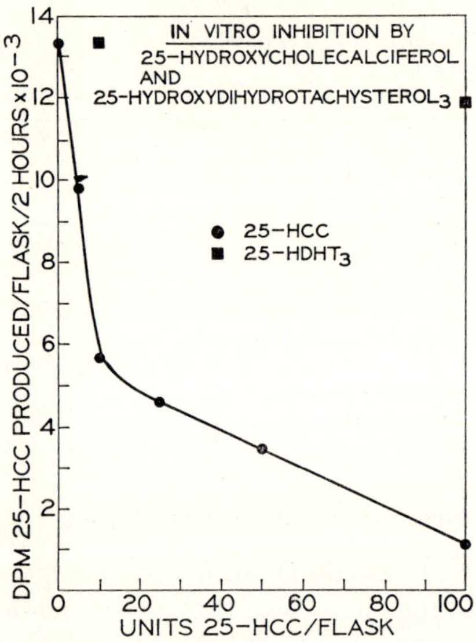

FIG. 3. Inhibition of vitamin D_3-25-hydroxylase of rat liver *in vitro* by 25-HCC. One unit of 25-HCC equals 65 pmoles. Note that even large concentrations of 25-OH-DHT do not inhibit. Incubations were carried out as described by Horsting & DeLuca (1969).

tions of vitamin D. Recently it has been possible to show with synthetic radioactive dihydrotachysterol$_3$ that this compound is hydroxylated in the liver by presumably the same systems which hydroxylate vitamin D in the 25 position (Hallick & DeLuca, 1971). This hydroxylation does not appear to be product inhibited. Clearly then, the 25-hydroxylation represents the first point at which the functional metabolism of vitamin D to its metabolically active forms is regulated. It is not known whether 25-HCC functions directly in any metabolic system. But it certainly represents the initial entry of vitamin D in the metabolic sequence which ultimately results in its function.

The regulation of 25-hydroxylation in the liver plays several important roles. It undoubtedly is responsible for maintaining vitamin D_3 in the storage form; thus active vitamin D, namely 25-HCC, is fed into the system as it is needed. On days when there are large intakes of vitamin D_3 it is not immediately hydroxylated to the active material, instead the regulation results in storage for days when there is little or no intake of vitamin D. This not only conserves vitamin D on the one hand but possibly protects the organism against vitamin D toxicity on the other during days of very great intake or exposure to ultraviolet light. It seems likely that massive amounts of vitamin D can overcome this regulation since it is well known that vitamin D_3 can be toxic at very high levels. If the bottleneck 25-hydroxylation reaction were not present, however, it may be that vitamin D would be toxic at even much lower doses.

1,25-DIHYDROXYCHOLECALCIFEROL—THE HORMONAL FORM OF VITAMIN D ACTIVE ON INTESTINE

Following the isolation and identification of 25-HCC it was possible to synthesize this compound chemically (Blunt & DeLuca, 1969) involving a series of reactions in which there could be introduced in the 26 and 27 positions the tritium label (Suda, DeLuca & Hallick, 1971). Thus tritium labelled 25-HCC became a useful tool for the study of its further metabolism. Certainly more polar metabolites of vitamin D_3 could be detected in a variety of tissues (Ponchon & DeLuca, 1969*b*; Lawson, Wilson & Kodicek, 1969; Haussler, Myrtle & Norman, 1968). Cousins *et al.*, (1970) using radioactive 25-HCC demonstrated that 25-HCC is metabolized very rapidly to more polar metabolites. This conversion took place prior to the initiation of intestinal calcium transport and bone mineral mobilization in response to 25-HCC. In parallel experiments, however, the metabolite fraction which was called fraction V on silicic acid columns was shown by Suda *et al.* (1970) to be extremely heterogeneous giving for example in the blood as many as five or six peak V metabolites. It soon became evident that the silicic acid column chromatographic methods originally devised for vitamin D metabolism studies had reached the limit of their usefulness. Although the liquid–liquid partition chromatography designed by Suda *et al.* (1970) presented a very efficient method of separation, the tediousness of its operations and the poor recoveries limited its usefulness. Holick & DeLuca (1971) described for the first time the use of liquid-gel chromatography for the separation of the metabolites of fraction V. Using this important chromatographic breakthrough, a new and important metabolite of 25-HCC could be demonstrated in the intestine and bone (Omdahl *et al.*, 1971). From the intestines of 1500 chickens given

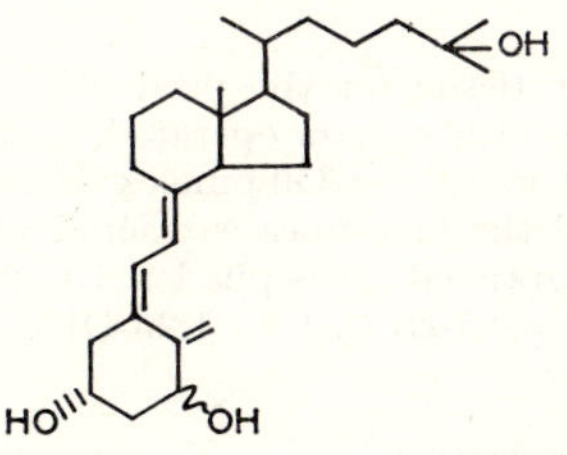

1, 25 DIHYDROXYCHOLECALCIFEROL

FIG. 4. 1,25-Dihydroxycholecalciferol (1,25-DHCC).

radioactive vitamin D_3, 2 μg of pure 25-trimethyl silyl ether derivative of peak V was obtained and this substance was unequivocally identified as 1,25-dihydroxycholecalciferol (1,25-DHCC) (Fig. 4) by Holick, Schnoes & DeLuca (1971) and Holick *et al.*, 1971). This structure was suggested but not established by Lawson *et al.* (1971) on the basis of mass spectrometer results using an impure preparation obtained from kidney homogenates. It must, however, be emphasized here that the conformation of the 1-hydroxyl position is as yet unknown. The speculation to date has been that it is α which puts it in the opposite plane from the 3-hydroxyl which is known to be β. However, it is important not to specify the configuration of the 1-hydroxy until it is unequivocally established.

25-HCC-1-HYDROXYLASE

Fraser & Kodicek (1970) made the important discovery that the intestinal metabolite which was subsequently identified as 1,25-DHCC in our laboratory is made exclusively in the kidney. We have confirmed this important observation and provided evidence that the metabolite identified by us in the intestine is identical with that produced by kidney homogenates and that it is the loss of kidney tissue in nephrectomized animals that prevents the conversion of 25-HCC to 1,25-DHCC and not the consequent uremia (Gray, Boyle & DeLuca, 1971) (Fig. 5).

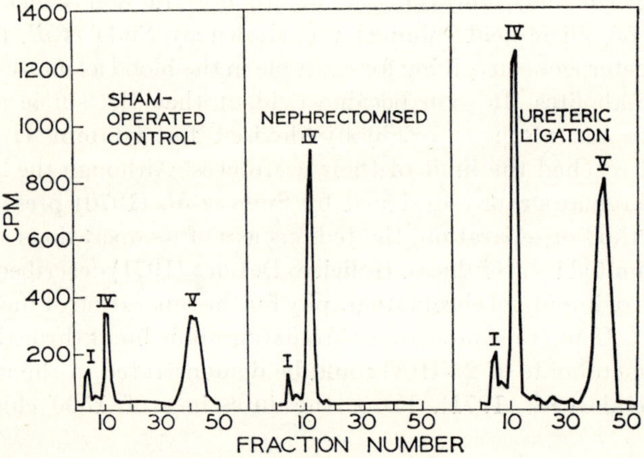

FIG. 5. Necessity of kidney tissue for the production of 1,25-DHCC from 25-HCC. Vitamin D-deficient rats were either sham operated, bilaterally nephrectomized or had their ureters ligated. Each rat was given 650 pmoles ^{3}H-25-HCC intravenously and 12 hr later each was sacrificed and the intestines extracted with methanol and chloroform. The extracts were chromatographed on Sephadex LH-20 according to the method of Holick & DeLuca (1971). IV = 25-HCC, V = 1,25-DHCC.

The 1-hydroxylase of 25-HCC is an enzyme located exclusively in the mitochondria of the kidney (Gray, Omdahl & DeLuca, unpublished results). In intact mitochondria, reduced pyridine nucleotide is not needed. Instead there is a requirement for malate or some Krebs cycle intermediate which generates intramitochondrial reduced pyridine nicleotide. Molecular oxygen is also required as are magnesium ions. The properties of this 1-hydroxylase have now been thoroughly studied and will be the subject of a forthcoming publication (Gray, Omdahl & DeLuca, unpublished results). At the present time the enzyme has not been solubilized and hence has not been substantially purified.

In 1965 Zull, Czarnowska-Misztal & DeLuca demonstrated that all physiological actions of vitamin D are blocked by a prior administration of actinomycin D, an antibiotic which binds to DNA and prevents transcription and protein synthesis. If actinomycin D is given after the vitamin it is no longer able to block the vitamin's action (Zull, Czarnowska-Misztal & DeLuca, 1966).

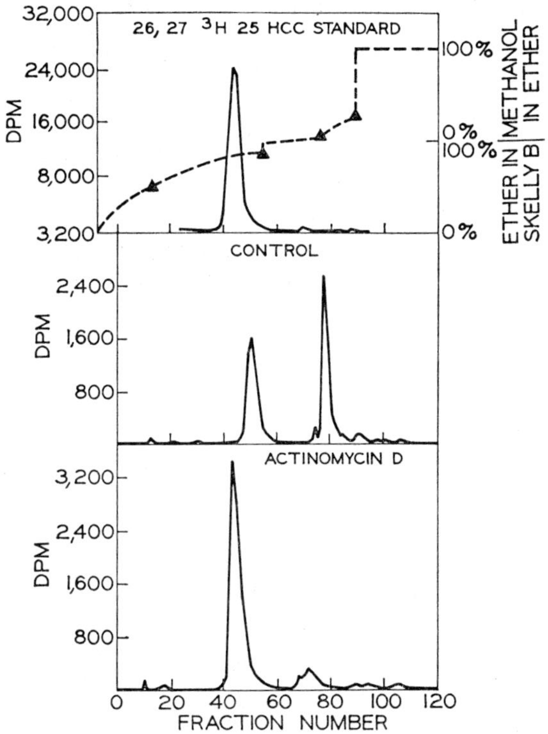

FIG. 6. Actinomycin D inhibition of 25-HCC metabolism to 1,25-DHCC. Deficient rats were given 1 μg/g body weight of actinomycin D intravenously, 2 hr later each was given 650 pmoles $_3$H-25-HCC and 8 hr later all rats were killed and the intestinal extracts chromatographed on Sephadex LH-20 (Holick & DeLuca, 1971). The first peak is 25-HCC and the second is 1,25-DHCC.

These results led to the conclusion that the metabolic function of vitamin D must involve retrieval of genetic information and synthesis of new protein. This conceptual advance provided a new impetus to vitamin D research, which is still continuing. Of great significance is the fact that actinomycin D given prior to radioactive 25-HCC to vitamin D-deficient animals prevents the conversion of 25-HCC to 1,25-DHCC (Tanaka & DeLuca, 1971*a*; Gray & DeLuca, 1971) (Fig. 6). If the antibiotic is given after radioactive 25-HCC, this conversion is not prevented. Cycloheximide, the well-known inhibitor of protein synthesis, also prevents the conversion of 25-HCC, to 1,25-DHCC. These results, therefore, suggest that at least one site of actinomycin D and cycloheximide sensitivity is the conversion of 25-HCC to 1,25-DHCC.

BIOLOGICAL ACTIVITY OF 1,25-DHCC

The 1,25-DHCC effects more rapid initiation of intestinal calcium transport than does even 25-HCC (Haussler *et al.*, 1971; Omdahl *et al.*, 1971) (Fig. 7). These results suggest that 1,25-DHCC or a further metabolite thereof is the

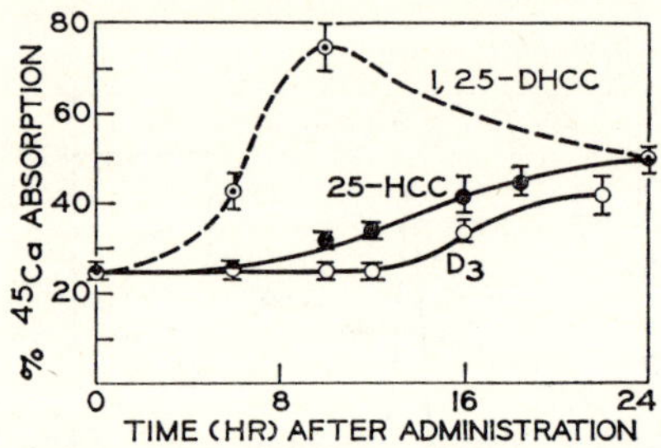

FIG. 7. Intestinal calcium absorption of rachitic chicks in response to 325 pmoles of either vitamin D_3, 25-HCC or 1,25-DHCC. Reproduced with the kind permission of the publisher (Omdahl *et al.*, 1971).

metabolically active form of vitamin D in the intestine. Nephrectomized animals which are unable to make 1,25-DHCC from 25-HCC are unable to respond to 25-HCC but are fully capable of responding to 1,25-DHCC in terms of intestinal calcium transport (Boyle *et al.*, 1972*a*) (Table 1). Furthermore, in

Table 1. *Calcium transport ratios in isolated everted duodenal sacs prepared* 12 *hr after parenteral administration of vitamin D metabolites to vitamin D-deficient rats subjected to nephrectomy, ureteric ligation or sham operation*

	No operation	Sham operation	Bilateral ureter ligation	Bilateral nephrectomy
50 μl 95% ethanol	1·7 ± 0·13 (4)	1·0 ± 0·08 (4)	1·6 ± 0·08 (4)	1·7 ± 0·18 (8)
325 pmoles* 25-HCC	3·4 ± 0·24 (7)	3·6 ± 0·22 (6)	2·6 ± 0·18 (5)	1·7 ± 0·15 (7)
650 pmoles 25-HCC	—	—	—	1·0 ± 0·15 (7)
65 nmoles 25-HCC	—	—	—	1·8 ± 0·15 (8)
325 pmoles 1,25-DHCC	3·3 ± 0·24 (13)	2·9 ± 0·20 (10)	2·8 ± 0·11 (12)	3·1 ± 0·26 (4)

Values are mean ± standard errors of the mean. Figures in brackets indicate number of rats.

* 1 μg 25-HCC is equivalent to 55–60 units of antirachitic activity in the rat.

animals given actinomycin D which also prevents the conversion of 25-HCC to 1,25-DHCC, at physiological doses 25-HCC has no effect on intestinal calcium transport whereas 1,25-DHCC is fully effective (Tanaka *et al.*, 1971) (Table 2). These results therefore tell us that 1,25-DHCC is unequivocally the metabolically active form of vitamin D in the intestine; and that 1,25-DHCC

Table 2. *Failure of actinomycin D to inhibit the rise in intestinal calcium transport induced by* 1,25-*DHCC*

Treatment		Transport
Actinomycin D	Vitamin D metabolite	^{45}Ca inside/^{45}Ca outside
—	—	1·7 ± 0·2*
—	25-HCC	2·7 ± 0·1†
+	25-HCC	1·8 ± 0·2
—	1,25-DHCC	2·9 ± 0·2†
+	1,25-DHCC	3·1 ± 0·3†

Rats were fed the 0·47% calcium 0·3% phosphorus vitamin D-deficient diet for 5 weeks. Where indicated each rat received 1 μg/g body weight of actinomycin D intrajugularly in 0·05 ml ethanol. Two hours later, where indicated, each rat received 65 pmoles 25-HCC intrajugularly in 0·05 ml ethanol. Control rats received the appropriate amount of ethanol in each case at the appropriate time. All rats were killed 12 hr after 25-HCC administration. Intestinal calcium transport was measured as described by Blunt, Tanaka and DeLuca (1968).

* Standard deviation. There were four rats in each group.

† Significantly different from control group (p. 0·01).

must function in the intestine by a mechanism not involving the retrieval of genetic information. Finally, it has been demonstrated that at the time radioactive 1,25-DHCC functions to promote intestinal calcium transport in the intestine, the only metabolite found in significant amounts is 1,25-DHCC (Frolik & DeLuca, 1971). Thus 1,25-DHCC must be the metabolically active form of vitamin D in the intestine.

The 1,25-DHCC also acts more rapidly to induce bone mineral mobilization than does 25-HCC when given in sufficient quantities (Tanaka & DeLuca, 1971*b*) (Fig. 8). However, it is important to realize that 25-HCC itself when

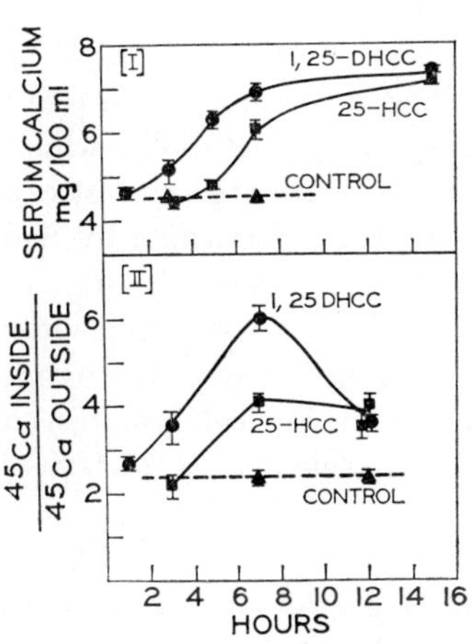

FIG. 8. (I) Rise in serum calcium in rats on a calcium-deficient, low vitamin D diet in response to 650 pmoles of 1,25-DHCC or 25-HCC. (II) Intestinal calcium transport response in rats on a low calcium, low vitamin D diet to 1,25-DHCC and 25-HCC. Reproduced with the kind permission of the publisher (Tanaka & DeLuca, 1971*b*).

added to cultures *in vitro* will induce bone mineral mobilization (Trummel *et al.*, 1969). Recently it has been shown that bone tissue cultures will respond to 1,25-DHCC much more readily than the 25-HCC (Raisz *et al.*, 1972). In fact, the results suggest that it is 100 times more effective than 25-HCC in inducing bone mineral mobilization in cultures of isolated bone. Although inconclusive at the present time it seems likely that 1,25-DHCC is the metabolic or hormonal form of vitamin D responsible for the mobilization of calcium from bone.

REGULATION OF 1,25-DHCC SYNTHESIS BY DIETARY AND PLASMA CALCIUM

It has long been known from the work of Rottensten (1938) and Nicolaysen (Nicolaysen & Eeg-Larsen, 1953) that animals maintained on a low calcium diet have a much more efficient intestinal calcium absorption mechanism than those maintained on a high calcium diet. Clearly then animals are able to adapt to low calcium intakes by increasing their intestinal calcium absorption ability. This adaptation requires the presence of vitamin D, is found predominantly in younger animals and has not been demonstrated to be related to other hormonal changes. In an extended study with human beings, Nicolaysen, Eeg-Larsen & Malm (1953) were able to show that this ability to absorb massive amounts of calcium remained with calcium-deprived animals or man until the needs of the bone are satisfied. This led Nicolaysen to postulate the existence of an endogenous or humoral factor presumably secreted from the skeleton which would inform the intestine to absorb calcium. The concept of the endogenous factor has remained fallow over many years but recently has received renewed attention. The results shown in Fig. 9 (Boyle, Gray & DeLuca, 1971; Boyle, Gray, Omdahl & DeLuca, 1972*b*) demonstrate that animals maintained on varied levels of dietary calcium show profound differences in their ability to synthesize 1,25-DHCC, the metabolically active form of vitamin D in the intestine. Clearly animals on a low calcium diet make predominantly 1,25-DHCC as shown from the profiles from the plasma of these animals. Animals maintained on high calcium diets make little or no 1,25-DHCC. Instead there appears a new metabolite which is termed metabolite V_a and which is likely to be 21,25-dihydroxycholecalciferol (21,25-DHCC), a metabolite previously isolated and identified by Suda *et al.* (1970). Of great interest is the fact that by feeding lactose plus high calcium diets it is possible to turn off the synthesis of 1,25-DHCC in even vitamin D-deficient animals. More recently it has been possible to demonstrate that kidney mitochondria or homogenates from chickens fed a low calcium diet containing small amounts of vitamin D are able to make exclusively 1,25-DHCC from 25-HCC, whereas homogenates or mitochondria from chicks fed a high calcium diet plus small amounts of vitamin D make exclusively metabolite V_a (Omdahl *et al.*, 1972*b*) (Fig. 10). Clearly then dietary calcium treatment can dictate the ability of the kidney to synthesize one metabolite of vitamin D or the other. The function of the V_a is not known at present. It may be a hormonal form of vitamin D which plays an important role in tubular phosphate reabsorption and tubular calcium excretion. This, however, remains to be investigated.

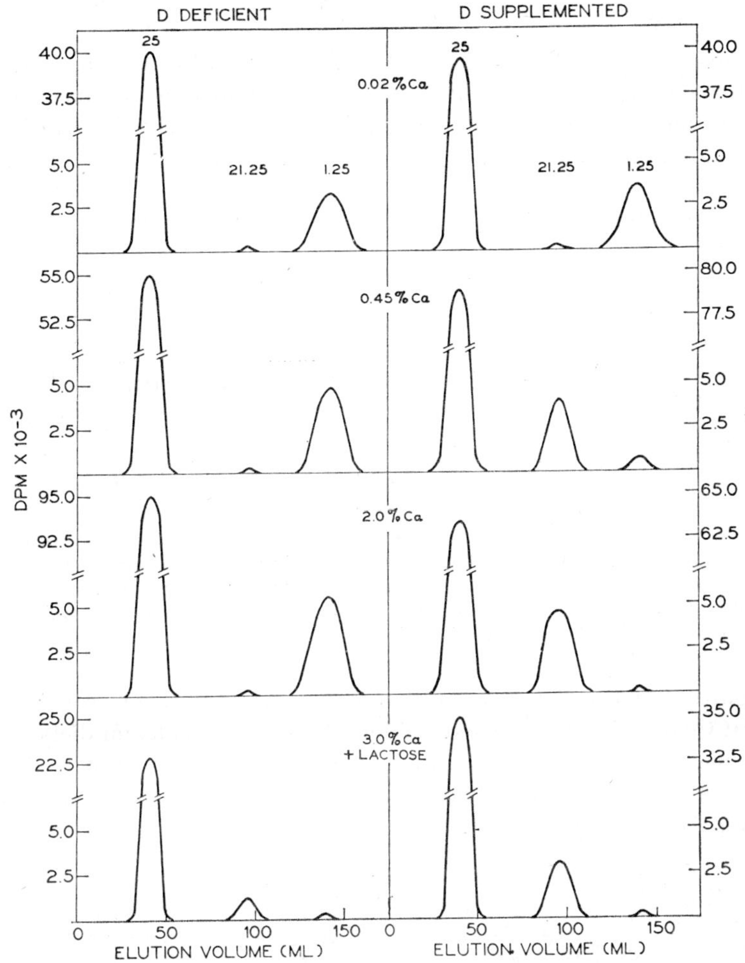

FIG. 9. Influence of dietary calcium on metabolism of ^{3}H-25-HCC by rats *in vivo*. Rats were maintained on the indicated diets for three weeks. One week prior to sacrifice one half of the rats received 1 IU daily of vitamin D_3. After the three week period all the rats were injected intravenously with 650 pmoles ^{3}H-25-HCC. Twelve hours later their blood plasma was collected, extracted with methanol-chloroform and chromatographed on Sephadex LH-20 (Holick & DeLuca, 1971). 25 = 25-HCC; 21,25 = 21,25-DHCC; 1,25 = 1,25-DHCC.

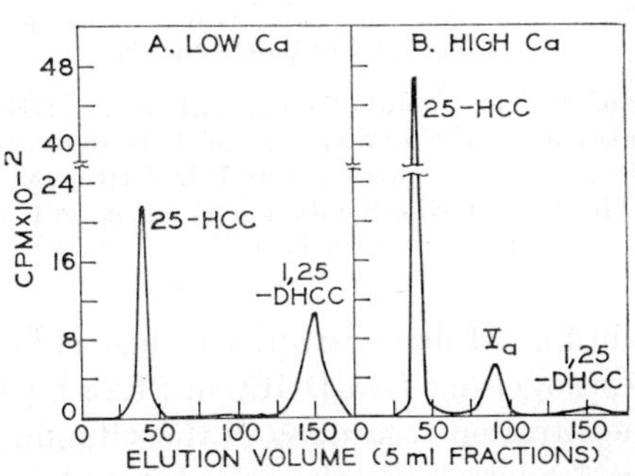

FIG. 10. Influence of dietary calcium on the *in vitro* hydroxylation of 25-HCC by kidney mitochondria. Chicks maintained on either a high calcium or a low calcium diet with a supplement of 5 IU of vitamin D_3 daily were killed after two weeks, their kidney mitochondria obtained and incubated with ^{3}H-25-HCC as described by Gray *et al.* (1971). The resultant extract was chromatographed on Sephadex LH-20.

Additional evidence that the 1,25-DHCC represents Nicolaysen's endogenous factor is shown in Table 3. Thus it is evident thatchicks maintained on 1,25-DHCC as their source of vitamin D are unable to adapt to low calcium or high calcium intakes. Instead their efficiency of calcium absorption remains

Table 3. *Failure of intestinal calcium absorption mechanism of chicks maintained on 1,25-DHCC to adapt to changes in dietary calcium*

Calcium in diet (%)	Ca^{45} absorbed %	
	25-HCC	1,25-DHCC
0·08	59·7 ± 2·9	74·9 ± 4·5
1·0	32·0 ± 5·1	71·0 ± 6·7
2·0	35·5 ± 3·4	78·6 ± 3·9

Chicks were maintained on synthetic diets containing the indicated level of calcium but with normal phosphate. One group of chicks on each diet was maintained on 325 pmoles of either 1,25-DHCC or 25-HCC for at least 8 days before determination of calcium absorption by the method of Omdahl *et al.* (1971).

high regardless of the dietary calcium content. In contrast, chicks maintained on 25-HCC adapt in the expected fashion, those on low calcium diets having higher efficiencies while those on high calcium diets have lower efficiencies. These results then provide evidence that 1,25-DHCC represents in part, Nicolaysen's endogenous factor.

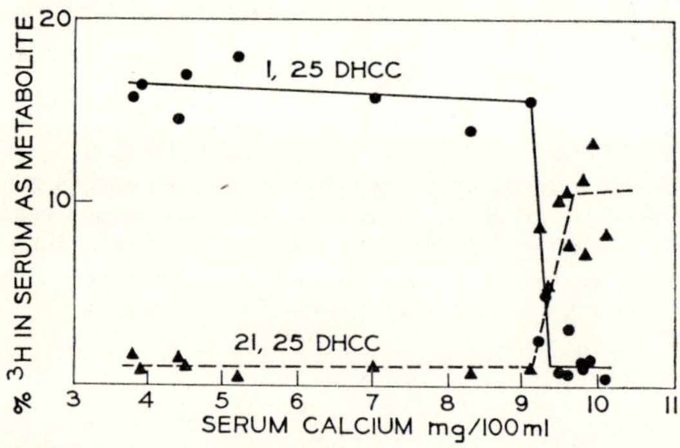

FIG. 11. Regulation of 25-HCC metabolism to either 1,25-DHCC or 21,25-DHCC by plasma calcium concentration. Rats on a variety of diets were given 325 pmoles ^{3}-H25-HCC and 12 hr later the plasma was extracted and the $CHCl_3$ extracts chromatographed on Sephadex LH-20. The % of radioactivity recovered as either 21,25-DHCC or 1,25-DHCC is plotted versus serum calcium concentration.

The results given in Fig. 11 demonstrate conclusively that the determinant in the *in vivo* synthesis of either 1,25-DHCC or 21,25-DHCC is the circulating plasma calcium concentration regardless of the vitamin D treatment of the animal. In this figure is plotted the plasma calcium concentration versus the synthesis of either 1,25-DHCC or 21,25-DHCC. Thus at calcium concentrations which are either normal or hypocalcemic, the predominant hormonal form of vitamin D made is the 1,25-DHCC. It is evident from the preceding discussion that this hormonal form is in turn responsible for the elevation of plasma calcium concentration by means of increased intestinal calcium absorption and increased bone mineral mobilization. This, therefore, provides the final segment in the feed back control loop of vitamin D metabolism which now no longer makes it possible to view vitamin D as a static dietary constituent but rather a substance whose functional metabolism is feed-back regulated much in the sense of a hormone.

ROLE OF 1,25-DHCC IN THE OVERALL FUNCTION OF VITAMIN D

It is worthy of note that the 1,25-DHCC is not as effective in the cure of rickets in rats as is 25-HCC (Omdahl *et al.*, 1971). In preliminary experiments, however, it does act almost as well as 25-HCC in curing rickets in chickens maintained on a diet containing adequate amounts of calcium and phosphate. It is possible that 1,25-DHCC cannot satisfy all the functions of vitamin D. Instead it may represent a specialized hormonal form which functions in the regulation of calcium metabolism. Once the function of 21,25-DHCC and perhaps other metabolites of vitamin D is known, it is possible that a more rational decision will be made concerning what segments of the role of vitamin D can be ascribed to its various metabolites.

SUMMARY

Vitamin D_3 is produced in the skin by means of ultraviolet irradiation of 7-dehydrocholesterol in the epidermis. It finds its way into the plasma where it is immediately picked up by the liver and under a controlled reaction is hydroxylated at the 25 position. This material is carried by means of a specific protein in the plasma to various tissues throughout the body. The 25-HCC proceeds to the kidney where it is hydroxylated further. Hydroxylation proceeds either in the 1 position or in the 21 position, depending upon the ambient plasma calcium concentration. On the low side of plasma calcium concentration a hormonal form of vitamin D, 1,25-DHCC, is made almost exclusively. At plasma levels above normal the 21,25-DGCC is made. The function of 1,25-DHCC is to mobilize calcium from bone to and increase intestinal calcium absorption. The increased intestinal calcium absorption is brought about either through the assembly of the calcium binding protein and the calcium dependent adenosine triphosphatase of the brush border or both or perhaps the formation of still some other factor by a process not involving transcription of DNA. The assembled transport material at the brush border surface brings about an increased rate of transfer of calcium from lumen to the cell. The mitochondria serve to take up the calcium, maintaining a cytoplasmic calcium concentration at around 10^{-6} M. At the basal side of the absorption cell, a downhill sodium gradient brings about the extrusion of calcium into the serosal medium. This depletes the cytoplasmic calcium concentration near the basement membrane surface which then is repleted by

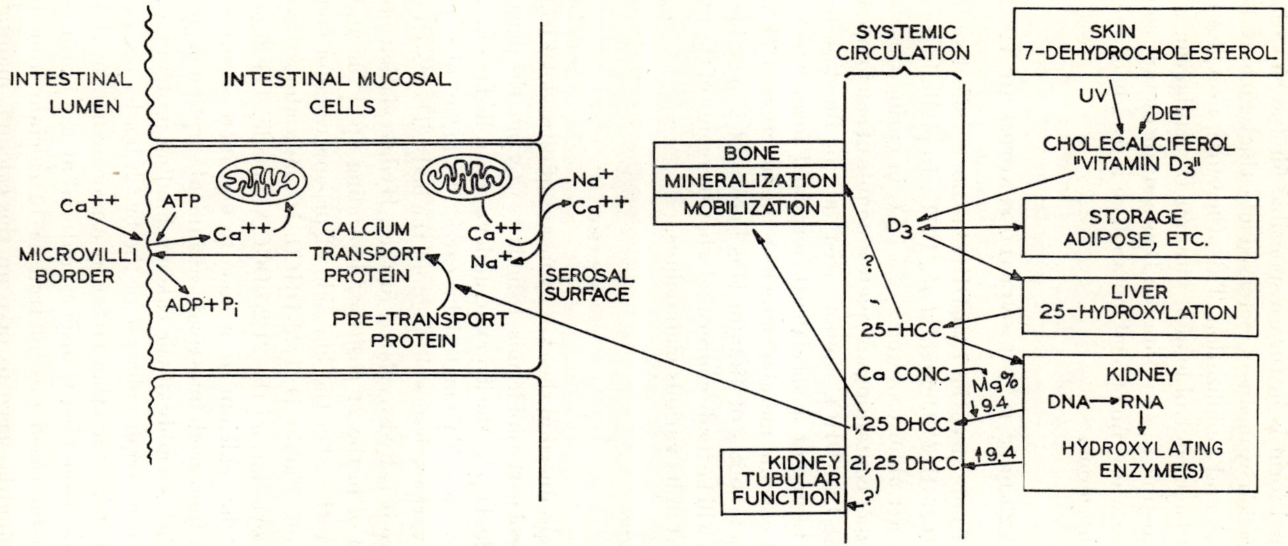

FIG. 12. Summary of metabolism and function of vitamin D.

calcium coming from the mitochondria. This postulated mechanism of vitamin D action merely serves to bring together those currently available bits of information and represents our best working hypothesis on how vitamin D brings about its important function in the intestine. Little is known concerning the vitamin D mechanisms in the mobilization of calcium from bone and possible renal tubular reabsorption mechanisms. Clearly a great deal has been learned not only concerning the functional metabolism of vitamin D but much in terms of conceptual framework within which vitamin D physiology is placed. It seems likely that the metabolism and function of vitamin D will be completely clarified in the next few years bringing to a close an era of some 50 years of investigation.

ACKNOWLEDGEMENT

This work was supported by the Harry Steenbock Research Fund and grants from the National Institutes of Health No. AM-14881 and No. AM-15512.

REFERENCES

Blunt, J. W. & DeLuca, H. F. (1969), *Biochemistry* **8**, 671.

Blunt, J. W., DeLuca, H. F. & Schnoes, H. K. (1968*a*). *Biochemistry* **7**, 3317.

Blunt, J. W., DeLuca, H. F. & Schnoes, H. K. (1968*b*). *Chem. Commun.* **14**, 801.

Blunt, J. W., Tanaka, Y. & DeLuca, H. F. (1968). *Proc. natn Acad. Sci. U.S.A.* **61**, 1503.

Boyle, I. T., Gray, R. W. & DeLuca, H. F. (1971). *Proc. natn Acad. Sci. U.S.A.* **68**, 2131.

Boyle, I. T., Miravet, L., Gray, R. W., Holick, M. F. & DeLuca, H. F. (1972*a*). *Endocrinology.* **90**, 605.

Boyle, I. T., Gray, R. W., Omdahl, J. L. & DeLuca, H. F. (1972*b*). *Proceedings of the Endocrinology 1971 Symposium, London, England.* pp. 468–476.

Cousins, R. J., DeLuca, H. F., Suda, T., Chen, T. & Tanaka, Y. (1970). *Biochemistry* **9**, 1453.

Fraser, D. R. & Kodicek, E. (1970). *Nature* **228**, 764.

Frolik, C. A. & DeLuca, H. F. (1971). *Archs Biochem. Biophys.* **147**, 143.

Gray, R. W., Boyle, I. T. & DeLuca, H. F. (1971). *Science* **172**, 1232.

Gray, R. W. & DeLuca, H. F. (1971). *Archs Biochem. Biophys.* **145**, 276.

Hallick, R. B. & DeLuca, H. F. (1971). *J. biol. Chem.* **246**, 5733.

Haussler, M. R., Myrtle, J. F. & Norman, A. W. (1968). *J. biol. Chem.* **243**, 4055.

Haussler, M. R. & Norman, A. W. (1967). *Archs Biochem. Biophys.* **118**, 145.

Haussler, M. R., Boyce, D. W., Littledike, E. T. & Rasmussen, H. (1971). *Proc. natn Acad. Sci. U.S.A.* **68**, 177.

Holick, M. F. & DeLuca, H. F. (1971). *J. Lipid Res.* **12**, 460.

Holick, M. F., Schnoes, H. K. & DeLuca, H. F. (1971). *Proc. natn Acad. Sci. U.S.A.* **68**, 803.

Holick, M. F., Schnoes, H. K., DeLuca, H. F., Suda, T. & Cousins, R. J. (1971). *Biochemistry* **10**, 2799.

Horsting, M. & DeLuca, H. F. (1969). *Biochem. biophys. Res. Commun.* **36**, 251.

Huldshinsky, K. (1919). *Deutsch Medizinische Wochenschrift* **45**, 712.

Kodicek, E. (1956). In *Ciba Foundation Symposium on Bone Structure and Metabolism* (Ed. by Wolstenholme, G. W. E. & O'Connor, C. M.), p. 161. Boston: Little, Brown and Company.

Lawson, D. E. M., Wilson, P. W. & Kodicek, E. (1969). *Biochem. J.* **115**, 269.

Lawson, D. E. M., Fraser, D. R., Kodicek, E., Morris, H. R. & Williams, D. H. (1971). *Nature* **230**, 228.

Lund, J. & DeLuca, H. F. (1966). *J. Lipid Res.* **7**, 739.

McCollum, E. V., Simmonds, N., Becker, J. E. & Shipley, P. G. (1922). *J. biol. Chem.* **53**, 293.

Mellanby, E. (1919). *J. Physiol* **52**, liii.
Morii, H., Lund, J., Neville, P. & DeLuca, H. F. (1967). *Archs Biochem. Biophys.* **120**, 508.
Neville, P. F. & DeLuca, H. F. (1966). *Biochemistry* **5**, 2201.
Nicolaysen, R. & Eeg-Larsen, N. (1953). *Vit. Horm.* **11**, 29.
Nicolaysen, R., Eeg-Larsen, N. & Malm, O. J. (1953). *Physiol. Rev.* **33**, 424.
Olsen, E. B. & DeLuca, H. F. (1969). *Science* **165**, 405.
Omdahl, J., Holick, M. F., Suda, T., Tanaka, Y. & DeLuca, H. F. (1971). *Biochemistry* **10**, 2935.
Omdahl, J. L., Gray, R. W., Boyle, I. T., Knutson, J. & DeLuca, H. F. (1971*b*). *Nature.* **237**, 63.
Ponchon, G. & DeLuca, H. F. (1969*a*). *J. clin. Invest.* **48**, 1273.
Ponchon, G. & DeLuca, H. F. (1969*b*). *J. Nutr.* **99**, 157.
Ponchon, G., Kennan, A. L. & DeLuca, H. F. (1969). *J. clin. Invest.* **48**, 2032.
Raisz, L. G., Trummel, C. L., Holick, M. F. & DeLuca, H. F. (1972). *Science.* **175**, 768
Rottensten, K. V. (1938). *Biochem. J.* **32**, 1285.
Steenbock, H. & Black, A. (1924). *J. biol. Chem.* **61**, 405.
Suda, T., DeLuca, H. F. & Hallick, R. B. (1971). *Anal. Biochem.* **43**, 139.
Suda, T., DeLuca, H. F., Schnoes, H. K. & Blunt, J. W. (1969). *Biochemistry* **8**, 3515.
Suda, T., DeLuca, H. F., Schnoes, H. K., Ponchon, G., Tanaka, Y. & Holick, M. F. (1970). *Biochemistry* **9**, 2917.
Tanaka, Y. & DeLuca, H. F. (1971*a*). *Proc. natn Acad. Sci. U.S.A.* **68**, 605.
Tanaka, Y. & DeLuca, H. F. (1971*b*). *Archs Biochem. Biophys.* **146**, 574.
Tanaka, Y., DeLuca, H. F., Omdahl, K. & Holick, M. F. (1971). *Proc. natn Acad. Sci. U.S.A.* **68**, 1286.
Trummel, C., Raisz, L. G., Blunt, J. W. & DeLuca, H. F. (1969). *Science* **163**, 1450.
Waddell, J. (1934). *J. biol. Chem.* **105**, 711.
Windaus, A., Schneck, F. & von Werder, F. (1936). *Hoppe-Seyler's Z. Phys. Chem.* **241**, 100.
Zull, J. E., Czarnowska-Misztal, E. & DeLuca, E. & DeLuca, H. F. (1965). *Science* **149**, 182.
Zull, J. E., Czarnowska-Misztal, E. & DeLuca, H. F. (1966). *Proc. natn Acad. Sci. U.S.A.* **55**, 177.

VITAMIN A TRANSPORT IN HUMAN AND RAT PLASMA AND ITS NUTRITIONAL REGULATION

By DeWITT S. GOODMAN

Department of Medicine, Columbia University College of Physicians and Surgeons, New York, New York, U.S.A.

The transport system for vitamin A in human plasma comprises two proteins: retinol-binding protein (RBP), the specific transport protein, and prealbumin (Kanai, Raz & Goodman, 1968; Peterson, 1971*a*). RBP has a molecular weight of approximately 21,000, α_1-mobility on electrophoresis, and a single binding site for one molecule of retinol. RBP interacts strongly with prealbumin, and circulates in plasma as a 1:1 molar RBP-prealbumin complex (Kanai *et al.*, 1968; Peterson, 1971*a*; Raz, Shiratori & Goodman, 1970). RBP normally circulates mainly as the holoprotein, containing a molecule of bound retinol. The usual level of RBP in plasma is about 40–50 μg/ml, and that of prealbumin 200–300 μg/ml (Smith, Raz & Goodman, 1970; Smith & Goodman, 1971).

In addition to its role in vitamin A transport, prealbumin has been recognized for some time as one of the three proteins responsible for the transport of thyroid hormone in plasma (Ingbar, 1963; Robbins & Rall, 1960; Oppenheimer, 1968). The prealbumin molecule appears to contain one major binding site for one molecule of thyroxine (Oppenheimer, Surks, Smith & Squef, 1965; Raz & Goodman, 1969). Estimates of the association constant for the binding of thyroxine to prealbumin have varied from $1 \cdot 6 \times 10^7$ (Raz & Goodman, 1969) to $2 \cdot 3$ or $3 \cdot 6 \times 10^8$ (Oppenheimer & Surks, 1964; Woeber & Ingbar, 1968). Recent studies from several laboratories have indicated that the prealbumin molecule is a tetramer, comprised of four very similar, and possibly identical, subunits (Morgan, Canfield & Goodman, 1971; Blake *et al.*, 1971; Branch, Robbins & Edelhoch, 1971; Rask, Peterson & Nilsson, 1971; Gonzalez & Offord, 1971). The amino terminal amino acid sequence of RBP is unrelated to that of prealbumin (Morgan *et al.*, 1971).

INTERACTIONS INVOLVING RETINOL, RBP AND PREALBUMIN

The transport system for vitamin A provides an interesting model for the study of both protein–protein interactions and the interactions of proteins with small molecules. The interaction of prealbumin with thyroxine appears to be independent of the RBP–prealbumin interaction (Raz & Goodman, 1969). In contrast, addition of prealbumin to RBP, to form the RBP–prealbumin complex, significantly affects the interaction of retinol with RBP. Thus, it has been shown that retinol is much less readily removed from RBP by gentle extraction with heptane, in the presence than in the absence of prealbumin (Peterson, 1971*b*; Goodman & Raz, 1972). This finding suggests that the interaction of retinol with RBP is stabilized by the formation of the RBP–prealbumin complex. This conclusion is supported by the results of recent studies employing fluorescence spectroscopy (Goodman & Leslie, 1972). Formation of the RBP–prealbumin complex increases the relative intensity of fluorescence of retinol bound to RBP by about 15%, and increases the polarization of fluorescence of retinol bound to RBP by approximately 80%. In addition, formation of the RBP–prealbumin complex substantially increases the thermal stability of the retinol–RBP complex (Goodman & Leslie, 1972). Thus, when RBP alone was progressively heated to 70° there occurred a partial disruption of the retinol–RBP complex, with loss of almost one-half of the retinol molecules from RBP. Addition of bovine serum albumin to RBP had very little effect on the thermal stability of the retinol–RBP complex. In contrast, in the presence of prealbumin the retinol–RBP complex was not at all disrupted by progressive heating to 70°, with no loss of retinol occurring from the RBP–prealbumin complex.

We have also recently used the technique of circular dichroism to examine the effects of the protein–lipid and protein–protein interactions involved in vitamin A transport on the secondary structures of both RBP and prealbumin (Gotto, Lux & Goodman, 1972). RBP appears to have a relatively high content of unordered conformation, a significant but small complement of β-structure, and little or no α-helix. Prealbumin appears to have a higher content of β-conformation (one-third to one-half the overall structure) and a lower content of unordered structure than does RBP; prealbumin also appears to have little or no α-helical structure. It is likely that the structures of both proteins are highly organized and specific, but mainly lack regular repeating characteristics. There was no evidence from the CD spectra to indicate that the interaction of either RBP with retinol or of prealbumin with thyroxine resulted in significant changes in the secondary structures of either protein. Moreover, the spectra of mixtures of RBP and prealbumin were additive, indicating that formation of the RBP-prealbumin complex results in very little if any alteration in the secondary structure of the two proteins.

The interaction of retinol with RBP is of considerable physiological importance, since this interaction serves to solubilize the water-insoluble retinol molecule and also to protect the unstable retinol molecule against chemical degradation. The interaction of RBP with prealbumin serves the physiological role of protecting the relatively small RBP molecule against glomerular filtration and subsequent renal catabolism and/or excretion. The physiological significance of the finding that the formation of the RBP–prealbumin complex significantly stabilizes the retinol–RBP complex is at present not clear.

RETINOL TRANSPORT IN RAT PLASMA

Recent studies in our laboratory have demonstrated that a similar transport system for vitamin A exists in rat plasma. The isolation and partial characterization of rat retinol-binding protein, the specific transport protein for vitamin A in rat plasma, has been achieved (Muto & Goodman, 1972). RBP was isolated from rat serum by a sequence of procedures which included: precipitation with ammonium sulphate between 30 and 50% saturation; chromatography on DEAE-Sephadex; gel filtration on Sephadex G-200 and G-100 and preparative polyacrylamide gel electrophoresis. These procedures resulted in RBP which had been purified approximately 2,300-fold, and which was completely pure by physical and immunological criteria. Purified rat RBP has a molecular weight of approximately 20,000, as estimated by sedimentation equilibrium analysis in the analytical ultracentrifuge, and also by gel filtration on standardized columns of Sephadex G-100 and by sodium dodecyl sulphate-disc gel electrophoresis. Rat RBP has α_1-mobility, a sedimentation constant ($S_{20,w}$) of 2·06 S, and apparently one binding site for one molecule of retinol. The properties of rat RBP resemble those of human plasma RBP in many ways. The two proteins have nearly identical ultraviolet absorption spectra (peak maxima at 280 nm and 330 nm) and fluorescence emission and excitation spectra. The amino acid composition of rat and human RBP are somewhat similar, both with a fairly high content of aromatic amino acids. Despite the many similarities between the two proteins, there was no immunological cross-reactivity between rat and human RBP, indicating that the two proteins are immunologically completely distinct.

RBP circulates in rat plasma in the form of a protein–protein complex, with an apparent molecular weight of approximately 60,000–70,000. The protein (prealbumin-2) with which rat RBP interacts to form a complex has an electrophoretic mobility slightly greater than that of rat serum albumin. Purified prealbumin-2 has an apparent molecular weight of approximately 45,000 to 50,000. Prealbumin-2 may also represent a major transport protein for thyroid hormone in the rat.

REGULATION OF RBP METABOLISM BY NUTRITIONAL VITAMIN A STATUS

A study was carried out in order to examine the role of dietary vitamin A in regulating the metabolism of RBP in the rat (Muto *et al.*, 1972). The study was designed to explore in detail the effects of vitamin A depletion and deficiency on the levels of serum RBP, and to determine the time course of such effects in relation to the levels of vitamin A in serum and liver. The effects of nutritional vitamin A status on the levels of immunoreactive RBP in the liver were also examined. Data were also obtained about the effects of repletion of deficient rats with vitamin A.

In order to be able to conduct this study it was necessary to develop an accurate method to measure the levels of RBP in rat plasma and/or tissues. A specific anti-rat RBP antiserum was prepared in a rabbit by the injection of purified rat RBP emulsified with Freund's adjuvant. A sensitive and specific radioimmunoassay was developed which accurately measures rat RBP in the amounts of 0·5 to 2 ng per assay tube. Figure 1 shows the standard curve for radioimmunoassay. As shown in Fig. 1, purified rat holo-RBP and apo-RBP (RBP without bound retinol) displayed identical immunoreactivities in the radioimmunoassay. In addition, RBP in whole serum from both control and vitamin A-deficient rats showed identical immunoreactivities in the radioimmunoassay. The immunoassay accurately measures the total amount of RBP present in the sample.

In the nutritional study (Muto *et al.* 1972), plasma RBP and vitamin A levels were compared in five groups of weanling male rats (a total of 131 rats), each fed a vitamin A-deficient diet with or without supplements as follows: (1) control: supplemented with vitamin A; (2) pair-fed control: supplemented with vitamin A but with food intake matched to that of the deficient rats;

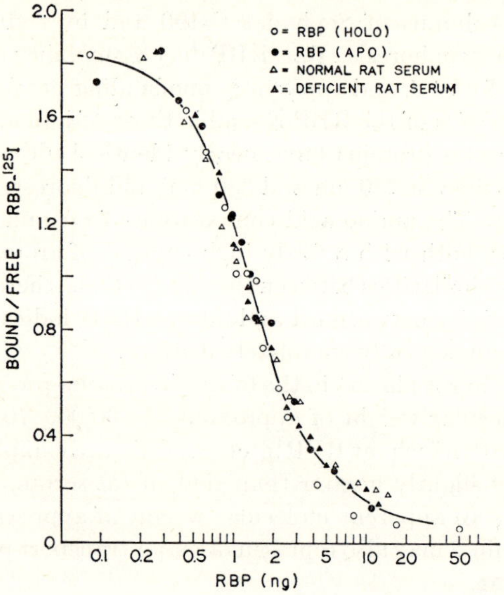

FIG. 1. The radioimmunoassay curve for rat RBP. The figure shows the displacement of rat RBP-^{125}I from antibody by purified holo-RBP, purified apo-RBP, and by dilutions of sera obtained from control and from vitamin A-deficient rats. (Reproduced by permission of *J. biol. Chem.*, from Muto *et al.*, 1972).

(3) deficient: no vitamin A supplementation; (4) and (5) retinoic acid: supplemented, after the initial depletion period, with a modest daily dose (14 μg or 28 μg for the two groups, respectively) of retinoic acid. The study was conducted for 75 days, during which time the food intake and growth of the rats were individually monitored. Some rats, from different groups, were sacrificed at intervals, in order to simultaneously determine the vitamin A and RBP levels in serum and liver. For most rats, serial blood samples were obtained by means of a highly effective venepuncture technique from the subclavian venous plexus. The effects of vitamin A repletion were studied in some of the deficient and some of the retinoic acid-supplemented animals. It was anticipated that this experimental design would permit us to distinguish the primary effects of vitamin A deficiency from the secondary effects due to reduced food intake (by comparison of control, pair-fed control and deficient groups), or due to bacterial infection and/or general ill-health (by comparison of control, and retinoic acid-fed groups, the latter displaying good health and normal growth rates).

In the rats fed the deficient diet, serum vitamin A levels decreased gradually during the first 25–30 days of the study, from 45 ± 6 (sem) μg/dl (on day 3) to levels of about 2 μg/dl. The gradual decrease in serum vitamin A level during depletion of liver stores (first 20 days) and for a few days thereafter was observed in every rat which was followed by sequential venepuncture. Figure 2 shows the serum RBP levels in the five groups of rats under study. In the rats fed the deficient diet, serum RBP levels declined substantially during the first 30 days of the study, particularly during the interval between days 15 and 25. The time course of the decline of RBP was fairly similar to that of vitamin A, but with a lag of approximately 3 days. Unlike vitamin A, however, RBP did not disappear from serum. After approximately 40 days on the deficient diet the level of RBP became relatively stable at about 12–15 μg/ml, representing about 25–30% of the level of RBP seen on the third day. Almost all of the RBP circulating at this time was present as apo-RBP.

The control rats maintained constant serum vitamin A levels throughout the entire study. After an initial small rise, the serum RBP levels of the control

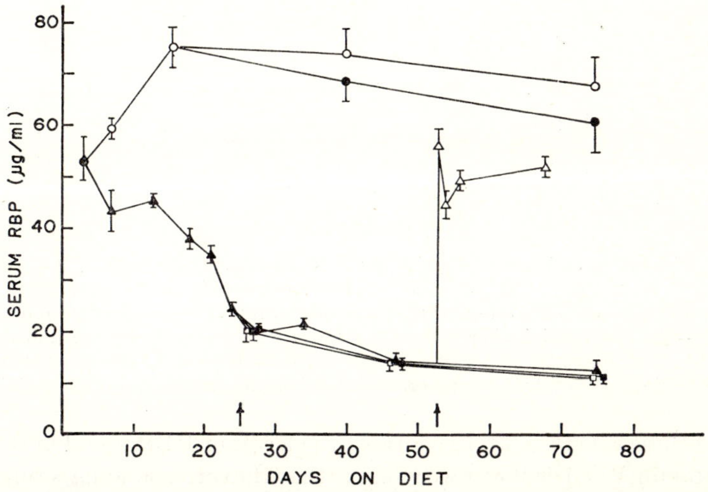

FIG. 2. The effects of dietary vitamin A status on the serum RBP level. Mean ± SEM values are shown for rats in each of the five dietary treatment groups. The first arrow at 25 days represents the point at which retinoic acid supplementation and pair-feeding were started. The second arrow (53 days) represents the time at which some of the deficient and retinoic acid-supplemented rats were first repleted with vitamin A. The symbols representing the several groups are: (1) control, ○; (2) pair-fed control, ●; (3) deficient, ▲; (4) retinoic acid, 14 μg/day, □; (5) retinoic acid, 28 μg/day, ■; (6) deficient rats repleted with vitamin A, △. (Reproduced by permission of *J. biol. Chem.*, from Muto *et al.*, 1972.)

rats remained approximately constant (65–75 μg/ml). The serum vitamin A and RBP levels in the pair-fed control group were similar to those in the *ad-lib* fed control group. In contrast, the results obtained with the retinoic acid-fed rats, who continued to grow normally, were identical with those of the deficient rats (Fig. 2).

When vitamin A was administered orally to the deficient rats on day 53, a very rapid increase was seen in both serum vitamin A and RBP levels.

Within 5 hr (the first time of sampling) the RBP level rose from a mean of approximately 14 μg/ml to 56 $\pm$ 3 μg/ml (see Fig. 2). With continued administration of vitamin A to these rats the RBP level decreased somewhat at 29 hr, and then rose and stabilized at about 50–52 μg/ml.

IMMUNOREACTIVE RBP IN RAT LIVER

When liver homogenates from control or deficient rats were tested in the radioimmunoassay for rat RBP, considerable amounts of immunoreactivity were found. Identical immunoassay curves were obtained with liver homogenates from the control and deficient rats, and these curves were indistinguishable from those generated by pure rat RBP or by whole serum. In the control rats the level of immunoreactive RBP in the liver remained low and constant throughout the study, at about one-half to one-third the level found in the

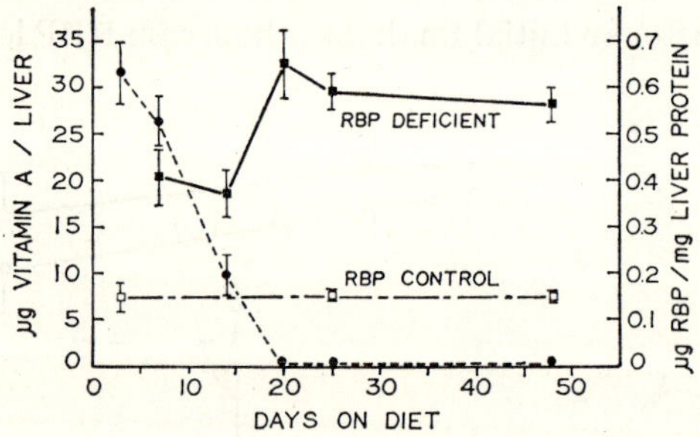

FIG. 3. The levels of immunoreactive RBP in the livers of control (□---□) and of vitamin A-deficient (■——■) rats. The values shown are the mean $\pm$ SEM for five rats which were slaughtered. Also shown are the values for total vitamin A content of deficient livers (●---●). The much higher values for vitamin A in control livers (with mean values of 37 μg/liver at 3 days, 259 μg at 14 days, 587 μg at 25 days, and 1704 μg at 48 days) are not shown. (Reproduced by permission of *J. biol. Chem.*, from Muto *et al.*, 1972).

serum of these rats (Fig. 3). In contrast, the levels of RBP found in the livers of the vitamin A-deficient rats were higher than the corresponding serum levels, and were much higher than the levels of RBP found in the control livers (Fig. 3). Prior to total depletion of the liver vitamin A stores the rats fed the deficient diet had liver RBP contents which were approximately 2–2·5-fold greater than those of controls ($p < 0{\cdot}01$). Coincident with the disappearance of liver vitamin A on about the 20th day of the study, the levels of RBP in the livers increased dramatically to a level about 4-times higher than that observed in the controls (see Fig. 3). After 20 days on the deficient diet the livers of the deficient rats contained about 750 μg more RBP per liver than did the livers of the control rats. This amount of RBP was equivalent to or somewhat greater than the amount of RBP which would have been needed to restore the plasma (and extracellular) compartment to a normal level.

In summary, these studies have shown that vitamin A deficiency results in a decreased level of RBP in serum together with an increased level of immunoreactive RBP in liver. The serum level of RBP can moreover be rapidly restored to normal by administration of vitamin A. These findings suggest that vitamin A deficiency primarily interferes in some way with the secretion, rather than with the synthesis, of RBP by the liver, and that the deficient liver contains a pool of preformed apo-RBP which can be released rapidly into the serum, as holo-RBP, when vitamin A becomes available. Further studies are planned to examine in detail the mechanisms whereby vitamin A regulates the production and secretion of RBP by the liver.

ACKNOWLEDGEMENTS

The studies dealing with rat RBP which are summarized here were carried out in collaboration with Doctors Y. Muto, J. E. Smith, and P. O. Milch. This work was supported by Grants AM-05968 and HL-14236 from the National Institutes of Health, Bethesda, Md., U.S.A.

REFERENCES

Blake, C. C. F., Swan, I. D. A., Rerat, C., Berthou, J., Laurent, A. & Rerat, B. (1971). *J. molec. Biol.* **61**, 217–224.

Branch, W. T., Jr., Robbins, J. & Edelhoch, H. (1971). *J. biol. Chem.* **246**, 6011–6018.

Gonzalez, G. & Offord, R. E. (1971). *Biochem. J.* **125**, 309–317.

Goodman, DeW. S. & Leslie, R. B. (1972). *Biochim. biophys. Acta.* **260**, 670–678.

Goodman, DeW. S. & Raz, A. (1972). *J. Lipid Res.* **13**, 338–347.

Gotto, A. M., Lux, S. E. & Goodman, DeW. S. (1972). *Biochim. biophys. Acta,* **271**, 429–435.

Ingbar, S. H. (1963). *J. clin. Invest.* **42**, 143–160.

Kanai, M., Raz, A. & Goodman, DeW. S. (1968). *J. clin. Invest.* **47**, 2025–2044.

Morgan, F. J., Canfield, R. E. & Goodman, DeW. S. (1971). *Biochim. biophys. Acta* **236**, 798–801.

Muto, Y. & Goodman, DeW. S. (1972). *J. biol. Chem.* **247**, 2533–2541.

Muto, Y., Smith, J. E., Milch, P. O. & Goodman, DeW. S. (1972). *J. biol. Chem.* **247**, 2542–2550.

Oppenheimer, J. H. (1968). *New Engl. J. Med.* **278**, 1153–1162.

Oppenheimer, J. H. & Surks, M. I. (1964). *J. clin. Endocrinol. Metab.* **24**, 785–793.

Oppenheimer, J. H., Surks, M. I., Smith, J. C. & Squef, R. (1965). *J. biol. Chem.* **240**, 173–180.

Peterson, P. A. (1971*a*). *J. biol. Chem.* **246**, 34–43.

Peterson, P. A. (1971*b*). *J. biol. Chem.* **246**, 44–49.

Rask, L., Peterson, P. A. & Nilsson, S. F. (1971). *J. biol. Chem.* **246**, 6087–6097.

Raz, A. & Goodman, DeW. S. (1969). *J. biol. Chem.* **244**, 3230–3237.

Raz, A., Shiratori, T. & Goodman, DeW. S. (1970). *J. biol. Chem.* **245**, 1903–1912.

Robbins, J. & Rall, J. E. (1960). *Physiol. Revs.* **40**, 415–489.

Smith, F. R. & Goodman, DeW. S. (1971). *J. clin. Invest.* **50**, 2426–2436.

Smith, F. R., Raz, A. & Goodman, DeW. S. (1970). *J. clin. Invest.* **49**, 1754–1761.

Woeber, K. A. & Ingbar, S. H. (1968). *J. clin. Invest.* **47**, 1710–1721.

CHOLESTEROL REGULATION OF HEPATIC HMG-CoA REDUCTASE*

By VICTOR W. RODWELL

Department of Biochemistry,
Purdue University, Lafayette, Indiana, USA

BEFORE discussing what is known about the physiological regulation of mammalian liver HMG-CoA reductase [mevalonate:NADP oxidoreductase (acylating CoA), EC 1.1.1.34] I shall comment briefly on HMG-CoA (3-hydroxy-3-methylglutaryl-CoA) as a branch-point compound. Some years ago, Lynen and his coworkers established that HMG-CoA functioned as an obligatory intermediate in hepatic ketogenesis. Since the work of Rudney, Lynen and others had previously established a role for HMG-CoA in polyisoprenoid biosynthesis, HMG-CoA is, at least in the liver, a true branch-point compound. By applying concepts evolved primarily in the areas of regulation of biosynthetic processes in microorganisms, one might therefore predict that enzymes catalysing conversion of HMG-CoA to products destined either for polyisoprenoid or for ketone body synthesis might be subject to stringent metabolic regulation. This prediction has been amply verified for the specialized case of hepatic cholesterol synthesis. The reduction of (S)-HMG-CoA to (R)-mevalonic acid, catalysed by liver microsomal HMG-CoA reductase, represents the first biosynthetic reaction unique to hepatic polyisoprenoid biosynthesis and is the rate-limiting and major regulated enzyme of hepatic cholesterol synthesis. However, while HMG-CoA is a branch point compound in liver, it remains to be established whether this is true for other tissues or for life forms other than vertebrates. In tissues or organisms where the sole known fate of HMG-CoA appears to be conversion to "active isoprene", HMG-CoA synthase [3-hydroxy-3-methylglutaryl-CoA acetoacetyl-CoA ligase (CoA acetylating), EC 4.1.3.5] rather than HMG-CoA reductase may catalyse the first reaction unique to polyisoprenoid synthesis. In tissues other than liver or in organisms other than vertebrates we might therefore predict regulation of HMG-CoA synthase or a prior reaction rather than of HMG-CoA reductase. The concepts and principles for regulation of hepatic cholesterol synthesis outlined below may therefore have little or no predictive value for other systems. Having sounded this cautionary note, I shall now proceed to discuss regulation of hepatic HMG-CoA reductase.

Despite an abundance of indirect evidence that cholesterol regulation of hepatic cholesterol synthesis occurs at the level of HMG-CoA reductase (Bucher *et al.*, 1959; Siperstein & Guest, 1959; Siperstein & Fagan, 1966),

* Not presented at the symposium.

evidence derived from studies of the activity of the reductase itself are of comparatively recent vintage. Various technical difficulties, among which may be cited the extremely low reductase activity of liver microsomal preparations, the presence of other enzymes utilizing NADPH, and the absence of a convenient method for isolation of trace quantities of mevalonate, greatly hindered early attempts to study the hepatic reductase. It is therefore not surprising that the first HMG-CoA reductases to be purified and studied in detail were those of yeast (Knappe, Ringelmann & Lynen, 1959; Knauss, Porter & Wasson, 1959; Durr & Rudney, 1960; Kirtley & Rudney, 1967; Rétey *et al.*, 1970) and of *Pseudomonas* (Fimognari & Rodwell, 1965; Bensch & Rodwell, 1970). The first real breakthrough with the mammalian reductase was the report of Linn (1967*a*) that rat liver microsomal MHG-CoA reductase could be successfully assayed by a thin-layer chromatographic method and was induced subject to regulation by dietary cholesterol (Linn, 1967*b*). Linn's major observations have been confirmed by several independent groups including our own, although we, among others, were unable to repeat the reported solubilization of the reductase from acetone-dried microsomal preparations. More recently, however, Kawachi & Rudney (1970) succeeded in solubilizing the microsomal reductase using bile salts and purified the enzyme 300-fold. Clearly, HMG-CoA reductase is not readily dissociated from the endoplasmic reticulum, and it is perhaps for this reason that the catalytic efficiency of the purified enzyme is only slightly higher than that of crude hepatic microsomes assayed under optimal conditions (Shapiro & Rodwell, 1971). Until this apparent discrepancy is overcome, there is much to be said for studying HMG-CoA reductase in a situation which more closely approaches the physiological state, i.e. in intact microsomes. This is the approach that we and others have followed.

An intriguing feature of liver HMG-CoA reductase is the presence of a diurnal rhythm. Although hepatic sterol synthesis from acetate was known to be low in the early morning and to rise significantly by late afternoon (Kandutsch & Saucier, 1969; Back, Hamprecht & Lynen, 1969). the presence of a rhythm in HMG-CoA reductase was first noted by Kandutsch & Saucier (1969) in mouse liver, and shortly thereafter confirmed for rat liver by Hamprecht, Nüssler & Lynen (1969) and by Shapiro & Rodwell (1969, 1971). When animals are maintained on a "normal" light–dark cycle (dark 6 p.m.–6 a.m.; light 6 a.m.–6 p.m.), a 4–10 fold rhythm in HMG-CoA reductase is observed with peak activity at 12 a.m.–2 a.m. and minimal activity at about noon. The cycle may be reversed by reversing the light schedule (Goldfarb & Pitot, 1971) and occurs both in animals fed *ad libitum* or only during restricted periods. It is worth noting that rhythmic changes are not observed for any of the other enzymes catalysing reactions between mevalonate and isopentenyl pyrophosphate (Dugan, Slakey & Porter, 1971).

A detailed study of the rhythm revealed that there are actually two peaks in HMG-CoA reductase activity which occur at midnight and at 1.45 a.m. in rats maintained on a "normal" light–dark schedule (Shapiro & Rodwell, 1971). Between these times there is a sharp decline in activity. Both the first (Shapiro & Rodwell, 1969; 1971) and the second (Shapiro & Rodwell, 1972) rises in activity are blocked by administration of cycloheximide, which suggests that both are dependent upon protein synthesis. Whether the cyclic rhythm indeed represents cyclic synthesis and degradation of HMG-CoA reductase itself remains, however, to be established. Although we believe this to be the case, this must ultimately be confirmed by techniques such as immunoassay or isotopic measurement of rates of synthesis and degradation of the reductase.* The double-peaked cyclic rhythm persists in fasted rats, although the reductase levels are at all times far below those present in animals fed a normal diet (Shapiro & Rodwell, 1972). The rise and fall in activity relative to the daytime activity nevertheless appears to be identical both in fed and in fasted rats. This suggests that, while the presence or absence of food can profoundly affect the level of HMG-CoA reductase in liver, dietary factors alone are not responsible for the double-peaked rhythm.

We have advanced the following possible explanations for this double-peaked rhythm.

1. *Transient appearance of a labile, HMG-CoA reductase-specific degradative system.* This might account for the rapid decline in reductase activity observed from midnight to 12.30 a.m. Following its disappearance, continued rapid synthesis of reductase would give rise to the second peak. Such a model would resemble that proposed by Boctor & Grossman (1970) for glucocorticoid regulation of hepatic tyrosine transaminase.

2. *Isozymes.* The double peak might represent asynchronous production of two isozymes of HMG-CoA reductase. Differential temporal regulation might result in one isozyme attaining peak levels at midnight and the other at 1.45 a.m. A related possibility is that dual peaks represent synthesis of reductase in different membrane fractions.

3. *Metabolic overshoot.* The dual peaks may represent an overcompensation as the cell attempts to stabilize reductase levels following a rapid increase. In several cases, enzymes whose levels change rapidly either in development (Moog, 1971) or following refeeding (Freedland & Szepesi, 1971) "overshoot" elevated levels and approach these via a series of damped oscillations.

4. *Periodic oscillations.* HMG-CoA reductase may undergo periodic oscillations which are detectable only at the peak of the cyclic rhythm. Stabilization of labile regulatory enzymes may involve control elements which sense the level of enzyme or of metabolites. When enzyme levels rise above the preprogrammed value, cellular controls would be activated which lower the enzyme level by altering the rate of enzyme synthesis or degradation. The level of enzyme would then decline until a second control element, which determines the lower limit, is engaged. This model suggests that the decline in reductase activity following a cyclic peak results from the action of activated cellular controls when levels of the reductase or a metabolite reach critical levels. The persistence of the cyclic rhythm in fasted rats nevertheless favours the alternate hypothesis that the entire cyclic rhythm is internally programmed and is independent of the level of enzyme or of its end product, cholesterol. Since HMG-CoA reductase activity in rats fasted for 36 hr is only about 5% of control activity, neither a high level of enzyme nor an elevated rate of cholesterol synthesis are required to initiate the cyclic decline in reductase levels.

* Note added in proof. This has recently been achieved (Higgins *et al.*, 1971).

Our ability to validate these or other hypotheses is hampered by the obscurity of the factors responsible for cyclic rhythms. Cyclic variations in food intake, with resulting variations in intracellular tryptophan levels, although cited by Wurtman (1969) as triggering cyclic variations in certain hepatic enzymes, would not appear to be the triggering factor here. Not only dietary cholesterol, but also food intake profoundly affect hepatic HMG-CoA reductase activity. Superimposed on this is the double-peaked cyclic rhythm whose causes and significance are at present obscure. What therefore can be stated with respect to cholesterol regulation of HMG-CoA reductase?

The first direct evidence for reduced reductase activity in microsomes from livers of cholesterol-fed rats was provided by Linn (1967*b*). Since reduced activity persisted after acetone and ether treatment, Linn inferred that the reduced activity was not due to the presence of an acetone- or ether-soluble inhibitor. Addition of various solubilized preparations of cholesterol did not inhibit reductase activity. Depression of reductase activity in the livers of cholesterol-fed rats was subsequently confirmed by others (Shapiro & Rodwell, 1969; White & Rudney, 1970). In addition to cholesterol, various other steroids can depress HMG-CoA reductase activity when present in the diet or when injected intraperitoneally (Kandutsch & Packie, 1970). Shapiro & Rodwell (1969; 1971) ascertained that the K_m for reductase from livers of cholesterol-fed and normal-fed rats did not differ significantly, while V_{max} decreased from 0·33 (normal-fed) to 0·04 nmoles mevalonate synthesized/min/mg microsomal protein (cholesterol-fed). Mixing experiments failed to suggest the presence of an inhibitor of HMG-CoA reductase in the livers of cholesterol-fed rats, and cholesterol-rich lipoproteins were without effect on HMG-CoA reductase activity (Shapiro & Rodwell, 1971). Finally, the conditions we employ for reductase assay (saturating levels of HMG-CoA and NADPH) should effectively reverse the inhibition by a classical allosteric inhibitor. No such effect was, however, noted.

In part because all attempts to demonstrate inhibition of reductase activity by cholesterol have failed, the view that cholesterol acts as a repressor of HMG-CoA reductase synthesis has attracted attention. The approximate half-life of HMG-CoA reductase in normal rats is sufficiently short to permit the rapid effects of dietary cholesterol to be mediated via processes involving protein synthesis. Cholesterol may thus act either by repressing synthesis of new reductase and/or by accelerating degradation of existing reductase. We have observed (Shapiro & Rodwell, 1971) that the fall in HMG-CoA reductase activity following cholesterol feeding is accompanied by a concommitant drop in overall synthesis of cholesterol from acetate. During the same time interval, cholesterol synthesis from mevalonate and HMG-CoA synthesis from acetyl-CoA are unaffected. We therefore propose that one can account for the short-term effects of dietary cholesterol on cholesterol synthesis solely on the basis of effects on HMG-CoA reductase, and that it is not necessary to postulate other points of control in order to explain the effects of cholesterol feeding on hepatic cholesterol synthesis.

The cyclic rhythm of HMG-CoA reductase activity closely resembles the idealized pattern for induction and degradation of a mammalian enzyme described by Schimke (1969). The cyclic rise in HMG-CoA reductase activity might be due to release of inhibition of the existing reductase, to activation of previously existing enzyme or proenzyme, or to *de novo* synthesis of new reductase protein. Both the similarity in basic kinetic properties of reductase from rats killed at noon and at midnight and the action of cycloheximide suggest that synthesis of new HMG-CoA reductase protein is responsible for the rise in activity.

The ability of cholesterol, as well as of cycloheximide, to suppress the cyclic rise in HMG-CoA reductase activity is consistent with the concept of cholesterol acting to repress reductase synthesis. During the 9 a.m. to 5 p.m. period reductase activity remains relatively constant. In this time period the rate of HMG-CoA reductase synthesis presumably equals the rate at which reductase is degraded. From 6 p.m. to midnight, when enzyme activity rises sharply, the number of molecules of reductase synthesized should greatly exceed the number being degraded or inactivated. During this period, the ratio of reductase activity in cholesterol-fed to that in control rats falls from 0·40 to 0·12. After midnight, reductase activity decreases in an exponential fashion, implying that the number of enzyme molecules degraded exceeds the number synthesized. From midnight to 2 a.m., while reductase activity in both normal and cholesterol-fed rats falls rapidly, the ratio of cholesterol-fed to normal-fed activity remains essentially constant (0·12 at midnight; 0·14 at 2 a.m.). This implies that the rate of reductase degradation or inactivation may be similar in normal and in cholesterol-fed rats.

The time course of changes in enzyme levels may be used to determine rates of enzyme synthesis and degradation. Using the simplified mathematical model of Berlin & Schmike (1965), we have estimated the half-life ($t_{\frac{1}{2}}$) and the rate of synthesis (k_s) for HMG-CoA reductase in control and in cholesterol-fed rats. Data for control rats was obtained during the cyclic fall in activity which occurs from 2 to 8 a.m. (Shapiro & Rodwell, unpublished data). Data for cholesterol-fed rats, taken from the work of Shapiro & Rodwell (1971), was from the daytime period immediately following cholesterol feeding. In both cases, $t_{\frac{1}{2}} = 2{\cdot}0 \pm 0{\cdot}1$ hr and $k_s =$ essentially zero (0·07 and 0·03 units/mg microsomal protein/hr in control and in cholesterol-fed rats, respectively). That these values are indeed negligible is seen by calculation of the k_s required to maintain a constant level of reductase by assuming a $t_{\frac{1}{2}}$ of 2·0 hr. These data are consistent with the view that the cyclic fall in activity and the decline seen following cholesterol feeding occur via the same mechanism, and that cholesterol acts by decreasing the rate of HMG-CoA reductase synthesis rather than by accelerating the rate of its destruction.

In intact cells, the end products of amino acid degradation enter oxidative metabolism and are frequently difficult to measure. This contributes to the lack of information concerning the physiological significance of alterations in enzyme levels in mammalian systems (Schimke & Doyle, 1970). Cholesterol, the end product of this pathway, is readily assayed. This permits us to correlate changes in HMG-CoA reductase activity with changes in the rate of sterol synthesis from acetate in a liver mince. The excellent agreement between these parameters, which decline in parallel following cholesterol feeding, indicates that the observed alterations in reductase activity reflect changes in the intracellular rate of cholesterol synthesis and represent a true physiological control mechanism. By contrast, Kim & Miller (1969) have reported that the induction of tyrosine transaminase and tryptophan pyrrolase was not associated with increased oxidation rates of ^{14}C amino acids *in vivo*.

It is of considerable interest that, although dietary cholesterol decreases cholesterol synthesis from acetate in liver tissue from normal or tumour-bearing rats, the tumour tissue itself appears to be insensitive to dietary cholesterol. The absence of regulation by cholesterol is common both to poorly-differentiated and to minimal deviation hepatomas (Siperstein & Fagan, 1964). That the site is indeed HMG-CoA reductase has recently been shown by Siperstein, Gyde & Morris (1971) and confirmed by Goldfarb & Pitot (1971) by direct assay of tumour reductase activity. These latter investigators showed, in addition, that the cyclic rhythm persists in tumour tissue. The "defect" in regulation of HMG-CoA reductase of hepatic tumour tissue may, however, be more apparent than real. The view has been advanced by several investigators that the apparent insensitivity of tumour HMG-CoA reductase to regulation by dietary cholesterol may reflect the failure of cholesterol or of its metabolites to reach their site of action in the tumour cell. The most convincing evidence in support of this view is that of Harry, Morris & McIntyre (1971) who showed that transplantable hepatomas exhibit impaired uptake and storage of free cholesterol. Additional evidence reported by Sabine (1969) suggests that while hepatic tumour tissue may be impermeable to bile acids, cholesterol synthesis from acetate is inhibited by 0·5 mM deoxycholate both in normal and in tumour tissue homogenates. Thus, while the apparent defect in regulation of HMG-CoA reductase in hepatic tumour tissue remains an intriguing observation, the possibility that this may reflect the inability of a metabolic regulator to reach its intracellular site of action merits serious consideration.

I should like to emphasize here that, quite apart from its significance for regulation of cholesterol synthesis, periods of rapid rise and fall in HMG-CoA reductase activity may provide a useful system for the study of enzyme and/or membrane synthesis and degradation. These processes are largely dissociated in time without the use of inhibitors of protein or RNA synthesis which frequently have undesirable secondary effects. I therefore anticipate a greatly increased interest in the regulation of this fascinating enzyme in the years ahead.

ACKNOWLEDGEMENTS

I should like to credit the Indiana and the American Heart Associations for funding our work on HMG-CoA reductase, and thank my colleagues and former students Drs Donald J. McNamara and Richard L. Imblum, and most particularly Dr David J. Shapiro, who executed the experiments described in our published work. Journal paper No. 4939, from the Purdue University Agricultural Experimental Station.

REFERENCES

Back, P., Hamprecht, B. & Lynen, F. (1969). *Archs Biochem. Biophys.* **133**, 11.

Bensch, W. R. & Rodwell, V. W. (1970). *J. biol. Chem.* **245**, 3755.

Berlin, C. M. & Schimke, R. T. (1965). *Molec. Pharmacol.* **1**, 149.

Boctor, A. & Grossman, A. (1970). *J. biol. Chem.* **245,** 6337.
Bucher, N. L. R., McGarrahan, K., Gould, E. & Loud, A. V. (1959). *J. biol. Chem.* **234,** 262.
Dugan, R. E., Slakey, L. L. & Porter, J. W. (1971). *Federation Proc.* **30,** 1229 abs.
Durr, I. F. & Rudney, H. (1960). *J. biol. Chem.* **235,** 2572.
Fimognari, G. M. & Rodwell, V. W. (1965). *Biochemistry* **4,** 2086.
Freedland, R. A. & Szepesi, B. (1971). In *Enzyme Synthesis and Degradation in Mammalian Systems* (Ed. by Rechcigl, M., Jr.), p. 104. Baltimore: University Park Press.
Goldfarb, S. & Pitot, H. (1971). *Cancer Res.* **31,** 1879.
Hamprecht, B., Nüssler, C. & Lynen, F. (1969). *FEBS Lett.* **4,** 117.
Harry, D. S., Morris, H. P. & McIntyre, N. (1971). *J. Lipid Res.* **12,** 313.
Higgins, M., Kawachi, T. & Rudney, H. (1971). *Biochem. Biophys. Res. Commun.* **45,** 138.
Kandutsch, A. A. & Packie, R. M (1970) *Archs Biochem. Biophys.* **140,** 122.
Kandutsch, A. A. & Saucier, S. E. (1969). *J. biol. Chem.* **244,** 2299.
Kawachi & Rudney (1970). *Biochemistry* **9,** 1700.
Kim, J. H. & Miller, L. L. (1969). *J. biol. Chem.* **244,** 1410.
Kirtley, M. E. & Rudney, H. (1967). *Biochemistry* **6,** 230.
Knappe, J., Ringelmann, E. & Lynen, F. (1959). *Biochem. Z.* **332,** 195.
Knauss, H. J., Porter, J. W. & Wasson, G. (1959). *J. biol. Chem.* **234,** 2835.
Linn, T. C. (1967*a*). *J. biol. Chem.* **242,** 984.
Linn, T. C. (1967*b*). *J. biol. Chem.* **242,** 990.
Moog, F. (1971). In *Enzyme Synthesis and Degradation in Mammalian Systems* (Ed. by Rechcigl, M., Jr.), p. 63. Baltimore: University Park Press.
Rétey, J., von Stetten, E., Coy, U. & Lynen, F. (1970). *Eur. J. Biochem.* **15,** 72.
Sabine, J. R. (1969). *Biochim. biophys. Acta* **176,** 600.
Schimke, R. T. (1969). In *Current Topics in Cellular Regulation* (Ed. by Horecker, B. L. & Stadtman, E. R.), p. 79. New York: Academic Press, Inc.
Schimke, R. T. & Doyle, D. (1970). *Ann. Rev. Biochem.* **39,** 929.
Shapiro, D. J. & Rodwell, V. W. (1969). *Biochem. biophys. Res. Commun.* **37,** 867.
Shapiro, D. J. & Rodwell, V. W. (1971). *J. biol. Chem.* **246,** 3210.
Shapiro, D. J. & Rodwell, V. W. (1972). Unpublished data.
Siperstein, M. D. & Fagan, V. M. (1964). *Cancer Res.* **241,** 1108.
Siperstein, M. D. & Fagan, V. M. (1966). *J. biol. Chem.* **241,** 602.
Siperstein, M. D. & Guest, M. J. (1959). *J. clin. Invest.* **38,** 1043.
Siperstein, M. D., Gyde, A. M. & Morris, H. P. (1971). *Proc. natn. Acad. Sci. U.S.A.* **68,** 315.
White, L. W. & Rudney, H. (1970). *Biochemistry* **9,** 2725.
Wurtman, R. J. (1969). *Adv. Enzyme Regulat.* **7,** 57.

EFFECT OF DIETARY CHANGES ON THE RATES OF SYNTHESIS AND DEGRADATION OF RAT LIVER FATTY ACID SYNTHETASE

BY MARGARET C. CRAIG, CARL M. NEPOKROEFF, M. R. LAKSHMANAN AND JOHN W. PORTER

Lipid Metabolism Laboratory,
Veterans Administration Hospital,
and
Department of Physiological Chemistry,
University of Wisconsin,
Madison, Wisconsin, U.S.A.

THE fatty acid synthetase of mammalian liver is one of several lipogenic enzymes whose activity is affected by the nutritional status of the animal. Fasting of the animal results in a reduction of the activity of this enzyme (Gibson & Hubbard, 1960; Numa, Matsuhashi & Lynen, 1961; Allmann, Hubbard & Gibson, 1965; Burton *et al.*, 1969). Upon refeeding the activity of the enzyme returns to normal, and if a fat-free diet is ingested the activity rises to levels far above normal (Allmann *et al.*, 1965; Gibson, Hicks & Allmann, 1966; Burton *et al.*, 1969).

Hicks, Allmann & Gibson (1965) have shown that the rise in the level of liver fatty acid synthetase upon refeeding can be prevented by administration of puromycin or antinomycin D at the beginning of the refeeding period. Their results suggest that the increase in fatty acid synthetase activity is primarily due to an increase in *de novo* synthesis of enzyme protein and not to activation of pre-existing enzyme. The work of Burton *et al.* (1969) confirmed this result by showing that the actual amount of enzyme isolated from the livers of fasted, normal and refed rats varied according to the nutritional state of the animal.

The net amount of enzyme present in any given state is a function of both synthetic and degradative parameters. In the present study we have investigated the roles of synthesis and degradation as determinants of the level of the fatty acid synthetase in the livers of rats subjected to various nutritional states. Isolation of this enzyme was carried out by an immunochemical procedure developed in this investigation or by purification of the enzyme protein. It will be shown that it is the rate of synthesis, and not that of degradation, that is the vital parameter in controlling the liver fatty acid synthetase in rats subjected to nutritional stress.

EXPERIMENTAL PROCEDURES

Materials

Experimental materials were obtained from the following U.S.A. sources: acetyl-CoA and malonyl-CoA from P-L Biochemicals, Milwaukee, Wisc.; NADPH from Sigma, St. Louis, Mo.; EDTA and dithiothreitol from Calbiochem, San Diego, Cal.; β-mercaptoethanol from Eastman Organic Chemicals, Rochester, N.Y.; L-[U-^{14}C]leucine from New England Nuclear, Boston, Mass.; NCS solubilizer from Amersham/Searle, Des Plains, Ill.; and DEAE-cellulose from Mann Research Laboratories, New York, N.Y. All other chemicals were of analytical grade.

Diets and animals

Male Holtzmann rats (Holtzmann Co., Madison, Wisc.) weighing approximately 200 g were used in all experiments. Animals were fed either a fat-free diet obtained from Nutritional Biochemicals, Cleveland, Ohio, or a normal rat diet, Wayne Lab-Blox, obtained from Allied Mills, Inc., Chicago, Ill.

Preparation of fatty acid synthetase

Rats were fasted for 48 hr and then re-fed a fat-free diet for 48 hr. Rat liver supernatant solutions were prepared from these animals by the method of Wakil, Porter & Gibson (1957), as modified by Hsu, Wasson & Porter (1965). The fatty acid synthetase was purified from the 105,000 g_{av} supernatant solution as described by Burton, Haavik & Porter (1968) and modified by Collins *et al.* (1971). The enzyme at this stage of purification will be referred to as DEAE-cellulose purified fatty acid synthetase.

Determination of protein

Protein concentrations of the rat liver soluble fractions or of the purified enzyme were determined by the biuret method of Gornall, Bardawill & David (1949).

Assay for fatty acid synthetase activity

The spectrophotometric assay method of Collins *et al.* (1971) for fatty acid synthetase activity was used, with the exception that 100 μg of 105,000 g_{av} supernatant protein or 10 μg of purified enzyme was assayed in a 1 ml volume. The activity of the fatty acid synthetase is expressed as nanomoles of NADPH oxidized or as nanomoles of palmitic acid (NADPH oxidized $\div$ 14) formed per minute.

Sucrose density gradient centrifugation

The results of Collins *et al.* (1971) showed that antisera to DEAE-cellulose purified fatty acid synthetase gave two bands on immunodiffusion, the major one due to the fatty acid synthetase and the minor one due to a contaminating 7S component. Hence, further purification of the enzyme was necessary in order to obtain antiserum specific for only fatty acid synthetase. This was accomplished by subjecting the DEAE-cellulose purified enzyme to two sequential sucrose density gradient centrifugations.

Sucrose density gradient centrifugations were carried out in a Spinco Model L ultracentrifuge with an SW 25·1 rotor. The centrifugations were performed for 46 hr at 2°C at 58,600 g_{av} in a 5–20%(w/v) sucrose gradient containing 0·5 M-potassium phosphate buffer, pH 7·0, 1 mM-EDTA and 1 mM-dithiothreitol. After centrifugation the bottom of the tube was punctured and the solution was collected dropwise. Profiles of protein in the gradient were monitored spectrophotometrically at 280 nm. The fractions from the first sucrose density gradient which contained fatty acid synthetase activity were pooled, concentrated and applied to a second gradient.

Preparation of antiserum

The enzyme purified by two successive sucrose density gradient centrifugations was mixed with an equal volume of incomplete Freund's adjuvant and injected into the foot pads of rabbits. Each rabbit received 2 mg of fatty acid synthetase protein. A second injection was given 10 days later. The rabbits were bled at 2-week intervals after the second injection. The sera were pooled and the γ-globulin fraction was isolated as described by Levy & Sober (1960). The serum from rabbits which were not injected with fatty acid synthetase was prepared in the same way and used as control serum. All sera were frozen at −15°C until used.

Immunological procedures

Ouchterlony micro-double diffusion was carried out on microscope slides in 0·5% agarose containing 1·0% (w/v) sodium chloride as described in the LKB Operation Manual, 6800A, Immunodiffusion Equipment. Equivalence point determinations and quantitative precipitin analyses were performed as outlined by Kabat & Mayer (1961). The protein content in antigen-antibody precipitates was determined by the Folin method of Lowry, Rosebrough, Farr & Randall (1951), after the precipitates were dissolved in 0·2 M-NaOH.

Incorporation of L-[U-^{14}C]leucine into fatty acid synthetase

Rats were fed the normal rat diet for at least a week and then divided into three groups. One group was fed *ad libitum* on the normal rat diet. Another group of animals was fasted for 48 hr then re-fed a fat-free diet for 72 hr. A third group was fasted for 48 hr. At the end of the respective experimental period each rat fed the normal or fat-free diet received a single intraperitoneal injection of 10 μCi of L-[U-^{14}C]leucine (specific activity = 100 mCi/mmol) in 0·9% (w/v) NaCl. Those in the fasted group received 20 μCi of radioactive leucine.

Assay for radioactive fatty acid synthetase

The rats were killed at various times after administration of radioactive leucine. The livers were removed and homogenized with a motor-driven teflon pestle with six to seven strokes in the buffer medium described by Hsu *et al.* (1965). An aliquot of the 105,000 g_{av} supernatant solution was used to determine the total radioactivity in the supernatant protein. The latter was precipitated with a final concentration of 10% (w/v) trichloroacetic acid and then washed with ethanol:diethyl ether, 2:1 (v/v) acidified with dilute hydrochloric acid. Both the trichloroacetic acid precipitable protein and immunoprecipitates (discussed later) were dissolved in NCS solubilizer before assaying for radioactivity in a toluene-based scintillation solution. All measurements of radioactivity were carried out in a Packard liquid scintillation spectrometer. Absolute d.p.m. values were calculated by using an automatic external standardization procedure.

Fatty acid synthetase was purified from the 105,000 g_{av} supernatant solution as described in a previous section. Because of the low levels of fatty acid synthetase in the livers of fasted rats, 5 ml of unlabelled 105,000 g_{av} supernatant solution from animals re-fed a fat-free diet was added as carrier prior to isolation of the enzyme.

An aliquot of the DEAE-cellulose purified enzyme was subjected to a single sucrose density gradient centrifugation as described above. Those fractions containing fatty acid synthetase activity were pooled, and the protein was precipitated with trichloroacetic acid as before and assayed for radioactivity.

Immunoprecipitation was performed on the 105,000 g_{av} supernatant solution obtained from rats re-fed a fat-free diet and on the DEAE-cellulose purified fraction obtained from livers of rats maintained in each nutritional state. In each case sufficient rabbit antisera was added to precipitate at least 1·25 times the amount of fatty acid synthetase present. A similar amount of control serum was added to determine the extent of nonspecific precipitation. There was no precipitation in tubes containing control serum. Nonspecific coprecipitation of radioactivity in the immunoprecipitates was determined by the method of Schimke, Sweeney & Berlin (1965). Unlabelled fatty acid synthetase equivalent to the amount that was present in the original preparation and a corresponding amount of antibody were added to the supernatant fraction from which the radioactive fatty acid synthetase had been removed by prior precipitation. Radioactivity, if any, in the precipitate of this second reaction was assumed to be due to nonspecific co-precipitation.

All immunoprecipitation reactions were incubated for 1 hr at 37°C and then placed at 4°C overnight. The precipitates were collected by centrifugation, washed twice with cold 0·9% (w/v) sodium chloride containing 0·01 M-potassium phosphate buffer, pH 7·0. They were then analysed for radioactivity.

RESULTS

Characterization of antibody to fatty acid synthetase

The Ouchterlony double diffusion pattern (Fig. 1*a,b*) show that only one precipitin band is obtained with antiserum prepared against fatty acid synthetase purified by sucrose density gradient centrifugation, whereas antiserum prepared against the DEAE-cellulose purified fatty acid synthetase gives a second minor band. Furthermore, when the antibody prepared against sucrose density gradient purified fatty acid synthetase is reacted with enzyme preparations derived from the livers of normal, fasted and re-fed rats, a single connecting band of precipitation is obtained (Fig. 1*d*,*e*). This indicates that the fatty acid synthetase from the livers of rats in the above nutritional states are immunologically identical. This result was found regardless of whether the antiserum reaction was performed with DEAE-purified enzyme (Fig. 1*d*) or with 105,000 g_{av} supernatant solution (Fig. 1*e*). Moreover, the Ouchterlony double diffusion patterns (Fig. 1*f*,*g*) indicate that the antibody reaction with the enzyme of the 105,000 g_{av} supernatant fraction is similar to the reaction with

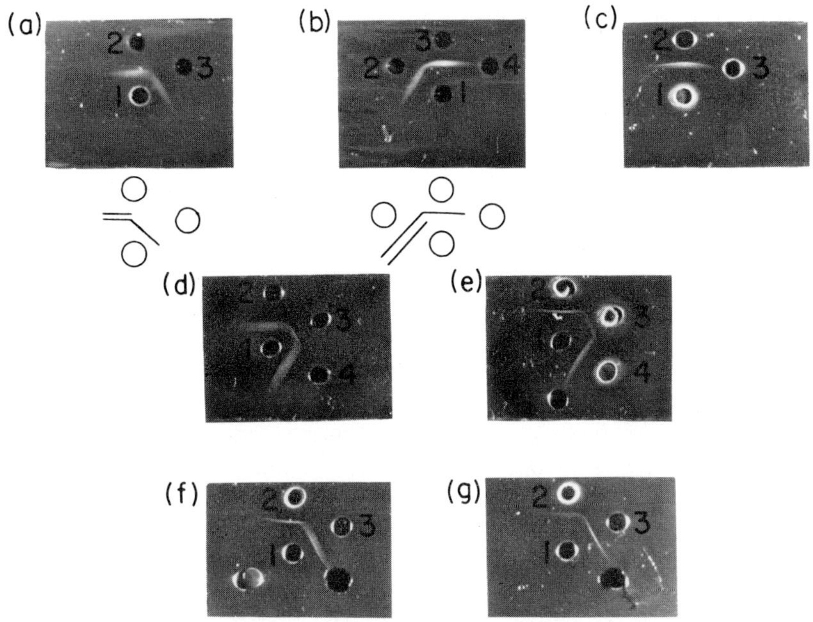

FIG. 1. Ouchterlony double diffusion analyses of fatty acid synthetase antibody and various preparations of the fatty acid synthetase. Agarose gels [0·5% (w/v)] in 1·0% (w/v) sodium chloride prepared on microscope slides were developed for 18–22 hr at 24°C before photographs were taken. The wells contained the following: (a) Well 1, antiserum prepared against DEAE-cellulose purified fatty acid synthetase; Well 2, DEAE-cellulose purified fatty acid synthetase; Well 3, sucrose density gradient purified fatty acid synthetase. (b) Well 1, DEAE-cellulose purified fatty acid synthetase; Well 2, antiserum prepared against DEAE-cellulose purified fatty acid synthetase; Well 3, antiserum prepared against sucrose density gradient purified fatty acid synthetase; Well 4, control serum. (c) Well 1, 105,000 g_{av} supernatant solution obtained from the livers of animals re-fed a fat-free diet; Well 2, antiserum prepared against sucrose density gradient purified enzyme; Well 3, control serum. (d) Well 1, antiserum prepared against sucrose density gradient purified enzyme; Wells 2, 3 and 4, DEAE-cellulose purified fatty acid synthetase from the livers of normal, fasted and re-fed animals, respectively. (e) Well 1, antiserum prepared against sucrose density gradient purified enzyme; Wells 2, 3 and 4, 105,000 g_{av} supernatant solution from the livers of normal, fasted and re-fed animals, respectively. (f) Well 1, antiserum prepared against sucrose density gradient purified enzyme; Well 2, 105,000 g_{av} supernatant solution from the livers of normal animals; Well 3, DEAE-cellulose purified fatty acid synthetase from the livers of normal animals. (g) Well 1, antiserum prepared against sucrose density gradient purified enzyme; Well 2, 105,000 g_{av} supernatant solution from the livers of fasted animals; Well 3, DEAE-cellulose purified fatty acid synthetase from the livers of fasted animals. Further details of enzyme purification procedures are described in the Experimental Procedures.

enzyme in the purified state. Neither the 105,000 g_{av} supernatant fraction nor the DEAE-cellulose purified enzyme showed any precipitin band when reacted against control serum (Fig. 1*b*,*c*).

The results of the quantitative precipitin reaction between anti-serum and DEAE-cellulose purified enzyme are shown in Fig. 2. Measurement of enzyme activity in each supernatant solution, after removal of the antigen-antibody precipitate, yielded an equivalence point of 33·5 mg of antiserum protein per mg of fatty acid synthetase protein. Control serum gave no precipitate nor did addition of control serum affect the activity of the enzyme. In the region of antibody excess, all fatty acid synthetase activity was removed. A similar curve for the 105,000 g_{av} supernatant solution gave an equivalence point of 75 mg of antiserum protein per mg of fatty acid synthetase protein. The content of fatty acid synthetase in the supernatant solution was estimated by enzyme analysis. Therefore, approximately 2·2 times as much antiserum is

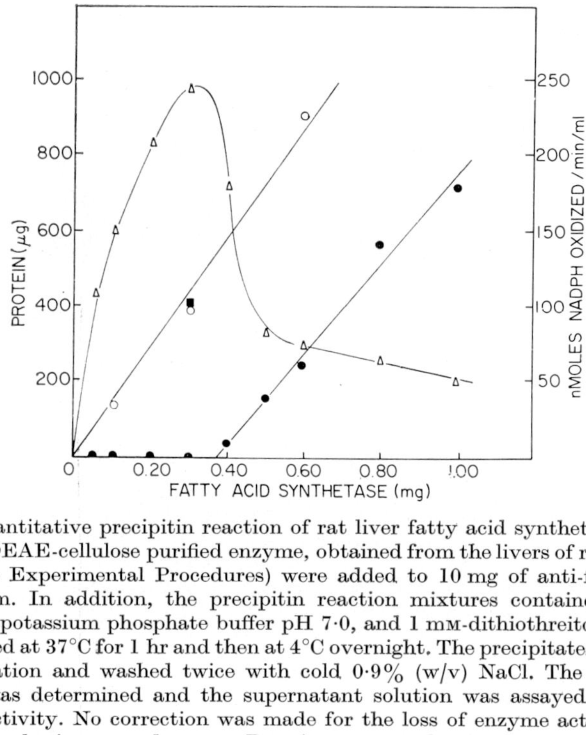

Fig. 2. Quantitative precipitin reaction of rat liver fatty acid synthetase. Increasing amounts of DEAE-cellulose purified enzyme, obtained from the livers of rats re-fed a fat-free diet (see Experimental Procedures) were added to 10 mg of anti-fatty acid synthetase serum. In addition, the precipitin reaction mixtures contained 0·9% (w/v) NaCl, 0·01 M-potassium phosphate buffer pH 7·0, and 1 mM-dithiothreitol. The samples were incubated at 37°C for 1 hr and then at 4°C overnight. The precipitates were collected by centrifugation and washed twice with cold 0·9% (w/v) NaCl. The protein in the precipitate was determined and the supernatant solution was assayed for fatty acid synthetase activity. No correction was made for the loss of enzyme activity occurring during the incubation procedure. △, Protein content of antibody-antigen precipitate; ■, fatty acid synthetase activity in the supernatant solution containing no serum; ○, fatty acid synthetase activity in supernatant containing control serum; ● fatty acid synthetase activity in the supernatant containing anti-fatty acid synthetase serum.

required to precipitate the same enzyme activity in the crude 105,000 g_{av} supernatant solution as required for the purified enzyme.

Immunochemical titrations of enzyme obtained from rats fed a normal diet or re-fed a fat-free one are shown in Fig. 3. Even though the specific activities of the enzyme preparations varied considerably, the equivalence point based on enzyme activity was found to be essentially the same for both nutritional states. Thus, the differences in the activity of DEAE-cellulose purified enzyme after changes in the diet of the animal are accompanied by proportionate changes in the quantity of immunoprecipitable fatty acid synthetase.

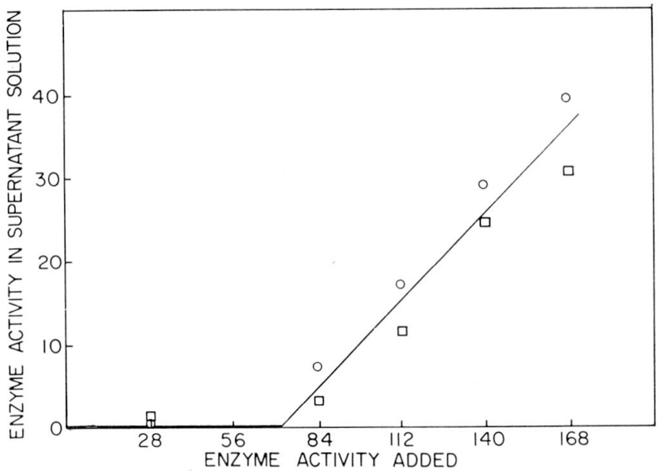

FIG. 3. Quantitative precipitin reactions of rat liver fatty acid synthetase obtained from animals fed a normal diet or re-fed a fat-free diet. Increasing amounts of DEAE-cellulose purified fatty acid synthetase obtained from the livers of rats re-fed a fat-free diet or fed a normal diet were added to 9·5 mg of antiserum. The specific activity of the enzyme, expressed as nanomoles of NADPH oxidized/min/mg of protein, was 616 for animals re-fed a fat-free diet and 280 for those fed a normal diet. Other conditions were the same as those described in Fig. 2, except that a different batch of antiserum was used. The enzyme activity added and the activity in the supernatant solution are expressed as nanomoles of NADPH oxidized/min. ○, fatty acid synthetase activity in the supernatant solution obtained after reaction of antiserum with enzyme prepared from animals re-fed a fat-free diet; □, fatty acid synthetase activity in the supernatant solution obtained after reaction of antiserum with enzyme prepared from animals fed a normal diet.

Isotopic leucine incorporation studies

Studies on the incorporation of L-[U-^{14}C]leucine into liver fatty acid synthetase were undertaken to determine whether the changes in enzyme level in various nutritional states are due to changes in the rate of enzyme synthesis or enzyme degradation.

The ratio of L-[U-^{14}C]leucine incorporated into protein precipitated by anti-fatty acid synthetase serum to that incorporated into total soluble liver protein was determined in order to measure the relative rates of enzyme synthesis (Table 1). The results of this experiment indicate that the relative rate of enzyme synthesis was increased 78-fold in rats re-fed a fat-free diet compared to animals in the fasting state, or 14·3-fold compared to animals in the normal state. The relative rate of enzyme synthesis was decreased 5·5-fold in fasting animals compared to those in the normal state. Thus, it appears that changes in the rate of synthesis are responsible for effecting the changes in the enzyme levels which occur when the nutritional state of an animal is altered.

Table 1. *Relative rates of synthesis of the fatty acid synthetase in the livers of rats in different nutritional states*

Nutritional state	Mean weight (g)	L-[U-^{14}C]leucine incorporation: A Fatty acid synthetase (d.p.m./liver)	B Total soluble protein (d.p.m./liver)	A/B × 10^5 Relative rates of synthesis
Normal	257	765	$3{\cdot}5 \times 10^5$	218
Fasted for 48 hr and subsequently re-fed a fat-free diet for 72 hr	222	15,900	$5{\cdot}1 \times 10^5$	3120
Fasted for 48 hr	204	332	$8{\cdot}3 \times 10^5$	40

Rats treated as indicated above were given a single intraperitoneal injection of L-[U-^{14}C]leucine (10 μCi per rat in animals fed a normal diet or re-fed a fat-free diet and 20 μCi per rat in the fasting group). Four hours later, two rats from each group were killed and the livers pooled. Radioactivity incorporated into the fatty acid synthetase was determined by immunoprecipitation of the DEAE-cellulose purified fatty acid synthetase. Further details of this experiment are given in the Experimental Section.

Rats in each of the various nutritional states were pulse-labelled with L-[U-^{14}C]leucine and the loss of radioactivity in the fatty acid synthetase with time was investigated in order to determine the rate of enzyme degradation. The results for rats re-fed a fat-free diet are shown in Fig. 4. A half-life of 44 hr was obtained (Fig. 4A) from the determination of radioactivity in the fatty acid synthetase precipitated from the 105,000 g_{av} supernatant solution by specific antiserum. However, a half-life of 69 hr was obtained when assays for radioactivity were carried out on the enzyme isolated by the standard purification procedure (Fig. 4B).

It has been shown that the DEAE-cellulose purified fatty acid synthetase from the livers of rats re-fed a fat-free diet consists almost entirely of fatty acid synthetase protein as indicated by analytical ultracentrifugation and sucrose density gradient centrifugation (Burton *et al.*, 1968). The near homogeneity of this preparation was confirmed by subjecting it to sucrose density gradient centrifugation. Between 75 and 90% of the radioactivity applied to the gradient appeared in the fractions containing fatty acid synthetase activity. Similar results were obtained when immunochemical analysis was performed on the DEAE-cellulose purified fatty acid synthetase. Thus, the same value for the half-life of the fatty acid synthetase (69 hr) was obtained by either direct measurements of the radioactivity in the purified enzyme fraction or by reaction of the purified enzyme fraction with antiserum.

The results of experiments with rats fed the normal diet are shown in Fig. 5. It should be pointed out that extremely high non-specific co-precipitation (approximately 50%) ruled out immunoprecipitation of the fatty acid synthetase from the 105,000 g_{av} supernatant fraction. A half-life of 69 hr was obtained (Fig. 5B) from determinations of the radioactivity incorporated into

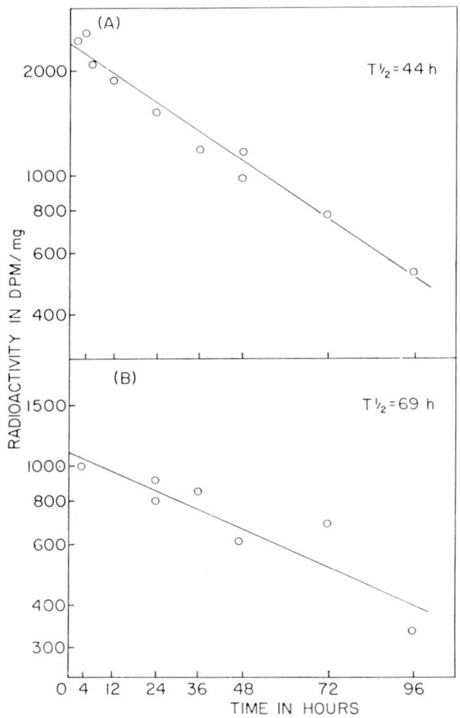

FIG. 4. Turnover of liver fatty acid synthetase in rats maintained on a fat-free diet. Rats were fasted for 48 hr then re-fed a fat-free diet for 72 hr before administration of 10 μCi of L-[U-^{14}C]leucine/rat. Animals were killed by decapitation at specified time intervals after administration of the labelled leucine. Analysis for radioactivity in the liver fatty acid synthetase was performed by either immunoprecipitation on the 105,000 g_{av} supernatant solution (Fig. 4A) or by enzyme isolation through the DEAE-cellulose step (Fig. 4B) as described in the Experimental Procedures. The amount of fatty acid synthetase in the 105,000 g_{av} supernatant was based on enzyme specific activity measurements. Two independent experiments were carried out and all values in the figure were normalized to a single time point that was common to both experiments. This was done in order to compensate for variation between experiments in ^{14}C-incorporation. The best fit for the decay curve was statistically computed by linear regression.

DEAE-cellulose purified enzyme. However, sucrose density gradient centrifugation of the DEAE-cellulose purified enzyme obtained from normal rats showed that only 25% of the radioactivity applied to the gradient sedimented with fatty acid synthetase activity. Therefore, immunoprecipitation was performed on the DEAE-cellulose purified enzyme obtained from normal rats, and the radioactivity in the immunoprecipitate was determined. The amount of radioactivity obtained by immunoprecipitation agreed well with that found associated with the fatty acid synthetase on sucrose density gradient centrifugation. A half-life of 68 hr was obtained from the radioactivity of the fatty acid synthetase that was precipitated by specific antiserum from the DEAE-cellulose purified fraction (Fig. 5A). Thus, similar half-life values were obtained when either purification or immunological procedures were used, even though only one-fifth of the radioactivity in the DEAE-cellulose fraction was immunoprecipitable.

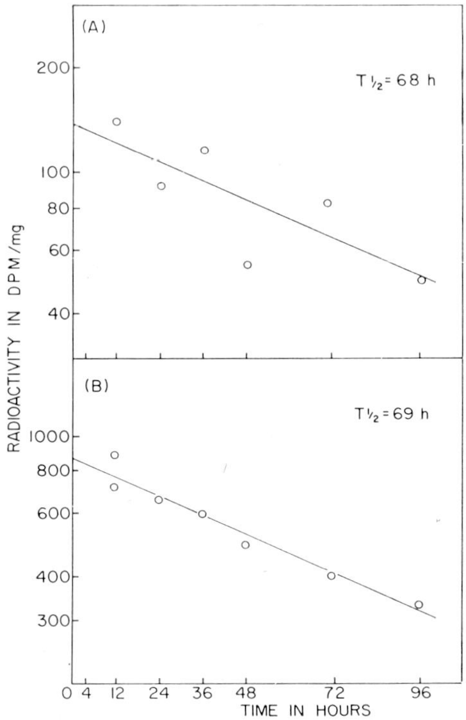

FIG. 5. Turnover of liver fatty acid synthetase in rats maintained on a normal diet. Conditions were the same as described in the legend for Fig. 4, except that analyses for radioactivity in fatty acid synthetase were performed either on the DEAE-cellulose purified enzyme (5B) or by immunoprecipitation of the DEAE-cellulose purified enzyme (5A).

High nonspecific co-precipitation (between 75 and 90%) was obtained when the 105,000 g_{av} supernatant solution from the livers of fasted animals was challenged with antiserum. Therefore, fatty acid synthetase was isolated by the standard purification procedure and then the labelled enzyme was immunochemically precipitated. However, high values for nonspecific adsorption were again obtained (approximately 50%). This was understandable since an excess of nonlabelled carrier fatty acid synthetase was added to the labelled 105,000 g_{av} supernatant fraction from fasted animals. This diluted the label in the fatty acid synthetase of fasted animals, thereby necessitating a relatively high correction for non-specific adsorption. Since the incorporation of radioactivity into fatty acid synthetase in fasted animals was very low (Table 1) and corrections for co-precipitation were high, especially for the longer time intervals, we were unable to obtain a reliable decay curve for the fatty acid synthetase in fasted rats.

Calculation of rate constants for synthesis and degradation

The change in the level of enzyme with respect to time is given by the expression:

$$\frac{dE}{dt} = k_s - k_d E \tag{1}$$

where E is the content of enzyme per mass; k_s is a zero order rate constant of synthesis per mass, and k_d is a first order rate constant of degradation expressed as time^{-1}. The half-life is related to the rate constant of degradation as follows: $T_{\frac{1}{2}} = \ln 2/k_d$. Under steady state conditions, i.e. $dE/dt = 0$, equation 1 reduces to:

$$E = \frac{k_s}{k_d} \tag{2}$$

Thus, knowing the half-life and the steady state level of the enzyme for a given state, the rate constant for synthesis can be calculated. The values for E, k_s and k_d are shown in Table 2. Since the rate constant of enzyme degradation was the same for the animals fed a normal diet or re-fed a fat-free diet, the change in

Table 2. *Rate constants for the synthesis and degradation of the fatty acid synthetase in livers of rats in different nutritional states*

Nutritional state	E (enzyme units* . liver^{-1})	$T_{\frac{1}{2}}$ (h)	k_d (h^{-1})	k_s (enzyme units* . liver^{-1} . h^{-1})
Normal	104	69	0·01	1·0
Refed a fat-free diet	1725	69	0·01	17·3
Fasted	14	(42–53)†	(0·013–0·016)†	(0·18–0·22)‡

* Enzyme units are given in nanomoles of palmitate formed per min at 30°C.
† k_d and $T_{\frac{1}{2}}$ values were calculated on the basis of the estimated value for k_s.
‡ The k_s value for the fatty acid synthetase of fasted animals was estimated on the assumption that the relative rates of synthesis of the enzyme as given in Table 1 would be proportional to the rate constant values for the synthesis of the fatty acid synthetase.

enzyme content is reflected in the rate constant of enzyme synthesis. The k_s for animals re-fed a fat-free diet was found to be 17·3 times that of the normal state. This value agrees well with the 14·3-fold difference observed in the relative rates of synthesis for the enzyme in animals re-fed a fat-free diet and those fed a normal diet (Table 1). Since the activity of the purified fatty acid synthetase is closely related to the quantity of enzyme protein present (Fig. 3) the E value, given in activity units, can be converted to the amount of enzyme protein, assuming an absolute specific activity of 70 units (nmol palmitate formed per min) per mg enzyme protein. Thus, the rate of synthesis is calculated to be 0·014 mg of fatty acid synthetase protein formed per hr per liver in the normal state and 0·246 mg of fatty acid synthetase per hr per liver in the re-fed state. Hence, the increase in the level of enzyme in the livers of animals re-fed a fat-free diet is attributed to a change in the rate of synthesis and not to a change in the rate of degradation.

Since the relative rates of synthesis of enzyme in the livers of animals fed a normal diet and those re-fed a fat-free diet (Table 1) are proportional to the values of k_s for these two states (Table 2), it was assumed that the k_s for fatty acid synthetase from fasted animals as compared to either the normal or the re-fed state would also be proportional to the relative rates of synthesis in these nutritional states (Table 1). When this assumption is made, a value of 0·18–0·22 units of enzyme activity per hr per liver is obtained for the k_s value for fatty acid synthetase in fasted animals. Assuming a steady state, one can then calculate values for k_d and $T_{\frac{1}{2}}$ (Table 2). From these results the half-life of the fatty acid synthetase in livers of fasted animals was estimated to be 42–53 hr. In the transition from the normal to the fasted state, a half-life of approximately 10 hr would be required if degradation were to completely account for the observed change in the level of fatty acid synthetase. Thus, it is evident that a change in the rate of synthesis, rather than in degradation, is the more probable cause for the change in the level of the enzyme in the fasted state.

In long-term turnover studies in which a labelled amino acid is used, the possibility exists that some of the incorporated radioactivity released as the result of protein degradation may be reutilized in protein synthesis. Under these conditions the observed half-life of the enzyme would be somewhat greater than the true value. In order to evaluate the effect of the reutilization of radioactivity we determined the half-life values for total soluble liver proteins. Values of 3·3, 3·8 and 3·0 days were obtained for the re-fed, normal and fasted dietary conditions, respectively; whereas the half-life reported for the normal state using a non-reutilizable amino acid was higher than the value we obtained (Arias, Doyle & Schimke, 1969). Moreover, Tweto *et al.* (1971) obtained a value of 71 hr for the half-life of liver fatty acid synthetase in normal rats using [guanidino-^{14}C]arginine, a non-reutilizable amino acid. Since this value is in good agreement with the half-life of 69 hr we obtained with L-[U-^{14}C]leucine, reutilization of isotope does not appear to be a significant factor in our experiments.

DISCUSSION

An antibody specific for rat liver fatty acid synthetase has been prepared and successfully employed for the accurate measurements of radioactivity incorporated into fatty acid synthetase in the livers of animals maintained in various nutritional states. Immunochemical methods have also enabled us to demonstrate that the rat liver fatty acid synthetase is homogeneous only after the DEAE-cellulose purified enzyme is subjected to sucrose density gradient centrifugation.

It is significant to note that the relative rates of synthesis of the fatty acid synthetase in livers of animals in the normal and re-fed states (Table 1) closely agree with the rate constants for enzyme synthesis (k_s values in Table 2) that were calculated from the results of the turnover studies of the enzyme in animals in these two nutritional states. Thus there was a 14·3 to 17·3-fold increase in the rate of synthesis of the fatty acid synthetase in the re-fed state as compared to the normal state, whereas there was little difference in the rates of degradation of the enzyme in these two states. Therefore, the dramatic increase in the level of rat liver fatty acid synthetase upon re-feeding a fat-free diet is due to an increase in the rate of *de novo* synthesis of the enzyme.

In addition to the fatty acid synthetase a number of other enzymes have been shown to exhibit similar changes in activity with a parallel change in hepatic lipogenesis in response to nutritional stress. Among these are acetyl-CoA carboxylase (Allmann *et al.*, 1965; Muto & Gibson, 1970), citrate cleavage enzyme (Kornacker & Lowenstein, 1965; Muto & Gibson, 1970), and malic enzyme (Tepperman & Tepperman, 1964; Muto & Gibson, 1970). No appreciable difference in the half-life of rat liver acetyl-CoA carboxylase between the normal and re-fed states was observed by either Majerus & Kilburn (1969) or Nakanishi & Numa (1970). Furthermore, the relative rate of synthesis of acetyl-CoA carboxylase in the re-fed state was 4-fold that observed in the normal state. A similar stimulation in enzyme synthesis for malic enzyme and citrate cleavage enzyme upon re-feeding a fat-free diet has been reported by Lyons & Gibson (1971). It appears, therefore, that the rate of enzyme synthesis is an important parameter in controlling the levels of these enzymes as well as that of the fatty acid synthetase in liver. Thus, the synthesis of these enzymes which are essential for the synthesis of fatty acids may be coordinately controlled, a mechanism which would obviously be of benefit to the organism. This conclusion is further supported by the recent finding of Muto & Gibson (1970) that the greatly stimulated activities of the lipogenic enzymes, namely, acetyl-CoA carboxylase, fatty acid synthetase, citrate cleavage enzyme, malic enzyme and glucose-6-phosphate dehydrogenase in the livers of rats maintained on a fat-free diet are progressively diminished on oral administration of polyunsaturated fatty acids. These results also suggest that polyunsaturated fatty acids are responsible for such adaptive changes in lipogenic enzymes and hence their mechanism of action deserves further study.

In contrast to the normal and re-fed states, fatty acid synthetase seems to be degraded more rapidly in the fasted state (Table 2), although the estimate of half-life in this state is based on certain assumptions. The degradation of other enzymes is also increased in the fasted state as in the case of acetyl-CoA carboxylase (Nakanishi & Numa, 1970) and malic enzyme (Silpananta & Goodridge, 1971). In general the degradation of total soluble liver proteins is increased on starvation. Even though there appears to be an increase in the rate of degradation of the fatty acid synthetase during starvation, this rate cannot account for the 7-fold change in the enzyme content from the normal to the fasted state. Thus, the 5·5-fold decrease in the relative rate of synthesis of the enzyme in the fasted state is the major parameter in determining the level of this enzyme in liver.

It is of interest to note that we have observed a discrepancy in the half-life of the fatty acid synthetase in re-fed animals, when determinations were made by immunochemical assay on the 105,000 g_{av} supernatant solution (44 hr) and on DEAE-cellulose purified fatty acid synthetase (69 hr). This difference could possibly be due to the presence of a species of fatty acid synthetase which is immunologically reactive but enzymatically inactive in the 105,000 g_{av} supernatant solution. Such a species would not co-purify with the fatty acid synthetase in our standard purification procedure. Evidence for the existence of such a protein(s) is obtained from our finding that the equivalence point for the 105,000 g_{av} supernatant solution is significantly higher than that for the

DEAE-cellulose purified fatty acid synthetase. It is tempting to speculate that this immunologically reactive species may be an incomplete fatty acid synthetase complex, either a precursor or a breakdown product, which has lost catalytic activity, but still retains the antigenic site.

On the basis of the high exchange rate of covalently bound 4′-phosphopantetheine with the pool of pantothenate compounds. Tweto *et al.* (1971) have proposed that this prosthetic group is alternately removed from, and recombined with, the fatty acid synthetase complex. The concept of coenzyme protection of the apo-enzyme against degradation has been reported (Schimke, 1970; Rivlin & Wolf, 1969). In fact it has recently been reported that specific degradative enzymes exist for pyridoxal phosphate enzymes as well as for pyridine nucleotide enzymes which attack only the "apo" form but not the "holo" form of these enzymes (Katunuma, Kominami & Kominami, 1971; Katunuma, Kito & Kominami, 1971). Hence, it is tempting to suggest that the 4′-phosphopantetheine confers protection to the "apo" fatty acid synthetase complex against degradation. If so, then the immunologically active, but enzymatically inactive, fatty acid synthetase might be an apoenzyme.

SUMMARY

Investigations were carried out to determine whether diet-induced changes in the level of rat liver fatty acid synthetase are due to changes in the rate of synthesis or degradation of this enzyme. 1. Assays for radioactivity in fatty acid synthetase were carried out through isolation of the enzyme by a standard purification procedure or by immunoprecipitation with antiserum prepared specifically against rat liver fatty acid synthetase. 2. Studies on the incorporation of radioactive leucine into the fatty acid synthetase showed that the relative rate of synthesis of this enzyme as compared to the normal state was decreased 5·5-fold on fasting and increased 14·3-fold by re-feeding a fat-free diet. 3. Studies on the degradation of the fatty acid synthetase, reported as half-life of the enzyme, yielded identical values, 69 hr for normal animals as well as for animals re-fed a fat-free diet. The rates of synthesis of enzyme (k_s) calculated from these half-life values and the enzyme content present in the liver in a given nutritional state agreed well with the relative rates of synthesis. 4. On the basis of these results it is concluded that changes in the level of liver fatty acid synthetase which are diet-induced are due to changes in the rate of synthesis rather than degradation of this enzyme.

ACKNOWLEDGEMENT

This investigation was supported in part by Research Grants AM 01383 from the National Institute of Arthritis and Metabolic Diseases of the National Institutes of Health, United States Public Health Service, G-68-38 from the Life Insurance Medical Research Fund, and 133-7563 from the Wisconsin Heart Association.

REFERENCES

Allmann, D. W., Hubbard, D. D. & Gibson, D. M. (1965). *J. Lipid Res.* **6**, 63.
Arias, I. M., Doyle, D. & Schimke, R. T. (1969). *J. biol. Chem.* **244**, 3303.
Burton, D. N., Haavik, A. G. & Porter, J. W. (1968). *Archs Biochem. Biophys.* **126**, 141.

Burton, D. N., Collins, J. M., Kennan, A. L. & Porter, J. W. (1969). *J. biol. Chem.* **244**, 4510.

Collins, J. M., Craig, M. C., Nepokroeff, C. M., Kennan, A. L. & Porter, J. W. (1971). *Archs Biochem. Biophys.* **143**, 343.

Gibson, D. M., Hicks, S. E. & Allmann, D. W. (1966). In *Advances in Enzyme Regulation*, Vol. 4 (Ed. Weber, G.), p. 239. Oxford: Pergamon Press Ltd.

Gibson, D. M. & Hubbard, D. D. (1960). *Biochem. biophys. Res. Commun.* **3**, 531.

Gornall, A. G., Bardawill, C. J. & David, M. M. (1949). *J. biol. Chem.* **177**, 751.

Hicks, S. E., Allmann, D. W. & Gibson, D. M. (1965). *Biochim. biophys. Acta* **106**, 441.

Hsu, R. Y., Wasson, G. & Porter, J. W. (1965). *J. biol. Chem.* **240**, 3736.

Kabat, E. A. & Mayer, M. M. (1961). *Experimental Immunochemistry*, p. 22. Springfield, Illinois: Charles C. Thomas.

Katunuma, N., Kominami, E. & Kominami, S. (1971). *Biochem. biophys. Res. Commun.* **45**, 70.

Katunuma, N., Kito, K. & Kominami, E. (1971). *Biochem. biophys. Res. Commun.* **45**, 76.

Kornacker, M. S. & Lowenstein, J. M. (1965). *Biochem. J.* **95**, 832.

Levy, H. B. & Sober, H. A. (1960). *Proc. Soc. exp. Biol. Med.* **103**, 250.

Lowry, O. H., Rosebrough, N. J., Farr, A. L. & Randall, R. J. (1951). *J. biol. Chem.* **193**, 265.

Lyons, R. T. & Gibson, D. M. (1971). *Fedn Proc. Fedn Am. Socs. exp. Biol.* **30**, p. 1080. Abs. No. 165.

Majerus, P. W. & Kilburn, E. (1969). *J. biol. Chem.* **244**, 6254.

Muto, Y. & Gibson, D. M. (1970). *Biochem. biophys. Res. Commun.* **38**, 9.

Nakanishi, S. & Numa, S. (1970). *Eur. J. Biochem.* **16**, 161.

Numa, S., Matsuhashi, M. & Lynen, F. (1961). *Biochem. Z.* **334**, 203.

Rivlin, R. S. & Wolf, G. (1969). *Nature, Lond.* **223**, 516.

Schimke, R. T. (1970). In *Mammalian Protein Metabolism*, Vol. 4 (Ed. Munro, H. N.), p. 177. New York: Academic Press.

Schimke, R. T., Sweeney, E. W. & Berlin, C. M. (1965). *J. biol. Chem.* **240**, 322.

Silpananta, P. & Goodridge, A. G. (1970). *J. biol. Chem.* **246**, 5754.

Tepperman, H. M. & Tepperman, J. (1964). *Am. J. Physiol.* **206**, 357.

Tweto, J., Liberati, M. & Larrabee, A. R. (1971). *J. biol. Chem.* **246**, 2468.

Wakil, S. J., Porter, J. W. & Gibson, D. M. (1957). *Biochim. biophys. Acta* **24**, 453.

IV

LIPIDS IN STRUCTURE AND FUNCTION OF MEMBRANES

CORRELATION OF CHEMICAL STRUCTURE AND BIOLOGICAL ACTIVITY OF MYCOBACTERIAL CELL WALLS AND DERIVED LIPIDS*

By E. LEDERER

Institut de Chimie des Substances Naturelles, C.N.R.S. and Institut de Biochimie, Centre d'Orsay, Université de Paris-Sud, France

THE study of the chemistry of bacterial cell walls is being actively pursued in many laboratories and is amply justified by its intrinsic importance. Mycobacterial cell walls have, however, an additional interest: they produce in experimental animals a series of biological events which lead to increased production of antibodies and, in general, an increase in resistance to infections not only by Mycobacteria, but also by agents quite unrelated to Mycobacteria.

The compounds responsible for these activities are mostly complex lipid-containing substances.

CHEMICAL STRUCTURE OF MYCOBACTERIAL CELL WALLS

Let us first have a puick look at the chemistry of the mycobacterial cell wall, which we have summarized in detail in a recent lecture (Lederer, 1971).**

Figure 1 shows the tentative structure of a "monomer" of the cell wall of Mycobacteria; mycolic acids are esterified to the primary hydroxyl group of an arabinofuranose which is part of an arabinogalactan bound to a phosphorylated mucopeptide.

What are the fundamental chemical differences of mycobacterial cell walls, in respect to those of "ordinary bacilli"?

The presence of the following structures is characteristic:

Mycolic acid,
an arabinogalactan,
N-glycolyl muramic acid (instead of *N*-acetyl muramic acid),
a diamidated tetrapeptide and
associated glycolipids.

* Not presented at the symposium.

** The author regrets that it has not been possible in this short review to give proper credit to many colleagues who have made significant contributions to this field in recent years: J. Asselineau, C. Asselineau, I. Azuma, A. H. Etémadi, M. B. Goren, P. Jollès, F. Kanetsuna, M. Kato, G. Lanéelle, G. Michel, N. Polgar, E. Vilkas to mention only the most important ones.

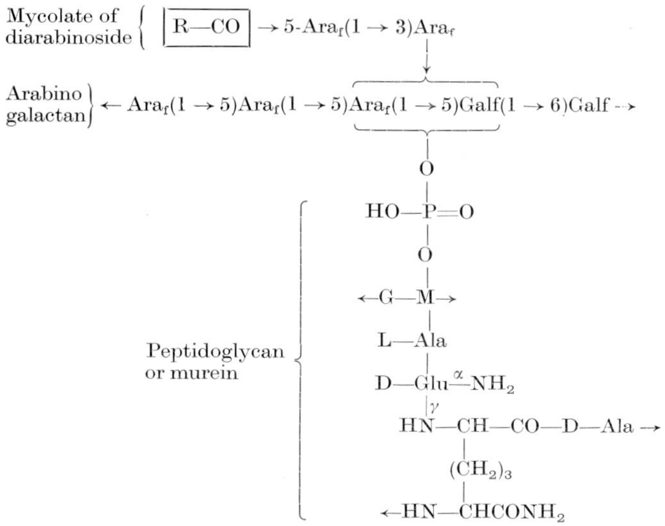

FIG. 1. Tentative structure of the "monomer" of the cell wall of Mycobacteria. RCOOH = mycolic acid G = *N*-acetyl glucosamine, M = *N*-glycolyl muramic acid.

Mycolic acids

These are "α-branched β-hydroxy acids" (Asselineau & Lederer, 1951). Three principal categories are now known.

(i) The corynomycolic acids ranging from C_{28} to C_{40}, found mostly in Corynebacteria, for instance (1) (Lederer & Pudles, 1951).

Corynomycolic acid $C_{32}H_{64}O_3$

$$CH_3(CH_2)_{14}CH(OH)\underset{\displaystyle C_{14}H_{29}}{\underset{|}{CH}}—COOH \tag{1}$$

(ii) The nocardic (or nocardomycolic) acids, ranging from C_{40} to C_{66} produced by strains of Nocardia for instance (2) (Michel, Bordet & Lederer, 1960; Lanéelle & Asselineau, 1970; Maurice, Vacheron & Michel, 1971).

Nocardic(nocardomycolic) acids $C_{56}H_{106}O_3$ and $C_{58}H_{110}O_3$

$$CH_3(CH_2)_aCH{=}CH(CH_2)_bCH{=}CH(CH_2)_cCH(OH)\underset{\displaystyle C_{14}H_{29} \text{ or } C_{16}H_{33}}{\underset{|}{CH}}(COOH) \tag{2}$$

$a + b + c = 32$ or 34

(iii) The mycobacterial mycolic acids, ranging from C_{60} to C_{90}. Two typical structures developed by Etémadi (1967) are shown below (3) and (4). (For reviews see Asselineau, 1966; Etémadi, 1967.)

α-Smegmamycolic acid from M. smegmatis
$C_{77}H_{154}O_3$

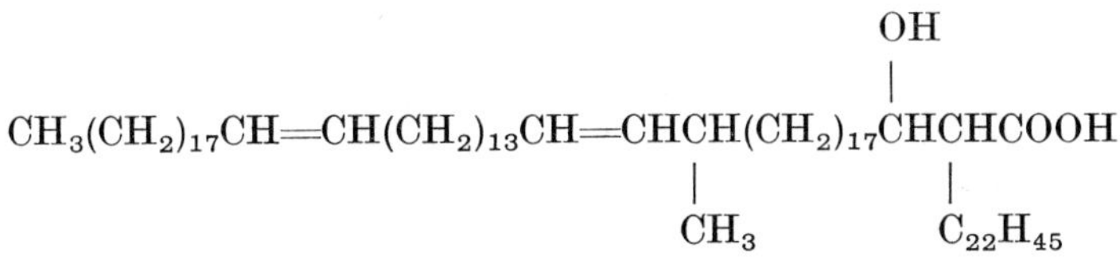

(3)

α-Kansamycolic acid from M. kansasii
$C_{80}H_{154}O_3$

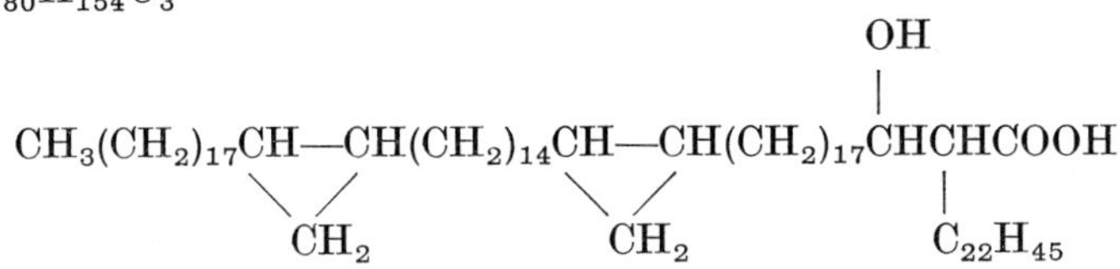

(4)

The structural differences of these three categories can be defined in terms of the increasing complexity of the enzyme system producing these compounds.

Corynebacteria have an enzyme capable of catalysing a "Claisen condensation" of two normal fatty acids

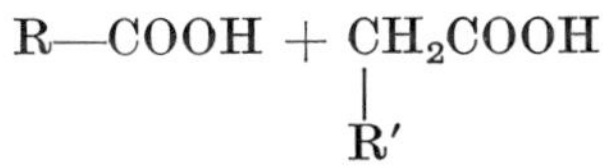

to produce the corynomycolic acids (1)

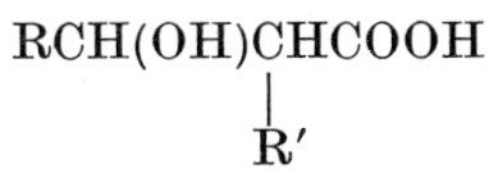

(1)

where R and R′ are normal C_{13} to C_{16} chains (Gastambide & Lederer, 1959).

Nocardiae have an additional enzyme system for lengthening the normal C_{16} or C_{18} fatty acids to C_{30}–C_{36}, with introduction of double bonds, thus producing the "meronocardic" chains which then are condensed with a normal short chain (C_{14}–C_{18}) acid to give the nocardomycolic acids (2).

Mycobacteria have an additional third enzyme system which C-alkylates the long "meromycolic" chains, thus producing the typical mycolic acids carrying cyclopropane or C-methyl groups in their "meromycolic" moiety (3,4) (Etémadi & Lederer, 1965; Jauréguiberry *et al.*, 1966). Lechevallier *et al.* (1971) have used these differences for classification of Nocardiae and Mycobacteria.

The arabinogalactan

This contains five D-arabinofuranose units for two galactose molecules, which Vilkas *et al.* (1971) have recently shown to be also in the furanose form. Acylated arabinogalactans have also been found in Nocardia strains (Lanéelle & Asselineau, 1970).

N-*glycolyl-muramic acid*

This seems to be restricted to Mycobacteria and Nocardiae (Adam *et al.*, 1969); it has not been found in Corynebacteria or *Streptomyces* (Azuma *et al.*, (1970).

The diamidated tetrapeptide

The structure of the diamidated tetrapeptide which can be seen in Fig. 1 has been defined by Wietzerbin-Falszpan *et al.* (1970) by mass spectrometry of the *N*-acetylated and permethylated compound; the same diamidated tetrapeptide is also present in cell walls of Corynebacteria.

Associated glycolipids

Mycobacterial cell walls are characterized by their association with a series of *glycolipids*, *peptidoglycolipids* and sometimes *glycosulfolipids*. Amongst these we have to mention especially *Wax D* and *Cord Factor* which have interesting biological properties, as mentioned below.

Wax D of human strains contains an arabinogalactan esterified by mycolic acids and linked to a typical mucopeptide containing *N*-acetyl glucosamine, *N*-glycolyl-muramic acid and an oligopeptide composed of L- and D-alanine, meso-α, ε-diaminopimelic acid (meso-DAP) and D-glutamic acid (Migliore & Jollès, 1969; Kanetsuna & San Blas, 1970; Markovits *et al.*, 1971). It is now considered to be an autolysis product of the cell wall, which explains the close chemical similarity (Markovits *et al.*, 1971). Azuma, Kimura & Yamamoto (1968) have found that the arabinogalactans of cell walls and wax D are immunologically identical. For more details on wax D see Lederer (1971).

Cord Factor is a 6,6′-dimycolate of trehalose (Noll *et al.*, 1956); Cornyebacteria and Nocardiae produce 6,6′-diesters of trehalose with shorter chain mycolic acids (Senn *et al.*, 1967; Ioneda, Lederer & Rozanis, 1970).

BIOLOGICAL ACTIVITY OF MYCOBACTERIAL CELL WALLS AND DERIVED LIPIDS

It is not possible, in the limited time available for this presentation to give a complete survey of the topic.

In a previous review (Lederer, 1971) we have mentioned three principal activities:

immunization against tuberculosis,
adjuvant activity and
stimulation of non-specific resistance to infections.

Immunization against tuberculosis

Anacker *et al.* (1969*a*) have described immunization against tuberculosis infection of experimental animals with oil-treated mycobacterial cell walls. More recently they have shown (1969*b*) that their cell wall preparations can be inactivated by organic solvent extractions and by alkali or lipase treatment: these inactivated preparations combined with a wax D fraction were again highly protective. It is not known how specific the immunization described by these authors is.

Bekierkunst *et al.* (1969) have described the granuloma formation induced in the lungs of mice by intravenous injection of cord factor. These mice were protected against an intravenous challenge with a virulent strain of *M. tuberculosis*. After intraperitoneal injection of cord factor no granuloma was found in the lungs and no protection against intravenous infection was observed. It thus seems that the formation of granuloma produces a local protective effect (see also Bekierkunst *et al.*, 1971*a*).

The immunizing effect obtained with "antigène méthylique" by Pigretti *et al.* (1965) may be ascribed to the presence of about 10% cord factor in their preparations.

Tsumita & Arita (1970) have isolated a protein fraction containing 5% carbohydrate and 3·3% lipid which protects mice and guinea pigs against infection by virulent Mycobacteria.

A specific immunization against tuberculosis infection has been described in a series of papers by Youmans & Youmans (1970), obtained by a fraction quite unrelated to the previous ones, an RNA preparation extracted from Mycobacteria.

The situation is thus far from being clear, various authors having claimed immunization effects with a rather wide variety of fractions isolated from Mycobacteria.

Smith *et al.* (1971) have clearly shown how difficult it is to compare the results of various groups of workers, using different animal test systems and different criteria of evaluation.

Immunological adjuvant activity

Coulaud (1935) observed that injection of an antigen in presence of killed Mycobacteria in mineral oil increases production of antibodies against the antigen; later, Freund (1956) developed the use of a water in oil emulsion containing killed Mycobacteria in paraffin oil and an antigen in the aqueous phase. This so-called "Freund's adjuvant" has been widely used in animal experiments for increasing antibody titers and for developing delayed hypersensitivity. In rats, Freund's adjuvant—without an antigen—induces an allergic arthritis (Pearson & Wood, 1959).

Wax D. While trying to isolate the component(s) of the mycobacterial cell responsible for their adjuvant activity, we could show (White *et al.*, 1958; 1964) that wax D of human strains of Mycobacteria could replace whole Mycobacteria in Freund's adjuvant (see also White, 1967).

Cell walls. As already mentioned the components of wax D are all typical cell wall constituents and it was thus not a surprise that purified cell walls, i.e. delipidated, trypsin and chymotrypsin treated preparations showed the same activity (Azuma *et al.*, 1971).

Quite recently, Adam *et al.* (1972) have reported a more precise evaluation of the adjuvant activity of cell walls of *M. smegmatis* and have found that a few μg in incomplete Freund's adjuvant* (FIA) injected into guinea pigs produces a significant stimulation of antibody production (Table 1).

Table 1. *Adjuvant activity of cell walls of* M. smegmatis. (*From Adam* et al., 1972)

Material injected	Dose (μg)	Antibody titers*
Freund's incomplete adjuvant	—	212
Freund's complete adjuvant	50	3000
Cell walls in Freund's incomplete adjuvant	2·5	2107
	10	3424
	25	3944
	75	4040
	200	4672

Hartley female guinea-pigs weighing 300 to 350 g were injected in both hind foot-pads with 0·1 ml of a water in oil emulsion made of equal parts of a solution of ovalbumin (egg albumin 5 X crystallized B grade, Calbiochem. Los Angeles, Calif.), 50 mg/ml in saline and either Freund's complete adjuvant or Freund's incomplete adjuvant Difco (Detroit, Michigan) completed as indicated with mycobacterial fractions. To obtain the sera, the guinea-pigs were sacrificed 21 days after the injection.

* Measured as μg/ml of antigen-antibody complex at the equivalence point. The complex is estimated with the Folin reagent, using ovalbumin as standard. Mean values for groups of six guinea-pigs.

A water soluble adjuvant. We could then go a step further by showing that lysozyme treatment of these cell walls liberates a hydrosoluble macromolecular compound which is strongly adjuvant active; we call this fraction "water soluble adjuvant" (WSA) (Adam *et al.*, 1972).

After lyophilization WSA is obtained as a fluffy, snow-white material easily soluble in water, $[\alpha]_D + 11{\cdot}3°$. It is insoluble in ether, chloroform, acetone and mixtures of methanol-chloroform.

In the ultracentrifuge, WSA behaves as a slightly polydisperse macromolecule with a sedimentation coefficient of $S_{20} = 2{\cdot}05$ corresponding approximately to a M.W. of 20,000 daltons.

WSA contains: amino sugars 12–15% (*N*-acetyl glucosamine and *N*-glycolylmuramic acid in equimolar proportions), neutral sugars 60–70% (arabinose and galactose in an approximate molar ratio of 2:1), amino acids 12–15% (Ala, Glu, DAP = α, ε-diaminopimelic acid in a molar ratio of 1·3:1:1). Only traces of other amino acids could be detected, thus excluding the presence of any other protein in the WSA fraction and less than 1% fatty acids.

All the constituents of WSA (aminosugars, neutral sugars, amino acids) are typical bacterial cell wall constituents; it can thus be considered to be a sort of "cell wall oligomer".

* I.e. Freund's complete adjuvant without mycobacteria.

Analogous WSA preparations have been obtained from cell walls of BCG (Bacille Calmette-Guérin) and from *M. kansasii*, differing apparently only by their molecular weight. It is probable that a whole series of adjuvant active "cell wall oligomers" can be prepared and it is expected that such compounds can be obtained from all Mycobacteria and probably also *Nocardia* strains.

Biological properties of WSA. When added to Freund's incomplete adjuvant (FIA) containing an antigen (e.g. ovalbumin) and injected into guinea-pigs, WSA increases the amount of precipitating antibodies, as shown by the precipitin tests (Table 2) (Adam *et al.*, 1972) and by passive haemagglutination (Chedid *et al.*, 1972). It also induces hypersensitivity to the same antigen

Table 2. *Adjuvant activity of the water soluble adjuvant (WSA). (From Adam* et al., 1972)

Material injected	Dose (μg)	Antibody titers*
Freund's incomplete adjuvant	—	640
Freund's complete adjuvant	50	3164
WSA in Freund's incomplete adjuvant	1	2120
	2·5	3380
	10	4680
	25	4828
	50	6880
	200	7500

* Mean values for 6 guinea pigs.
Experimental conditions same as in Table 1.

(ovalbumin), and the biosynthesis of γ_2 type of precipitating antibodies. It does not, however, induce hypersensitivity to tuberculin.

WSA has a stronger adjuvant action than equal amounts of whole bacteria, cell walls or wax D.

To our knowledge this is the first time that a water soluble adjuvant active fraction has been obtained from Mycobacteria.

WSA also stimulates antibody production in mice infected with influenza and Columbia SK virus (Chedid *et al.*, 1972). It has no undesirable effects such as induction of hyperreactivity to endotoxin, pyrogenicity, production of lung granuloma and polyarthritis and is thus a promising candidate for clinical experiments (Chedid *et al.*, 1972).

Cord factor. In experiments with *guinea pigs*, where WSA (and wax D) were active, cord factor (and mycoside C) were inactive. Bekierkunst *et al.* (1971*a*) have, however, shown that 10 μg of cord factor, injected into the foot pad of mice increased the antibody response to sheep red blood cells subsequently injected into the same site. Furthermore, administration of sheep red blood cells emulsified in incomplete Freund's adjuvant containing 5 μg of cord factor induced a very strong antibody response, as compared to that after injection of the same amount of antigen in incomplete Freund's adjuvant containing wax D or mycobacteria.

Wax D was also found to be inactive in chickens (White, personal communication). There is thus very definite evidence that different animal species react in a very different way to the same glycolipid fractions.

Stimulation of non specific resistance to infections

Antibacterial action. Dubos & Schaedler (1957) and Williams & Dubos (1959) have shown that injection of intact mycobacterial cells into mice can produce immunity against heterologous infections.

Howard *et al.* (1959) have reported that BCG produces a stimulatory effect on the phagocytic activity of the reticuloendothelial system in mice and thus increases resistance to infection with *Salmonella enteridis*.

Weiss, Bonhag & Parks (1964) have shown that this heterologous immunogenicity of BCG was retained in the insoluble residue obtained after methanol extraction (see also Steinkuller, Krigbaum & Weiss, 1969). Fox *et al.* (1966) investigated the stimulation of non-specific resistance to infection by crude cell wall preparations from *M. phlei*, and Misaki *et al.* (1966) showed that a mucopeptide preparation of BCG had a protective effect against staphylococcal infection in mice, which was as high as that of intact BCG cells.

We have reported similar results with cell wall preparations which stimulated resistance of mice against infection with *Klebsiella pneumonia* (Lederer 1971, experiments of L. Chedid & A. Lamensans). BCG vaccination has recently been shown to have a beneficial effect on malaria also (Brown, 1971).

Antiviral and antitumor action

An increasing amount of evidence shows that living BCG can protect experimental animals against viral infections and against experimental tumours.

Effect of whole bacilli. Favourable effects of BCG infections on viral leukaemia have been reported by Lemonde & Clode (1962) and Lamensans, Mollier & Laurent (1968), on viral tumours by Bergman, Allison & Pereira (1967) and on Friends disease virus in mice by Larson *et al.* (1971).

Immunotherapy of various types of tumours in experimental animals by living BCG has recently been reported by Amiel & Berardet (1971), Bartlett (1971), Nathanson (1971), Piessen *et al.* (1971) and by Zbar & Tanaka (1971).

Clinical experiments with successful application of BCG for the treatment of acute lymphoblastic leukaemia have been reported by Mathé *et al.* (1968) and the statistical study of Davignon *et al.* (1970) seems to show that BCG vaccination exerts significant beneficial effects against leukaemia in children.

Effects of mycobacterial fractions. Cell walls of mycobacteria (BCG, or *M. kansasii*, or *M. smegmatis*) have similar effects on lymphoid leukaemia in mice, as shown by unpublished experiments of L.Chedid & A. Lamensans mentioned recently (Lederer, 1971).

Wax D has also been reported to shorten the delay of production of interferon (Gorhé, Asso & Paraf, 1968) and to stimulate non-specific resistance against tumours (Amiel & Hiu, 1971).

Cord factor has been shown by Bekierkunst *et al.* (1971*c*) to suppress the development of urethan induced lung tumours in mice, to a similar degree as living BCG and inhibition observed was ascribed to the reactions of host cells caused locally by cord factor, injected intravenously. In the mycobacterial field this is the first instance of a chemically defined compound having a definite (if only local) effect on tumour growth.

A methanol soluble fraction of *M. butyricum* has been found to counteract the immuno suppressive effect of tumour growth (Esber, Bogden & Taylor, 1971). The chemical nature of the active compound is not known (it might be cord factor).

In conclusion, the two main recent advances in the field of biologically active mycobacterial products are: the discovery of WSA, a water-soluble strongly adjuvant active "cell wall oligomer" (see Table 2) which might find interesting applications in therapy, and the demonstration of the various cellular reactions caused by cord factor, which can lead to local immunity to tuberculosis or other infections and also local resistance to tumours (see Table 3). These observations have a general significance in relation to the mode of action of lipid and glycolipid adjuvants in general, such as vitamin A, sugar esters, etc., which interact with the membranes of lyosomes and other cellular structures.

Table 3. *Biological activities of Cord Factor in mice*

Activity	Reference
Toxicity: kills mice by intravenous injection of 20 μg	Bloch (1950)
Produces hyperreactivity to endotoxins	Suter *et al.* (1958)
Inhibits *in vitro* mitochondrial respiration and phosphorylation	Kato (1968)
Damages mitochondrial and lysosomal membranes	Kato (1970)
Produces lung granuloma after intravenous injection	Bekierkunst *et al.* (1969)
Induces local immunity against infection by tubercle bacilli	Bekierkunst *et al.* (1969)
Causes local cellular reactions very similar to those produced by living BCG bacilli	Bekierkunst *et al.* (1971*a*)
Increases antibody response to sheep red blood cells injected at the same site	Bekierkunst *et al.* (1971*b*)
Suppresses urethan-induced lung adenomas	Bekierkunst *et al.* (1971*c*)

The cord factor and its analogues may well prove useful in further studies on the mechanisms of cellular immunity and differentiation.

ACKNOWLEDGEMENT

The work of the author was supported, in part by grants from the World Health Organization (Geneva), the Ligue Nationale Française contre le Cancer and the Fondation de la Recherche Médicale.

REFERENCES

Adam, A., Ciorbaru, R., Petit, J. F. & Lederer, E. (1972). *Proc. natn Acad. Sci. USA*. **69**, 851.

Adam, A., Petit, J.-F., Wietzerbin-Falszpan, J., Sinay, P., Thomas, D. W. & Lederer, E. (1969). *FEBS Lett.* **4**, 87.

Anacker, R. L., Barclay, W. R., Brehmer, W., Goode, G., List, R. H., Ribi, E. & Tarmina, D. F. (1969*a*). *Am. Rev. Resp. Dis.* **99**, 242.

Anacker, R. L., Bickel, W. D., Brehmer, W., Niwa, M., Ribi, E. & Tarmina, D. F. (1969*b*). *Proc. Soc. exp. Biol. Med.* **130**, 723.

Amiel, J. L. & Berardet, M. (1971). *Experientia* **8**, 941.

Amiel, J. L. & Hiu, I. J. (1971). *Proc. Am. Ass. Cancer Res.* **12**, 32 (Abstr. 128).

Asselineau, J. (1966). *The Bacterial Lipids*. Paris: Herman. San Francisco: Holden Day.

Asselineau, J. & Lederer, E. (1951). *Biochim. biophys. Acta* **7**, 126.

Azuma, I. Kimura, H. & Yamamura, Y. (1968). *J. Bacteriol.* **96**, 567.

Azuma, I., Thomas, D. W., Adam, A., Ghuysen, J.-M., Bonaly, R., Petit, J.-F. & Lederer, E. (1970). *Biochim. biophys. Acta* **208**, 444.

Azuma, I., Kishimoto, S., Yamamura, Y. & Petit, J.-F. (1971). *Jap. J. Microbiol.* **15**, 193.

Bartlett, G. L. (1971). *Proc. Am. Ass. Cancer Res.* **12**, 41 (Abstr. 164).

Bekierkunst, A., Levij, I. S., Yarkoni, E., Vilkas, E., Adam, A. & Lederer, E. (1969). *J. Bacteriol.* **100**, 95.

Bekierkunst, A., Levij, I. S., Yarkoni, E., Vilkas, E. & Lederer, E. (1971). *Infect. Immun.* **4**, 245.

Bekierkunst, A., Levij, I. S., Yarkoni, E., Flechner, I., Morecki, S., Vilkas, E. & Lederer, E. (1971*a*). *Infect. Immun* **4**, 256.

Bekierkunst, A., Yarkoni, E., Vilkas, E. & Lederer, E. (1971*b*). *Sciences* **174**, 1240.

Bergmann, L. D., Allison, A. & Pereira, H. G. (1967). *Int. J. Cancer*, **2**, 539.

Bloch, H. (1950). *J. exp. Med.* **91**, 197; **92**, 507.

Bordet, C. & Michel, G. (1969). *Bull. Soc. Chim. biol.* **51**, 527.

Brown, K. N. (1971). *Nature* **230**, 166.

Chedid, L., Parant, M., Parant, F., Gustafson, R. H. & Berger, R. M. (1972). *Proc. natn Acad. Sci. U.S.A.* **69**, 855.

Coulaud, E. (1935). *C. r. hebd. Séanc. Soc. Biol. Paris*, **119**, 368.

Davignon, L., Lemonde, P., Robillard, P. & Frappier, A. (1970). *Lancet*, p. 638.

Dubos, R. J. & Schaedler, R. W. (1957). *J. exp. Med.* **106**, 703.

Esber, H. J., Bogden, A. & Taylor, D. J. (1971). *Proc. Am. Ass. Cancer Res.* **11**, 23.

Etémadi, A. H. (1967). *Exposés Annuels de Biochimie Médicale*, **28**, 77. Paris: Masson.

Etémadi, A. H. & Lederer, E. (1965). *Biochim. biophys. Acta* **98**, 160.

Fox, A. E., Evans, G. L., Turner, F. J., Schwartz, B. S. & Blaustein, A. (1966). *J. Bacteriol.* **92**, 1.

Freund, J. (1956). *Adv. Tuberc. Res.* **7**, 130. Basle: Karger.

Gastambide, M. & Lederer, E. (1959). *Nature, Lond.* **184**, 1563.

Gorhé, D. S., Asso, J. & Paraf, A. (1968). *Annls Inst. Pasteur, Paris*, **115**, 446.

Howard, J. G., Biozzi, G., Halpern, B. N., Stiffel, C. & Mouton, D. (1959). *Brit. J. exp. Pathol.* **40**, 281.

Ioneda, T., Lederer, E. & Rozanis, J. (1970). *Chem. Phys. Lipids* **4**, 375.

Jauréguiberry, G., Lenfant, M., Das, B. C. & Lederer, E. (1966). *Tetrahedron*, Suppl. 8, part I, 27.

Kanetsuna, F. & San Blas, G. (1970). *Biochim. biophys. Acta* **208**, 434.

Kato, M. (1968). *Am. Rev. Resp. Dis.* **98**, 260.

Kato, M. (1970). *Archs Biochem. biophys.* **140**, 379.

Lanéelle, M. A. & Asselineau, J. (1970). *FEBS Lett.* **7**, 64.

Lamensans, A., Mollier, M. F. & Laurent, M. (1968). *Revue Fr. Etudes Clin. Biol.* **13**, 871.

Larson, C. L., Ushijima, A. R., Florey, M. J., Baker, R. E. & Baker, M. B. (1971). *Nature, New Biol.* **229**, 243.

Lechevalier, M. P., Horan, A. C. & Lechevalier, H. (1971). *J. Bacteriol.* **105**, 313.

Lederer, E. (1971). *Pure Appl. Chem.* **25,** 135.
Lederer, E. & Pudles, J. (1951). *Bull. Soc. Chim. biol.* **33,** 1003.
Lemonde, P. & Clode, M. (1962). *Proc. Soc. exp. Biol. Med.* **111,** 739.
Markovits, J., Vilkas, E. & Lederer, E. (1971). *Eur. J. Biochem.* **18,** 287.
Mathé, G., Amiel, J. L., Schwarzenberg, L., Schneider, M., Cattan, A., Schlumberger, J. R., Hayat, M. & de Vassal, F. (1968). *Revue Fr. Etudes Clin. Biol.* **13,** 454.
Maurice, M. T., Vacheron, M. J. & Michel, G. (1971). *Chem. Phys. Lipids* **7,** 9.
Michel, G., Bordet, C. & Lederer, E. (1960). *C. r. hebd. Séanc. Acad. Sci. Paris,* **250,** 3518.
Migliore, D. & Jollès, P. (1969). *C. r. hebd. Séanc. Acad. Sci. Paris,* **269,** sér. D., 2268.
Nathanson, L. (1971). *Proc. Am. Ass. Cancer Res.* **12,** 99 (Abstr. 393).
Noll, H., Bloch, H., Asselineau, J. & Lederer, E. (1956). *Biochim. biophys. Acta* **20,** 299.
Pearson, C. M. & Wood, F. D. (1959). *Arth. Rheum.* **2,** 440.
Piessens, W. F., Heimann, R., Legros, N. & Heuson, J. C. (1971). *Cancer Res.* **31,** 1061.
Pigretti, M., Vilkas, E., Lederer, E. & Bloch, H. (1965). *Bull. Soc. Chim. biol.* **47,** 2039.
Senn, M., Ioneda, T., Pudles, J. & Lederer, E. (1967). *Eur. J. Biochem.* **1,** 353.
Smith, D., Harding, G. E., Wiegeshaus, E. H., Grover, A. A. & McMurray, D. N. (1971). Intern. Symp. on B.C.G. Vaccine, Frankfurt (Main). (1970). *Symp. Series immunobiol. Standard* Vol. 17, p. 259. Basel, Minchen; New York: Karger.
Steinkuller, C. B., Krigbaum, L. G. & Weiss, D. W. (1969). *Immunology* **16,** 255.
Suter, E., Ullman, G. E. & Hoffman, R. G. (1958). *Proc. Soc. exp. Biol. Med.* **99,** 167.
Tsumita, T. & Arita, M. (1970). *Japan J. exp. Med.* **40,** 283.
Vilkas, E., Markovits, J., Amar-Nacasch, C. & Lederer, E. (1971). *C. r. hebd. Séanc. Acad. Sci. Paris,* **273,** sér. C. 845.
Weiss, D. W., Bonhag, R. S. & Parks, J. A. (1964). *J. exp.* **119,** 53.
White, R. G. (1967). In Modern Trends in Immunology. Ed. by Cruickshank & Weir, D. M. Vol. II, p. 28. London: Butterworths.
White, R. G., Bernstock, L., Johns, R. G. S. & Lederer, E. (1958). *Immunology* **1,** 54.
White, R. G., Jollès, P., Samour, D. & Lederer, E. (1964). *Immunology* **7,** 158.
Wietzerbin-Falszpan, J., Das, B. C., Azuma, I., Adam, A., Petit, J.-F. & Lederer, E. (1970). *Biochem. biophys. Res. Comm.* **40,** 57.
Williams, C. A. & Dubos, R. J. (1959). *J. exp. Med.* **110,** 981.
Youmans, A. S. & Youmans, G. P. (1970). *Infect. Immun.* **2,** 659.
Zbar, B. & Tanaka, T. (1971). *Science* **172,** 271.

CO-ORDINATED SYNTHESIS OF MEMBRANE PHOSPHOLIPIDS WITH THE FORMATION OF RIBOSOMES AND ACCELERATED PROTEIN SYNTHESIS DURING GROWTH AND DEVELOPMENT

By J. R. TATA

National Institute for Medical Research,
Mill Hill, London, England

As cells differentiate, grow and mature the changing pattern of biosynthetic activity is accompanied by an increasing complexity of intracellular structures. Relatively little is known about the relationship between the mechanisms that regulate growth and development via a control of RNA and protein synthesis on the one hand, and the cellular structural requirements for such regulations to be effective, on the other. In this paper I wish to discuss the question of how changes in cellular structure, particularly in the endoplasmic reticulum are co-ordinated with those of protein synthesis during growth and development. Much of the experimental work described here has been performed in our laboratory in which hormones were used to initiate or accelerate growth and development; however the phenomena described are common to most late embryonic growth and developmental processes.

It is well known that free polyribosomes with endogenous or synthetic messenger will effect the incorporation of amino acids into protein. However, it seems that in intact cells some form of structural or topological deployment of ribosomes is necessary for the synthesis of proteins which may involve an attachment of ribosomes to membranes of the endoplasmic reticulum (Hendler, 1968; Campbell, 1970). It is of some interest that such membranes are themselves continuously formed and broken down according to the demands for protein synthesis (Siekevitz *et al.*, 1968; Tata, 1968, 1970*a*). Unfortunately, much of the work done so far on the proliferation of the endoplasmic reticulum has been carried out on predominantly protein secreting tissues such as the liver and pancreas. The well known participation of both rough and smooth endoplasmic reticulum (see Palade, 1966) in the process of transport of secretory proteins has therefore tended to mask other possible roles for the attachment of ribosomes to endoplasmic membranes, especially during growth. However, it is important to note that in non-secretory animal cells the attached ribosomes appear to be functionally very active in intact cells (Hendler, 1968; Tata, 1971*a*; Andrews & Tata, 1971*a*,*b*).

RNA AND PROTEIN SYNTHESIS DURING GROWTH AND DEVELOPMENT

It is not necessary to stress here that the protein synthetic machinery of the cell is activated during growth and differentiation. In most instances the actual increase in protein synthetic activity is anticipated by changes in the rate and nature of RNA synthesized in the nucleus and its transfer into the cytoplasm. It has been well established that a large majority of growth-promoting hormones, when administered *in vivo*, markedly affect protein synthesis *in vivo* or *in vitro* at the level of the ribosomes of the target cell but fail to influence amino acid incorporation when added *in vitro* to the ribosomes (see Korner, 1965; Tata, 1967*a*, 1970*b*). This finding led to the observations that RNA synthesis in target cells is very sensitive to growth and developmental hormones, and experiments with inhibitors of RNA synthesis revealed that part of the RNA synthesized under hormonal influence is essential for the biological expression of hormonal activity (Tata, 1966). In non-hormone induced growth and development also one observes an important involvement of RNA synthesis.

A rapid stimulation of nuclear RNA synthesis, anticipating stimulation of protein synthesis in the target cells is now a well known feature of the administration of growth and developmental hormones to immature or hormone-deprived animals, or at the onset of rapid growth. (See Tata, 1966, 1967*a*, 1970*b*; Hamilton, 1968). Evidence for the regulation of RNA synthesis, independent of hormonal effects on precursor pools *in vivo*, can be produced by a stimulated RNA polymerase activity in isolated nuclear preparations or when chromatin template activity is measured (see Tata, 1966). In rapidly growing or maturing tissues the stimulation of RNA polymerase assayed at low ionic strength is, in every instance, more rapid and more pronounced than when the polymerase is assayed at high ionic strength or when the isolated chromatin is transcribed by a bacterial polymerase. The significance of this finding lies in the fact that the product of the first reaction is rich in ribosomal RNA and that of the other two is DNA-like or perhaps messenger RNA so that the transcription of ribosomal RNA is more sensitive to the stimulus for growth than that of other types of RNA.

As regards the nature of additional RNA, the synthesis of all classes of RNA is accelerated in rapidly growing tissues, but with most growth and developmental hormones, at the time of a peak shift in the rate of RNA synthesis, the major effect is on the synthesis of ribosomal RNA (see Tata, 1966, 1967*a*, 1970*b*; Hamilton, 1968). However, at early time intervals after hormone administration, or at the onset of development when the overall rate of RNA synthesis is unaffected, it is likely that there may be a subtle qualitative change in the nature of RNA synthesized without much effect on ribosomal RNA synthesis. It is important to recall here that the nature of highly hybridizable RNA from the nucleus is still in doubt. Much of it may not be m-RNA while some of it may represent the products of highly repetitive sequences of genes whose function is as yet unknown.

CYTOPLASMIC RNA AND RIBOSOMES

Soon after the activation of nuclear RNA synthesis by the stimulus of growth there occurs a gradual build-up of cytoplasmic RNA and ribosomes. In immature tissues or in tissues of animals deprived of the appropriate hormone the increase is very rapid and often most impressive. For example, growth-hormone treatment can almost double the population of hepatic polyribosomes in hypophysectomized rats, while androgen and oestrogen can cause a 4- to 8-fold increase in cytoplasmic RNA of their target accessory sexual tissues of castrated rats. The appearance of newly formed RNA in the cytoplasm following hormonal induction is accompanied by a gradual change in the polyribosomal profile. In virtually every instance of a growth and developmental system it has been shown that withdrawal of the hormone or cessation of growth causes a shift to smaller aggregates and an increase in the monomeric ribosomal peak whereas hormone administration causes a shift in the opposite direction (see Korner, 1965; Tata, 1966, 1967*b*; Teng & Hamilton, 1970). There is good evidence now to show that hormone deficiency leads to an inherent defect in the capacity of the ribosome itself and that hormone replacement corrects the defect, either directly or through the generation of new and more active population of ribosomes (see Tata, 1967*a*).

The relevance of the accumulation or alteration of cytoplasmic RNA to the question of hormonal regulation of protein synthesis becomes most obvious when the timing of changes in protein synthetic activity is taken into consideration. In case of many hormones, including androgens, estrogens, thyroid hormones and growth hormones, during metamorphosis or during liver regeneration or kidney hypertrophy the rate of protein synthesis increased rather abruptly following a relatively long lag period after hormone replacement, hepatectomy or unilateral nephrectomy (Tata, 1967*a*,*b*; Tsukada *et al.*, 1968; Malt, 1969). The onset of increased protein synthetic activity coincides with the accumulation in the cytoplasm of the hormone-induced newly formed ribosomes. It appears therefore that a large part of the effect of the hormone or growth stimulus on protein synthesis is incorporated into the new ribosomes formed after hormone treatment; but it seems likely that hormones control protein synthesis merely by increasing the number of ribosomes although this would certainly be important for a sustained growth action.

CYTOPLASMIC DISTRIBUTION OF RIBOSOMES DURING GROWTH

The importance of structural considerations in an evaluation of the action of growth and developmental hormones first became obvious from studies on the timing of hormone-induced acceleration of protein synthesis and the accumulation of newly formed RNA in the cytoplasm. Simultaneously with the observation of the presence of additional ribosomes accompanying the stimulation of protein synthesis, we found that the polysomes from mammalian and amphibian cells were more firmly bound to microsomal membranes in hormonally stimulated tissues, as judged by the ease with which the particles were released from membrane complexes with detergents. A firmer attachment of this type also caused a shift in the distribution of ribosomes from the free polysome fraction to those attached to rough endoplasmic reticulum membranes as the predominant type. Such changes were most marked in bullfrog tadpole liver following thyroid hormone-induced metamorphosis and also in the liver of young, thyroidectomized, hypophysectomized or adrenalectomized rats whose rate of growth was stimulated by growth hormone, thyroid hormone and corticosteroid, respectively, (Tata & Widnell, 1966; Tata & Williams-Ashman, 1967; Tata, 1967*b*; Cox & Mathias, 1969). When the activities *in vitro* of bound and free ribosomes from tissues of animals deprived of or treated with growth-promoting hormones were compared it was found that it is not necessary to have membrane-attached ribosomes *in vitro* to reflect a hormonal stimulation *in vivo* (see Korner, 1965; Tata, 1967*a*; Hamilton, 1968). However, if hormonal stimulation caused ribosomes to be re-distributed in the intact cell between free and membrane-bound fractions, it became imperative to prepare sub-microsomal fractions *without the use of detergents* and under very mild conditions so as to cause minimal damage to bound ribosomes. Table 1 summarizes results of some experiments on the effect of hypophysectomy and combined treatment with growth hormone and tri-iodothyronine on the capacity of sub-microsomal fractions to incorporate amino acids into protein, both with endogenous and synthetic m-RNA (Tata & Williams-Ashman, 1967). When the ability to incorporate ^{14}C-phenylalanine and ^{14}C-isoleucine was measured, in the presence or absence of synthetic messengers (polyuridylic acid and copolymer of uridylic and adenylic acids), it was found that hormone treatment corrected the lowered capacity for incorporation caused by hypophysectomy in all three types of sub-microsomal fractions. The effect of hormone deprivation or replacement was of greater magnitude in the heavy rough membranes than in the light rough membranes. To a large extent the response to synthetic messengers of both membrane-bound and free ribosomes paralleled that to endogenous messenger, whether growth had been retarded or accelerated. Under the conditions of rather prolonged hormonal treatment as used in these experiments, the population of hepatic ribosomes would be 50–60% higher than in the control hypophysectomized rats. Thus there is a distinct possibility that the newly formed ribosomes upon hormone administration were inherently more efficient in handling synthetic or natural messengers. At the same time, the relative stimulation by synthetic polynucleotides was slightly higher in the less active preparations which indicated that a small part of hormonal effect could be due to an increase in the polysomal n-RNA content. This type of result raises the possibility that the relatively slow anabolic and developmental actions of hormones may be restricted to a small population of newly generated ribosomes rather than modify the activity of all the ribosomes in the cytoplasm.

CO-ORDINATION BETWEEN THE PROLIFERATION OF MICROSOMAL MEMBRANES AND RIBOSOMES, AND ACCELERATED PROTEIN SYNTHESIS *in vivo*

Above observations on the distribution of ribosomes and their amino acid incorporating activity *in vitro* posed an important question. Are the new ribosomes formed following growth stimulation merely re-distributed on existing membranes or whether there also occurred simultaneous alterations in the rate or type of membranes formed? To answer this question, the proliferation of membranes of the endoplasmic reticulum was followed in our laboratory by measuring the incorporation of ^{32}P-phosphate and ^{14}C-labelled choline and glycerol into membrane phospholipids. In order to correlate the

Table 1. *Incorporation of ^{14}C-phenylalanine and ^{14}C-isoleucine into protein by "free" polysomes and polysomes in light and heavy rough membrane fractions of liver microsomes from normal and hypophysectomized rats with or without hormonal treatment: response to poly U and poly UA*

Hormonal treatment consisted of 3 injections of 75 μg of human growth hormone (HGH) at 21, 47 and 74 hr and 2 injections of tri-iodothyronine (T_3) at 47 and 74 hr before killing. Light and heavy rough membrane fractions consist of ribosomes attached to membranes of the endoplasmic reticulum and differ from one another by the density of ribosome packing. (From Tata & Williams-Ashman, 1967.)

Rats	Polysomes	mg RNA/g	^{14}C-Phe incorporated (c.p.m./10 min/mg RNA)		^{14}C-Ileu incorporated (c.p.m./10 min/mg RNA)	
			−Poly U	+Poly U	−Poly UA	+Poly UA
Normal	Free	0·57	13,590	38,090	5,320	6,383
	Light membranes	0·51	18,620	41,840	3,590	5,748
	Heavy membranes	1·62	58,800	128,400	23,800	29,980
Hx	Free	0·44	6,700	30,520	3,120	6,150
	Light membranes	0·46	14,550	44,250	3,560	5,036
	Heavy membranes	0·59	40,100	102,900	18,550	21,190
Hx + HGH + T_3	Free	0·66	26,700	86,300	10,900	20,720
	Light membranes	0·47	22,630	101,150	6,980	11,590
	Heavy membranes	1·40	66,000	144,000	32,100	45,700

Hx = hypophysectomized; HGH = human growth hormone; T_3 = 3,3′,5-tri-iodothyronine.

effect of hormones or hepatectomy on this process with those on biosynthetic activity of the tissue we studied the rate of incorporation of radioactive amino acids into nascent proteins of the different sub-microsomal fractions simultaneously with the accumulation of newly synthesized microsomal RNA. This has now been done in a variety of systems (see Tata, 1970*a,b*) which include: (a) growth of liver induced by a single injection of growth hormone, tri-iodothyronine and testosterone in hypophysectomized, thyroidectomized and castrated rats, respectively; (b) growth and maturation of the seminal vesicle of castrated rat when stimulated by testosterone; (c) precocious initiation of

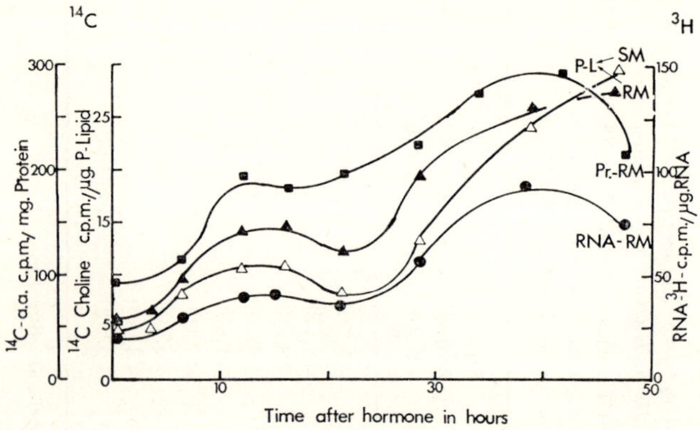

FIG. 1. Co-ordination of the appearance of additional newly formed hepatic microsomal phospholipids (△, ▲) and RNA (●) and the recovery of nascent microsomal protein formed *in vivo* (■), following administration to hypophysectomized rats of human growth hormone and tri-iodothyronine together. The rate of formation of microsomal RNA and phospholipids was determined from the incorporation of [^{3}H]-orotic acid and [^{14}C]choline, respectively, whereas the rate of protein synthesis *in vivo* was derived from the incorporation of ^{14}C-labelled amino acids at 10 min after their administration. The results are expressed as per cent increase in specific radioactivities in hormone-treated animals over the values obtained in untreated controls. The first burst of synthetic activity is in response to growth hormone, whereas the second burst is due to the slower effect of tri-iodothyronine. Phospholipid synthesis was measured in both smooth (△) and rough (▲) e.r. (From Tata, 1970*a*.)

enzyme and protein synthesis in bullfrog tadpole liver during thyroid hormone-induced metamorphosis; (d) accelerated synthesis of thyroglobulin or iodoprotein by thyroid tissue under the influence of thyrotrophic hormone (TSH) and (e) regenerating rat liver.

Figure 1 compares the rates of formation of liver microsomal phospholipids with those of microsomal RNA and of nascent protein in microsomes following the combined administration, at maximal growth promoting levels, of growth hormone and tri-iodothyronine to hypophysectomized rats. Since the two hormones have very different lag periods it was quite easy to detect a temporally co-ordinated enhancement of microsomal phospholipid formation by the hormones with that of the rates of microsomal RNA and protein synthesis. Thus, growth hormone stimulated the protein synthetic capacity of hepatic ribosomes several hours earlier than did tri-iodothyronine and the same pattern was observed for the accelerated formation of polysomes (much of the additional microsomal RNA appeared as polysomes).

Such co-ordination of proliferation of cellular structures and biosynthetic activity seems to be a feature common to other growth and developmental situations. A good co-ordination in the onset of accelerated synthesis of phospholipid, RNA and protein was also observed in microsomes isolated from the seminal vesicles of castrated rats following growth stimulation with testosterone. In another system, although ribosomes and membranes were not actually isolated and studied in the same laboratory, a co-ordinated synthesis of phospholipid and RNA in their respective target tissues may underline the same process as in the case of oestrogen (Teng & Hamilton, 1970; Nishigori & Aizawa, 1968). In a growth system not directly involving hormones, e.g. in regenerating rat-liver, one also notices a simultaneous build-up of rough endoplasmic reticulum and enhanced protein synthesis *in vivo* (Tables 2 and 3). Tsukada *et al.* (1968) and Fex (1970) have separately shown that regeneration involves rapid build-up of ribosomes and membrane phospholipids. This co-ordination thus seems to be a common feature of cells undergoing rapid development in general, irrespective of whether the stimulus is hormonal or not, in which during this period new biochemical functions tend to be concentrated in cytoplasmic structures of increasing complexities (see Siekevitz *et al.*, 1968; Dallner, Siekevitz & Palade, 1966). In some cases the appearance of rough endoplasmic reticulum at the time of rapid growth and maturation has been corroborated by electron microscopy (see Table 4).

An indirect suggestion regarding some sort of specificity in the proliferation of microsomal structures is provided by systems in which the hormone induced the synthesis *de novo* of specific enzymes or proteins. Two examples of this specificity are the following systems which have been studied in our laboratory: (a) induction of urea cycle enzymes and serum albumin during tri-iodothyronine induced metamorphosis in bullfrog tadpoles; (b) enhancement of thyroglobulin formation by thyroid cells under the influence of thyrotrophic hormone where the hormone causes an almost simultaneous increase in the rate of whole thyroid phospholipid and RNA synthesis *in vivo*. What is particularly interesting is that the onset of this change coincides with the formation of increasing amounts of a thyroglobulin-like protein (Kerkof & Tata, 1969). Since one of the multiple actions of TSH is to stimulate secretion of thyroglobulin, it could be argued that additional membrane synthesis would be necessary to perform the vectorial discharge of extra iodoprotein.

In the system of artificial induction of metamorphosis in the tadpole, there occurs a simultaneous increase in the synthesis of ribosomes and membranes of the endoplasmic reticulum as judged by ribosomal RNA and membrane phospholipid synthesis in the rough endoplasmic reticulum (Fig. 2a). It is interesting to note that this simultaneous increase in the synthesis of the two rough endoplasmic reticulum components is accompanied by an enhanced rate of protein synthesis, among which are proteins (Fig. 2b), synthesized *de novo*, such as urea cycle enzymes and serum albumin. These biochemical changes are so marked that it is easy to observe structural changes by electron microscopy. These consist of a shift in the distribution of ribosomes from around the simple vesicular membrane structures of the immature larvae to the more complex double lamellar structures more commonly seen in mature

Table 2. *Co-ordinated increase in formation of rough microsomal phospholipid and RNA after partial hepatectomy*

Each group consisted of four sham-operated or partially hepatectomized rats. They were given 52 μCi of [^{3}H]-orotic acid or 65 μCi of [^{32}P]-orthophosphate 3·2 and 2·0 hr before death. The uptake of the label and acid-soluble fraction of radioactivity in the whole liver homogenate were measured to allow for the correction of specific radioactivities for increases in uptake of label after hepatectomy. (From Tata, 1970*a*.)

Time after operation (hr)	Operation	Wt. of liver (g)	Specific radioactivity	
			^{3}H-labelled RNA (c.p.m./mg)	^{32}P-labelled phospholipid (c.p.m./μg of P)
4	Sham	6·8	12,100	270
4	Hepatectomy	1·5	18,480	339
9	Sham	6·9	11,633	465
9	Hepatectomy	1·6	21,070	808
18	Sham	6·7	10,450	366
18	Hepatectomy	1·6	24,600	543
24	Sham	7·0	9,335	297
24	Hepatectomy	1·8	15,870	413

Table 3. *Increase in amino acid incorporation into rough microsomal protein after partial hepatectomy*

The experiment was performed in the same way as that in Table 2, except that 8 min before death hepatectomized rats received 4 μCi of a mixture of ^{14}C-labelled amino acids and sham-operated animals were given 10 μCi. Different amounts of radioactivity were administered to the two groups of rats to allow for the different amounts of tissue and corrections were also applied for uptake. (From Tata, 1970*a*.)

Time after operation (hr)	Operation	Wt. of liver (g)	Specific radioactivity (c.p.m./mg of protein)	Protein (mg/m of liver)
2	Sham	6·4	435	9·2
	Hepatectomy	1·6	398	8·8
4	Sham	6·6	490	8·9
	Hepatectomy	1·9	410	9·3
8	Sham	6·3	385	8·7
	Hepatectomy	1·7	588	9·1
16	Sham	6·3	361	8·8
	Hepatectomy	1·9	639	9·3
24	Sham	6·6	343	8·9
	Hepatectomy	2·0	605	9·6

Table 4. *Co-ordination between the appearance of newly formed ribosomes and membranes (phospholipid) of the endoplasmic reticulum accompanying the onset of additional or new protein synthesis*

References to individual findings are not given but may be found in Tata (1970*a*,*b*) and Andrews & Tata (1971*a*).

Tissue	Species	Growth or Induction Stimulus	Onset of increase in microsomal*			Supporting electron microscopic data
			Protein synthesis	Ribosomes or RNA	P-lipid	
Thyroid (*in vivo*)	Guinea pig	Thyrotrophic hormone	2	1	1	+
Uterus	Rat	Oestrogen	4	3	2	—
Liver	Rat	Growth hormone	6	5	4	—
		Thyroid hormone	20	20	18	—
		Testosterone	14	12	15	—
		Phenobarbital	18	10	14	+
		Hepatectomy	9	10	8	+
Wing bud	Cecropia (Silk moth)	Ecdysone	12–16	9	12(?)	+
Mammary gland in culture (Epithelial cells)	Rat	Corticosteroids	20–24	20–24	20–24	+
Liver	Bullfrog tadpole	Thyroid hormone	48	50	58	+

* Earliest time after the stimulus is applied when a 10–20% change over control values is observed. A rapid and transient increase in amino acid incorporation rate, often due to changes in levels of available amino acids, is not included.

tissues. This shift in structural organization of the endoplasmic reticulum coincided temporarily with the changes in the formation of microsomal RNA and phospholipids, and with the synthesis of proteins by polyribosomes. It is also interesting that much of the newly formed rough endoplasmic reticulum is localized in the perinuclear region during the initial stages of induction.

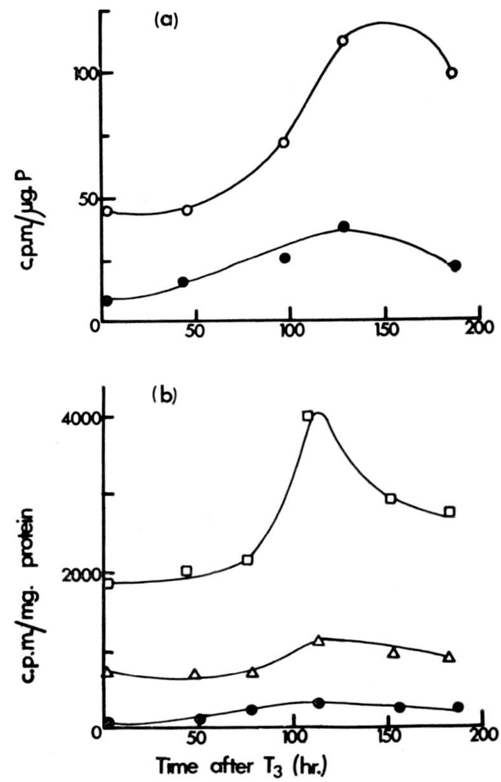

FIG. 2. Co-ordinated proliferation of endoplasmic reticulum and the increase in protein synthesis *in vivo* in the heavy rough membranes (densely packed ribosomes attached to membranes) during tri-iodothyronine-induced metamorphosis of young bullfrog tadpoles. (a) Appearance of newly synthesized (^{32}P-labelled) heavy rough membrane phospholipids (○) and RNA (●); (b) recovery of labelled nascent proteins formed *in vivo* after a short pulse of radioactive amino acids in heavy rough membranes (□), light rough membranes (△) and free polysomes (●). For details see Tata (1967*b*).

In the studies mentioned above it is quite possible that the hormonal stimulus leads to both accelerated formation and retarded degradation of those membranes for which demand is most acute, as for example the rough endoplasmic reticulum. The question of membrane stability is an important one and ought to be studied along the lines recently laid down by the groups of Palade and Schimke in studying drug-induced proliferation of the endoplasmic reticulum (Siekevitz *et al.*, 1968; Omura, Siekevitz & Palade, 1967; Arias, Doyle & Schimke, 1969). How the phospholipid and protein components of membranes are assembled in an orderly fashion to yield precise structures is not known. It may be that the initial response to a growth stimulus may be to accelerate the assembly or "crystallization" of rigid structures from existing components whose depletion in turn would trigger off the synthesis of additional membrane phospholipids and proteins.

SIGNIFICANCE OF CO-ORDINATED PROLIFERATION OF RIBOSOMES AND MEMBRANES DURING GROWTH AND DEVELOPMENT

It can be concluded from the work described above that there exists some mechanism in the cell that tightly co-ordinates the formation of different components of rough endoplasmic reticulum in response to an increased demand for protein synthesis following stimulation with a variety of growth and developmental hormones or other stimuli. In order to deduce the significance of the co-ordination between membrane and ribosome proliferation, on the one hand, and protein synthesis, on the other, it is necessary to know the functions of membrane-ribosome interactions. Until now the only well established function is facilitation of the export of secretory proteins. But this seems to be a most unlikely reason for the proliferation of the rough endoplasmic reticulum during growth because even in secretory cells there is relatively little increase in protein secretion during the period of active growth. We have recently found in our laboratory that bound ribosomes in non-secretory and secretory tissues show a major difference in the vectorial discharge of protein newly synthesized on ribosomes bound to membranes (Andrews & Tata, 1971*a*,*b*). It is often said that proteins for export are made on bound ribosomes and conversely those for intracellular use are made on free ribosomes. While there is no doubt that proteins for secretion or storage are made on the rough endoplasmic reticulum (see Palade, 1966; Campbell, 1970), there is little reason to accept that some intracellular proteins are not synthesized on bound ribosomes as well. It is important to recall in this context that recently Pitot *et al.* (1969) have identified, by radio-immunological techniques, bound and not free ribosomes of the liver as the site of synthesis of serine dehydrase, an exclusively intracellular protein. Further attempts to identify the site of synthesis of intracellular constituents may indeed reveal a greater involvement of bound ribosomes in the synthesis of intracellular proteins. In this context it would be well worth investigating the proteins synthesized on membrane-bound ribosomes of predominantly non-secretory tissues.

The significance of a co-ordinated membrane-ribosome assembly may be that a firm attachment would in effect lead to a segregation of different populations of ribosomes. During rapid developmental or adaptive changes populations of ribosomes formed at different periods would be pre-coded with different spectra of messengers, so that operationally it would mean a topological segregation of ribosomes synthesizing different classes of proteins. There are two important pre-requisites for a consideration of such a topographical segregation as a likely physiological process: (a) there should be only limited or no exchange between different groups of membrane-bound ribosomes or between bound and free ribosomes and (b) there are qualitative differences in the composition of products formed by bound and free ribosomes. As regards the first point, while it is technically not possible to study populations of bound polysomes that would be kept apart *in vivo*, a relatively extensive exchange of ribosomes between the bound and free compartments is generally accepted. A critical appraisal of the evidence on which this acceptance is based will however show that there is indeed no convincing proof for or against such an exchange occurring. Bound and free ribosomes could well have similar half-lives without the two pools necessarily being exchangeable, or the exchange may be minimal and may perhaps be topographically restricted within small regions of the cytoplasm (see Tata, 1971).

There is also some evidence for chemical differences between membrane-bound and free ribosomes and a number of papers have recently reported on the identification of different products synthesized on bound and free ribosomes. For example, Fridlender & Wettstein (1970) find by polyacrylamide gel electrophoresis of ribosomal proteins of chick embryo absence of two components in the bound particles which strongly suggests the unlikelihood of an exchange between bound and free ribosomes. Dingman, Kakefuda & Aronow (1970) found that removal of the cytoplasmic supernatant from microsomes or treatment of microsomes with sodium deoxycholate caused a partial degradation of ribosomal RNA. By isotopic labelling of RNA it was found that newly formed RNA and ribosomes were preferentially broken down suggesting the presence of some mechanism in the cytoplasm that allows a discrimination between ribosomes formed at different periods of a cell's lifetime. As regards different classes of proteins synthesized by bound and free ribosomes, there is considerable evidence from microbial, yeast and *E. coli* cells or structurally uncomplicated animal cells, such as the reticulocyte, in which the ribosomes may be attached to the cell membrane or structures other than the endoplasmic reticulum (see Tata, 1971).

Thus the idea of a topographical segregation of different classes of ribosomes synthesizing different groups of proteins may not be too far from reality. What is holding up a convincing demonstration of such a segregation *in vivo* are technical limitations, but a solution is not impossible if immunochemical, histochemical, electron microscopic and isotope labelling methods are combined. Until such time ribosomal segregation will remain an intriguing concept. The advantages of functional segregation of ribosomes would be principally felt in cells undergoing rapid development or adaptation. It would enable such cells to respond (based on selective synthesis of proteins) to hormonal and non-hormonal stimuli without perturbing the synthesis of proteins that are not initially involved in growth and development.

REFERENCES

Andrews, T. M. & Tata, J. R. (1971*a*). *Biochem. J.* **121,** 683.
Andrews, T. M. & Tata, J. R. (1971*b*). *Biochem. J.* **124,** 883.
Arias, I. M., Doyle, D. & Schimke, R. T. (1969). *J. biol. Chem.* **244,** 3303.
Campbell, P. N. (1970). *FEBS Lett.* **7,** 1.
Cox, R. F. & Mathias, A. P. (1969). *Biochem. J.* **115,** 777.
Dallner, G., Siekevitz, P. & Palade, G. E. (1966). *J. Cell Biol.* **30,** 73.
Dingman, C. W., Kakefuda, T. & Aronow, A. (1970). *Biochim. biophys. Acta* **224,** 114.
Fex, G. (1970). *Biochem. J.* **119,** 743.

Fridlender, B. R. & Wettstein, F. O. (1970). *Biochem. biophys. Res. Commun.* **39,** 247.
Hamilton, T. H. (1968). *Science,* **161,** 649.
Hendler, R. W. (1968). *Protein Biosynthesis and Membrane Biochemistry.* New York: John Wiley.
Kerkof, P. R. & Tata, J. R. (1969). *Biochem. J.* **112,** 729.
Korner, A. (1965). *Rec. Progr Hormone Res* **21,** 205.
Malt, R. A. (1969). In *Renal Hypertrophy.* (Ed. by Nowinski, M. W. and Goss, R. J.), pp. 131–154. New York: Academic Press.
Nishigori, H. & Aizawa, Y. (1968). *Endocrinol. Japan* **15,** 209.
Omura, T., Siekevitz, P. & Palade, G. E. (1967). *J. biol. Chem.* **242,** 2389.
Palade, G. E. (1966). *J. Am. med. Ass.* **198,** 815.
Pitot, H., Sladek, N., Ragland, W., Murray, R. K., Mayer, G., Soling, H. D. & Jost, J. P. (1969). In *Microsomes and Drug Oxidations.* (Ed. by Gillette, J. R., Conney, A. H., Cosmides, G. J., Estabrook, R. W., Fonts, J. R. and Mannering, G. J.), pp. 59–79. New York: Academic Press.
Siekevitz, P., Palade, G. E., Dallner, G., Ohad, I. & Omura, T. (1968). In *Organizational Biosynthesis.* (Ed. by Vogel, H. J., Lampen, J. O. and Bryson, V.), pp. 331–373. New York: Academic Press.
Tata, J. R. (1966). *Progr. Nucleic Acid. Res. Molec. Biol.* **5,** 191.
Tata, J. R. (1967*a*). *Biochem. J.* **104,** 1.
Tata, J. R. (1967*b*). *Biochem. J.* **105,** 783.
Tata, J. R. (1968). *Nature,* **219,** 331.
Tata, J. R. (1970*a*). *Biochem. J.* **116,** 617.
Tata, J. R. (1970*b*). In *Biochemical Actions of Hormones.* (Ed. by Litwack, G.), pp. 89–133. New York: Academic Press.
Tata, J. R. (1971). *Subcellular Biochemistry* **1,** 83.
Tata, J. R. & Widnell, C. C. (1966). *Biochem. J.* **98,** 604.
Tata, J. R. & Williams-Ashman, H. G. (1967). *European J. Biochem.* **2,** 366.
Teng, C. S. & Hamilton, T. H. (1970). *Biochem. J.* **118,** 341.
Tsukada, K., Moriyama, T., Umeda, T. & Lieberman, J. (1968). *J. biol. Chem.* **243,** 1160.

STUDIES ON THE MEMBRANES OF *BACILLI*. III. DISAGGREGATION AND REAGGREGATION OF THE MEMBRANE COMPONENTS OF *BACILLUS MEGATERIUM* KM

By R. A. MAKULA, K. A. DEVOR AND W. J. LENNARZ

Department of Physiological Chemistry, The Johns Hopkins University School of Medicine, Baltimore, Maryland, U.S.A.

INTRODUCTION

As part of an extensive study of the membranes of *Bacilli* (Patterson & Lennarz, 1970; Patterson & Lennarz, 1971; Devor *et al.*, 1972) we have investigated the possibility that a membrane completely dissociated into its individual molecular components might, under suitable conditions, reassemble into a structure with enzymatic and morphological properties similar, if not identical, to those of the original membrane.

The phospholipids of *Bacillus megaterium* constitute 25% of the total weight of the cytoplasmic membrane. We have previously shown that the enzymes involved in the synthesis of these phospholipids are associated with the cytoplasmic membrane (Patterson & Lennarz, 1971). Since these enzymes are responsible for the synthesis of over 25% of the weight of the membrane, it seems logical to assume that they are an integral part of it and play a key role in its structural organization. For this reason we have initially utilized these enzymes to assess the results of our attempts to reassociate membranes that have been dissociated with urea-sodium dodecylsulphate. In this paper the current state of our studies on dissociation and reassociation of the membrane of *Bacillus megaterium* are reported.

EXPERIMENTAL PROCEDURE

Materials

Urea was purchased from J. T. Baker Chemical Co., Phillipsburg (N.J.) and was purified and stored as described by Duesberg & Reuckert (1965). Sodium dodecyl sulphate (SDS) was obtained from Eastman Kodak Co., Rochester (N.Y.). AG-1 X-2 ion exchange resin, 200–400 mesh, acetate form, was purchased from Biorad Lab. Richmond (Cal.). L-chloramphenicol, blue dextran 2,000, bovine thyroglobulin (Type I), and Trizma base were obtained from

Sigma Chemical Co., St. Louis, Missouri. Bovine serum albumin was purchased from Pentex, Inc., Kankakee, Illinois. Sepharose 4B and 6B were obtained from Pharmacia, Uppsala, Sweden. Dialysis tubing purified by boiling in a solution containing 5% Na_2CO_3, 0·05 M disodium ethylene diamine tetra-acetate was purchased from Arthur H. Thomas Co., Philadelphia (Pa.). Dithiothreitol (DTT) was obtained from Calbiochem, Los Angeles (Cal.). Glycerol-^{14}C, uniformly labelled was purchased from International Chemical and Nuclear Corp., Long Island City (N.Y.). L-leucine labelled with tritium in the 4 and 5 positions was purchased from New England Nuclear, Boston (Mass.). The ^{14}C-sn-glycerol-3-phosphate was prepared in this laboratory by Dr F. Albright as described by Chang & Kennedy (1967). ^{32}P labelled T-7 phage DNA was a generous gift from Dr Paul Englund. The sources of other radio chemicals and counting techniques have been previously described (Patterson & Lennarz, 1971; Gould & Lennarz, 1970).

Radioactive [1-^{14}C]-SDS (sp. activity 2·92 μCi/mmole) was prepared by reduction of methyl laurate-1-^{14}C with lithium aluminium hydride followed by sulphation of the resulting alcohol with chlorosulphonic acid.

METHODS

Preparation of ghosts

Bacillus megaterium KM ATCC 13632 was grown as described by Weibull, Beckman & Bergstron (1959) with agitation at 37°. The cells were harvested at mid-log phase and were collected by centrifugation at 10,000 × g for 10 min. The cells were washed with 20 volumes of 50 mM sodium phosphate pH 7·0 and recollected by centrifugation at 10,000 × g for 10 min. Lysozyme treatment of the washed cells was carried out in 40 volumes of a solution containing 0·25 M sucrose, 50 mM sodium phosphate, pH 7·0, 5 mM $MgCl_2$ and 10 mM NaCl (sucrose buffer) for 60 min at 23°. The resulting protoplasts were collected by centrifugation at 10,000 × g for 15 min and were lysed in 20 volumes of solution containing 50 mM sodium phosphate, pH 7·0 and 5 mM $MgCl_2$ (Mg-buffer). The ghosts resulting from lysis were then treated with DNase and RNase (2 mg of each per 10 g of wet cell weight) at 0° for 30 min after which time an equal volume of Mg-buffer was added. The ghosts were washed twice with 40 volumes Mg-buffer and once with 40 volumes 50 mM Tris-HCl, pH 7·0. The ghosts were resuspended for storage at 4° in 50 mM Tris-Cl, pH 7·0 containing 10 μg/ml L-chloramphenicol at a final protein concentration of 10–15 mg/ml. Ghosts were stored no longer than a week under these conditions.

Solubilization procedure

The membranes were solubilized in a final concentration of 2 mg of protein/ml at 0° in 50 mM Tris-acetate (pH 7·0) containing 1 mM DTT, 7·5 M urea, and 0·2% sodium dodecyl sulphate. In the final procedure employed, SDS was added to the membranes 15 min after the addition of urea. Solubilization, as judged visually, occurred immediately upon addition of SDS, but the solution was routinely stirred for 30 min at 0°.

Removal of SDS by dialysis

The solubilized membranes, after centrifugation at 105,000 for 20 hr to remove any insoluble residue, were dialysed against 50 mM Tris-HCl (pH 7·0), 5 M urea with stirring for 24 hr to remove the SDS. The membranes were then reaggregated by dialysis against 2000 volumes of 50 mM Tris-HCl (pH 7·0) containing 20 mM $MgCl_2$ for 24–48 hr without stirring. The reaggregated membranes were collected by centrifugation at 105,000 × g for 1 hr.

Removal of SDS by ion exchange resin

The procedure of Weber & Kuter (1971) was modified for removal of the SDS from the solubilized membranes. The solution containing solubilized membrane (2 mg protein ml) was added to Biorad AG-1 X-2 acetate ion exchange resin, 200–400 mesh (50 mg) and the mixture stirred for 30 min at 0°. The resin was then pelleted by low speed centrifugation (2000 rpm for 1 min), and the resulting supernatant solution dialysed against 50 mM Tris-acetate (pH 7·0) overnight. When reaggregation was desired, 20 mM $MgCl_2$ was added to the Tris-acetate buffer.

Preparation of labelled ghosts

Radioactively labelled ghosts were prepared by growth to mid-log phase in 15 ml of media as described above except that casamino acids were omitted and either ^{3}H-leucine (60 μCi, spec. act. 55 Ci/mmole) or ^{14}C-leucine (22·5 μCi, spec. act. 0·262 Ci/mmole) was added. Cultures were protoplasted with 200 μg lysozyme in 4 ml of sucrose buffer (see above). The protoplasts were lysed in 2 ml Mg-buffer and the resultant ghosts, resuspended in 2 ml of Mg-buffer, were treated with 20 μg RNase and DNase. The ghosts were then washed twice with Mg-buffer and once with 50 mM Tris-Cl, pH 7·0 and were resuspended for storage in 0·5 ml Tris-Cl, pH 7·0. The branched amino acids serve as precursors for the branched chain fatty acyl groups of the lipids as well as the protein of the membrane (Lennarz, 1966). Lipid extraction of the ghosts showed that 12–20% of the radioactivity was in lipid.

Enzyme assays and analytical methods

L-*Serine:CDP-diglyceride Phosphatidyltransferase*. The incubation mixture consisted of: 5 μmoles $MgCl_2$, 40 nmoles cytidine diphosphate diglyceride, 120 nmoles ^{14}C-L-serine, 100 μmoles Tris-Cl, pH 9·0, 100 μg ghost protein and 0·25% cutscum in 0·4 ml. Incubation was for 60 min at 37°.

Phosphatidylglycerol synthetase. The incubation mixture consisted of: 10 μmoles $MgCl_2$, 40 nmoles cytidine diphosphate diglyceride, 100 nmoles ^{14}C-sn-glycerol-3-phosphate, 10 μmoles Tris-maleate, pH 7·0, 100 μg ghost protein and 0·1% cutscum in 0·4 ml. Incubation was for 30 min at 25°.

CTP:phosphatidic acid cytidyltransferase. The incubation mixture consisted of: 2·5 μmoles $MgCl_2$, 80 nmoles phosphatidic acid, 200 nmoles ^{3}H-CTP,

10 μmoles Tris-Cl, pH 7·0, 50 μmoles KCl, 100 μg ghost protein and 0·4% cutscum in 0·4 ml. Incubation was for 30 min at 25°.

The products of enzymic reactions were extracted from the incubation mixtures as described by Patterson & Lennarz (1971). Protein was determined by the method of Lowry *et al.* (1951). Lipid phosphate was determined by the method of Bartlett (1959). The procedure for lipid extraction has been described previously (Patterson & Lennarz, 1971).

RESULTS

Criteria of solubilization

Our first concern was to devise conditions that led to complete solubilization of the cytoplasmic membrane using a minimal amount of SDS. When the membrane protein was held constant (2 mg/ml) and the concentration of SDS varied, the percent protein solubilized was found to be maximal when the ratio of SDS to membrane protein was 1:1 (Fig. 1). When the membrane protein concentration was varied and the SDS held constant at 0·2% (2 mg/ml),

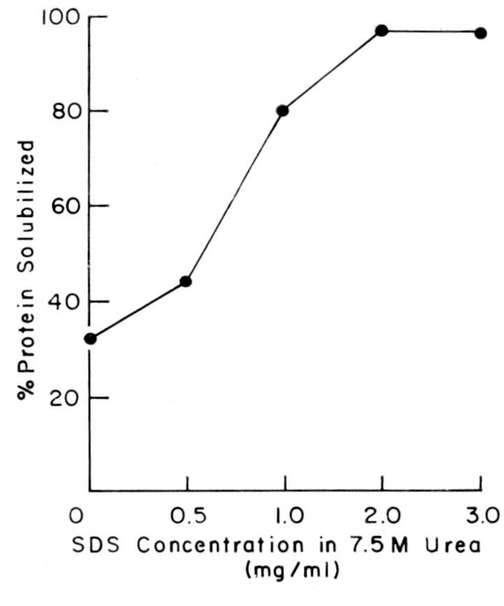

FIG. 1. The effect of SDS concentration on solubilization of ghosts. The concentration of ghost protein was held constant and the SDS concentration varied as indicated.

maximal solubilization again occurred when the ratio of SDS to membrane protein was 1:1 (data not shown). The criterion of solubilization for these experiments was the amount of protein remaining in the supernatant solution after centrifugation of the samples in urea-SDS at 105,000 × g for 20 hr.

To further study the effect of time of centrifugation on the amount of membrane protein solubilized, membranes were solubilized in either 0·1% or 0·2% SDS in 7·5 M urea and centrifuged for varying lengths of time. The results of these experiments are shown in Fig. 2. After 1 hr of centrifugation, virtually none of the membrane sediments from either the 0·1% or 0·2% SDS solution. However, with increased time of centrifugation, an increasing fraction of the membranes solubilized in 0·1% SDS was sedimentable. In contrast, no increase in the sedimentable protein was observed when membranes were solubilized in 0·2% SDS; only 3% of the protein sedimented after 20 hr of centrifugation at 105,000 × g.

A second criterion of solubilization was the use of gel filtration column chromatography. The solubilized membranes (in SDS-urea), labelled with

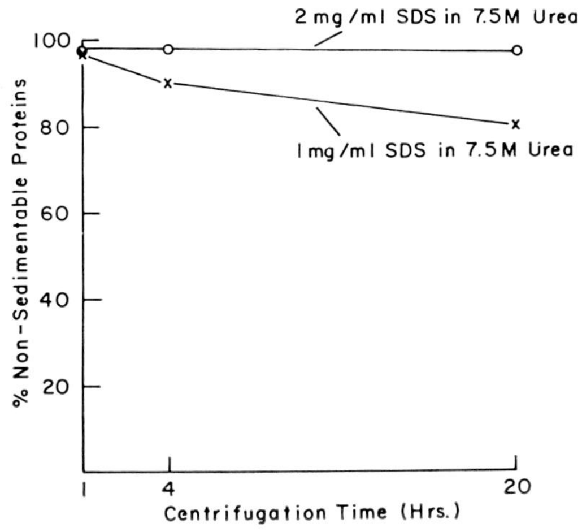

FIG. 2. Effect of time of centrifugation on the sedimentation of SDS-urea solubilized membranes.

^{3}H-leucine, were centrifuged for 105,000 × g for 1 hr and the supernatant fraction was applied to a Sepharose 6B column equilibrated in 50 mM Tris-Cl (pH 7·0) containing 7·5 M urea and 0·2% SDS and eluted with the same solvent. Thyroglobulin, which is reported to have a molecular weight of 250,000 in urea-SDS (Steiner & Edelhoch, 1961), was used to determine the exclusion volume. The results in Fig. 3A demonstrate that essentially all the labelled proteins and lipids (leucine serves to label both the proteins and the fatty acyl groups) were included in the column. It was also found that over 50% of the labelled protein was less than 70,000 M.W. Another interesting observation was that the membrane lipid was markedly retained in the column, suggesting that either the lipid was dissociated from the bulk of the protein or that the lipid was associated with a particular group of low molecular weight proteins. Polyacrylamide gel electrophoresis in urea-SDS of pooled fractions was used to determine if there had been a fractionation of the membrane proteins. Figure 3B shows the results obtained from these gels. The first gel depicts the protein pattern of the solubilized membrane applied to the column. The other four gels represent pooled fractions of the column. A definite fractionation of the proteins was observed, those of higher molecular weight being eluted from the column earlier than the lower molecular weight proteins. Because the resolution on the Sepharose 6B column was not maximal and it was important for subsequent

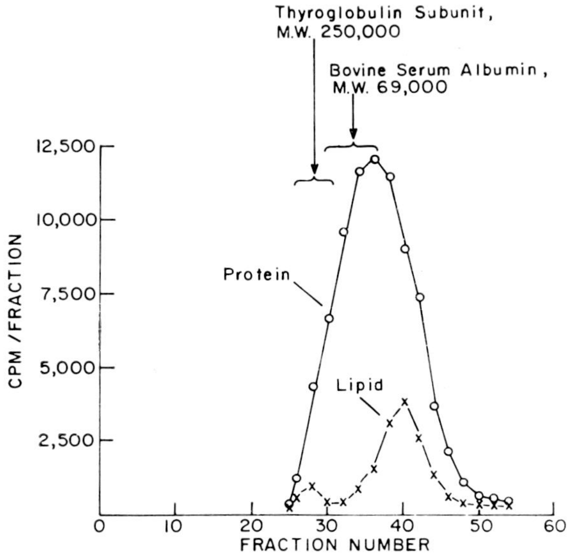

FIG. 3A. Sepharose 6B gel filtration of membrane components in SDS-urea.

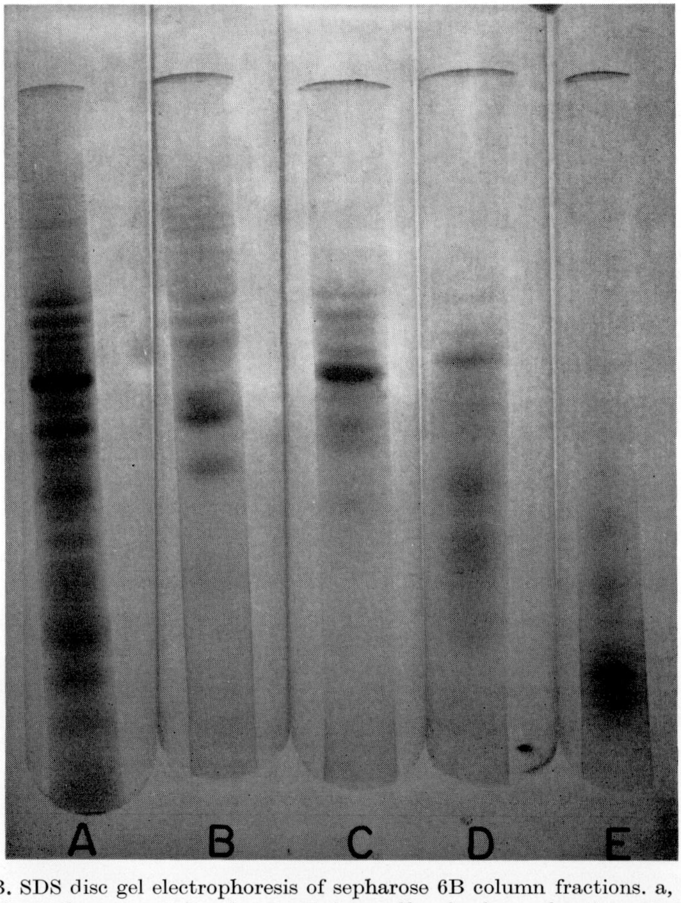

FIG. 3B. SDS disc gel electrophoresis of sepharose 6B column fractions. a, SDS-urea solubilized membrane proteins; b–e, protein profile of column fractions 21–31, 32–35, 36–39, and 40–43, respectively.

enzymatic studies to have definitive information on the size distribution of the membrane components in urea-SDS, another sample of membranes solubilized in urea-SDS, but not subjected to centrifugation was applied to a Sepharose 4B column. The results shown in Fig. 4 indicate the presence of a

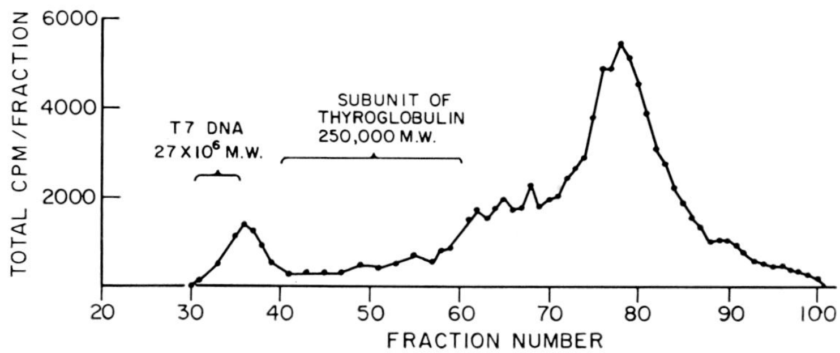

FIG. 4. Sepharose 4B column chromatography of membrane components in urea-SDS.

small fraction of the membrane components (*ca* 8%) that are of very high molecular weight. However, the majority of proteins and lipids are disaggregated to components with a molecular weight less than 250,000.

Removal of SDS

Having fulfilled two criteria for solubilization of the membranes, conditions that would lead to maximal reassociation of the solubilized lipids and proteins were investigated. Of primary concern was development of a rapid method for complete removal of SDS. In order to study the dialysis method of SDS removal, the kinetics of ^{14}C-SDS efflux from the solubilized membranes upon dialysis was studied (Fig. 5). The data indicated that after 20 hr of dialysis

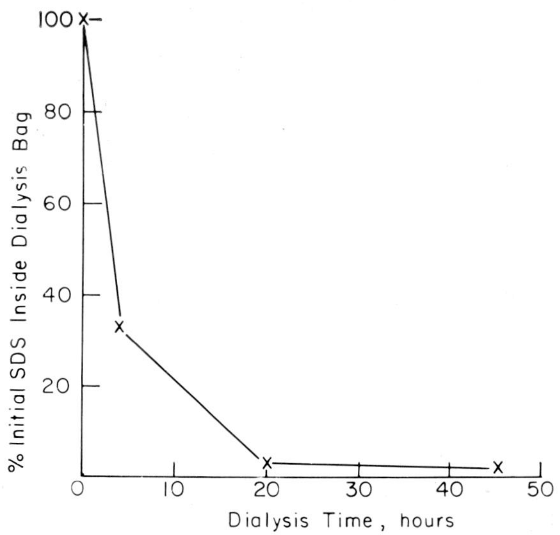

FIG. 5. Kinetics of SDS removal from solubilized membrane components by dialysis.

against 5 M urea, only 3·3% of the SDS remained associated with the membrane components. After an additional 24 hr of dialysis against $MgCl_2$-buffer, less than 2% of the labelled SDS remained. This figure represents less than 15 nmoles of SDS per 700 nmoles of lipid phosphorus per 2 mg of membrane protein.

Although the resulting SDS-depleted membrane components could be effectively reaggregated, the resulting reaggregated material was devoid of enzymatic activity (see below). Consequently, a rapid method that would ensure virtually complete removal of the SDS was employed, namely, treatment of the solubilized membranes with AG-1 X-2 ion exchange resin. Experiments in which the resin to SDS ratio was varied indicated that treatment of 2 mg of solubilized membranes with 50 mg of AG-1 X-2 effected removal of 99% of the SDS without significant loss of membrane proteins or membrane lipids (Fig. 6). Since there was little difference in the amount of sodium dodecyl sulphate removed upon increasing the amount of AG-1 X-2 to 200 mg,

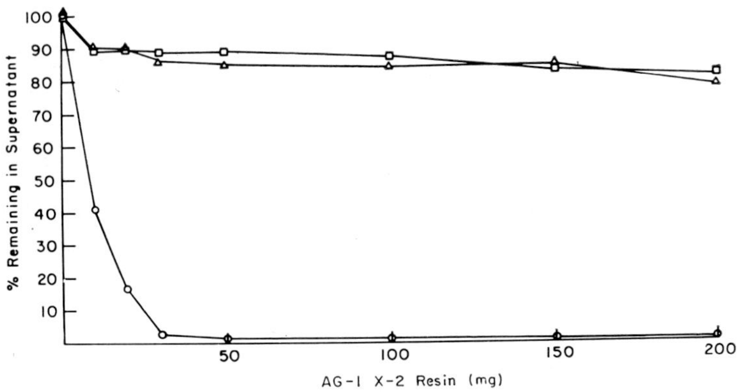

FIG. 6. Removal of ^{14}C sodium dodecyl sulphate by ion exchange resin. *Bacillus megaterium* KM ghosts at a concentration of 2 mg/ml protein were mixed in a final volume of 10 ml containing 50 mM Tris-acetate pH 7·0, 7·5 M urea and 0·2% sodium dodecyl sulphate containing 400,000 cpm ^{14}C sodium dodecyl sulphate. This mixture was stirred 15 min at 0° and 1 ml aliquots of this mixture were added to tubes containing the indicated amounts of AG-1 X-2 acetate resin. The resin and the solubilized membrane mixture were stirred at 0° for 20 min and then centrifuged at 2000 × g for 3 min to sediment the resin. Aliquots of each tube were then taken for analysis of ^{14}C-SDS, protein, and lipid phosphate. ○——○, ^{14}C-SDS; △——△, protein; □——□, lipid phosphate.

the ratio of resin to ghost protein used in further studies was 25:1. Under these conditions over 85% of the membrane proteins and membrane lipids was recovered in the supernatant upon removal of the resin.

Reaggregation of membrane components

Reaggregation of the membrane proteins occurred during the second step of removal of SDS by dialysis, namely, during the step to remove urea. Reaggregation was found to be dependent upon the presence of magnesium ion and the degree of reaggregation was a function of the magnesium ion concentration.

The extent of reaggregation as a function of $MgCl_2$ concentration is shown in Fig. 7. In the absence of magnesium there was little reaggregation, that is, there was little material which was sedimentable by centrifugation for 1 hr at 105,000 × g. The optimal magnesium concentration for reaggregation was found to be approximately 20 mM. Higher concentrations of magnesium did

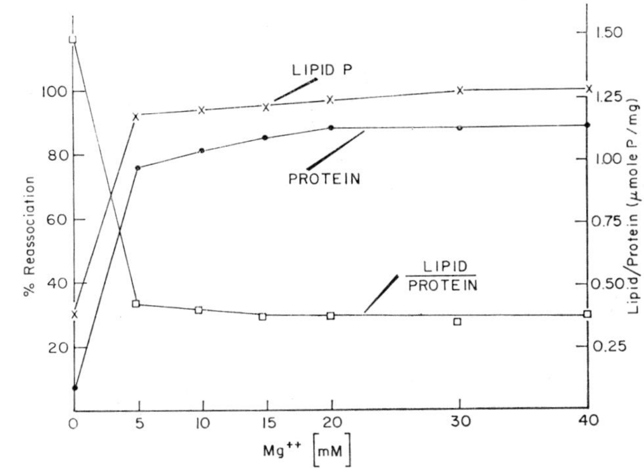

FIG. 7. Dependence of reassociation of urea-SDS solubilized membrane components on Mg^{++} ion concentration.

not result in an increase in the level of reaggregation. The extent of reaggregation of lipids is slightly greater than that of proteins, resulting in a pellet that has a higher ratio of lipid phosphorus to protein than that found in the original membranes (0·31). As shown in Fig. 7, the ratio of lipid phosphorus to proteins changes as the magnesium ion concentration increases, and approaches that of the original membranes.

In order to investigate the degree of lipid-protein interactions in the reaggregated pellet, the behaviour of the original ghost membrane and the reaggregated pellet was studied by means of sucrose density centrifugation. As shown in Fig. 8, not unexpectedly the buoyant density of the reaggregated pellet, which is relatively lipid rich, was slightly less than that of the original membranes. Of particular interest was the finding that virtually all of the lipid co-sedimented with the membrane proteins, thus indicating that during the reaggregation process the membrane proteins are interacting with the lipids as well as with each other. If this were not the case, and lipids and proteins were reaggregating independently, one would observe two peaks, one for lipids with a buoyant density of approximately 1·02 gm/cc and one for proteins with a buoyant density of 1·35 gm/cc.

The morphology, as well as the functional activity of the reaggregated pellet in phospholipid biosynthesis, was investigated. Electron microscopic examinations using both negative staining and thin sectioning procedures failed to reveal a meaningful proportion of structures with morphology typical of the original membranes. Thus, although during the reaggregation process the lipids and proteins interact, they do not do so in a manner that leads to formation of structures with the trilaminar appearance of the original membrane of the ghosts. Furthermore, despite the fact that a great number of variables in the basic dialysis procedure were studied (temperature, pH, ionic

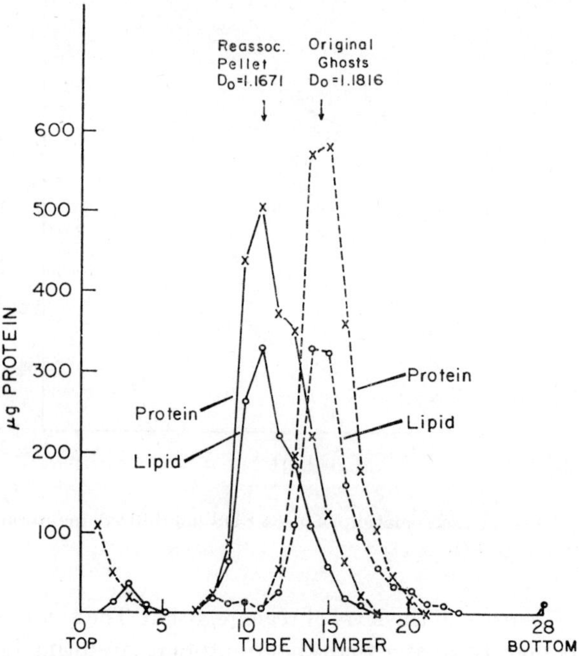

FIG. 8. Sucrose density centrifugation of reaggregated pellet and of ghosts. Into separate centrifuge tubes containing 17 ml of sucrose in a concentration gradient from 30–60% a suspension of reaggregated pellet (2·5 ml, 2·15 mg) or ghosts (2·5 ml, 2·5 mg) was layered. The tubes were centrifuged for 18 hr at 26,000 rpm in a Spinco SW-27 rotor. Fractions of 20 drops each were collected starting at the top of the gradient. Protein and lipid was assayed in each fraction as indicated in the "Experimental Procedure."

strength), in all cases the reaggregated pellet was found to be devoid of any enzymatic activity for the synthesis of phosphatidylglycerol, phosphatidylserine, or CDP-diglyceride.

Recovery of phosphatidylglycerol synthetase activity

On the assumption that these negative results might be caused by the presence of residual SDS in the reaggregated pellet, the alternative method for SDS removal was employed by using AG-1 X-2 acetate resin. Initial experiments revealed that, after dissociation and reassociation of the membrane components, up to 50% of phosphatidylglycerol synthetase activity could be recovered. Resin treatment of intact membranes did not inhibit phosphatidylglycerol synthetase activity (a slight increase in specific activity was observed). Varying the concentration of resin from 25 to 200 mg per 2 mg solubilized membrane protein also had no effect on the recovery of enzymatic activity.

The optimal concentration of urea necessary for maximal recovery of phosphatidylglycerol synthetase activity after solubilization and removal of SDS was studied (Fig. 9). In 5·0 to 6·5 M urea, the percent recovery of activity was low (about 4%). Compared to the activity in 6·5 M urea, there was a 2·5-fold increase

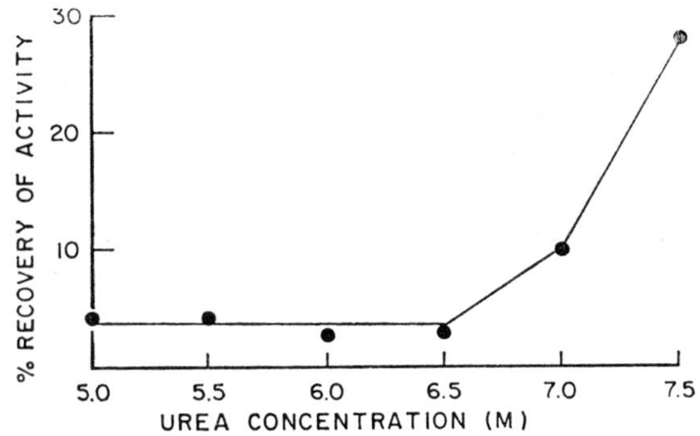

FIG. 9. Effect of urea concentration during solubilization on recovery of PG synthetase activity after reaggregation.

in 7 M urea and a 7-fold increase in 7·5 M urea. Concentrations of urea greater than 7·5 M could not be employed because of solubility limitations.

The order of addition of the solubilizing components, urea and SDS, to the membranes was found to be crucial for maximal recovery of enzymatic activity (Table 1). When membranes were added to urea-SDS, the recovery of

Table 1. *Effect of the order of addition of urea in SDS on phosphatidylglycerol synthetase activity in reaggregated pellet*

Order of addition	Relative enzyme activity
1. Untreated membranes	(100)
2. Urea, SDS, 0 min, membranes	4
3. Membranes, SDS, 0 min, urea	8
4. Membranes, urea, 0 min, SDS	20
5. Membranes, urea, 15 min, SDS	43
6. Membranes, urea, 30 min, SDS	39
7. Membranes, urea, 60 min, SDS	32

Components of the solubilization solution were added in the order indicated. In experiment 2, urea and SDS were pre-mixed and then membranes added. In experiments 4–7, the membrane solution in urea was allowed to stir for the indicated time before SDS was added. In all experiments the mixtures were then allowed to stir for 30 min at 0°, and then subjected to SDS removal by resin treatment and urea removal by dialysis against 0·05 M Tris·HCl.

enzymatic activity was only 4%. Upon addition of SDS to the membranes followed by urea, the recovery was still low, although somewhat increased. The highest recovery was obtained when urea was first added to the membranes, followed by SDS. Moreover, the recovery could be increased significantly if membranes were stirred in the presence of urea for some time prior to the addition of SDS. As seen from the data in Table 1, the highest recovery of enzymatic activity was obtained if membranes were treated with urea for 15 min prior to the addition of SDS.

Having established that enzymatic activity for one of the enzymes of phospholipid synthesis, phosphatidylglycerol synthetase, was detectable after urea-SDS treatment it was essential to show that this activity did not reside exclusively in the small fraction (3–8%) of membranes that resisted solubilization to components of relatively low molecular weight. The two most direct means of investigating this possibility, long term centrifugation in urea-SDS or gel filtration in urea-SDS presented serious experimental difficulties. For reasons not yet understood, when the membranes solubilized in urea-SDS were subjected to centrifugation followed by dialysis, *neither* the supernate *nor* the pellet had any enzymatic activity. This was true even if the supernate and the pellet were recombined prior to the dialysis step. In contrast, when the membranes solubilized in urea-SDS were centrifuged after SDS removal, 35 to 50% of the original phosphatidylglycerol synthetase activity was recovered following removal of urea by dialysis. Thus, apparently high speed centrifugation in the presence of SDS somehow leads to enzyme inactivation. The other direct method, gel filtration in urea-SDS, was precluded because using this solvent it was necessary to elute the column at room temperature. These conditions lead to inactivation of the enzyme.

GHOSTS

[15·4 mg protein (100%); Total activity 100%]

↓

1. Solubilization
2. AG-1 X-2 treatment
3. Dialysis

↓

MEMBRANE SOLUTION

[13·2 mg (86%); T.A. 35%]

↓

100,000 × g

↓ ↓

PELLET [3·2 mg (21%); T.A. 15%]

SUPERNATANT FRACTION [9·6 mg (62%); T.A. 22%]

FIG. 10. Fractionation procedure for study of the distribution of phosphatidylglycerol synthetase.

For this reason, a less direct method for the study of the distribution of enzymatic activity was employed. The results of this study are outlined in Fig. 10. Membranes were solubilized in urea-SDS, treated with AG-1 X-2 resin, and subjected to dialysis (in the absence of Mg^{++} ion) to remove urea. As shown in Fig. 10, this preparation contained 35% of the original phosphatidylglycerol synthetase activity. When this solution was subjected to centrifugation a pellet comprising 21% of the total proteins was obtained. These results substantiated the earlier study on the effect of Mg^{++} upon reaggregation (Fig. 7) that showed that a relatively small fraction of the total membrane proteins reaggregated in the absence of Mg^{++} ion. It is noteworthy that this particulate fraction also contained the phosphatidylglycerol synthetase activity.

Of particular interest, however, was the distribution of enzymatic activity in the supernatant fraction. This fraction contained two-thirds of the total proteins and more than half of the total recovered enzymatic activity. The molecular size distribution of the proteins of the supernatant was studied by

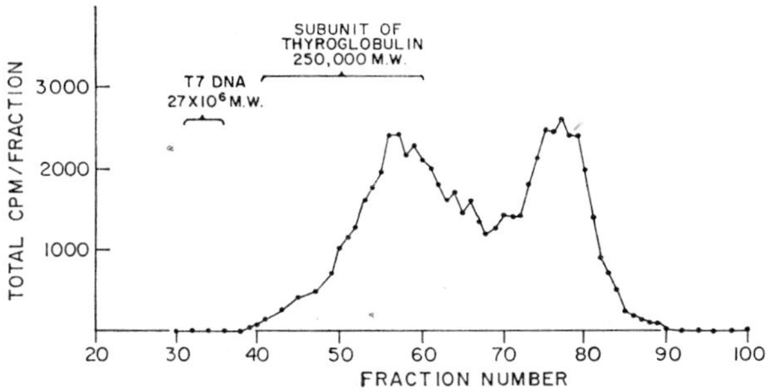

FIG. 11. Sepharose 4B chromatography in SDS-urea of supernatant membrane components obtained after SDS-urea removal. The supernatant fraction was prepared as described in Fig. 10. It was then dissolved and chromatographed in the presence of urea-SDS.

gel-filtration in urea-SDS, that is, under the same conditions under which they were originally solubilized. The results (Fig. 11) indicate that in urea-SDS, these proteins, from which phosphatidylglycerol synthetase activity can be recovered, have a molecular weight of 100,000 or less. This finding has been confirmed by SDS-polyacrylamide gel electrophoresis (data not shown).

Since upon dialysis, even in the absence of Mg^{++} ion, some of the protein aggregated to an extent sufficient to allow it to be sedimented at $105{,}000 \times g$, it seemed likely that in an aqueous environment the supernatant proteins might also aggregate to sizes greater than 100,000. Therefore, a sample of the supernatant fraction was subjected to gel filtration in 50 mM Tris buffer as shown in Fig. 12. It is evident that the protein and phospholipid profiles are coincident and very broad, indicating the presence of molecular species ranging from over 70,000 to at least 700,000. In contrast, the distribution of enzymatic activity is skewed towards the high molecular weight region, although a significant fraction of it is of molecular weight 700,000 or less.

DISCUSSION

During the past decade it has become clear that a variety of supramolecular structures can be resolved into their individual molecular components and that these components can, under appropriate conditions, self assemble *in vitro* to produce an aggregate that is morphologically and functionally identical with the original structure. Perhaps the most complex structure (in terms of numbers of different proteins) yet shown to undergo self assembly is the ribosome. One of the subunits of the ribosome (30 S) contains at least 20 proteins and these have been dissociated from the core RNA and shown to reassemble on to it, producing a biologically active supramolecular aggregate (Kurland, 1970).

There is increasing evidence that a large proportion of the lipids in membranes exist in the form of a bilayer (cf. Hendler, 1971). If this proves to be the case, then a reasonable, albeit oversimplified, analogy can be drawn between the 30 S ribosome subunit, with a core of RNA, and the membrane, with a "backbone" of lipids. It is clear that the thermodynamically most favourable

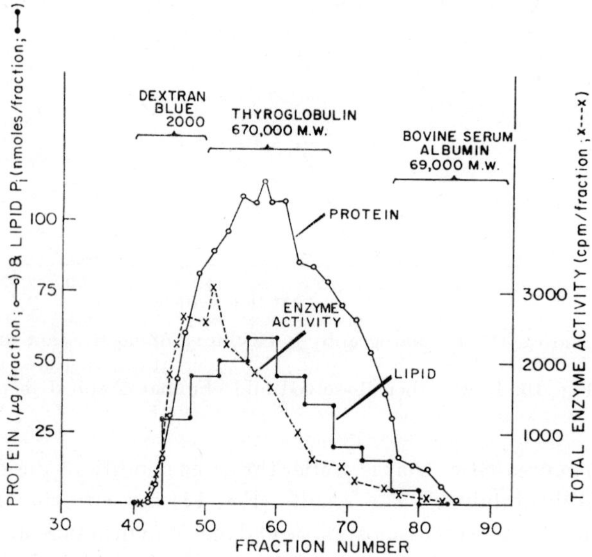

FIG. 12. Sepharose 6B chromatography of supernatant membrane components after SDS-urea removal. The supernatant fraction was prepared as described in Fig. 10 and chromatographed in 0·05 M Tris·HCl.

form of lipids in an aqueous environment is the bimolecular leaflet, and that lipids isolated from natural membranes can readily be reformed into bilayers. What is not clear, however, is how the many proteins of the membrane can be isolated in forms that will be amenable to renaturation and reassociation with themselves and with phospholipids. Clearly, here the analogy with the ribosome is weak because in the case of membranes one is dealing with a number of protein species that differ greatly in their enzymatic function and perhaps in their structure, whereas in the ribosome all the proteins are, in one way or another, involved in the single function of protein biosynthesis.

In any event, despite the paucity of knowledge of the physical properties of bacterial membrane proteins and the forces that are involved in their interaction we are investigating the possibility that the membrane, or at least some functional components of the membrane, might be reassembled *in vitro* after complete dissociation of the original structure.

After unsuccessful preliminary attempts with non-ionic detergents we chose SDS as the primary disaggregating reagent because of a wealth of information on its efficacy in disrupting membranes (cf. Bishop, Rutberg & Samuelsson,

1967) and on the nature of its interaction with proteins (Fish, Reynolds & Tanford, 1970). However, in most earlier studies on the effect of SDS on proteins it was not possible to obtain native, enzymatically active proteins upon removal of SDS (Visser & Blout, 1971). Based on the fact that enzymes have been renatured following denaturative unfolding in urea or guanidine·HCl (Teipel & Koshland, 1971), we developed a procedure in which urea and SDS were both present during the membrane dissociation but in which the SDS was removed prior to removal of urea. After initiation of our studies such a procedure was used by Weber & Kuter (1971) successfully to renature a number of soluble enzymes not associated with membranes.

Throughout these studies we have been concerned with two aspects of this problem. On the one hand, we wished to establish conditions for the quantitatively complete disaggregation and reassociation of the membrane components and, on the other, we wished to use techniques that did not lead to irreversible inactivation of the enzymes of phospholipid synthesis. Ultimately conditions for disaggregation and reassociation must be established that fulfil both requirements; that is, the disaggregation-reassociation process must be quantitative and this reaggregated membrane must have full enzymatic activity. However, at the present stage of our studies, for lack of information we must, for the most part, discuss these two aspects independent of each other.

Using SDS in 7·5 M urea it was possible to establish that the solubilization of ghosts was maximal when the membrane protein and SDS concentrations were equivalent on a weight basis. Under these conditions (2 mg membrane protein and 2 mg SDS per ml) less than 3% of the membrane protein was sedimentable by centrifugation for 20 hr at 105,000 $\times$ g. Moreover, gel filtration on Sepharose 6B or 4B in urea-SDS showed that over 90% of the membrane proteins were in a form with molecular weight less than 250,000. The remaining 8–10% was very large in molecular weight and may represent residual membrane fragments.

As noted above, the basic rationale of the reassociation process was initial removal of the SDS from the membranes solubilized in urea-SDS, followed by removal of urea. Dialysis was found to be an effective procedure to remove greater than 97% of the SDS, although the process was slow. The more serious limitation of this technique, however, was that the enzymatic activities of phospholipid synthesis that we studied could not be detected in the membrane components after subsequent reaggregation. However, using an ion exchanger, AG-1 X-2, it was possible to achieve rapid and virtually complete removal of SDS. Under these conditions up to 50% of the original phosphatidylglycerol synthetase activity was recovered after subsequent removal of urea.

A number of parameters in the reaggregation process subsequent to removal of the SDS have been studied. In the absence of Mg^{++} ion only 10 to 20% of the protein and 30–40% of the lipid are sufficiently aggregated to be sedimentable. Upon addition of Mg^{++} ion over 85% of the protein and 95% of the lipid are readily sedimentable. The resulting aggregate has a ratio of lipid to protein somewhat greater than that of the original ghosts and, upon sucrose gradient centrifugation the aggregate bands at a buoyant density slightly less than that of the original ghosts. Of particular interest is the fact that the lipid and protein in the aggregate co-sediment, indicating that there is, in some manner, either direct or indirect interaction of all of the lipids with all of the proteins. It is clear, however, that whatever the mode of interaction, it is not identical to that found in the original membrane because electron microscopy of the reaggregate failed to reveal any orderly, bilayer-containing structures. Rather, most of the preparation appeared to lack any recognizable regularity in structure.

With regard to the second aspect of this study, namely, the retention of enzymatic activity of phospholipid synthesis, it was found that, if ion exchange resin was used to remove the SDS, the reaggregate contained up to 50% of the original activity of the ghosts in phosphatidylglycerol synthesis. The other enzymes of phospholipid synthesis known to be present in the ghosts have not yet been studied in the reaggregate. Interestingly, the urea concentration in the urea-SDS was important for recovery of activity, and only when it was 7 M or greater was there significant recovery of phosphatidylglycerol synthetase activity. This suggests that it may be essential to have the protein in the random coil form for renaturation. Another important factor in recovery of activity was the order of addition of membrane, SDS and urea. Only when the membranes were first treated with urea, followed by SDS, was there maximal recovery of activity.

Because a small fraction (3–8%) of the total membrane protein remains insoluble upon treatment with urea-SDS it was essential to show that the phosphatidylglycerol synthetase had indeed been solubilized and was not concentrated in this minor insoluble fraction. The most direct proof of this, by centrifugation of the solubilized membrane in urea-SDS followed by assay of the supernatant fraction and the pellet, failed for unknown reasons because *both* the supernate and the pellet, alone or combined, were enzymatically inactive. Therefore, the distribution of enzymatic activity in the fractions resulting from sequential removal of SDS and urea in the absence of Mg^{++} ion was studied. From these results, as well as gel filtration studies on the supernatant fraction, it is clear that the majority of the phosphatidylglycerol synthetase activity is solubilized by urea-SDS and is not present in the small fraction of insoluble residue.

From these chemical and enzymatic studies our current view on the state of the proteins and lipids during the disaggregation process is as follows: Upon treatment with urea-SDS, *Bacillus megaterium* ghosts disaggregate to components of molecular weight less than 250,000. No greater than 8% of the proteins are found in larger molecular weight aggregates. In urea-SDS the lipids interact with either a select class of relatively low molecular weight proteins or, more likely, exist as independent aggregates. Upon removal of SDS and urea in the presence of Mg^{++}, greater than 85% of the protein and lipid co-reaggregate into a readily sedimentable form. In the absence of Mg^{++}, one-fourth of the protein aggregates to a sedimentable form, whereas the remaining three-fourths is found in a "soluble" form with all protein and lipid species having a molecular weight less than 700,000. Phospholipid appears to interact uniformly with these aggregates of M.W. 700,000 or less.

Thus, there appear to be two stages of aggregation. The majority of the completely solubilized, low molecular weight proteins aggregate with the lipids upon removal of urea-SDS in the absence of Mg^{++} to form aggregates of intermediate molecular weight; in the presence of Mg^{++} large aggregates of protein and lipid that are readily sedimentable are formed. Clearly, the present studies do not provide any clear-cut evidence on whether or not membranes of *Bacillus megaterium* can be reassembled *in vitro* but hopefully they do provide a foundation for further investigations.

SUMMARY

This study was undertaken in order to explore the possibility that under appropriate conditions the ghosts of *Bacillus megaterium* could be completely disaggregated and then reassociated to produce a membranous structure with the morphology and functional properties of the original membrane. Using mixtures of urea and sodium dodecyl sulphate, conditions for the essentially complete disaggregation of the membrane were established. Under these conditions virtually all the proteins of the membranes were less than 250,000 daltons. Upon the sequential removal of SDS and urea from the solubilized membrane, reaggregation was observed. In the absence of Mg^{++} ion the reaggregation was only partial and the majority of the proteins and phospholipids associated to form aggregates of less than 700,000 daltons. In the presence of Mg^{++} ion greater than 85% of the phospholipids and proteins coaggregated into a readily sedimentable form. Although this aggregate lacked the trilaminar morphology typical of the original membrane it did contain up to 50% of the activity of the original ghosts in the biosynthesis of phosphatidylglycerol.

ACKNOWLEDGEMENTS

It is a pleasure to acknowledge the expert technical assistance of Mrs B. J. Earles. We are also indebted to Mr G. Decker and Dr J. W. Greenawalt for electron microscopic studies.

This work was supported by National Institutes of Health Grant No. 5 RO1 AI-06888-07 to W.J.L. K.A.D. and R.A.M. are recipients of N.I.H. Postdoctoral Fellowships, 5 FO 2 GM 35573-02, and 1 FO 2 GM 50658-01, respectively.

REFERENCES

Bartlett, G. R. (1959). *J. biol. Chem.* **234**, 466.

Bishop, D. G., Rutberg, L. & Samuelsson, B. (1967). *Eur. J. Biochem.* **2**, 454.

Chang, Y. Y. & Kennedy, E. P. (1967). *J. Lipid Res.* **8**, 447.

Devor, K. A., Makula, R. A., Patterson, P. H., Kennedy, M., Decker, G., Greenawalt, J. W. & Lennarz, W. J. (1972). *J. biol. Chem.* **297**, 1288.

Duesberg, P. H. & Rueckert, R. R. (1965). *Analyt. Biochem.* **11**, 342.

Fish, W. W., Reynolds, J. A. & Tanford, C. (1970). *J. biol. Chem.* **245**, 5166.

Gould, R. M. & Lennarz, W. J. (1970). *J. Bacteriol.* **104**, 1135.

Hendler, R. W. (1971). *Physiol. Rev.* **51**, 66.

Kates, M. (1964). *Adv. Lipid Res.* **2**, 17.

Kurland, C. G. (1970). *Science*, **169**, 1171.

Lennarz, W. J. (1966). *Adv. Lipid Res.* **4**, 175.

Lowry, O. H., Rosebrough, N. J., Farr, A. L. & Randall, R. (1951). *J. biol. Chem.* **193**, 265.
Patterson, P. H. & Lennarz, W. J. (1970). *Biochem. biophys. Res. Commun.* **40**, 408.
Patterson, P. H. and Lennarz, W. J. (1971). *J. biol. Chem.* **246**, 1062.
Steiner, R. F. & Edelhoch, H. (1961). *J. Am. chem. Soc.* **83**, 1435.
Teipel, J. W. & Koshland, D. E. (1971). *Biochem.* **10**, 792.
Visser, L. & Blout, E. R. (1971). *Biochem.* **10**, 743.
Weber, K. & Kuter, D. J. (1971). *J. biol. Chem.* **246**, 4504.
Weibull, C., Beckman, H. & Bergstron, L. J. (1959). *J. gen. Microbiol.* **20**, 519.

SYNTHESIS AND EXCHANGE OF PHOSPHOLIPIDS WITHIN BRAIN AND LIVER CELLS

By R. M. C. DAWSON, F. B. JUNGALWALA, E. MILLER and W. C. McMURRAY

Biochemistry Department, Institute of Animal Physiology, Babraham, Cambridge, England

The demonstration by Wilgram & Kennedy (1963) and Schneider (1963) that phosphatidylcholine synthesis from CDP choline or deoxy CDP choline in liver homogenates was selectively localized in the microsomal fraction, led to the suggestion (Dawson, 1966) that newly synthesized phospholipids are transferred to other cellular membranes and organelles by a process of exchange diffusion of lipoproteins. It was postulated that this exchange could occur either through the intermediary role of soluble cytoplasmic lipoproteins or through a continuum of membranous lipoproteins throughout the cell. The whole "network could then be considered to be in dynamic equilibrium; utilization of phospholipid at a given location, e.g. for the formation of new membrane or by phospholipid hydrolysis would result in a new flow of phospholipid towards that area". The most convenient model for experimental verification of such a postulate was clearly the liver cell, since it is known that in marked contrast to the *in vitro* results obtained by Wilgram & Kennedy (1963) with isolated subcellular fractions, the mitochondrial phosphatidylcholine in the intact liver becomes labelled at a rate which is only slightly less than that of the endoplasmic reticulum (McMurray & Dawson, 1969). In addition, it can be shown that the slow synthesis of phosphatidylcholine from CDP choline by liver mitochondria observed by Wilgram & Kennedy (1963) can be entirely attributed to the microsomal contamination of these organelles (McMurray & Dawson, 1969). Thus, in a system in which incremental amounts of microsomal protein were added to a constant amount of mitochondria, the plot relating synthesis of phosphatidylcholine from CDP choline to microsomal protein extrapolated to zero providing due allowance was made for the microsomal contamination of the mitochondrial fraction as determined by NADPH-cytochrome c reductase assays (Fig. 1). It was shown by three groups working independently (Wirtz & Zilversmit, 1968; McMurray & Dawson, 1969; and Akiyama & Sakagami, 1969) that rat liver microsomes could exchange phospholipids with liver mitochondria when incubated together providing a soluble "macromolecular" component of the supernatant fraction was present. In our own experiments microsomal phosphatidylcholine was specifically labelled by incubation with CDP-[Me-^{14}C]choline and after incubation with unlabelled mitochondria, the labelling of the phosphatidylcholine

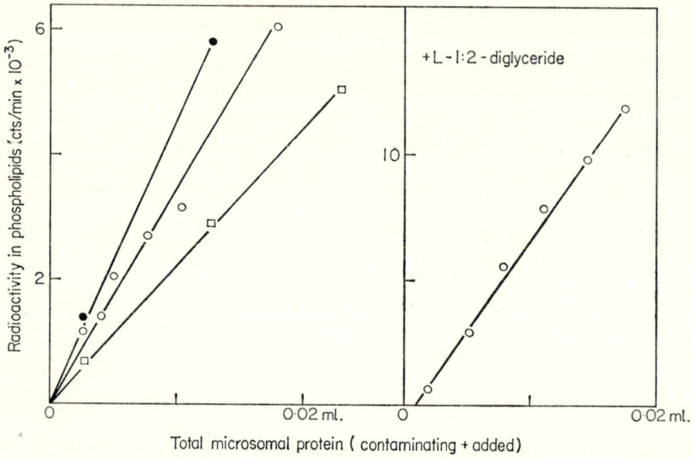

FIG. 1. The uptake of radioactivity from CDP-[1,2-^{14}C]choline into phospholipids on adding increments of microsomes to a constant amount of liver mitochondria. The figure on the left represents three independent experiments performed in the absence of added L-1:2 diglyceride whereas that on the right shows an experiment in its presence. The first point on all lines represents the synthesis by the mitochondrial fraction alone where the contaminating microsomal protein has been computed from NADPH-cytochrome c reductase assays (McMurray & Dawson, 1969).

in both subcellular fractions was followed after these had been reisolated by centrifugation. It is clear that there had been a rapid and substantial transfer of phospholipid from the microsomes to the mitochondria so that by the end of two hours the phosphatidylcholines were almost in radioactive equilibrium (Fig. 2). This transfer was not accompanied by a corresponding movement of the endoplasmic reticulum marker enzyme NADPH-cytochrome c reductase to the mitochondria or of cytochrome oxidase from the mitochondria to the reisolated microsomes so the movement of phospholipid cannot be ascribed to heterocoagulation of the two particulate fragments. Moreover, when labelled mitochondria were incubated with unlabelled microsomes there was a corresponding transfer of phosphatidylcholine showing that the process involved is one of true exchange. All workers have found that a non-dialysable supernatant fraction is obligatory to such transfers and Wirtz (1971) has gone on to partially purify one of the protein factors involved. It seems to specifically enhance the exchange of phosphatidylcholine and phosphatidylinositol, which accounts for the observation that the exchange of these phospholipids between ^{32}P-labelled liver microsomes and unlabelled mitochondria is faster than that of phosphatidylethanolamine (McMurray & Dawson, 1969). Does such an exchange occur *in vivo*? It can be argued that this may not be so since the spatial relationship between membrane fractions in free suspension is different to that existing between the endoplasmic reticulum and mitochondria in the intact cell. To investigate this, Jungalwala & Dawson (1970) isolated hepatic cells from rat liver and conducted pulse-chase experiments with labelled phospholipid precursors. It was found that after labelling the cells and washing, incubation with inactive precursor chase caused a decrease in the specific radioactivity of the microsomal phospholipid. This was of a magnitude which could not be accounted for either by a selective loss of highly labelled phospholipid by phospholipase action or by its dilution with newly synthesized phospholipid. Moreover, there was generally an increase in the specific radioactivity of the same phospholipid in the mitochondrial fraction when the cells were incubated with inactive precursor. These observations are entirely consistent with the idea of an exchange of phospholipid between the two membranes in the intact structure of the liver cell.

It cannot, of course, be assumed that similar phospholipid exchanges can occur in all tissues and between all diverse types of membranes. In some ways the liver is perhaps an atypical organ regarding phospholipid turnover since *de novo* phospholipid synthesis is required not only for membrane turnover but also for supplying materials, especially phosphatidylcholine, for secretion of the plasma liproproteins and of bile. It is conceivable that phospholipid exchange processes might be connected with such functions. On the other hand the brain is an organ which, as far as is known, does not secrete phospholipids and has a variety of membranes which can now be separated in a state of comparative purity by differential and gradient density centrifugation techniques (Gray & Whittaker, 1962; Eichberg, Whittaker & Dawson, 1964). Marker enzyme assays for endoplasmic reticulum contamination (we have found that in brain NADPH-cytochrome c reductase and NAD nucleosidase are satisfactory while RNA estimation gives an approximate indication of rough endoplasmic reticulum) of the isolated organelles suggests that the mitochondria

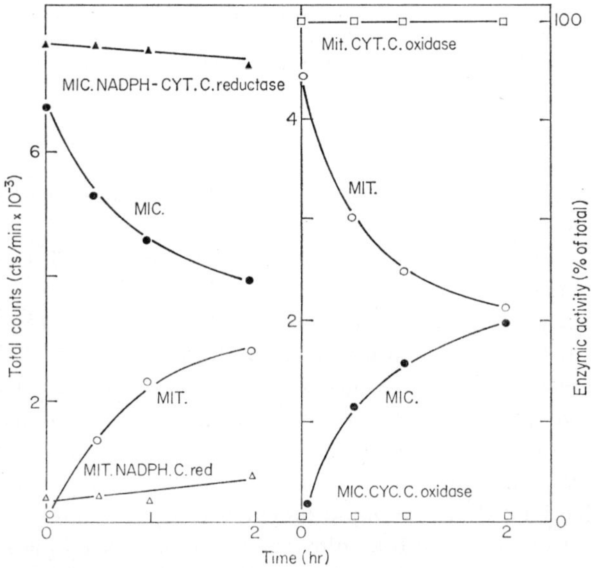

FIG. 2. Exchange of ^{14}C-choline labelled phospholipid between liver microsomes and mitochondria incubated together with a supernatant fraction. After incubation the two fractions were reisolated and assessed for radioactivity and marker enzymes. In the left-hand figure choline-labelled microsomes (●) were incubated with unlabelled mitochondria (○), whereas on the right, choline-labelled mitochondria (○) were incubated with unlabelled microsomes (●) (McMurray & Dawson, 1969).

are contaminated with about 15–20% microsomal protein while the synaptosomes or nerve ending fraction contains as much as 25–30%. Myelin isolated by these methods is heavily contaminated (50–60% contaminating protein) although this is highlighted by the low protein content of this membrane. Successive washing of mitochondria with sucrose solution can be used to reduce

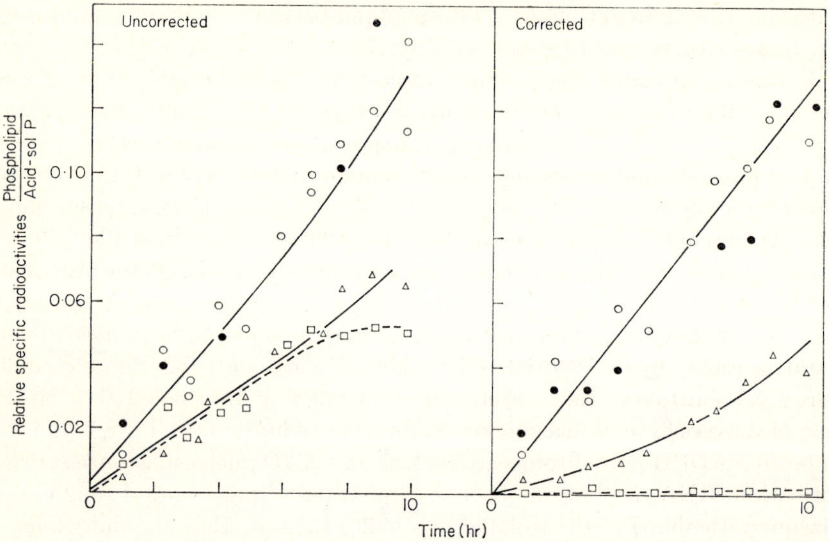

FIG. 3. The incorporation of ^{32}P into the phospholipids of some subcellular fractions of guinea-pig brain after intraperitoneal injection of [^{32}P]Pi: ○ microsomes, ● mitochondria; △ synaptosomes; □ myelin. In the right-hand figure the values for mitochondria, synaptosomal and myelin turnover have been corrected for contamination by labelled microsomal phospholipids as computed from NADPH-cytochrome c reductase assays.

this contamination, but with synaptosomes this results in appreciable fragmentation.

An example of the influence of this contamination on investigations of phospholipid turnover can be seen in Fig. 3. After intraperitoneal injection of ^{32}P into guinea pigs, the incorporation of ^{32}P into the phospholipids of the isolated myelin fraction was appreciable and equal to that of the synaptosomes. On correction for the presence of endoplasmic reticulum, the incorporation into the synaptosomes was greatly reduced and that into myelin became virtually nil. Recent studies in our laboratory have shown that although the myelin membrane does not turn over appreciably at short periods after intraperitoneal injections of phospholipid precursors, a well-defined incorporation of isotope is seen even in the adult animal when the isotope is injected intracerebrally (Jungalwala & Dawson, 1971). Thus, like the liver, all cerebral membranes examined seem to be constantly turning over the phospholipid complement of their membranes even though at somewhat different rates. Although mitochondria isolated from brain appear to have a limited capacity for *de novo* phospholipid synthesis on incubation with labelled ^{32}Pi, it can be seen that on adding microsomal increments and extrapolating to zero microsomal concentration (contaminating + added) the incorporation into phosphatidylinositol, phosphatidylcholine, and phosphatidylethanolamine can be entirely ascribed to contamination of the organelles with endoplasmic reticulum (Fig. 4). On the other hand, cardiolipin the mitochondrial phospholipid, can be synthesized as can phosphatidic acid. Using the same method of assessment it has been shown that brain mitochondria are incapable of forming phosphatidylcholine from [Me-^{14}C]-choline (either the nucleotide mediated pathway or the Ca^{2+} catalysed exchange reaction) phosphoryl [Me-^{14}C]-choline or CDP [Me-^{14}C]-choline: all the synthesis in the isolated mitochondrial fraction can be ascribed to contamination with microsomes. Osmotic shocking of the mitochondria to improve their permeability to substrates did not change this picture.

As regards synaptosomes the position is not so clear cut. The incorporation into most of the phospholipids from ^{32}Pi is very sluggish and the system shows little response to added microsomes, possibly because the synaptosomes

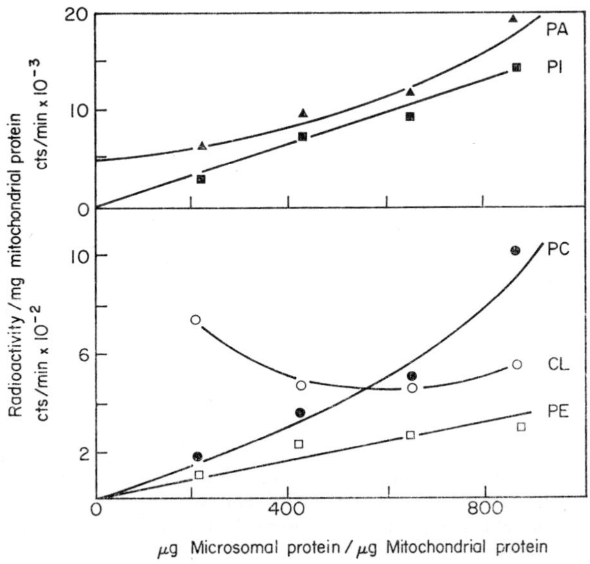

FIG. 4. Uptake of radioactivity from [^{32}P]Pi into individual phospholipids on adding increments of brain microsomes to a constant amount of brain mitochondria (Miller & Dawson, 1972). ▲ = phosphatidic acid; ● = phosphatidylcholine; □ = phosphatidylethanolamine; ■ = phosphatidylinositol; ○ = cardiolipin. The first point on each curve represents microsomal contamination of the mitochondria computed from NADPH-cytochrome c reductase assays.

cannot produce sufficient external ATP (γ^{32}P). The main phospholipid labelled by the preparation is phosphatidylinositol and by the same tests as we have used for mitochondria the synthesis can be entirely attributed to microsomal contamination of the synaptosomes. Using the already phosphorylated substrates, phosphoryl [^{14}C-Me]choline or CDP-[^{14}C-Me]choline, most of the synthesis of synaptosomal phosphatidylcholine can be attributed to contaminating microsomes, but because of the extent of the contamination the extrapolation to zero microsomal concentration is too uncertain to eliminate the possibility of some limited synthesis by the synaptosomes themselves. However, if the synaptosomes, after labelling from CDP-[Me^{14}C]choline, were disrupted and subfractionated on a sucrose density gradient, then maximum labelling occurred in the soluble fraction and in the membrane fractions of the synaptic vesicles rather than the microsomes (Fig. 5). If the synaptosomal

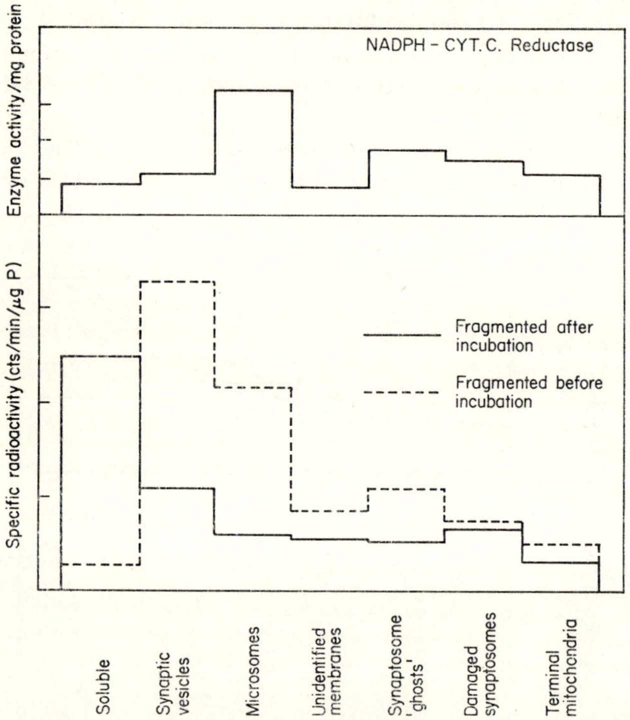

FIG. 5. Labelling of phospholipids of synaptosomal membranes on incubation of synaptosomes with CDP-[Me-^{14}C]choline either before or after disruption of the synaptosomes by osmotic shocking. Upper figure shows the distribution of NADPH-cytochrome c reductase in the membrane fractions.

fractions were osmotically shocked before incubation with CDP-[^{14}C]choline followed by a subfractionation then little labelling was observed in the phospholipids of the soluble fraction, but again maximal-specific activity was observed in the membrane fragments of the synaptic vesicles (Fig. 5). The ability of synaptic vesicles to form labelled phospholipid from CDP-[Me-^{14}C]-choline was again clearly indicated by experiments in which graded increments of microsomes were added to a constant amount of synaptic vesicles. These show that synthesis in the latter cannot all be accounted for by microsomal contamination as assessed by NADPH-cytochrome c reductase assays. In addition, the microsomal fraction isolated from the synaptosomes can synthesize phospholipids at a rate above that which can be computed from contaminating external microsomes. It has often been speculated that nerve endings contain a rudimentary endoplasmic reticulum and in this respect we have shown that synaptosomes contain an occluded form of NADPH-cytochrome c reductase which can be revealed only by detergent treatment and osmotic shocking. No synthesis is shown by the other synaptosomal subfractions including the intraterminal mitochondria and plasma membrane above that which can be accounted for by microsomal contamination. Thus, apart from synaptic vesicles, the only membrane present in cerebral homogenates which can synthesize phosphatidylcholine by the cytidine mediated pathway appears to be that associated with the NADPH-cytochrome c reductase activity or in other words the smooth endoplasmic reticulum. In addition, isolated myelin particles when adequately purified appear to be quite devoid of most anabolic enzymes including those synthesizing phospholipids although the phosphorylation of diphosphoinositide may be an exception (Iacobelli, 1969).

It seems likely that some cerebral membranes may be able to replace molecular moieties in their phospholipids by exchange processes which do not involve *de novo* synthesis. The outer membrane of liver mitochondria can exchange phospholipids by a deacylation-reacylation cycle (Sarzala, van Golde, De Kruyff & van Deenen, 1970). It is not known whether this process occurs in brain mitochondria, although brain tissue contains active phospholipases (Al + A2) and a reacylation system using lysolecithin as acceptor (Webster, 1970). A Ca^{2+}-catalysed non-energy requiring incorporation of labelled ethanolamine and serine into chick brain microsomal phospholipids has been demonstrated by Porcellati & di Jeso (1971) but an exchange of choline was not observed. With guinea-pig brain we have been able to demonstrate massive incorporation of choline into phosphatidylcholine which is catalysed by Ca^{2+} and does not require nucleotides or energy generation. The incorporation by this mechanism into the isolated mitochondrial fraction can all be ascribed to contamination with endoplasmic reticulum, but that into synaptosomes cannot. If the synaptosomes are subfractionated after labelled choline incorporation it is found, rather surprisingly, that the choline has been incorporated maximally into the intraterminal mitochondria whereas uptake by the microsomes is quite low. The reason for this remarkable difference between the behaviour of intra- and extra-terminal mitochondria and microsomes is unknown.

It is clear therefore, that the ability of various isolated cerebral membranes to synthesize phospholipids, which we and others have shown, would not account for the isotopic turnover of phospholipids demonstrated by experiments *in vivo*. We have therefore examined the ability of cerebral membranes to exchange phospholipid molecules under conditions which allowed such an exchange to be demonstrated using subcellular fractions of liver. One of the problems encountered was that myelin had a pronounced inhibitory effect on exchange processes even when added to a liver system. Another was that problems of non-specific aggregation were more severe and could vary with different incubation conditions, e.g. a change of pH. Nevertheless, a well-defined exchange process between a ^{32}P-labelled brain microsomal fraction and purified mitochondria could be demonstrated (Fig. 6). It is apparent from these results that assessing the contaminating endoplasmic reticulum by NADPH-cytochrome c reductase assays and assuming that its phospholipids immediately equilibrate with the added microsomes provide a reasonable exchange time-curve which starts at zero. There is no transfer of the microsomal marker enzyme to the mitochondria of a magnitude which could account for the phospholipid exchange. As in the liver system, the transfer is greatly enhanced by adding a brain supernatant fraction and the active factor is non-dialysable and heat-labile. Unfortunately, it is not yet known whether, as in

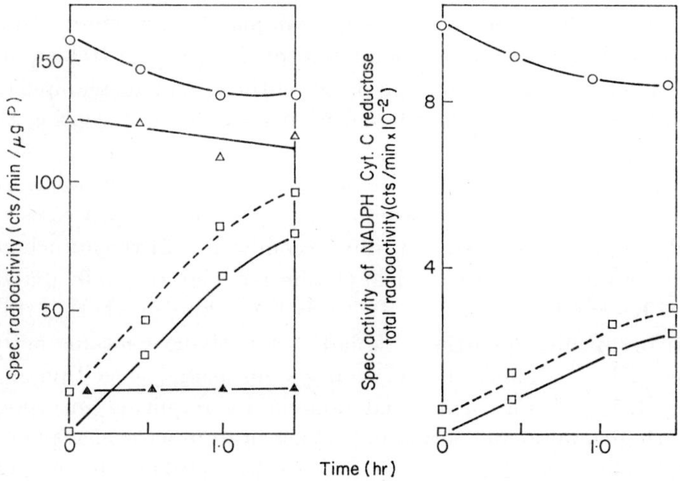

FIG. 6. Transfer of phospholipid from ^{32}P-labelled brain microsomes (○) to unlabelled brain mitochondria (□) in the presence of a supernatant fraction (Miller & Dawson, 1972). △ shows the level of NADPH-cytochrome c reductase in the reisolated microsomes and ▲ the level in the reisolated mitochondria. With the interrupted line the radioactivity in the mitochondrial phospholipids has not been corrected for contamination by microsomal phospholipids.

liver, this is due to a protein factor which specifically enhances the transfer of phosphatidylcholine and phosphatidylinositol. The protein carrier may be different since unlike liver, the brain system shows little difference between the exchange rate of these lipids and that of phosphatidylethanolamine. The concentration of the soluble supernatant factor seems to be the rate-limiting determinant of the exchange *in vitro* (Fig. 7). Its intracellular concentration must be many times the concentration achieved in *in vitro* incubations because of the dilution attendant on the preparation of the supernatant. The rate of phospholipid exchange between microsomes and mitochondria did not increase on increasing the concentration density of the particulate preparation while maintaining the concentration of the supernatant fraction constant. Therefore, the exchange appears not to depend on thermal collisions between the two particles which would presumably be more numerous in the more concentrated suspensions. This result tends to favour the idea of a protein factor functioning as a carrier in the transfer of phospholipid molecules. In this connection it can be shown that phospholipids in the supernatant become labelled on incubation

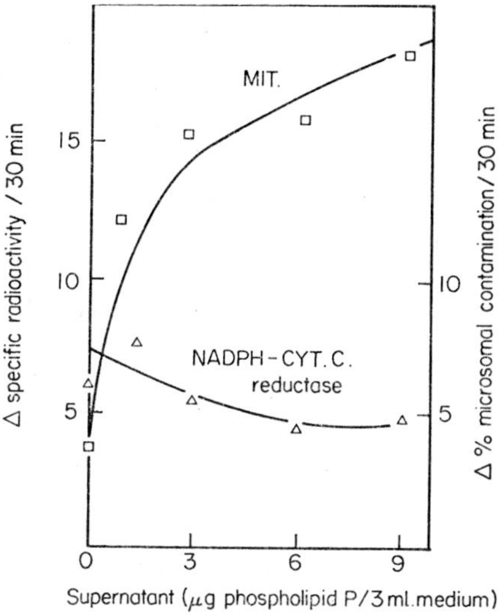

FIG. 7. The effect of a brain supernatant fraction on the transfer of phospholipids from ^{32}P-labelled microsomes to unlabelled brain mitochondria (Miller & Dawson, 1972). □ = specific radioactivity of mitochondria reisolated from incubation medium; △ = NADPH-cytochrome c reductase present in reisolated mitochondria.

with labelled microsomes (Fig. 8) and that such a labelled supernatant will transfer the label to mitochondrial phospholipids. The process involved is one of true exchange since labelled phospholipids in brain mitochondria will exchange quite readily with those in unlabelled microsomes in the presence of a supernatant fraction. Furthermore, mitochondria isolated from synaptosomes showed the same ability to exchange phospholipids as those isolated directly from an homogenate.

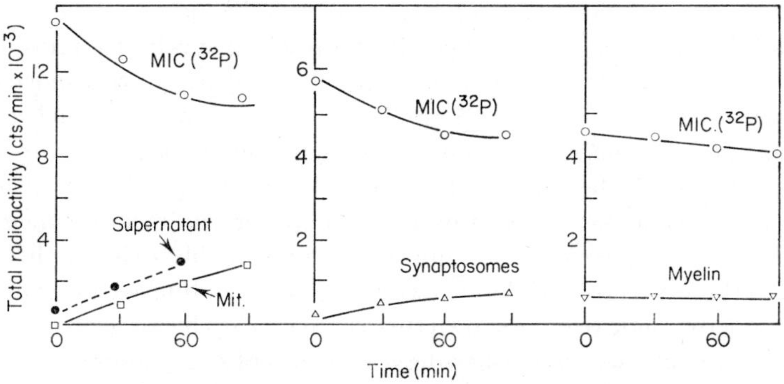

FIG. 8. Transfer of phospholipids from ^{32}P-labelled brain microsomes (○) to unlabelled mitochondria (□), supernatant (●), synaptosomes (△) and myelin (▽), (Miller & Dawson, 1972). With the reisolated mitochondria and synaptosomes the radioactivities have been corrected for contamination by microsomal phospholipids. This has not been done with the myelin fraction since the phospholipid present in the added microsomal preparation does not come into rapid equilibrium with all the microsomal phospholipid contaminating the reisolated myelin.

When synaptosomes were incubated with labelled microsomes, an exchange of phospholipids could be demonstrated although this tended to be slower than that shown by the cerebral mitochondria (Fig. 8). Subfractionation of the synaptosomes after phospholipid transfer from microsomes had occurred showed that at zero incubation time the phospholipids of the smaller particulate matter, e.g. synaptic vesicles, synaptic "microsomes" were heavily labelled. This means that either they were heavily contaminated with the added ^{32}P-microsomes or alternatively that exchange had occurred extremely rapidly before reisolation of the synaptosomes from the incubation medium was achieved. None of this smaller particulate matter appeared to increase its labelling on further incubation suggesting that equilibrium had been reached. Although the synaptosomal outer membranes showed some increase in labelling during incubation, the fraction showing the most response appeared to be the intraterminal mitochondria. Clearly, the membranes of the intact synaptosomes can show a completely variable ability to act as a receptor of phospholipid transferred from externally added microsomes. If the synaptosomes have not become "leaky" to the added microsomes during the incubation and the mitochondria are truly intraneural, this must mean that only a certain fraction of the phospholipids in the synaptosomal plasma membrane (ghosts) can act as carriers of the microsomal phospholipid transferred.

The inability of certain isolated cerebral membranes to accept phospholipids from ^{32}P-labelled microsomes was also seen in experiments in which isolated myelin was incubated with the latter—no transfer at all could be demonstrated during a ninety minute incubation period above that which could be accounted for by heterocoagulation of the two particles (Fig. 8). We are therefore left with enormous gaps in our understanding of the way in which the continual *de novo* synthesis and breakdown of membrane phospholipids in the nervous system observed in experiments *in vivo* is brought about. It seems reasonable to assess that the bulk of the *de novo* synthesis of phospholipid occurs in the endoplasmic reticulum of both glial and neuronal cells. This newly synthesized phospholipid can rapidly exchange and replenish the membranes of organelles such as mitochondria which cannot by themselves synthesize phospholipid. Some phospholipid synthesized in the neuronal body may be transported along the axon to the nerve terminal (Miani, 1964), but the kinetics of such a process based on figures available for protein movement seem to be too slow to account for the rapidity of labelling of synaptosomal phospholipid *in vivo*. The nerve ending probably makes most of its phospholipid in rudimentary membranes equivalent to endoplasmic reticulum which presumably exist in this region. Although these are not well-defined when the nerve ending is viewed under the electron microscope, we have shown that there is an occluded form of NADPH-cytochrome c reductase and others have shown that synaptosomes are capable of independent protein synthesis (Morgan & Austin, 1968). These synaptosomal membranes equivalent to endoplasmic reticulum can presumably supply the phospholipid needed for the replacement of the phospholipids present in other synaptosomal membranes and the intraterminal mitochondria. Both these latter membranes, although incapable of independent phospholipid synthesis, can rapidly acquire labelled phospholipid when incubated with a labelled brain supernatant fraction. Some doubt must remain as to whether the mitochondria can also acquire phospholipid from intraneural sources but it seems unlikely that such phospholipids can pass through both the plasma membrane of glial cells and the synaptosomal membrane. It seems more likely that the exchange observed between ^{32}P-labelled microsomes and synaptosomes is due to mitochondria contaminating the synaptosomal preparation or released by fragmentation of the latter during incubation. It is possible that the rapid *de novo* synthesis of phosphatidylcholine observed with synaptic vesicles could be connected with the resynthesis of their membranes after the release of neurotransmitter substances at the presynaptic membrane. Certain evidence suggests that a phospholipase may be active in breaking down the vesicular membrane (Heilbronn & Nilsson, 1971). However, Whittaker (1971) has obtained evidence from *in vitro* studies of the labelling of synaptical vesicles by ^{3}H-choline that, if this is so, there must be considerable reutilization of the intact phosphatidylcholine in the membrane during the regeneration of hormone-charged vesicles.

As to the mechanism of turnover of myelin phospholipids in the adult brain, we have at present to real idea. Since the myelin membrane possesses no capacity for phospholipid synthesis, newly synthesized phospholipids could be derived from either the oligodendria cells whose plasma membrane is wound round the nerve axon or in the neurone itself. In the latter case, the axolemma would presumably have to act as an intermediary membrane for the exchange since the synaptosomal membrane which is really a continuation of the axolemma is incapable of independent phospholipid synthesis. Both origins would involve an exchange process through a hydrated lipoprotein structure with minimal contact with cytoplasm. It is perhaps therefore asking a little too much to expect to demonstrate an exchange of phospholipids through a cytoplasmic intermediary as occurs between brain microsomes and mitochondria.

REFERENCES

Akiyama, M. & Sakagami, T. (1969). *Biochim. biophys. Acta* **187**, 105.

Dawson, R. M. C. (1966). In *Essays in Biochemistry*, Vol. 2. (Ed. by Campbell, P. N. & Greville, G. D.), p. 69. London: Academic Press.

Eichberg, Jun., J., Whittaker, V. P. & Dawson, R. M. C. (1964). *Biochem. J.* **92**, 91.

Gray, E. G. & Whittaker, V. P. (1962). *J. Anat. Lond.* **96**, 79.

Heilbronn, E. & Nilsson, M. G. (1971). *Proc. 3rd Int. Meet. Int. Soc. Neurochem. Budapest*, p. 392.

Iacobelli, S. (1969). *J. Neurochem.* **16**, 909.

Jungalwala, F. B. & Dawson, R. M. C. (1970). *Biochem. J.* **117**, 481.

Jungalwala, F. B. & Dawson, R. M. C. (1971). *Biochem. J.* **123**, 683.

McMurray, W. C. & Dawson, R. M. C. (1969). *Biochem. J.* **112**, 91.

Miani, N. (1964). *Prog. Brain Res.* **13**, 115.

Miller, E. K. & Dawson, R. M. C. (1972). *Biochem. J.* (In press.)

Morgan, I. G. & Austin, L. (1968). *J. Neurochem.* **15**, 41.

Porcellati, G. & di Jeso, F. (1971). *Membrane Bound Enzymes: Adv. exp. Med. Biol.* **14**, 111.

Sarzala, M. G., van Golde, L. M. G., De Kruyff, B. & van Deenen, L. L. M. (1970). *Biochim. biophys. Acta* **202**, 106.

Schneider, W. C. (1963). *J. biol. Chem.* **238**, 3572.
Webster, G. R. (1970). *Biochem. J.* **117**, 10P.
Whittaker, V. P. (1971). *Ann. N.Y. Acad. Sci.* (In press.)
Wilgram, G. F. & Kennedy, E. P. (1963). *J. biol. Chem.* **238**, 2615.
Wirtz, K. W. A. (1971). Ph.D. Thesis. University of Utrecht.
Wirtz, K. W. A. & Zilversmit, D. B. (1968). *J. biol. Chem.* **243**, 3596.

RELATIONS BETWEEN LIPID COMPOSITION AND PERMEABILITY OF MEMBRANES

By L. L. M. VAN DEENEN, J. DE GIER AND R. A. DEMEL

State University of Utrecht, Department of Biochemistry, Utrecht, The Netherlands

RENEWED interests in the architecture and transport mechanisms of biological membranes have greatly stimulated research in phospholipids. The combination of lipophilic and hydrophilic groups makes these molecules highly suitable for interfaces. The hydrocarbon chains present an apolar barrier by lipid-lipid interaction, and in addition are considered to interact with hydrophobic regions of membrane proteins. Electrostatic interactions between lipids and proteins, binding of cations and formation of ternary complexes depend on the polar headgroups of phospholipids. It is now agreed that phospholipids are essential for the structure of all biological interfaces. Many investigators believe that significant portions of a number of membranes contain phospholipid bilayers, but a certain degree of heterogeneity in membrane architecture is most likely. Without minimizing the crucial role of proteins in membranes, it can be shown that barrier properties of membranes are dictated to a great extent by the nature of the lipids. In this brief review we will report that the permeability of both artificial and natural membranes depends on the following lipid parameters.

The nature of the hydrocarbon chains. Because of the less compact packing membranes containing unsaturated phospholipids when compared with the saturated ones the *former barrier* will be more permeable.

Interaction with sterols. The so-termed condensing effect of sterols like cholesterol limits the mobility of the fatty acid chains. The effect depends not only the chemical nature of the phospholipid, but also on structural details of the sterol partner.

The chemical nature of the polar headgroup of the phospholipid. Wide variations in structure occur and the net charge of phospholipids can range from extremely negative to positive, thereby influencing ion permeation.

FATTY ACID CONSTITUENTS OF PHOSPHOLIPIDS

A simple phosphoglyceride such as lecithin from a given membrane may contain some ten different fatty acid constituents. Theoretically these can be combined in a great number of pairs. The members of this lecithin family can be recognized in a quantitative manner and prepared by chemical synthesis.

Conspicuous differences in physical properties have been detected between various lecithin species. In our laboratory research was concentrated on the behaviour of the synthetic phospholipids in monomolecular layers and in bilayers with the hope of obtaining information about the contributions of phospholipids to the barrier properties of biological membranes.

In a monomolecular layer at the air-water interface the area occupied by a lecithin molecule appears to depend to a great extent on the nature of the hydrocarbon chains. As the saturated hydrocarbon chains become shorter, there is a decrease in the London–van der Waals forces allowing a greater mobility of the chains. The same is true for the introduction of unsaturated fatty acid constituents demonstrating that, in general, the space occupied by a phospholipid molecule increases with increasing unsaturation. The implications of making a phospholipid mixed-acid are very pronounced. The fluidity of the film is greatly promoted although the values for some of the diunsaturated species are not attained (van Deenen *et al.*, 1962; Demel, van Deenen & Pethica, 1967; Demel, van Kessel & van Deenen, 1972*b*).

By extrapolating from the monolayer results one can expect that the permeation through a lipid bilayer will be dependent on the species of lecithin present. For verifying this assumption use was made of liposomes which are known to be made up of concentric, closed bimolecular membranes and to behave as osmometers. Both the penetration rates across these bilayers and the leakage from liposomes were measured (Demel, Kinsky & van Deenen, 1968; Demel, van Kessel & van Deenen, 1972*b*; de Gier, Mandersloot and van Deenen, 1968). Using liposomes made of various synthetic lecithin species it could be demonstrated that a decrease in chain length caused a higher permeability of the lipid bilayer for compounds such as glycerol and glucose. The permeation of these non-electrolytes through the lipid-barrier becomes easier when the degree of unsaturation of the lecithin species increases. There is an obvious correlation between the packing of molecules in the monolayer and artificial bilayer. It can also be noted that some lecithin species which do not occur as major membrane constituents display an extremely different behaviour. The (distearoyl)-lecithin forms rather impermeable barriers while those of (dilinoleoyl)-lecithin are very leaky. Mixed-acids species which are so abundant in natural membranes show an intermediate behaviour. While the phospholipid species in the bilayer dictate the barrier properties as expected from their behaviour in monolayers, the crucial question is, whether we can extrapolate them to the biological interface. In other words, is the permeability behaviour of natural membranes dependent on the nature of the fatty acid constituents.

For several membranes investigated to date the answer is yes. A good example is provided by *Mycoplasma laidlawii*. In this micro-organism the fatty acid composition can be varied within rather wide limits by supplementing different fatty acids to the growth medium. Measurements of the permeability were carried out on intact cells and on the liposomes prepared from the lipids extracted from these cells and both natural membranes and artificial bilayers showed an increase in permeability with increasing unsaturation (McElhaney, de Gier & van Deenen, 1970). It should also be noted that in this organism saturated and unsaturated fatty acids are located mainly at the 1- and 2-positions of the phospholipids, respectively. This specificity is achieved during the first step of the biosynthesis, namely the formation of phosphatidic acid. Thus when palmitic acid and oleic acid are supplied to the medium, a major molecular species of a phospholipid is (1-palmitoyl-2-oleoyl)-phosphatidylglycerol. However, with an increase in the number of double bonds of the fatty acid, the rate of incorporation decreases. The order is monoenoic > dienoic > trienoic > tetraenoic, while a hexaenoic acid is hardly incorporated. We can then conclude that the membrane properties are controlled here at the level of phospholipid biosynthesis in order to prevent the formation of leaky membranes. The mechanism of this control system has not been elucidated yet.

The preceding experiments on passive diffusion raise the question as to whether the lipid composition in a membrane may also regulate facilitated diffusion. This problem was studied by using valinomycin as a model for carrier transport. It has been well established that this depsipeptide specifically transports K^+ and Rb^+, but not Na^+ through lipid bilayers and natural membranes. It has also been shown that valinomycin can form charged complexes with the first mentioned cations and that the exterior of the complex is rather apolar. The action of valinomycin was studied on bilayers made up of different lecithins. The spontaneous leak of radioactive Rb^+ was very low. But the addition of very small amounts of valinomycin gave a significant stimulus to the transport and the effect turned out to be highly dependent on the nature of the fatty acid chains of the phospholipids. The more unsaturated lipids allowed a higher mobility and efficiency of the carrier (de Gier *et al.*, 1970). Very recently it was demonstrated that this phenomenon occurs in a natural membrane as well. Mycoplasma containing more unsaturated fatty acid constituents appeared to allow a more efficient exchange of K^+ by valinomycin. These results suggest that the nature of the lipid is not only important for simple diffusion but may also control the rate of carrier mediated transport. In a complex biological membrane carriers with a high and low working-gear should be located in an unsaturated and a saturated lipid locus respectively.

PHOSPHOLIPID-STEROL INTERACTION

Yet another factor which appears to be important for the stability and permeability of membranes is the sterol content. This lipid parameter varies widely. In erythrocyte membranes the molecular ratio of phospholipid to free cholesterol is close to unity. However, in most bacterial membranes sterols are absent. The role of sterols has been extensively studied in mono- and bilayers and it has been well established that a sterol like cholesterol, when spread together with a natural lecithin at the air-water interface, leads to, as an overall effect, a reduction in the mean molecular area of the lecithin molecules. This observation raises several questions.

Is the condensing effect dependent on the nature of the lecithin species?

Which structural features of the sterol are important for this action? Is this phenomenon relevant to biological membranes?

Using synthetic lecithin species, it was found that indeed the so-termed condensing or ordering effect of cholesterol and its magnitude depend on the fatty acid constituents and their pairing within the lecithin molecule (Demel *et al.*, 1967). Without going into details it can be concluded that at least all natural phospholipid species having one unsaturated fatty acid constituent reveal reduction in their molecular area in a monolayer in the presence of cholesterol. No condensation was observed with either dipolyunsaturated lecithins or certain disaturated lecithins (this depends very much on the chain length). One would expect that those lecithin species which are at the air-water interface limited in the mobility of their hydrocarbon chains by cholesterol would display similar behaviour in a lipid bilayer. That this was indeed the case was demonstrated by a decrease in the permeability of liposomes in which cholesterol was inserted (de Gier *et al.*, 1970). In the natural barriers also cholesterol reduced the permeability and this was proved in a different manner. Cells of *Mycoplasma laidlawii* grown in the presence of cholesterol appeared to have a lower permeability than those in the absence of the sterol (McElhaney, de Gier & van Deenen, 1970; de Kruyff, Demel & van Deenen, 1972).

In fact there is a similarity in behaviour between the membranes of different lipid compositions as found in artificial barriers made of these respective lipids. A second approach was quite different. Erythrocytes can be made to loose a part of their membrane cholesterol by simple incubation with liposomes. It was found that those cells which are depleted to the extent of some 30%–40% of their membrane cholesterol become more permeable and also less stable (Bruckdorfer *et al.*, 1969). Relations between altered permeability, osmotic stability and cholesterol contents of erythrocytes can also be found in patients with different diseases.

The common role of cholesterol in these biological membranes appears to be a reduction in the mobility of the hydrocarbon chains of the unsaturated phospholipid species, thereby leading to a reduction in the permeability of the lipid barrier. Not only the structural details of the phospholipids, but those of the sterol are also important. This can be demonstrated by replacing the cholesterol of the erythrocyte by its analogues which can be effected by incubation with liposomes containing the sterol to be inserted into the membrane. At the same time one can study the behaviour of the particular sterol in monolayers and bilayers with an appropriate phospholipid partner. The results so far have been very encouraging. It turned out that a number of sterols mimic cholesterol in the natural and artificial barrier successfully while others cause deviations. For instance, some keto steroids cause an increase in the permeability (Bruckdorfer *et al.*, 1969). Agreement between natural and artificial systems was, in general, excellent (Demel, Bruckdorfer & van Deenen, 1972). At this stage of the investigations it was concluded that the reduction in membrane permeability caused by cholesterol is dependent on a planar sterol nucleus, an intact side chain and the presence of a 3-β-hydroxy group in the sterol. Regarding the last conclusion, it was of interest that various sterol analogues were also incorporated into *Mycoplasma laidlawii*. It was found that the 3-α-hydroxy isomer of cholesterol, epicholesterol, is incorporated to the same extent as cholesterol but does not reduce the permeability as is done by cholesterol (de Kruyff *et al.*, 1972). At the same time it was observed that epicholesterol fails to mimic cholesterol in both monolayers and bilayers. These results again confirmed that relations between model systems and natural membranes exist. They also indicated that various structural details of the sterol are extremely important, and that continuation and further extension of these studies are necessary for arriving at a proper understanding of phospholipid-sterol interactions at the molecular level.

THE POLAR HEADGROUP OF PHOSPHOLIPIDS

That the nature of the polar headgroup can affect the permeability properties of a natural membrane could be demonstrated in bacteria (Haest *et al.*, 1972). Two important phospholipids of gram-positive bacteria are phosphatidylglycerol and a derivative denoted lysylphosphatidylglycerol. The first has a net negative charge and the second is positively charged. The ratios between the two phospholipids in the membrane can vary significantly depending upon the environmental conditions. In agreement with the previous studies on lipid bilayers showing that the net charge of the interface has an enormous effect, it was found that artificial barriers of phosphatidylglycerol were cation selective whereas those of lysylphosphatidylglycerol were anion selective. As for non-electrolytes, the permeation through bilayers of lysylphosphatidylglycerol was easier than through phosphatidylglycerol. The former has a more bulky headgroup as revealed on monolayers and prevents close packing of the chains which are similar in both phospholipids. Thus the properties of ion permeability are dependent upon the charge of the phospholipid. It was found that valinomycin is highly active as a carrier through lipid bilayers of negatively charged phosphatidylglycerol. But, its action appears to be immobilized with a bilayer of lysylphosphatidylglycerol. The formation of the charged complex at or in the membrane may be affected by the presence of positive groups in the lipid.

These observations on liposomes could be extended to membranes of *Staphylococcus aureus* (Haest *et al.*, 1972). The ratio of phosphatidylglycerol to lysylphosphatidylglycerol was varied, depending upon the time of exposure to a glucose medium of pH 5 and was found that cells with a dominant lysylphosphatidylglycerol content were more permeable to non-electrolytes, just as in the case of liposomes. Measurements of the action of valinomycin showed that there is a striking reduction in the ability of this carrier with an increase in the ratio of the positively charged lysylphosphatidylglycerol to the negatively charged phosphatidylglycerol. Again, this carrier facilitated transport appears to reflect the significance of the lipid composition just as in artificial membranes.

In several cases we have seen that not only the permeability of a natural membrane depends on the details of the lipid constituents but also that there is a similarity between the natural membrane and artificial bilayers. Within the limitations of our present experimental techniques, this certainly should not be interpreted to mean that membranes are built up exclusively of bilayers of lipids. The approximations made here were for the purpose of illustrating only some of the progress which has been made for establishing the fact that the chemical structures of phospholipids and sterols are of consequence for membrane properties. These and many still unknown details are necessary for understanding the interactions of lipids and proteins so as to arrive in the future at better molecular descriptions of biological membranes.

SUMMARY

The lipid composition controls barrier properties of artificial and natural membranes. Diffusion processes and carrier-mediated transport may depend on the nature of the hydrocarbon chain of lipids, the interaction of phospholipids with sterols and the chemical nature of the polar headgroups of phospholipids.

REFERENCES

Bruckdorfer, K. R., Demel, R. A., Gier, J. de & Deenen, L. L. M. van (1969). *Biochim. biophys. Acta* **181**, 334.

Deenen, L. L. M. van, Houtsmuller, U. M. T., Haas, G. H. de & Mulder, E. (1962). *J. pharm. Pharmacol.* **14**, 429.

Demel, R. A., Bruckdorfer, K. R. & Deenen, L. L. M. van (1972). *Biochim. biophys. Acta* **255**, 321.

Demel, R. A., Deenen, L. L. M. van & Pethica, B. A. (1967). *Biochim. biophys. Acta* **135**, 11.

Demel, R. A., Kessel, W. S. M. van & Deenen, U. M. van (1972). *Biochim. biophys. Acta* **266**, 26.

Demel, R. A., Kinsky, S. C. & Deenen, L. L. M. van (1968). *Biochim. biophys. Acta* **150**, 655.

Gier, J. de, Mandersloot, J. G. & Deenen, L. L. M. van (1968). *Biochim. biophys. Acta* **150**, 666.

Gier, J. de, Mandersloot, J. G. & Deenen, L. L. M. van (1969). *Biochim. biophys. Acta* **173**, 143.

Gier, J. de, Haest, C. W. M., Mandersloot, J. G. & Deenen, L. L. M. van (1970). *Biochim. biophys. Acta* **211**, 373.

Haest, C. W. M., Gier, J. de, Op den Kamp, J. A. F., Bartels, P. & Deenen, L. L. M. van (1972). *Biochim. biophys. Acta* **255**, 720.

Kruyff, B. de, Demel, R. A. & Deenen, L. L. M. van (1972). *Biochim. biophys. Acta* **255**, 331.

McElhaney, R. N., Gier, J. de & Deenen, L. L. M. van (1970). *Biochim. biophys. Acta* **219**, 245.

INOSITOL LIPID METABOLISM AND THE CELL MEMBRANE

By J. N. HAWTHORNE

*Department of Chemistry, University of Calgary, Alberta, Canada**

PHOSPHATIDYLINOSITOL (I) is widely distributed in the animal and plant kingdoms and is a constituent of most subcellular membranes in mammalian tissues, where it accounts for between 2% and 10% of the total lipid P. The phosphorylated derivatives diphosphoinositide (II) and triphosphoinositide

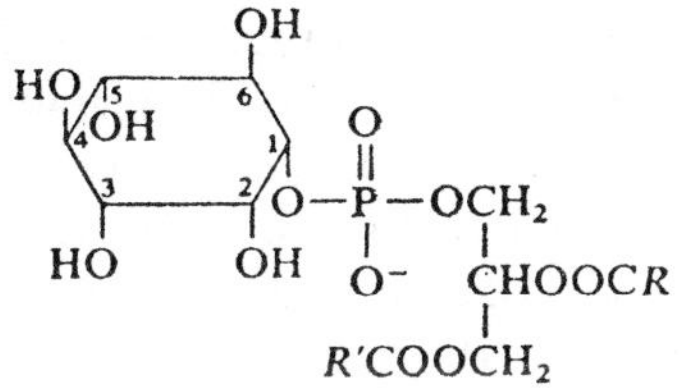

(I)

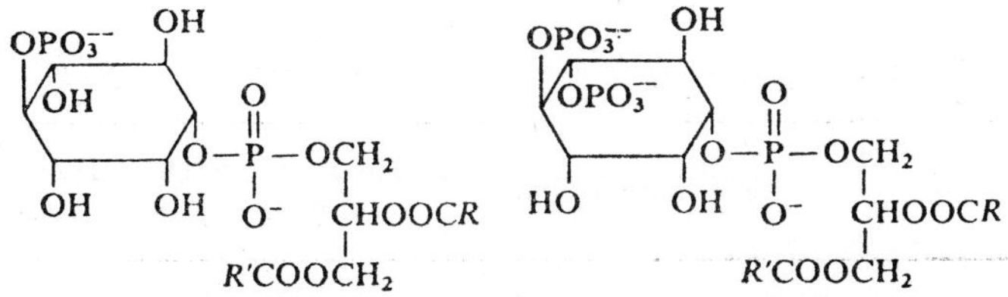

(II) (III)

FIG. 1. Structures of the phosphoinositides.

(III) have a more limited distribution and are associated particularly with a number of plasma membrane structures. The monoesterified phosphate groups of these two phospholipids have a considerable affinity for divalent cations, particularly calcium (Hendrickson & Fullington, 1965; Hawthorne & Kai, 1970).

This review represents a progress report on three of the topics currently being studied in my laboratory, together with some background information from the literature. Dr J. T. Buckley, Mr P. H. Cooper and Dr G. L. White are responsible for the experimental work, and their names appear with the appropriate section titles.

* Present address, Department of Biochemistry, Medical School, University Park, Nottingham, England.

ERYTHROCYTE POLYPHOSPHOINOSITIDES AND CALCIUM (J. T. BUCKLEY)

Some years ago it was suggested that the diphosphoinositide and triphosphoinositide levels in the axolemma might control the permeability of that membrane by a calcium-binding mechanism (Hawthorne & Kemp, 1964). Erythrocyte ghosts form the two polyphosphoinositides from ATP and phosphatidylinositol (Hokin & Hokin, 1964; Schneider & Kirschner, 1970; Peterson & Kirschner, 1970) and Harrison & Long (1968) have suggested that the tightly bound calcium of the erythrocyte membrane might be attached to triphosphoinositide. It seemed therefore that the red cell would provide a model system for a study of the relation, if any, between polyphosphoinositides and calcium.

Polyphosphoinositide synthesis

Pig erythrocyte ghosts were prepared by the method of Schneider & Kirschner (1970) and washed twice with 50 mM-tris chloride buffer (pH 7·4) to remove EDTA. They were then incubated for various lengths of time at 37° in the same 50 mM-tris chloride buffer containing 40 mM-$MgCl_2$ and 5 mM-sodium ATP. Previous experiments had shown that optimum polyphosphoinositide synthesis was obtained with $MgCl_2$ concentrations between 20 and 40 mM and ATP concentrations between 1 and 6 mM. After incubation the membranes were centrifuged down and extracted with chloroform-methanol-HCl according to Schneider and Kirschner (1970). Diphosphoinositide and triphosphoinositide were separated by the t.l.c. procedure of Gonzalez-Sastre & Folch (1968) and the phosphorus content of each spot was estimated.

Table 1. *Phospholipid composition of pig erythrocyte ghosts*

	Fresh ghosts	Incubated without $MgCl_2$	Incubated with 40 mM-$MgCl_2$
Material at origin	5·3 ± 1·2 (4)	3·5	9·8
Phosphatidylinositol	3·7 ± 0·3 (4)	3·4	not detected
Phosphatidylserine	19·3 ± 0·5 (5)	19·3	19·0
Sphingomyelin	22·0 ± 1·2 (5)	22·6	21·7
Phosphatidylcholine	22·4 ± 2·0 (5)	23·1	22·7
Phosphatidylethanolamine	27·2 ± 1·2 (5)	28·2	26·8
		n moles per 2 ml membrane suspension	
Diphosphoinositide	—	17·7	114·7
Triphosphoinositide	—	87·6	123·0
Polyphosphoinositide increase	—	132·4	
Initial phosphatidylinositol	—	155·1	

Figures represent per cent recovered lipid phosphorus with S.D. and number of preparations in parentheses. Ghosts were incubated for 30 min at 37° as described in the text, with or without $MgCl_2$. Phospholipids were separated by two-dimensional t.l.c. (Turner & Rouser (1970)). Polyphosphoinositides remain at the origin.

In a typical experiment diphosphoinositide concentration (nmoles/mg protein) rose from 2·3 to 15·0 and triphosphoinositide concentration rose from 11·4 to 16·3 over a 30 min period. These figures are similar to those of Schneider & Kirschner (1970), though the diphosphoinositide change is greater. Table 1 shows the phospholipid composition of fresh and incubated membranes. Only the inositol lipid concentrations changed during incubation, almost all the phosphatidylinositol being converted to polyphosphoinositide.

Though optimum conditions may not yet have been reached, particularly in relation to the lipid substrate, phosphatidylinositol kinase is relatively active in erythrocyte membranes. The initial reaction rate, which was constant for 2–3 min, represented the formation of 3·10 nmoles diphosphoinositide/min/mg protein. The corresponding figure for whole brain is 0·12 and is much less for other tissues (Harwood & Hawthorne, 1969). A rate of 3·3 nmoles diphosphoinositide/min/mg protein was only obtained with a brain microsomal fraction in the presence of the detergent Cutscum (Colodzin & Kennedy, 1965).

Calcium content of membranes

In parallel experiments the firmly bound calcium of the erythrocyte membranes was compared after incubation in the presence or absence of ATP, using the same 40 mM-$MgCl_2$/50 mM-tris chloride medium. After incubation the membranes were mixed with ice-cold $CaCl_2$ solution to give a final concentration of 5 mM, and centrifuged down. The sediment was washed once with

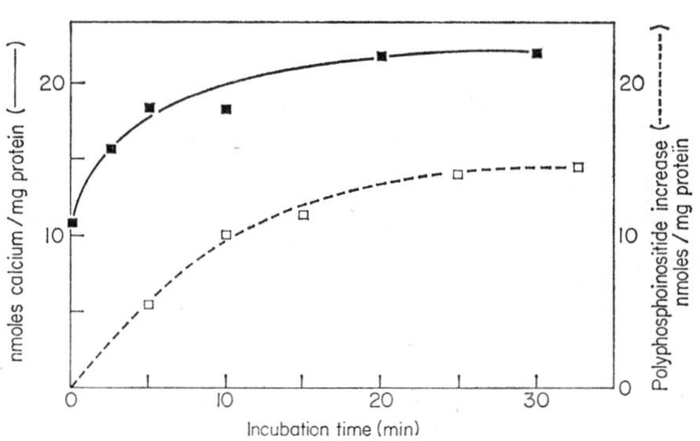

FIG. 2. Changes in calcium and polyphosphoinositide contents of pig erythrocyte membranes during incubation as described in the text.

33 mM-NaCl/2 mM $CaCl_2$ and then twice with 33 mM-NaCl. Bound calcium was extracted with 153 mM-NaCl/5 mM-EDTA and determined by atomic absorption. The values obtained were within the range quoted by Harrison & Long (1968) for human erythrocyte membranes. After incubation the calcium content approximately doubled and the increase followed a similar time-course to that of the polyphosphoinositides (Fig. 2).

Calculating from these and other more recent experiments (Buckley, 1972) the molar ratio of increased phosphoinositide monophosphate to increased calcium was 1·2 : 1·0. How calcium is bound to the phosphoinositides is not yet clear. It is likely that the 4- and 5-phosphates of the polyphosphoinositides are also bound to some other membrane constituents, probably basic groups in a protein. The failure of chloroform-methanol to extract the lipids unless HCl is added also indicates that there is ionic bonding between membrane protein and polyphosphoinositide. It was also significant that EDTA treatment did not render the lipids extractable.

Sodium and potassium fluxes

In another series of experiments re-sealed pig erythrocyte ghosts were used to study the possible effect of increased membrane polyphosphoinositide levels on permeability to sodium and potassium ions. Whole erythrocytes incubated for 3 hr with ATP to increase polyphosphoinositide concentrations (Peterson & Kirschner, 1970) were compared with cells which had not been incubated. Each batch of cells was then haemolysed in 4 mM-$MgCl_2$/4 mM-sodium ATP at pH 7·4 (tris chloride) and re-sealed by addition of 3 M-NaCl to give a final concentration of 154 mM. Two volumes of 140 mM-NaCl/10 mM-KCl/5 mM-tris chloride pH 7·4/1 mM-$MgCl_2$ were added at 0° and the cells were centrifuged. They were then washed with a similar solution containing less NaCl (120 mM) and more KCl (30 mM) and finally suspended in 75 mM-NaCl/75 mM-KCl/5 mM-tris chloride pH 7·4/1 mM-$MgCl_2$/1 mM-$CaCl_2$. After incubation at 30° for various lengths of time (0–45 min) samples were mixed with 5 volumes of isotonic choline chloride at pH 7·4, centrifuged and washed once with isotonic choline chloride. The cells were again centrifuged and re-suspended in water. Trichloroacetic acid was added to give a final concentration of 5% (w/v), the resultant precipitate centrifuged down and the trichloroacetic acid removed by ether extraction. Sodium and potassium were then determined by flame photometry. There were no significant differences in sodium or potassium flux between cells with high and cells with low polyphosphoinositide concentrations.

Calcium-dependent ATPase

Various ATPase activities of the pig erythrocyte membranes were measured after incubation for 30 min at 37° in 50 mM-tris chloride (pH 7·4) containing 20 mM-$MgCl_2$, with or without the addition of 5 mM-ATP. As shown above, polyphosphoinositide levels increased during the incubation with ATP. After incubation the membranes were centrifuged down and washed with 50 mM-tris chloride. ATPase activities were then determined in the presence of various cations and 2 mM-[^{32}P]ATP labelled in the terminal phosphate by the method of Glynn & Chappell (1964). Radioactive inorganic phosphate was determined after extraction according to Martin & Doty (1949).

Table 2 shows that there was a significant increase in the Ca^{2+}-activated ATPase when polyphosphoinositide levels were high. The Mg^{2+}-ATPase and $Na^+/K^+/Mg^{2+}$-ATPase were not affected.

Table 2. *ATPase activities of erythrocyte membranes*

Activators (mM)				ATPase (nmoles inorganic P/min/ml ghosts)	
$MgCl_2$	$CaCl_2$	NaCl	KCl	Polyphosphoinositide concentration normal	increased
2	0	0	0	1·25	2·04
2	2	0	0	1·28	6·16
2	0	50	12·5	3·36	4·25

Pig erythrocyte ghosts were incubated with or without ATP as described in the text. Each figure represents the mean of four determinations. 50 mM-tris chloride was always present.

Conclusions

It seems significant that the firmly bound calcium of erythrocyte membranes is bound to diphosphoinositide and triphosphoinositide and that these lipids affect the activity of the membrane calcium-dependent ATPase. It is not yet certain whether this enzyme activity is related to the calcium pump which maintains low levels of Ca^{2+} within the red cell. The possibility that polyphosphoinositides are associated either directly or indirectly with this pump is being investigated.

POLYPHOSPHOINOSITIDE METABOLISM IN KIDNEY (P. H. COOPER)

Previous work had suggested that synthesis of diphosphoinositide by phosphatidylinositol kinase (1) was associated with both nuclear and

$$\text{phosphatidylinositol} + \text{ATP} \rightarrow \text{diphosphoinositide} + \text{ADP} \qquad (1)$$

microsomal fractions of kidney (Harwood & Hawthorne, 1969). The microsomal enzyme was markedly activated by the detergent Cutscum and the subcellular distribution of the kinase suggested that some of the enzyme was associated with the various types of kidney plasma membrane. Tou, Hurst & Huggins (1969) on the other hand, considered that the kinase was primarily bound to the endoplasmic reticulum, though their assays appear to have been done in the presence of Cutscum. Subsequent studies (Tou *et al.*, 1970) indicated that kidney diphosphoinositide kinase (2) was associated with the plasma membrane.

$$\text{diphosphoinositide} + \text{ATP} \rightarrow \text{triphosphoinositide} + \text{ADP} \qquad (2)$$

Triphosphoinositide phosphomonoesterase (3) was most active in the supernatant fraction of rat brain (Salway, Kai & Hawthorne, 1967) but appears to

$$\text{triphosphoinositide} \rightarrow \text{phosphatidylinositol} + 2P_i \qquad (3)$$

be associated with the microsomal fraction of kidney (Lee & Huggins, 1968*a*). However, the presence of an active alkaline phosphatase in the kidney microsomal fraction makes assays difficult (Lee & Huggins, 1968*b*) since this enzyme can also release inorganic phosphate from triphosphoinositide (Prottey, Salway & Hawthorne, 1968). With these problems in mind, the enzymes catalysing reactions 1–3 in rat kidney have been studied in some detail.

Polyphosphoinositide purification

In the course of this work a new column method for the isolation of diphosphoinositide and triphosphoinositide was developed. It is based on the t.l.c. procedure of Gonzalez-Sastre & Folch (1968). Silica gel (30 g, 100–200 mesh) was slurried with 70 ml 1% (w/v) aqueous potassium oxalate and then dried overnight at 110°. It was made into a column with 7:3 (v/v) chloroform-methanol and crude brain diphosphoinositide fraction was applied to the column in the same solvent with the addition of a little chloroform as necessary, to give a clear solution. The loading ratio was 1 mg lipid P/g silica gel. Gradient elution followed, using chloroform-methanol-water (9:7:2, by vol.) in the mixing vessel and chloroform-methanol-4N-ammonia (9:7:2, by vol.) in the reservoir. Fractions were collected and monitored by the t.l.c. method above, which showed that phosphatidylinositol was eluted first, followed by diphosphoinositide and triphosphoinositide with little overlap. Suitable fractions were combined, neutralized with acetic acid and the solvents were removed by rotary evaporation at 40°. The resulting aqueous dispersion of lipid was then dialysed against sodium EDTA and either lyophilized for storage or used directly as a substrate.

Phosphatidylinositol kinase

This enzyme was about twice as active in cortex as in the medulla of rat kidney. Cortex homogenates in the presence of Cutscum showed a specific activity of 0·09 nmoles diphosphoinositide formed/min/mg protein. The subcellular distribution of the kinase followed that of arylesterase (E.C. 3.1.1.2) most closely, suggesting localization in the microsomal fraction. However, there was also kinase activity in the light mitochondrial fraction and the activity there was relatively higher than that of arylesterase. The plasma membrane enzymes 5′-nucleotidase (E.C. 3.1.3.5) and alkaline phosphatase (E.C. 3.1.3.1) were also very active in this fraction. The presence of a plasma membrane phosphatidylinositol kinase cannot be excluded and work continues on this topic.

In liver (Harwood & Hawthorne, 1969) phosphatidylinositol kinase was most active in the nuclear fraction in the absence of Cutscum. In the presence of Cutscum, microsomal activity was greatly increased, but nuclear activity was unaffected. This did not apply to kidney, where Cutscum had much the same effect on all subcellular fractions.

The diphosphoinositide kinase was much more active; cortex homogenate in the absence of Cutscum synthesized 0·15 nmoles triphosphoinositide/min/mg protein. Of the divalent cations tested, Mg^{2+} was most effective, but Mn^{2+} and Co^{2+} also activated the enzyme. Optimum magnesium levels had still not been reached at 50 mM, but this may be a feature of the homogenate. Maximum activity was reached with 0·5 mM-diphosphoinositide and 4 mM-ATP. An optimum concentration of 20 mM-$MgCl_2$ was obtained for this kinase in plasma membrane by Tou *et al.* (1970).

Triphosphoinositide phosphomonoesterase

Because of the possible interference by alkaline phosphatase or acid phosphatase, an assay system was developed which excluded these enzymes. This took advantage of the fact that fluoride completely and specifically inhibits kidney acid phosphatase, and cysteine completely and specifically inhibits the alkaline phosphatase. Hydrolysis of triphosphoinositide or diphosphoinositide was not affected by fluoride or cysteine added to the kidney homogenates. The final assay system involved pre-incubation at 0° for 10 min of 0·2–0·4 mg homogenate protein in 45 mM-tris chloride buffer pH 7·2, 1 mM-NaF, 3 mM-cysteine and 1 mM-$MgCl_2$. Substrate triphosphoinositide (1 mM) was then added with 2 mM-cetyl trimethylammonium chloride and the mixture was incubated for 30 min at 37°.

Slightly different optimum conditions were arrived at with diphosphoinositide as substrate. The $MgCl_2$ concentration was higher (1·5 mM), bovine serum albumin (8 mg/ml) was required to avoid an initial lag in the rate curve and substrate was optimum at 2 mM with 1 mM cetyltrimethylammonium chloride.

It is still not known whether triphosphoinositide monoesterase and diphosphoinositide monoesterase are separate enzymes. With kidney homogenate the pH optimum for the release of inorganic phosphate was 7·5 whether triphosphoinositide or diphosphoinositide was used as substrate.

Conclusions

As Huggins and his colleagues have pointed out, phosphoinositide metabolism is active in kidney. Studies are in progress to determine more precisely the subcellular localization of the various enzymes involved. It will be interesting to see whether there is any relationship between active cation transport and the inositol lipids.

NERVE CONDUCTION AND TRIPHOSPHOINOSITIDE (G. L. WHITE)

Ganglionic transmission seems to be associated with specific changes in the turnover of phosphatidylinositol as indicated by incorporation of ^{32}P. It has long been known that the 4- and 5-phosphate groups of triphosphoinositide in nervous tissue have the most rapid turnover of all the phospholipid phosphate groups. When the incorporation of radioactive phosphate into the three inositol lipids was compared in rat vagus nerve and sympathetic ganglia, triphosphoinositide was most rapidly labelled in the vagus, and phosphatidylinositol in ganglia (Yagihara, Salway & Hawthorne, 1969). It is possible that most of the highly labelled triphosphoinositide was in the axonal membrane, since rat vagus fibres are chiefly unmyelinated (Larrabee, 1970). Through its ability to complex with calcium, triphosphoinositide may be able to control ion permeabilities in this membrane (Hawthorne & Kai, 1970). Changes in bound calcium and ligand charge resulting from conversion of triphosphoinositide to diphosphoinositide in such a model system have been calculated by Henrickson & Reinertsen (1971). At pH 7·0 such a change would result in loss of 70% of bound calcium and a 25% decrease in average ligand charge.

Neuronal activity and phospholipid labelling

Larrabee and his colleagues (Larrabee, Klingman & Leicht, 1963; Larrabee & Leicht, 1965) have developed an excellent system for studying changes in turnover of phospholipids due to electrical stimulation of sympathetic ganglia or nerve trunks. This seemed to provide one approach to the possible involvement of triphosphoinositide in membrane changes associated with excitation. Studies of this type were therefore made in Dr M. G. Larrabee's laboratory at Johns Hopkins University. Rat vagus nerves, phrenic nerves and superior cervical ganglia were pre-incubated at 37° for 60 min in Earle's balanced salt solution (Grand Island Biological Company, New York) containing approximately 0·25 mCi/ml [^{32}P]orthophosphate and then stimulated for 3 hr in the apparatus described by Larrabee *et al.* (1963), the compound action potentials being recorded. Control nerves and ganglia were similarly treated, being placed in the apparatus alongside the experimental tissues but not stimulated. At the end of the 3 hr stimulation period the tissue was extracted three times with chloroform/methanol/conc. HCl, 800:400:3 (by vol.), carrier Folch diphosphoinositide fraction was added and the combined extracts were washed twice with 0·2 vol. N-HCl. The lower layer was then evaporated down and the lipids applied to a t.l.c. plate for chromatography according to Gonzalez-Sastre & Folch (1968). The plate was scanned for ^{32}P and each spot was counted. Alternatively formaldehyde-impregnated paper chromatography was used.

Table 3 gives the results showing that the only significant changes were the expected increase in phosphatidylinositol labelling of ganglia and a reduction

Table 3. *Neuronal activity and phospholipid labelling**

	Vagus nerve	Phrenic nerve	Sympathetic ganglion
Impulses/sec	5	200	10
Triphosphoinositide	−24 ± 2† (7)	6 ± 11 (4)	−10 ± 10 (11)
Diphosphoinositide	3 ± 11 (8)	0 ± 8 (4)	11 ± 8 (11)
Phosphatidylinositol	−5 ± 5 (7)	−10 ± 8 (4)	69 ± 5† (10)
Phosphatidylethanolamine/ phosphatidylserine	−2 ± 4 (8)	1 ± 23 (4)	−14 ± 4 (11)

Pre-incubation and stimulation are described in the text. The figures are based on ^{32}P (counts/sec) in the phosphoinositide spots (t.l.c.) divided by the equivalent figure for the phosphatidylcholine plus phosphatidate spot. The per cent change in this ratio due to stimulation is given, with the S.E.M. and number of experiments.

* G. L. White and M. G. Larrabee, *J. Neurochem.* (In press.)

† $P < 0·0001$.

in triphosphoinositide labelling in vague nerve. The incorporation of ^{32}P into the phosphoinositides was normalized with respect to the phosphatidylcholine/phosphatidate spot, radioactivity of which was unaffected by stimulation (Larrabee & Leicht, 1965).

The results of Yagihara *et al.* (1969) suggest that the 4- and 5-phosphates of vagus triphosphoinositide reach equilibrium with the pool of labelled ATP in 2 hr, assuming no net breakdown of the lipid. Once this equilibrium is reached, changes in turnover of these phosphate groups due to stimulation of the nerve would not be detected. Experiments were therefore done with a shorter period of stimulation (Table 4). Under these conditions excitation caused a reduction

Table 4. *Nerve conduction and phospholipid labelling*

Impulses/sec	Vagus 5	Phrenic 200
Triphosphoinositide	-72 ± 8†	-49 ± 19
Diphosphoinositide	-76 ± 7†	-57 ± 16*
Phosphatidylinositol/phosphatidylserine	-74 ± 4†	-57 ± 14*
Phosphatidylcholine/phosphatidylethanolamine	-68 ± 7†	-35 ± 26

The conditions were similar to those in Table 3, except that pre-incubation was for 15 min and stimulation for 30 min. The figures represent per cent changes in ^{32}P incorporation (nmoles phosphate incorporated/g protein/45min) with S.E.M. Four experiments were done with vagus and five with phrenic nerve.

* P 0·01–0·02. † P $<$ 0·003.

in labelling, but this was much the same in all the phospholipids studied. One explanation might be that stimulation channels radioactive ATP away from phospholipid synthesis, presumably most of it being used in the restoration of sodium and potassium ion gradients.

Effects of ouabain and tetrodotoxin

Similar experiments were done in the presence of various concentrations of these two drugs, both of which block conduction of nerve impulses, though by different mechanisms. In the unmyelinated vagus nerve the decrease in polyphosphoinositide labelling (incorporation of ^{32}P/mg protein) caused by stimulation (Table 4) was somewhat smaller after the addition of a concentration of ouabain (10^{-3} M) or tetrodotoxin (10^{-7} M) sufficient to abolish the action potential during the 45 min incubation. The phosphatidylinositol and phosphatidylcholine decreases, however, almost disappeared in the presence of either drug. With the myelinated phrenic nerve the addition of ouabain in sufficient concentrations to markedly affect the compound action potential did not appear to affect ^{32}P incorporation (per mg protein) into any of the phospholipids.

Thus the metabolism of the higher inositides in vagus nerve or phrenic nerve does not appear to be affected greatly by ouabain, which inhibits the Na^+/K^+-dependent ATPase of the sodium pump. Glynn *et al.* (1965) have also concluded that these lipids are not involved in sodium transport in electric organ.

Tetrodotoxin blocks the sodium pores of the axonal membrane from the outside (Narahashi, Anderson & Moore, 1966) and like ouabain, it does not seem to affect polyphosphoinositide metabolism. This is the expected result, since if triphosphoinositide synthesis and breakdown do regulate sodium ion movement through such pores, this metabolism would take place at the inner surface of the membrane where ATP and triphosphoinositide phosphomonoesterase are available.

The restoration of phosphatidylcholine and phosphatidylinositol labelling to normal may be due to the increased availability of ATP when the action potential is blocked.

Conclusions

Though most of the above results can be incorporated into a theory linking trophosphoinositide metabolism with the permeability changes accompanying the action potential, direct evidence is still lacking. One problem in all this work is the variability in the response, which entails many experiments and statistical treatment of the results. Using lobster leg nerves, Birnberger *et al.* (1971) claim that stimulation *increases* the turnover of triphosphoinositide phosphate and that this increase is not affected by cyanide poisoning. Their technique was quite different from ours, the nerves being labelled *in vivo* and then removed for stimulation (5 min) in a medium containing no ^{32}P. However, it is equally possible to argue from these results, as Birnberger *et al.* (1971) do, that triphosphoinositide is involved in permeability changes during excitation.

SUMMARY

The amount of tightly-bound calcium in the pig erythrocyte membrane is proportional to the polyphosphoinositide content. The molar ratio of phosphoinositide monoesterified phosphate to calcium is approximately 1:2:1. The Ca^{2+}-ATPase activity is higher in polyphosphoinositide-rich membranes.

The properties of phosphatidylinositol kinase, diphosphoinositide kinase and triphosphoinositide phosphomonoesterase in rat kidney are described.

Electrical stimulation of vagus nerve depresses the incorporation of ^{32}P into triphosphoinositide. Some effects of ouabain and tetrodotoxin on phospholipid labelling during stimulation are discussed. The results are consistent with the theory that triphosphoinositide controls the permeability of the axonal membrane during excitation.

ACKNOWLEDGEMENTS

This work is being supported by the Medical Research Council of Canada, grant MA 3989 and the National Research Council of Canada, grant A 6217.

REFERENCES

Birnberger, A. C., Birnberger, K. L., Eliasson, S. G. & Simpson, P. C. (1971). *J. Neurochem.* **18**, 1291–1298.

Buckley, J. T. & Hawthorne, J. N. (1972). *J. biol. Chem.* (In press.)

Colodzin, M. & Kennedy, E. P. (1965). *J. biol. Chem.* **240**, 3771–3780.
Glynn. I. M. & Chappell, J. B. (1964). *Biochem. J.* **90**, 147–149.
Glynn, I. M., Hayman, C. W., Eichberg, J. & Dawson, R. M. C. (1965). *Biochem. J.* **94**, 692–699.
Gonzalez-Sastre, F. & Folch, J. (1968). *J. Lipid Res.* **9**, 532–533.
Harrison, D. G. & Long, C. (1968). *J. Physiol.* **199**, 367–381.
Harwood, J. L. & Hawthorne, J. N. (1969). *Biochim. biophys. Acta* **171**, 75–88.
Hawthorne, J. N. & Kai, M. (1970). In *Handbook of Neurochemistry*, Vol. 3. (Ed. by Lajtha, A.), pp. 491–508. New York: Plenum Press.
Hawthorne, J. N. & Kemp, P. (1964). *Adv. Lipid Res.* **2**, 127–166.
Hendrickson, H. S. & Fullington, J. G. (1965). *Biochemistry* **4**, 1599–1603.
Hendrickson, H. S. & Reinertsen, J. L. (1971). *Biochem. biophys. Res. Commun.* **44**, 1258–1264.
Hokin, L. E. & Hokin, M. E. (1964). *Biochim. biophys. Acta* **84**, 563–575.
Larrabee, M. G. (1970). *Fedn Proc. Fedn Am. Socs. exp. Biol.* **29**, 1919–1928.
Larrabee, M. G., Klingman, J. D. & Leicht, W. S. (1963). *J. Neurochem.* **10**, 549–570.
Larrabee, M. G. & Leicht, W. S. (1965). *J. Neurochem.* **12**, 1–13.
Lee, T.-C. & Huggins, C. G. (1968*a*). *Archs Biochem. Biophys.* **126**, 206–213.
Lee, T.-C. & Huggins, C. G. (1968*b*). *Archs Biochem. Biophys.* **126**, 214–220.
Martin, J. B. & Doty, D. M. (1949). *Anal. Chem.* **21**, 965–967.
Narahashi, T., Anderson, N. C. & Moore, J. W. (1966). *Science* **153**, 765–767.
Peterson, S. C. & Kirschner, L. B. (1970). *Biochim. biophys. Acta* **202**, 295–304.
Prottey, C., Salway, J. G. & Hawthorne, J. N. (1968). *Biochim. biophys. Acta* **164**, 238–251.
Salway, J. G., Kai, M. & Hawthorne, J. N. (1967). *J. Neurochem.* **14**, 1013–1024.
Schneider, R. P. & Kirschner, L. B. (1970). *Biochim. biophys. Acta* **202**, 283–294.
Tou, J.-S., Hurst, M. W. & Huggins, C. G. (1969). *Archs Biochem. Biophys.* **131**, 596–602.
Tou, J.-S., Hurst, M. W., Huggins, C. G. & Foor, W. E. (1970). *Archs Biochem. Biophys.* **140**, 492–502.
Turner, J. D. & Rouser, G. (1970). *Anal. Biochem.* **38**, 423–426.
Yagihara, Y., Salway, J. G. & Hawthorne, J. N. (1969). *J. Neurochem.* **16**, 1133–1139.

THE ROLE OF LIPIDS IN THE STRUCTURE AND FUNCTION OF *E. COLI* MEMBRANE

By SALIH J. WAKIL AND MOJTABA ESFAHANI

Department of Biochemistry,
Baylor College of Medicine,
Houston, Texas, U.S.A.

CURRENTLY there is much interest in the relationship between structure and function of biological membranes. A possible approach to the understanding of the structural requirements for the lipids and proteins which are essential for the expression of membrane function is, alteration of these constituents and investigation into the effects of such alterations on membrane properties. In this paper we shall present an account of our studies on the modification of the apolar portions of membrane lipids in unsaturated fatty acid auxotrophs of *Escherichia coli*, and the effects of such modifications on membrane properties.

FATTY ACIDS OF *E. coli*

Extraction of *E. coli* with organic solvents removes all the cellular lipids except those that are constituents of cell envelope lipopolysaccharides (Kaneshiro & Marr, 1961). The lipids isolated in this way consist of membrane phospholipids, neutral lipids, and free fatty acids (Okuyama, 1969). Palmitic acid is the major saturated fatty acid in these lipids, while palmitoleic and *cis*-vaccenic acids are the only unsaturated fatty acids and account for approximately 55% of the total fatty acids (Norris, Matsumara & Bloch, 1964; Pugh *et al.*, 1966; Silbert, Ruch & Vagelos, 1968). Approximately 85% of the total fatty acids are found in the phospholipids. Cyclopropane fatty acids are detected in trace amounts in exponentially growing cells. As the culture approaches the stationary phase of growth, the unsaturated fatty acids are progressively converted to their cyclopropane analogs (Cronan, 1968). Substantial amounts of lauric, myristic and β-hydroxymyristic acids are found in the cell envelope lipopolysaccharides and can be extracted with organic solvents only after the cells have been subjected to acid or alkali hydrolysis (Kaneshiro & Marr, 1961; Burton & Carter, 1964).

The relative proportions of saturated and unsaturated fatty acids in *E. coli* vary with growth temperature. Lowering of growth temperature results in an increase in the levels of unsaturated fatty acids in the lipids (Marr & Ingraham, 1962) indicating that the organism attempts to counterbalance the condensing effect of low temperature by maintaining higher levels of unsaturated fatty acids in membrane lipids.

In wild type strains of *E. coli* exogenous fatty acids have no effects on cellular fatty acid composition (Silbert, Ruch and Vagelos, 1968 and Table 1). This indicates that the fatty acid synthetase of this organism is not repressible.

Table 1. *Fatty acid composition of phospholipids of* E. coli *strain* CR_{34}*

Fatty acids in phospholipids	Cells grown on	
	amino acids	amino acids plus oleate
	(%)	(%)
$C_{14:0}$	1	1
$C_{16:0}$	43	42
$C_{16:1}$	11	10
$C_{17:0}$†	9	4
$C_{18:0}$	2	3
$C_{18:1}$	29	34
$C_{19:0}$†	3	5
Unknown	2	1
Saturated	46	45
Unsaturated‡	52	53

* Cells were harvested at stationary phase of growth.
† Cyclopropane acids.
‡ Includes cyclopropane acids.

In contrast, fatty acid synthesis by *Lactobaccilus plantarum* is repressed by the addition of a number of unsaturated and cyclopropane fatty acids to the growth medium (Henderson & McNeil, 1966; Weeks & Wakil, 1970).

FATTY ACID REPLACEMENTS IN FATTY ACID AUXOTROPHS OF *E. coli*

A wide variety of unsaturated fatty acids can meet the growth requirements of unsaturated fatty acid auxotrophs of *E. coli* (Silbert, Ruch and Vagelos, 1968; Esfahani, Barnes & Wakil, 1969; Schairer & Overath, 1969; Esfahani, Ioneda & Wakil, 1971). However, the extent to which the exogenous fatty acid is incorporated into the phospholipids of the auxotrophs is tightly regulated and depends on the physico-chemical properties of the fatty acid. When each member of a structurally homologous series of *cis*-unsaturated acids serves as a growth factor for the unsaturated fatty acid auxotroph strain OL_2^-, the percentage of saturated fatty acids present in the lipids decreases with increasing chain length ($C_{14:1}$, $C_{16:1}$, $C_{18:1}$) or with decreasing the degree of unsaturation ($C_{18:3}$, $C_{18:2}$, $C_{18:1}$) in the apolar group of the supplement (Table 2). *trans*-Octadecenoic acids are incorporated at levels higher than the corresponding *cis*-isomers (the same table). Thus, it appears that the organism attempts to compensate for the condensing effects of *trans* acids by minimizing the levels of saturated fatty acids in the lipids, and for the expanding effects of short chain or polyunsaturated fatty acids by effecting an elevated level of saturated acids in the membrane lipids. This regulation presupposes the existence of a control mechanism for the synthesis of lipids having physicochemical properties within limits.

Table 2. *Effect of exogenous unsaturated fatty acids on the percentage of saturated and unsaturated fatty acids of phospholipids of* OL_2^- *strain*

Fatty acid added to growth medium	Fatty acids in phospholipids*			Satur./Unsatur.
	Saturated	Unsaturated	Unknown	
	(%)	(%)	(%)	
cis-Δ^9-$C_{14:1}$	70	29	1	2·41
cis-Δ^9-$C_{16:1}$	52	46	2	1·13
cis-Δ^9-$C_{18:1}$	45	53	2	0·84
All *cis*-$\Delta^{9,12}C_{18:2}$	60	37	3	1·62
All *cis*-$\Delta^{9,12,15}C_{18:3}$	66	30	4	2·20
cis-Δ^{11}-$C_{18:1}$	45	53	2	0·84
trans-Δ^{11}-$C_{18:1}$	29	69	2	0·42
trans-Δ^9-$C_{18:1}$	27	65	7	0·41

* Fatty acids with eighteen carbon atoms were partly degraded to the corresponding C_{16} acids, but no chain elongation, or *cis*, *trans*-isomerization or change in the degree of unsaturation was noted (Esfahani *et al.*, 1969).

RELATIONSHIP BETWEEN FATTY ACID SYNTHETASE AND MEMBRANE LIPID SYNTHESIS

A direct relationship is observed between the activity of the fatty acid synthetase *in vivo* and the amount of fatty acid required for membrane phospholipid synthesis (Esfahani, Ioneda & Wakil, 1971). The amount of fatty acids synthesized from ^{14}C-acetate by resting celis of strain *civ*-2 *fao*-6 (a β-oxidation negative derivative of OL_2^-) (Esfahani, Ioneda & Wakil, 1971) was highest in those cells which contain the highest level of saturated fatty acids in the phospholipids (Table 3). Furthermore, when this mutant was grown on *cis*-Δ^{11}-eicosenoic acid, its phospholipids contained a high level of

Table 3. *Incorporation of* ^{14}C*-acetate into fatty acids by resting cells of civ*-2 *fao*-6

Fatty acid added to growth medium	^{14}C-fatty acid synthesized	Saturated fatty acids in phospholipids
	(cpm/mg cells (dry weight))	(% Total)
Elaidic	10,500	13
Oleic	17,800	41
Linolenic	42,000	63

myristic acid (cf. Table 4), indicating that the cells are also able to regulate the synthesis of the type of fatty acid in response to growth conditions. This qualitative response of the synthetase to requirements for phospholipid synthesis was further demonstrated by determining the composition of the fatty

Table 4. *Effect of exogenous fatty acids on the fatty acid composition of phospholipids of strain civ-2 fao-6*

Fatty acids in phospholipids	Fatty acid added to growth medium	
	cis-Δ^9-$C_{18:1}$	*cis*-Δ^{11}-$C_{20:1}$
	(%)	(%)
$C_{14:0}$	4	25
$C_{16:0}$	37	46
$C_{18:0}$	1	—
$C_{18:1}$	58	—
$C_{20:1}$	—	30

acids synthesized by resting cells from ^{14}C-acetate. As shown in Table 5, the proportions of lauric and myristic acids were much higher in the *in vivo* products of the fatty acid synthetase of cells containing eicosenoate than in those of the synthetase of cells grown on oleate. Soluble extracts of the eicosenoate-grown cells contained the usual fatty acid synthetase which yields

Table 5. *Composition of fatty acids synthesized from ^{14}C-acetate by resting cells and extracts of civ-2 fao-6 mutant*

System	Fatty acid added to growth medium	Fatty acid synthesized			
		$C_{12:0}$	$C_{14:0}$	$C_{16:0}$	$C_{18:0}$
		(% Total)			
Resting cells	*cis*-Δ^9-$C_{18:1}$	8	11	79	2
	cis-Δ^{11}-$C_{20:1}$	20	24	53	3
Soluble extracts	*cis*-Δ^{11}-$C_{20:1}$	2	7	65	27

palmitate and stearate but very little myristate or laurate (Table 5), suggesting that either cellular organization is important for this type of regulation or that a factor controlling chain length is lost during the preparation of the soluble enzymes. The alteration of the fatty acid synthetase so that a high proportion of myristic acid is synthesized suggests that a feedback type regulation exists between fatty acid synthesis and membrane lipid synthesis. This type of regulation presupposes a direct interaction of the fatty acid synthetase and the lipid-synthesizing enzymes and requires their close localization within or near the cell membrane where lipid synthesis occurs. Experiments of Vagelos and co-workers (van den Bosch, Williamson & Vagelos, 1970) conclusively support this by demonstrating the localization of ACP (acyl carrier protein) (and presumably the fatty acid synthetase) on the cell membrane of *E. coli*.

IMPORTANCE OF THE STRUCTURE OF APOLAR GROUP OF MEMBRANE LIPIDS

The unsaturated fatty acid auxotrophs of *E. coli* have been useful in the investigations on the effect of alterations of lipid composition on membrane-associated properties (Fox, 1969; Esfahani, Barnes & Wakil, 1969; Overath, Schairer & Stoffel, 1970; Esfahani, Ioneda & Wakil, 1971; Wakil & Esfahani, 1971). Fox (1969) has shown that induction and maintenance of lactose

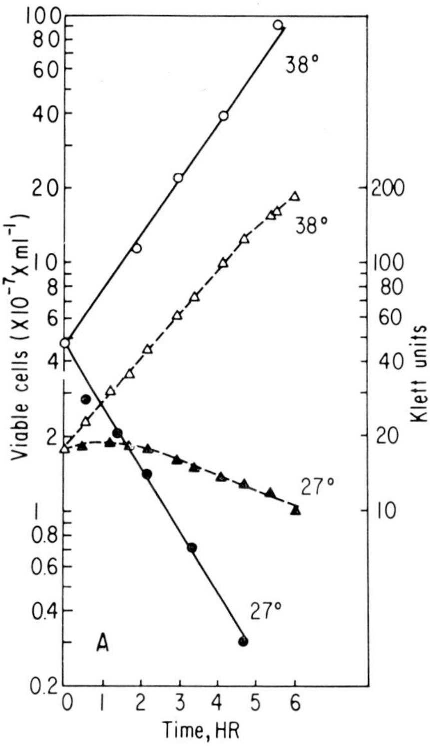

FIG. 1. Effect of temperature on growth of OL_2^- supported by elaidic acid. Cells growing exponentially on elaidic acid were incubated at the indicated temperatures. Viability was determined by serial dilution and plating on agar plates. △, ▲, Klett unit (blue filter); ○, ●, viability (Esfahani, Barnes & Wakil, 1969).

permease requires an appropriate unsaturated fatty acid. Cells growing exponentially on *trans*-octadecenoic acid or petroselenic acid at 37°, lyse when incubated at 27° or 18°, respectively (Esfahani, Barnes & Wakil, 1969; Esfahani, Ioneda & Wakil, 1971, and Fig. 1). Cell death at low temperature is prevented by addition of oleic acid to the culture which is incorporated into all major phospholipids of *E. coli* (Esfahani, Ioneda & Wakil, 1971).

Wide-angle X-ray diffraction techniques have been employed to investigate the physical state of lipids in isolated membranes of strain OL_2^- (Esfahani *et al.*, 1971). Lipids in these membranes undergo thermotropic phase transitions. The transition is characterized by a gradual change from a diffuse band centered at a Bragg spacing of 4·6 Å to a sharp band at 4·2 Å, as the temperature is lowered (Fig. 2). The temperature at which phase changes are detected depends on the extent of incorporation of the exogenous unsaturated fatty acid into the membrane lipids as well as its physico-chemical properties (Table 6). In all these instances the transition temperatures were at or below the temperature at which the cells were grown.

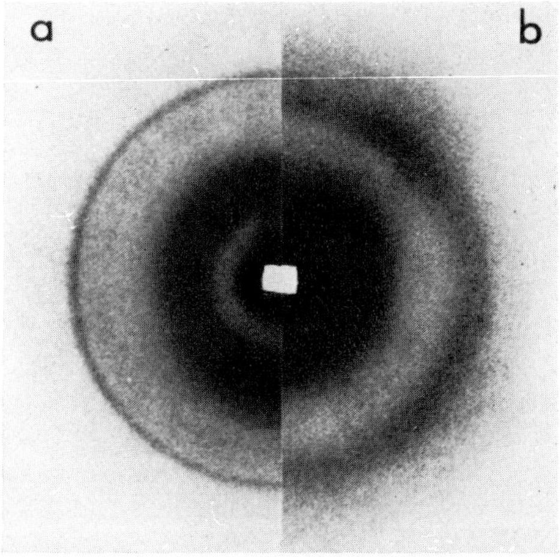

FIG. 2. X-ray diffraction pattern from *E. coli* strain OL_2^- membranes. (a) Pattern from specimens below the transition showing (from left to right) a diffuse band due to water at 3·4 Å, a sharp band due to lipids at 4·2 Å, and a diffuse band at 10·5 Å. (b) Pattern from specimen above the transition temperature showing (from right to left) a diffuse band at 3·4 Å, a diffuse band at 4·6 Å, and a diffuse band at 10·5 Å (from Esfahani *et al.*, 1971).

Table 6. *Effect of supplemental fatty acids on transition temperatures in membrane vesicles derived from cells supplemented with various unsaturated acids*

Fatty acid added to growth medium	Saturated fatty acids in phospholipids	Transition temperature (°C) as determined by		
		X-ray diffraction*	Proline transport	Succinic Dehydrogenase
	(%)			
Oleate	50	19–29 (24)	19	11
Linolenate	70	36–46 (41)	14	23
Myristoleate	70	36–46 (41)	—†	—†
Elaidate	20	30–40 (35)	26	22

* The values given here represent the range of temperatures during which transition occurs. The values in parentheses represent the mid-point transition temperature.
† The transition temperatures for these membranes were not determined.

The similarity between the broad 4·6 Å band obtained from the membranes and that from liquid paraffins (Warren, 1933; Schmitt, Bear & Clark, 1935; Engelman, 1970) indicates that the apolar part of the membrane lipids is in a liquid state. The sharp 4·2 Å band is a diffraction feature of long chain paraffins in a closely packed hexagonal state (Muller, 1932). On this basis the gradual thermal transition observed in the diffraction patterns of the membrane is attributed to a phase change in the fatty acyl chains from a fluid state to a closely packed condensed state.

The Arrhenius plots of several membrane-associated activities exhibit discontinuities at temperatures that are unique for each membrane. The discontinuities in the Arrhenius plots are indicative of a phase change in the lipids involved in the expression of the given function (Overath, Schairer & Stoffel, 1970; Wilson & Fox, 1971). That is, the increase in activation energy below the temperature at which discontinuity occurs is a reflection of the closer packing of the hydrocarbon chains of the membrane lipids. For example, Fig. 3 shows that an Arrhenius plot of the initial rate of proline uptake is biphasic. The temperature at which discontinuity in the plot was detected varied depending upon the type of fatty acids present in membrane lipids. In membranes derived from oleate-grown cells, the transition temperature of proline transport is 19° whereas the transition temperature in membranes containing linolenate or elaidate is 14° or 26°, respectively (cf. Table 6). This finding suggests a sensitivity of the proline transport system to the degree of unsaturation in the supplemental fatty acids. Similar conclusions have been

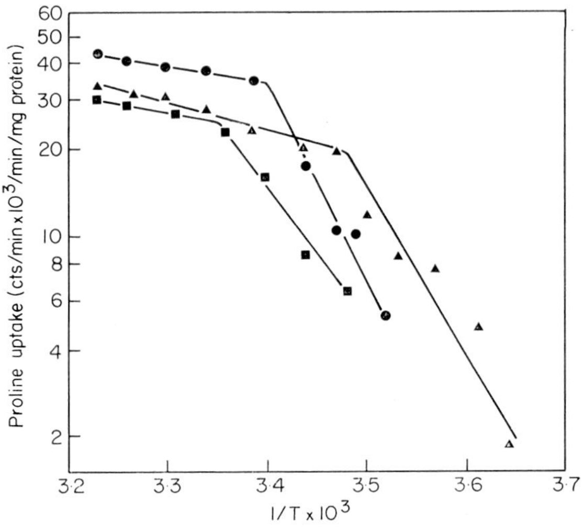

FIG. 3. Arrhenius plots for proline uptake by membranes of strain OL_2^-. Membranes were prepared from cells supplemented with oleate (●), elaidate (■), and linolenate (▲), (Esfahani *et al.*, 1971).

drawn by other investigators for sugar transport (Wilson, Rose & Fox, 1970; Overath, Schairer & Stoffel, 1970; Wilson & Fox, 1971). These observations indicate that the lipid sites in the membrane associated with proline uptake undergo a phase change from a less ordered state above the transition to a more ordered state below it.

The succinic dehydrogenase activity of the membranes also exhibited discontinuities in the Arrhenius plots which are unique for each membrane type (Fig. 4). However, the order at which the discontinuities were observed does not follow the order in the degree of unsaturation in the supplemental fatty acids (cf. Table 6). This is indicative of an association of the enzyme with lipids structurally distinguishable from those associated with the sites involved in the transport of proline.

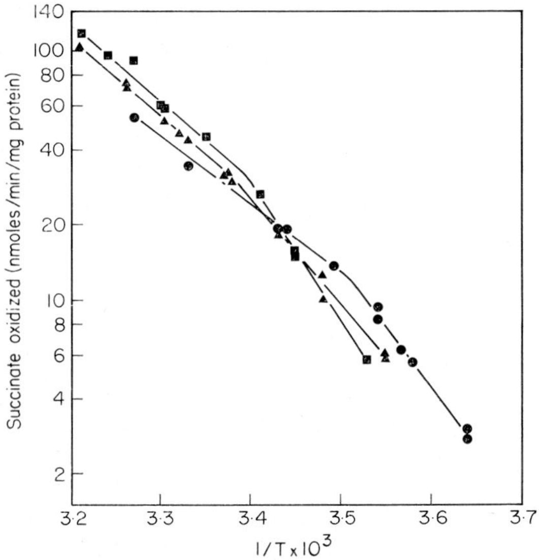

FIG. 4. Arrhenius plots for succinic dehydrogenase activity of membranes derived from strain OL_2^- grown on oleate (●), elaidate (■), and linolenate (▲) (Esfahani *et al.*, 1971).

This conclusion is based primarily on the assumption that transitions in activation energies were due to changes in enzymic activities in response to phase changes in lipids at different temperatures. In other words, organization of the lipids within the membrane and packing of the hydrocarbon chains of these lipids are responsible for this transition and alterations of this organization would affect this transition. Data presented in Fig. 5 show that the Arrhenius plot of succinic dehydrogenase is continuous in deoxycholate treated vesicles but discontinuous in normal membranes. This finding is consistent with the aforementioned interpretation but by all means is not proof for it. Further information is necessary in order to determine the exact relationship between lipids and membrane enzyme activities.

For each membrane the temperatures at which discontinuities were detected in the Arrhenius plots of proline transport or succinic dehydrogenase activities are different from each other, and occur below the temperature at which physical changes were noted in the lipids with X-ray diffraction. These observations are most readily explained in terms of heterogeneity in the molecular organization of the membrane. That is, within a given membrane, a specific class of lipids having a physical character determined by its apolar as well as polar groups, is associated with a specific enzymic activity.

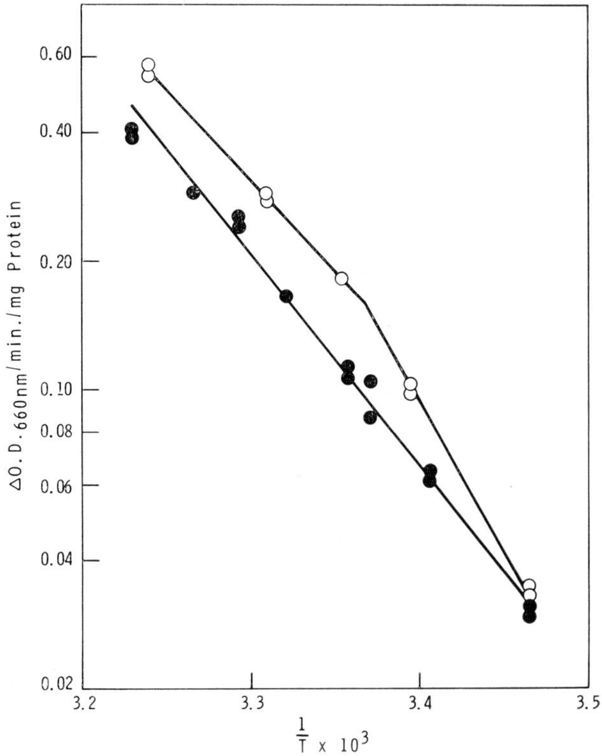

FIG. 5. Arrhenius plots for succinic dehydrogenase activity of membranes derived from strain OL_2^- grown on elaidate. Assay conditions were the same as described previously (Esfahani *et al.*, 1971). ●, in the presence of 0·05% sodium deoxycholate; ○, in the absence of deoxycholate.

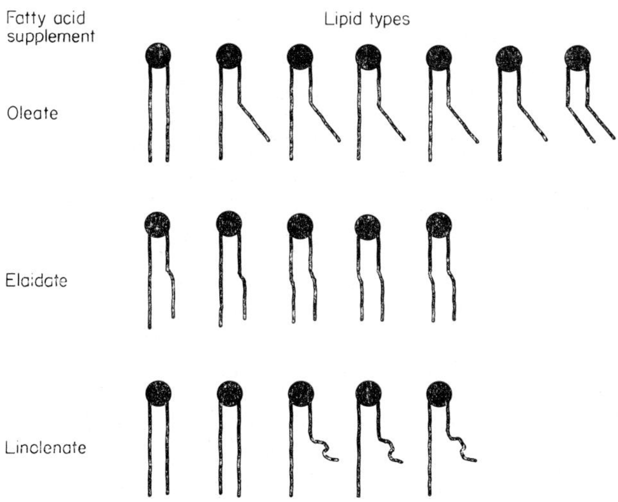

FIG. 6

Considering only the apolar groups, the possible phospholipid species available in membranes from cells grown on oleate, elaidate, or linolenate are shown in Fig. 5. This presentation was based on the analysis of the fatty acid composition of membrane phospholipids and on positional distribution of fatty acids in each phospholipid of the unsaturated fatty acid auxotroph of *E. coli* (Silbert, 1970). Since in elaidate-grown cells 80% of the fatty acids of membrane phospholipid is elaidic acid and since the physical properties of elaidate are close to those of palmitate (Chapman, Owens & Walker, 1966), the fluid character of the different lipid components of these cells is *more similar* to each other than that of the lipids of linolenate-enriched membranes. Lipids of oleate-rich membranes are intermediates in their fluid characteristics. The fact that the lipids of linolenate-enriched membranes contain 30% linolenate tends to provide a wider range in the fluid character of the different phospholipids as compared to those of elaidate-containing membranes. The near homogeneous nature of the lipid components in elaidate membranes is further reflected by the observation that the transition temperatures for proline transport and succinic dehydrogenase activity are closer to each other and to the phase transition detected by X-ray diffraction than is the case for membranes containing oleic or linolenic acids. While the order of the transition temperatures for proline transport follows the order in the degree of fluidity of the fatty acid supplement, this is not the case for succinic dehydrogenase activity. This observation suggests that proline transport has more stringent requirements for highly fluid membrane lipid. A similar conclusion was reached by other workers (Overath, Schairer & Stoffel, 1970; Wilson, Rose & Fox, 1970, and Wilson & Fox, 1971) who found that sugar transport is associated with the most liquid sites of the membrane. Altogether these results suggest that lipids are heterogeneously distributed within the membrane and that a specific lipid(s) is required for the optimal activity of membrane enzymes.

ACKNOWLEDGEMENTS

This investigation was supported in part by Grant GM-06242 from the National Institutes of Health, and by a grant from the American Cancer Society.

REFERENCES

Burton, A. J. & Carter, H. E. (1964). *Biochemistry* **3**, 411.

Chapman, D., Owens, N. F. & Walker, D. A. (1966). *Biochim. biophys. Acta* **120**, 148.

Cronan, T. E., Jr. (1968). *J. Bact.* **95**, 2054.

Engelman, D. M. (1970). *J. molec. Biol.* **47**, 115.

Esfahani, M., Barnes, E. M., Jr. & Wakil, S. J. (1969). *Proc. natn Acad. Sci. U.S.A.* **64**, 1057.

Esfahani, M., Ioneda, T. & Wakil, S. J. (1971). *J. biol. Chem.* **246**, 50.

Esfahani, M., Limbrick, A. R., Knutton, S., Oka, T. & Wakil, S. J. (1971). *Proc. natn Acad. Sci. U.S.A.* **68**, 3180.

Fox, C. F. (1969). *Proc. natn. Acad. Sci. U.S.A.* **63**, 850.

Henderson, T. O. & McNeil, J. J. (1966). *Biochem. biophys. Res. Commun.* **25**, 662.

Kaneshiro, T. & Marr, A. G. (1961). *J. biol. Chem.* **236**, 2615.

Marr, A. G. & Ingraham, J. L. (1962). *J. Bact.* **84**, 1260.

Muller, A. (1932). *Proc. R. Soc. A*, **138**, 514.

Norris, A. T., Matsumara, S. & Bloch, K. (1964). *J. biol. Chem.* **239**, 3653.
Okuyama, H. (1969). *Biochim. biophys. Acta* **176**, 125.
Overath, P., Schairer, H. U. & Stoffel, W. (1969). *Proc. natn Acad. Sci. U.S.A.* **64**, 606.
Pugh, E. L., Sauer, F., Waite, M. B., Toomey, R. E. & Wakil, S. J. (1966). *J. biol. Chem.* **241**, 2635.
Romeo, D., Binckley, A. & Rothfield, L. (1970). *J. molec. Biol.* **53**, 491.
Rothfield, L. & Finkelstein, A. (1968). *Ann. Rev. Biochem.* **37**, 463.
Schairer, H. U. & Overath, P. (1969). *J. molec. Biol.* **44**, 209.
Schmitt, F. O., Bear, R. S. & Clark, G. L. (1935). *Radiology* **25**, 131.
Silbert, D. F., Ruch, F. & Vagelos, P. R. (1968). *J. Bact.* **95**, 1658.
Van den Bosch, H., Williamson, J. R. & Vagelos, P. R. (1970). *Nature, Lond.* **228**, 338.
Wakil, S. J. & Esfahani, M. (1971). In *Biochemical Responses to Environmental Stress.* (Ed. by Bernstein, I. A.), p. 15. New York: Plenum Press.
Warren, B. E. (1933). *Phys. Rev.* **44**, 969.
Weeks, G. & Wakil, S. J. (1970). *J. biol. Chem.* **245**, 1913.
Wilson, G. & Fox, C. F. (1971). *J. molec. Biol.* **55**, 49.
Wilson, G., Rose, S. P. & Fox, C. F. (1970). *Biochem. biophys. Res. Commun.* **38**, 617.

NOTE ADDED IN PROOF

Recent work in our laboratory clearly demonstrates that the transitions noted in the Arrhenius plots of succinic dehydrogenase (or more precisely, succinic-ubiquinone reductase activity) is dependent on the phospholipids, and that the temperature at which this transition occurs is determined by the nature of the apolar group of the lipids. (Esfahani, M., Crowfoot, P. D. & Wakil, S. J. (1972). *J. biol. Chem.* **247**, in press.) These observations, therefore, corroborate the conclusions drawn in this presentation. In addition, recent studies on mixtures of leithins have given evidence for cluster formation and phase segregation (Phillips, M. C., Leadbrooke, B. D. & Chapman, D. (1970). *Biochim. biophys. Acta* **196**, 35; Oldfield, E., Keough, K. M. & Chapman, D. (1972). *Fed. Eur. Biochem. Soc. Lett.* **20**, 344; Phillips, M. C., Hauser, H. & Paltauf, F. (1972). *Chem. Phys. Lipids* **9**, 68). This segregation can account for the formation of heterogeneous gel systems whenever the temperature is intermediate between the transition temperatures of the component lipids, as for example, in the membrane of cells growing on linolenic acid at 37°C. When such a membrane is warmed to a temperature above that of the gel to liquid-crystalline phase transition of the saturated phospholipid components, the bulk of the lipids may attain a homogeneous state due to latteral diffusion (Kornberg, R. D. & McConnel, H. M. (1971). *Proc. natn Acad. Sci. U.S.A.* **68**, 2561) and phase mixing of lipids. Conversely it is conceivable that the membranes with uniform viscosity at growth temperature, when cooled to temperature of the highest "melting" lipid component, may attain domains of gel coexisting with liquid crystals through lateral diffusion and phase segregation. It is also likely that some components of lipids are immobilized by proteins through apolar and/or polar interactions. This then would give rise to a protein-dependent heterogeneity in the molecular organization of the lipids.

LIPIDS OF SEMLIKI FOREST VIRUS AND OF HOST CELL MEMBRANES

By O. RENKONEN, L. KÄÄRIÄINEN, C. G. GAHMBERG AND K. SIMONS

Department of Biochemistry, University of Helsinki, Helsinki, Finland

WE have chosen Semliki Forest virus as a model for the study of biological membranes at the cellular level. We shall compare the viral membranes with those of the host cell, and we intend to show that the viral membrane is relatively simple. Similar findings have been reported for Sindbis virus, a close relative of Semliki Forest virus (Strauss *et al.*, 1968).

In our work BHK21 cells are used for culturing the virus. These cells are derived from baby hamster kidney and are grown in monolayer cultures. Electron microscopy has revealed that these cells contain all the organelles typical of animal systems, such as plasma membrane, endoplasmic reticulum, mitochondria, Golgi body, and nucleus (Fig. 1). All these organelles contain lipoprotein membranes.

Some of these membranes have been isolated in our laboratory (Gahmberg & Simons, 1970), and their chemical composition has been studied. Figure 2 shows the proteins of plasma membrane and endoplasmic reticulum membrane in gel electrophoresis in the presence of sodium dodecyl sulphate. It is evident that the two membranes contain many different protein species (Gahmberg, 1971), and that their protein patterns are also rather different.

Phospholipids are the major lipid components of these two membranes (Renkonen *et al.*, 1971*a*). A number of different phospholipid classes are found in the two membranes by two-dimensional thin-layer chromatography (Figs. 3 and 4). The relative amounts of individual phospholipid classes are different in them. For instance, the plasma membrane is relatively rich in GPS-lipids, and poor in GPI-lipids, whereas the reverse is true for the endoplasmic reticulum membrane.

Our experience with the BHK21 membranes suggests that they are fairly difficult to isolate in a satisfactory state of purity and that there are very complex structures containing many protein and lipid species. Both these factors are general and major problems in contemporary membrane research. These problems are considerably reduced, however, when the viral membrane is selected as the object of study.

Semliki Forest virus is an RNA virus. Its natural hosts are insects and mammals. It was first isolated from mosquitoes of Semliki Forest in Africa. This virus is a convenient object of study, because it is harmless to man

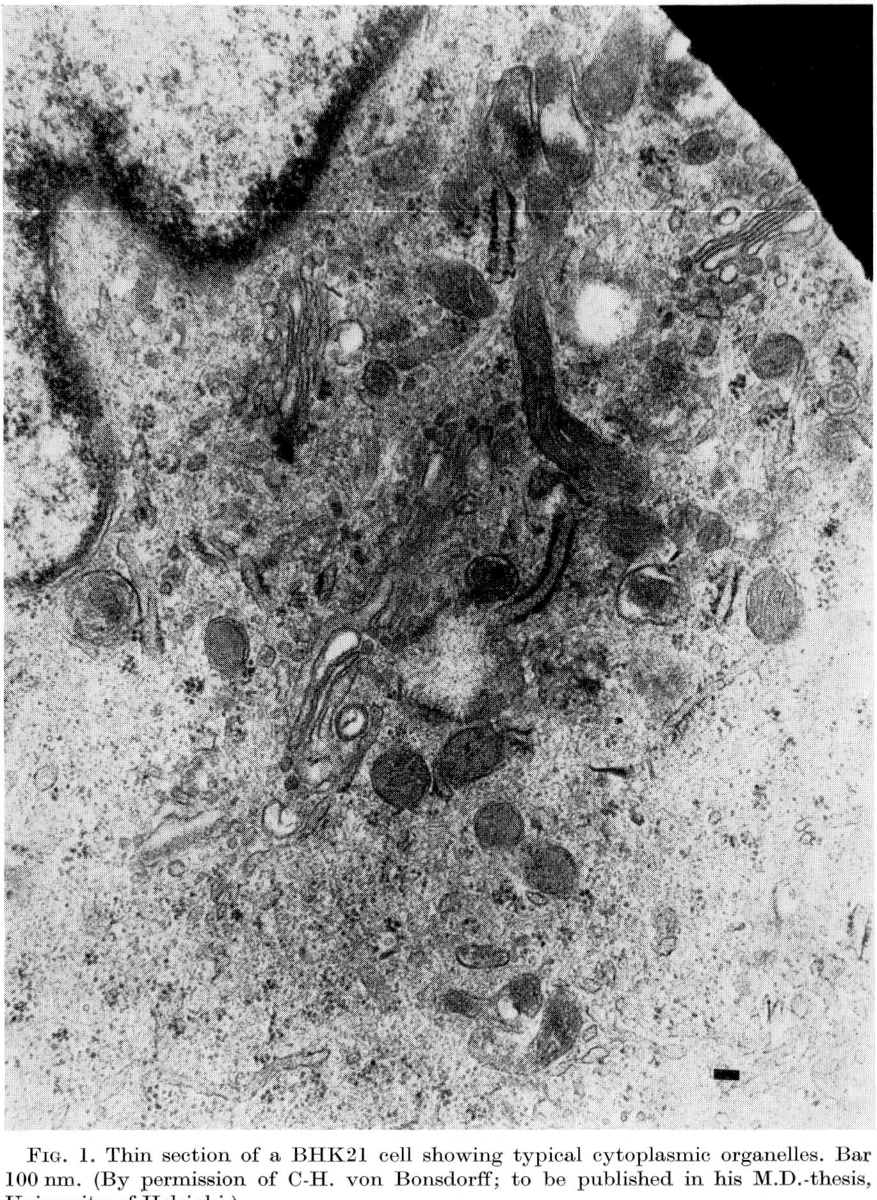

FIG. 1. Thin section of a BHK21 cell showing typical cytoplasmic organelles. Bar 100 nm. (By permission of C-H. von Bonsdorff; to be published in his M.D.-thesis, University of Helsinki.)

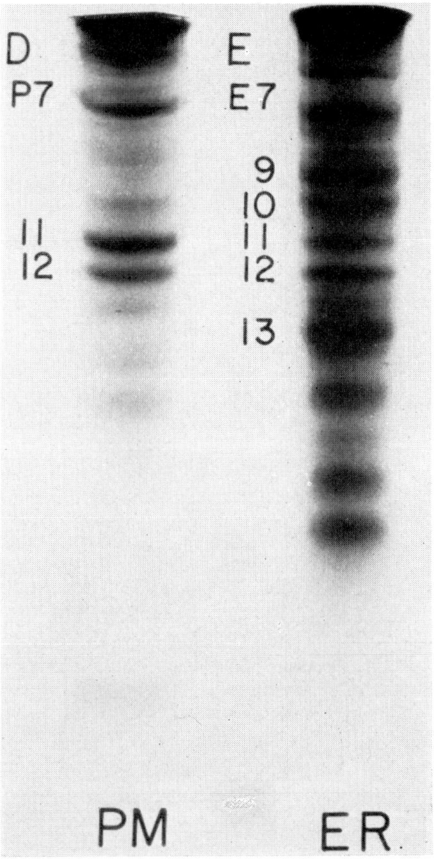

FIG. 2. Disc electrophoresis of plasma membrane and endoplasmic reticulum of BHK21 cells. (PM = Plasma membrane; ER = Endoplasmic reticulum). (From Gahmberg, 1971.)

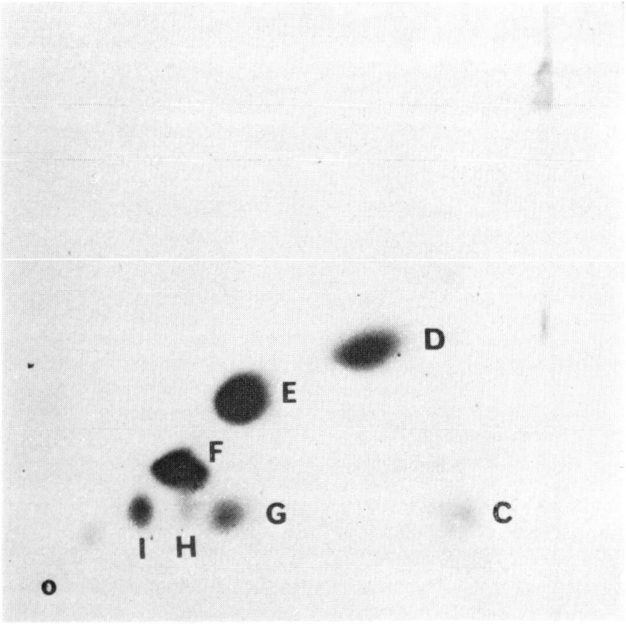

FIG. 3. Two dimensional-TLC of phospholipids from plasma membrane of BHK21 cells. (G = GPS-lipids; H = GPI-lipids.)

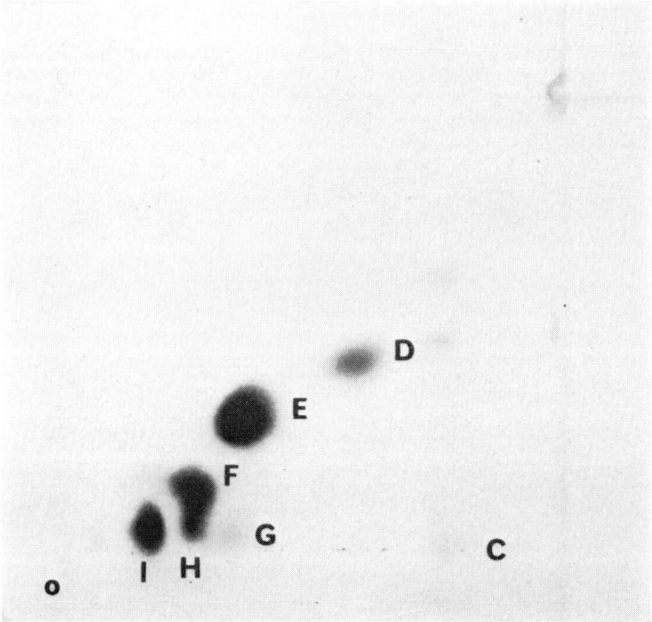

FIG. 4. Two dimensional-TLC of phospholipids from endoplasmic reticulum of BHK21 cells. (G = GPS-lipids; H = GPI-lipids.)

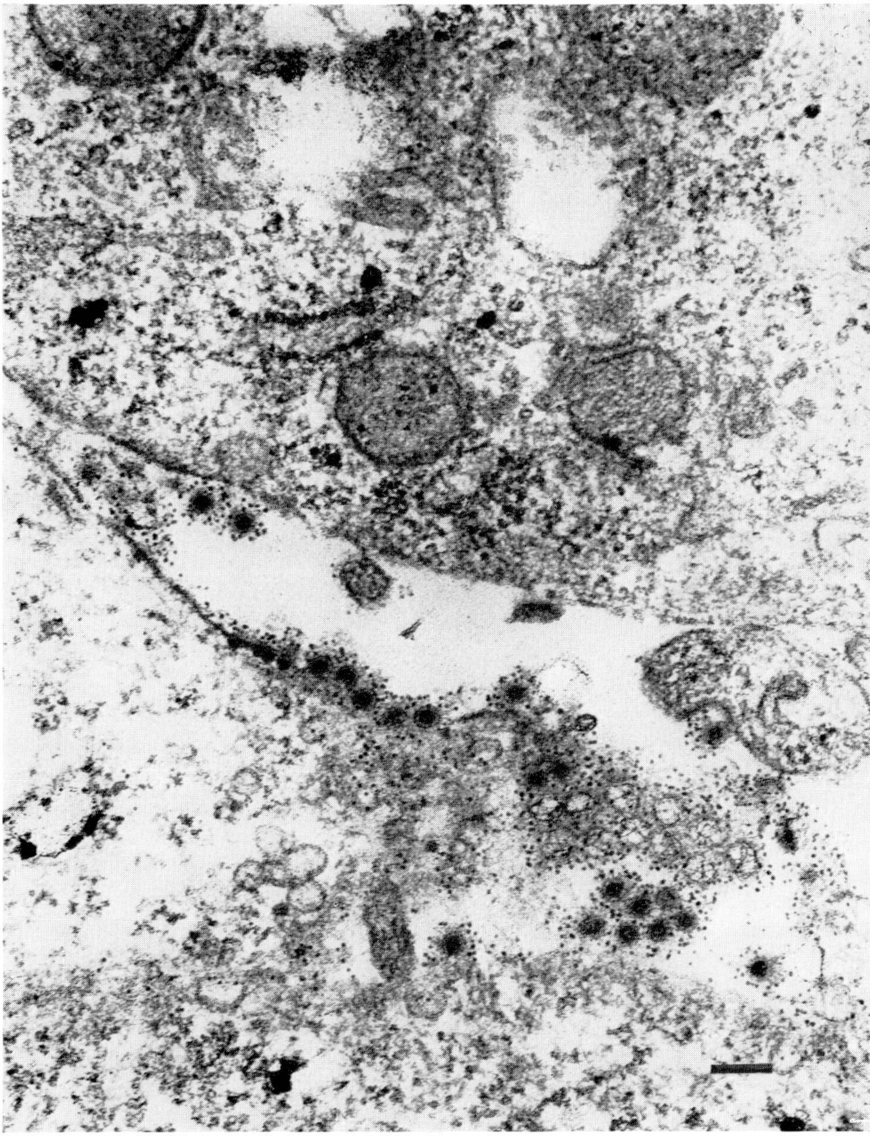

FIG. 5. Part of the plasma membrane of a BHK21 cell infected with Semliki Forest virus. Virus particles at different stages of the budding process are visible. Anti-Semliki Forest virus-ferritin conjugate was used to identify the virus. Bar 100 nm. (By permission of C-H. von Bonsdorff; to be published in his M.D. thesis, University of Helsinki.)

although it kills mice. It can also be easily cultured in various cells including the BHK21 cells. Each infected cell produces in one cycle a few thousand viruses which are released into the culture medium.

Electron microscopic evidence suggests that Semliki Forest virus matures by budding through the host plasma membrane (Acheson & Tamm, 1967). Figure 5 shows an electron micrograph of this event in our cells. A number of virus particles are seen at different stages of the budding through the plasma membrane. It looks as if the core particles of the virus become surrounded by the cellular plasma membrane during the process.

We isolate the virus from the culture medium in the following three steps: (1) the cells and debris are removed by low speed centrifugation, (2) the virus

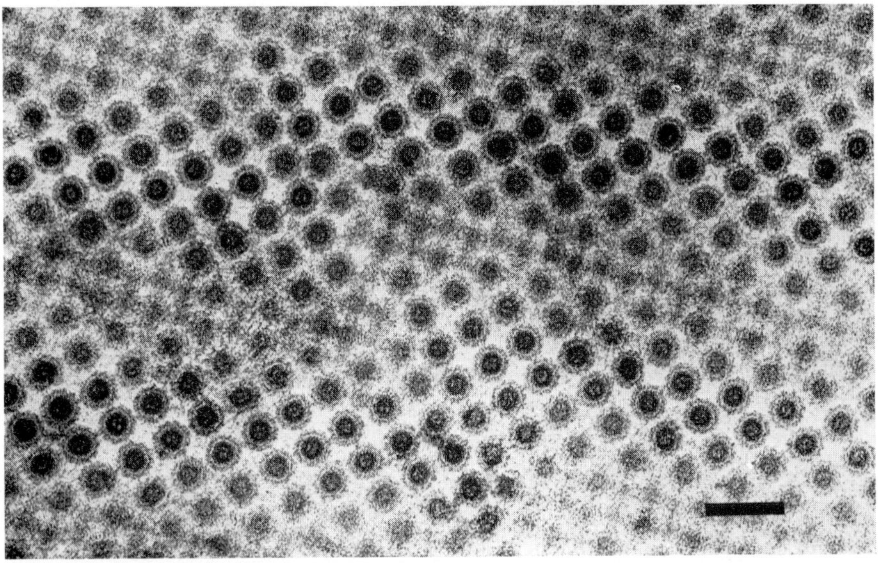

FIG. 6. Thin section of pelleted Semliki Forest virus. Bar 100 nm. (By permission of C-H. von Bonsdorff; to be published in his M.D. thesis, University of Helsinki.)

is concentrated and purified by continuous flow zonal centrifugation in sucrose, and (3) final purification is achieved by a combined velocity and isopycnic centrifugation. The purified virus looks very homogeneous on electron microscopy (Fig. 6). In fact it is so pure that it produces crystal-like structures. The high state of purity of the virus preparation is revealed also by immunodiffusion and by a number of quantitative analytical data which are described below. Preparations of the cellular membranes cannot be isolated in similar state of purity and this is the first great advantage of the virus as a membrane model.

The virus can be split by detergent treatment (Kääriäinen, Simons & von Bonsdorff, 1969). Incubation with detergent and subsequent gradient centrifugation gives two fractions: a fast sedimenting fraction containing RNA and protein, and a slow sedimenting fraction containing protein and lipid. The former fraction contains the intact ribonucleoprotein, i.e. the nucleocapsid of the virus, while the latter contains pieces of broken envelope membrane which surrounds the nucleocapsid in the virus.

When the whole virus and its two fractions are submitted to gel electrophoresis in the presence of sodium dodecylsulphate, the picture that emerges is shown in Fig. 7. The virus contains two protein species: the larger being in the membrane and the smaller in the nucleocapsid (Kääriäinen *et al.*, 1969; Simons & Kääriäinen, 1970). The fact that Semliki Forest virus envelope contains only one protein species is the second great advantage in considering the virus as a membrane model.

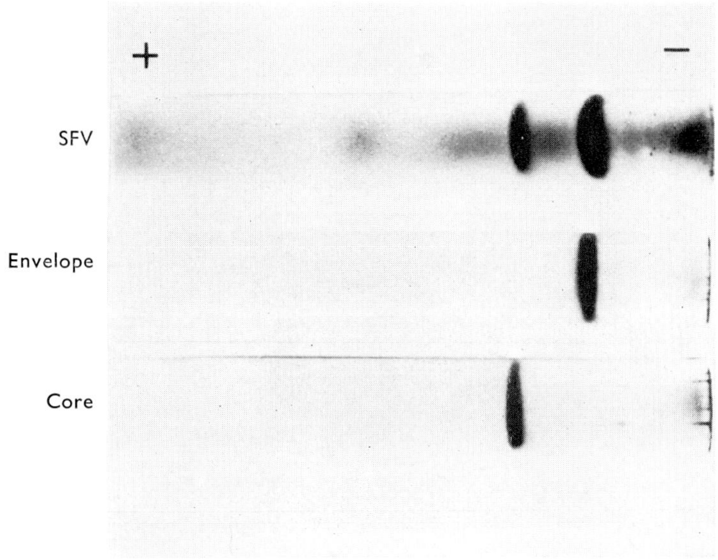

FIG. 7. Disc electrophoresis of Semliki Forest virus and its nucleocapsid and membrane proteins (Simons & Kääriäinen, 1970). Schlesinger *et al.* (1972) have succeeded in separating the membrane protein in two components.

The overall composition of Semliki Forest virus has been determined adequately. From our unpublished data we can estimate that it consists of 6·4% RNA, 56% protein, 31% lipid and 6·6% protein-bound carbohydrate. The viral membrane contains about 80% of the total protein, and all of the lipid and carbohydrate; the lipids consist of 32% neutral lipids, 61% phospholipids, and 7% glycolipids (Renkonen *et al.* 1971*b*).

Thin-layer chromatography showed that the neutral lipid fraction of Semliki Forest virus consists almost exclusively of free cholesterol (Fig. 8, Renkonen *et al.*, 1971*b*).

Two-dimensional thin-layer chromatography of the phospholipids showed that the main components are sphingomyelin and GPE-, GPC- as well as GPS-lipids, either in diacyl, or in alkenyl-acyl, or in alkyl-acyl form (Fig. 9; Renkonen *et al.*, 1971*b*). The viral phospholipid components are thus qualitatively similar to those in the host cell membranes.

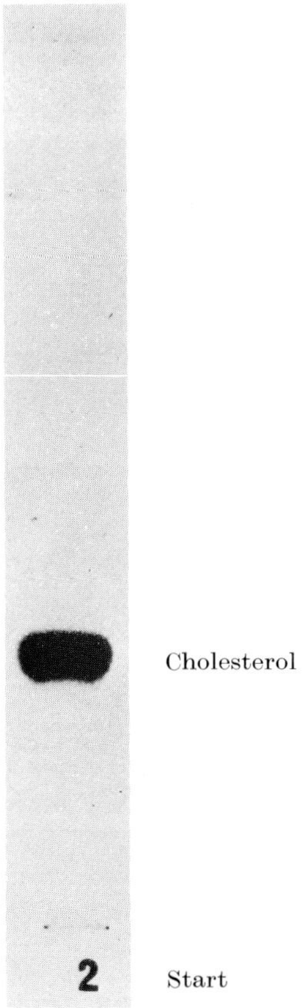

FIG. 8. TLC of neutral lipids from Semliki Forest virus.

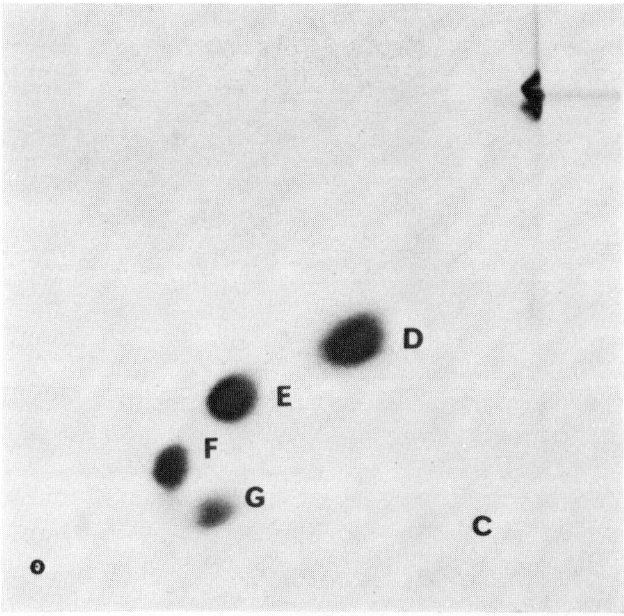

FIG. 9. Two dimensional-TLC of phospholipids from Semliki Forest virus. D = GPE-lipids; E = GPC-lipids; F = sphingomyelin; G = GPS-lipids; O = start.

The glycolipid fraction of the virus consists almost exclusively of sialolactosyl ceramides (Fig. 10; Renkonen *et al.*, 1971*b*). These lipids also predominate in the glycolipid fraction of the host cell membranes.

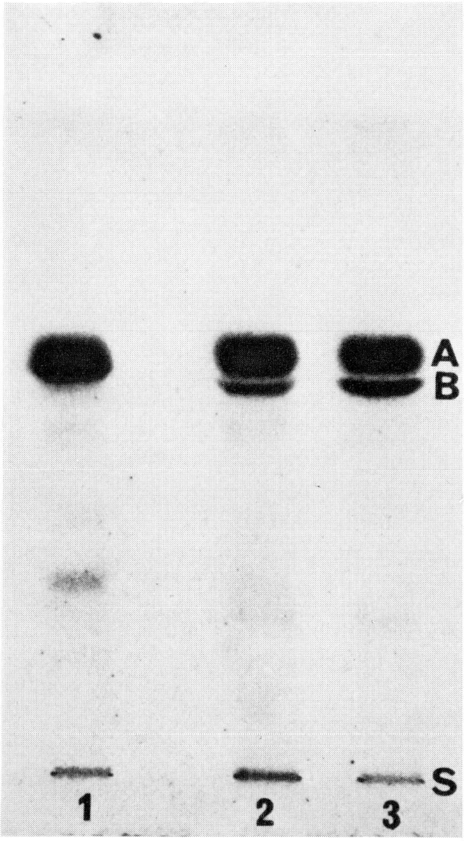

FIG. 10. TLC of gangliosides. 1. N-acetyl neuraminyl lactosyl ceramide. 2. Gangliosides of Semliki Forest virus. 3. Gangliosides of plasma membrane of BHK21 cells. A = N-acetyl neuraminyl lactosyl ceramide; B = N-glycolyl neuraminyl lactosyl ceramide; S = start. (From Renkonen *et al.*, 1971*b*.)

The virus contains 0·99 moles of free cholesterol to every mole of total phospholipid (Table 1) and 0·08 moles of sphingoglycolipids, i.e. sialolactosyl ceramides, to every mole of total phospholipid (Table 1). It also contains 0·20 moles of sphingomyelin, 0·23 moles of GPE-lipids, 0·33 moles of GPC-lipids, 0·13 moles of GPS-lipids and 0·02 moles of GPI-lipids to every mole of total phospholipids (Table 1; Renkonen *et al.*, 1971*b*).

When the distribution of the viral lipid classes is compared with that of the host cell membranes, some interesting correlations become evident (Table 1). Thus the viral lipids are remarkably similar to those of the host plasma membrane, but quite dissimilar to those of endoplasmic reticulum membrane. The most significant observations include: (1) high cholesterol values in the virus and plasma membrane, but low values in the endoplasmic reticulum; (2) high sphingoglycolipid values in the virus and plasma membrane but low values in

Table 1. *Lipid class composition (mole ratios) of Semliki Forest virus and the host membranes*

	Semliki Forest virus	Host plasma membrane	Host endoplasmic reticulum
Cholesterol	0·99	0·70	0·29
Glycolipids	0·08	0·10	0·03
Phospholipids	1·00	1·00	1·00
GPE-lipids	0·23	0·18	0·14
GPC-lipids	0·33	0·38	0·55
GPS-lipids	0·13	0·09	0·04
GPI-lipids	0·02	0·02	0·04
Sphingomyelin	0·20	0·18	0·10

the endoplasmic reticulum membrane; (3) high sphingomyelin values in the virus and plasma membrane but low values in the endoplasmic reticulum; (4) high values for the GPS-lipids in the virus and plasma membrane, but low values in the endoplasmic reticulum; (5) low values for the GPC-lipid in the virus and plasma membrane and high values in the endoplasmic reticulum; (6) low values for the GPI-lipids in the virus and the plasma membrane and high values in the endoplasmic reticulum (Renkonen *et al.*, 1971*b*).

These findings, together with the electron microscopic evidence, suggest that the lipids of the viral membrane are derived from the host plasma membrane. If this were the case, the fatty acid compositions should be similar in each individual lipid class of the virus and of the host plasma membrane. Analyses of rat liver by Keenan & Morré (1970) have indicated that the given phospholipid classes may have different fatty acid compositions in the plasma membrane and in the other organelles of the cell. The fatty acid compositions of the four major glycerophospholipid classes of Semliki Forest virus as given in Table 2 show that all four classes have different fatty acid compositions

Table 2. *Fatty acid composition (%) of major glycerophospholipid classes of Semliki Forest virus, of plasma membrane and of endoplasmic reticulum from BHK21-Wi2 cells**

Fatty acid	GPC-lipids			GPS-lipids			GPI-lipids			GPE-lipids		
	SFV	PM	ER	SFV	PM	ER	SFV	PM	ER	SFV	PM	ER
16:0	40	32	22	4	4	4	8	6	3	5	5	5
16:1	7	7	7	3	3	1	3	2	1	2	3	4
18:0	7	8	6	32	28	27	42	39	37	5	8	12
18:1	33	44	50	48	55	43	16	31	29	47	52	56
18:2	7	5	8	8	5	4	4	3	4	13	7	9
C_{20}–C_{24}	2	2	2	2	4	9	27	17	25	16	8	6

* One virus sample and 3–4 samples of host membranes were analysed. GPC = glyceryl phosphoryl choline; GPS = glyceryl phosphoryl serine; GPI = glyceryl phosphoryl inositol; GPE = glyceryl phosphoryl ethanolamine. SFV = Semliki Forest virus; PM = plasma membrane; ER = endoplasmic reticulum.

(Laine *et al.*, 1972). The GPC-lipids contain more palmitic than stearic acid; they are also very poor in long chain polyenoic acids. The GPS-lipids in turn are characterized by relatively low palmitic acid and high stearic acid contents. The GPI-lipids have high stearic acid content like the GPS-lipids, but they are also rich in the long chain polyenoic acids. Finally, the GPE-lipids contain a far lower amount of palmitic and stearic acids than do the other classes.

All four glycerophospholipid classes in the virus and plasma membrane are similar in their fatty acid compositions (Table 2). The lipid classes of the endoplasmic reticulum, too, have the same fatty acid compositions as their counterparts in the plasma membrane (Laine *et al.*, 1972).

In addition to the sphingomyelins the viral phospholipids contain other alkali stable components. Analysis after mild alkaline treatment (Renkonen, 1963) revealed that the viral phospholipids contain 40% of alkali stable compounds. Since there is only 21% of sphingomyelins, the virus contains, by difference, 19% ether phospholipids, i.e. either alkenyl-acyl or alkyl-acyl phospholipids (Table 3). Similar analysis of the host cell membranes showed

Table 3. *Percentage of alkali stable phospholipids in Semliki Forest virus and BHK21-Wi2 cell membranes**

	Semliki Forest virus	Whole BHK21 cells	Plasma membranes	Endoplasmic reticulum
Alkali stable phospholipids	40	25	40	28
Sphingomyelins†	21	7	18	10
Ether phospholipids (by difference)	19	18	22	18

* Results of one analysis.
† Taken from Renkonen *et al.* (1971*a*,*b*).

that their phospholipids, too, contain about 20% of the ether phospholipids (Laine *et al.*, 1972). Analysis of the phosphatides after mild acid treatment (Renkonen, 1963) indicated that the viral and the cellular phospholipids contain about 14% alkenyl-acyl and 5–8% alkyl-acyl phospholipids (Laine *et al.*, 1972).

The amount of alkenyl-acyl phospholipids in the different glycerolipid classes was determined by comparing the viral phospholipid composition before and after mild acid methanolysis (Renkonen, 1963). During the acid treatment the alkenyl-acyl compounds are transformed into lyso compounds, and their amount is reflected by the decrease of the parent phospholipid and simultaneous increase of the corresponding lyso phospholipid. Two-thirds of the viral GPE-lipids are of the alkenyl-acyl type, but the GPC-lipids contain only 3% of these molecules (Table 4; Laine *et al.*, 1972). The host plasma membrane as also the whole host cell are similar to the virus in that 50–60% of their GPE-lipids and only a few per cent of their GPC-lipids are of the

Table 4. *Alkenyl-acyl lipid content in different glycerophospholipid classes of Semliki Forest virus and BHK21-Wi2 cell membranes*

	GPE-lipids	GPC-lipids
	(%)	(%)
Semliki Forest virus	63	3
Plasma membranes of BHK21-Wi2 cells	59	2
BHK21-Wi2 cells	48	4

alkenyl-acyl type (Table 4; Laine *et al.*, 1972). The GPS- and the GPI-lipids of the virus and the host revealed only traces of alkenyl-acyl lipids. The presence of large amounts of alkenyl groups in the GPE-lipids may explain the presence of low amounts of palmitic and stearic acid in these lipids of the virus and the host (Table 2); these acids appear to be substituted by the alkenyl-groups in the GPE-lipids. Gas-liquid chromatography showed that the dimethyl acetals derived from the alkenyl-acyl phospholipids of the virus and the host are quite similar (Table 5; Laine *et al.*, 1972). The major components

Table 5. *Composition of dimethyl acetals derived from the alkenyl groups of plasmalogenic phospholipids of Semliki Forest virus and BHK21-Wi2 cells**

Major aldehyde	Semliki Forest virus	BHK21-Wi2 cells
	(%)	(%)
Total aldehydes	100	100
16:0	33	31
17:0	6	6
18:0	22	24
18:1	38	39

* Results of one analysis.

are saturated and monoenoic straight chain molecules of 16–18 carbon atoms. These are the components commonly found in animal systems.

The alkyl-acyl lipids of the virus and the host are also concentrated in the GPE- and GPC-lipids. Thin-layer chromatography of those phospholipids which resisted mild acid and subsequent mild alkaline treatments revealed the presence of alkyl-GPE and alkyl-GPC, in addition to the sphingomyelins. Quantitative determination of these compounds showed that the viral GPE-lipids contain 7% and the GPC-lipids 10% of the alkyl-acyl lipids (Table 6; Laine *et al.*, 1972). The host cell lipids had similar compositions.

Table 6. *Alkyl-acyl lipid contents of different glycerophospholipid classes of Semliki Forest virus and BHK21-Wi2 cell membranes**

	GPE-lipids	GPC-lipids
	(%)	(%)
Semliki Forest virus	7	10
BHK21-Wi2 cells	9	13

* Results of one analysis. Abbreviations as in Table 2.

The sphingolipids of Semliki Forest virus have a rather different set of component fatty acids than the glycerophospholipids. The sphingolipid fatty acids are characterized by their high contents of saturated and monoenoic acids, and of C_{20}–C_{24} acids, and by the absence of polyenoic acids (Table 7; Laine *et al.*, 1972). These features are common in animal sphingolipids.

Table 7. *Fatty acid composition (%) of major sphingolipid classes of Semliki Forest virus, of plasma membrane and of endoplasmic reticulum from BHK21 cells**

Fatty acid	Gangliosides			Sphingomyelins		
	SFV	PM	ER	SFV	PM	ER
16:0	35	41	46	76	69	78
16:1	6	4	6	1		1
18:0	10	6	8			
18:1	10	6	5			
20:0	1	1	1			
22:0	6	5	7	6	10	7
23:0	2	2	4			
24:0	7	8	7	6		6
24:1	14	16	10	11	22	8

* Results of one analysis. Abbreviations as in Table 2.

The viral gangliosides have a similar fatty acid composition as the gangliosides of host plasma membranes (Table 7), and the gangliosides of the endoplasmic reticulum also have a similar fatty acid composition as those of the plasma membrane.

The viral sphingomyelins also have the same fatty acid composition as the plasma membrane sphingomyelins (Table 7). Even the sphingomyelins of the endoplasmic reticulum are similar in their fatty acids. The data of Table 7 show some differences between the ganglioside and sphingomyelin fatty acids; these may or may not be significant.

The other long chain compound of the sphingolipid molecules, namely the long chain base must be the classical sphingosine, or Δ^4-*trans*-sphingenine, in the virus as well as in all host membranes, because this was the only long chain base that could be found in the gangliosides and in the sphingomyelins of the whole BHK21 cells (Laine *et al.*, 1972).

When the results of our lipid analyses are integrated we see that the lipids of Semliki Forest virus and the host plasma membrane are similar in the following important aspects: (1) composition as defined by the polar groups; (2) composition as defined by the hydrocarbon chains; (3) composition as defined by the linkages between the hydrocarbon chains and the polar groups. On the other hand, the viral lipids are quite different from those of the other membranes of the host. These facts lend further support to the hypothesis that the viral lipids are derived from the host cell plasma membrane. Similar but less extensive analyses have suggested that parainfluenza virus, vesicular stomatitis virus, Rous sarcoma virus and Sindbis virus also resemble the host plasma membrane in their lipid composition (Klenk & Choppin, 1969, 1970; McSharry & Wagner, 1971; Quigley, Rifkin & Reich, 1971; David, 1971).

Our data also allow description of an individual Semliki Forest virus particle. The molecular weights of the components are known with some degree of certainty. They are 4×10^6 daltons for the RNA (Levin & Friedman, 1971); 34,000 daltons for the nucleocapsid protein and 55,000 daltons for the membrane protein (Simons & Kääriäinen, 1970); 775 daltons for the phospholipid and 387 daltons for cholesterol. The appropriate molecular ratios are obtained when the weight ratios of the different constituents are supplemented with these data. The actual number of the different molecules in a virus particle is obtained if it is assumed that there is only one RNA-molecule per virion. Each virion may thus contain about 200 core protein molecules and 550 membrane protein molecules, about 15,000 cholesterol molecules and about 16,000 phospho- and glycolipid molecules (Table 8). Each virus particle thus contains about 28 pairs of cholesterol and polar lipid for every membrane protein.

Table 8. *The number of different molecules in a Semliki Forest virus particle*

Constituent	Number of molecules per virion	
RNA		1
Core proteins		200
Membrane proteins		550
Cholesterol		15,000
Polar lipids		16,000
GPE-lipids	3,500	
GPC-lipids	6,400	
GPS-lipids	2,000	
GPI-lipids	200	
Sphingomyelins	2,400	
Gangliosides	1,000	

Recent work of Compans (1971) suggests that in Sindbis virus, which is very similar to Semliki Forest virus, proteins may be the outermost components of the membrane. In agreement with this concept, Harrison *et al.* (1971) have shown the presence of a lipid bilayer in the Sindbis virus. Assuming that Semliki Forest virus has the same basic structure we end up with the quantitative model shown in Fig. 11.

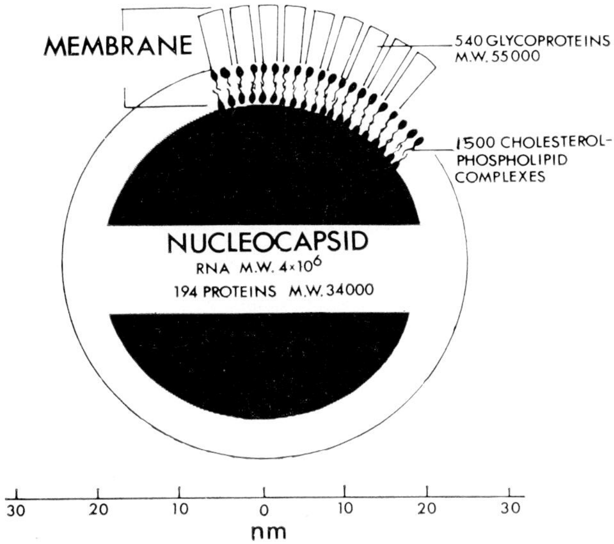

FIG. 11. Model of the Semliki Forest virus particle.

We have good reasons to hope that Semliki Forest and other enveloped viruses may prove to be fine models in the dynamic studies which are currently in progress in our laboratory and also in other laboratories.

ACKNOWLEDGEMENTS

The work reported here has been supported in part by Sigrid Jusélius Foundation, the Finnish State Council for Sciences and Medical Research, and the Association of Finnish Life Assurance Companies.

REFERENCES

Acheson, N. H. & Tamm, I. (1967). *Virology* **32**, 128.
Compans, R. W. (1971). *Nature New Biol.* (*London*) **229**, 114.
David, A. (1971). *Virology* **46**, 711.
Gahmberg, C. G. (1971). *Biochim. biophys. Acta* **249**, 81.
Gahmberg, C. G. & Simons, K. (1970). *Acta path. microbiol. scand.* Section B, **78**, 176.
Harrison, S. C., David, A., Jumblatt, J. & Darnell, J. E. (1971). *J. molec. Biol.* **60**, 523.
Keenan, T. W. & Morré, D. J. (1970). *Biochemistry* **9**, 19.
Klenk, H-D. & Choppin, P. W. (1969). *Virology* **38**, 255.
Klenk, H-D. & Choppin, P. W. (1970). *Virology* **40**, 939.
Kääriäinen, L., Simons, K. & von Bonsdorff, C-H. (1969). *Ann. Med. exp. Fenn.* (*Helsinki*) **47**, 235.
Laine, R., Kethunen, M. L., Gahmberg, C. G., Kääriäinen, L. & Renkonen, O. (1972). *Virology* **10**, 433.
Levin, J. G. & Friedman, R. M. (1971). *J. Virol.* **7**, 504.
McSharry, J. J. & Wagner, R. R. (1971). *J. Virol.* **7**, 59.

Quigley, J. P., Rifkin, D. B. & Reich, E. (1971). *Virology* **46**, 10.
Renkonen, O. (1963). *Acta chem. scand.* **17**, 646.
Renkonen, O., Gahmberg, C. G., Simons, K. & Kääriäinen, L. (1971*a*). *Biochim. biophys. Acta* **255**, 66.
Renkonen, O., Kääriäinen, L., Simons, K. & Gahmberg, C. G. (1971*b*). *Virology* **46**, 318.
Schlesinger, M. J., Schlesinger, S. & Burge, B. W. (1972). *Virology* **45**, 539.
Simons, K. & Kääriäinen, L. (1970). *Biochem. biophys. Res. Comm.* **38**, 981.
Strauss, J. H., Burge, B. W., Pfefferkorn, B. R. & Darnell, J. E. (1968). *Proc. natn Acad. Sci. U.S.* **59**, 533.

THE ROLE OF LIPIDS IN MEMBRANE STRUCTURE

By F. L. CRANE, T. F. CHUANG AND Y. C. AWASTHI*

Department of Biological Sciences, Purdue University, Lafayette, Indiana, U.S.A.

INTRODUCTION

CYTOCHROME, oxidase, in association with phospholipid, forms membranous vesicles which may be used as a model for the study of the directive influence of protein in membrane structure (McConnell *et al.*, 1966; Jacobs *et al.*, 1966; Sun *et al.*, 1968; Chuang, Awasthi & Crane, 1970; Seki & Oda, 1970; Vanderkooi, 1972; Green & Perdue, 1966). In this paper we will present evidence to show that the membrane formed is a model mosaic membrane in which the oxidase protein is associated with hydrophobic fatty acid chains inside the lipid bilayer.

METHODS

Preparation of cytochrome oxidase was carried out as described by Sun *et al.* (1968). Lipid-depleted detergent-free cytochrome oxidase was prepared by dilution of the lipid-depleted oxidase ten-fold with 0·02 M *tris*/HCl buffer pH 7·4, followed by centrifugation at 270,000 × g in the Spinco L2-65 for 30 min. Four or five further washes were carried out by suspending the green pellet in *tris*/HCl buffer followed by centrifugation at 103,000 × g. The final pellet had less than 0·03 mg Triton per mg protein. Triton was determined according to Chuang, Sun & Crane (1970).

Negative staining was with 2% potassium phosphotungstate pH 6·8 (Cunningham & Crane, 1966). Membranes were embedded in Epon for sectioning. Fixation was with glutaraldehyde and osmium tetroxide, poststained with uranylacetate and lead citrate (Hall & Crane, 1971). Infra red spectra were obtained from aqueous suspensions dried on an Irtran disk *in vacuo*. Cytochrome c oxidase activity was determined using an oxygen electrode (Chuang *et al.*, 1970).

Membrane formation was carried out by mixing the oxidase and phospholipid micelles in 0·25 M sucrose in 0·02 M *tris*/HCl buffer of pH 7·4. The mixture was sonicated for five 30 sec periods at maximum output (Branson Sonifier). Membranes were separated by centrifugation at 108,000 × g for 30 min.

* Present address: National Botanic Gardens, Lucknow, India.

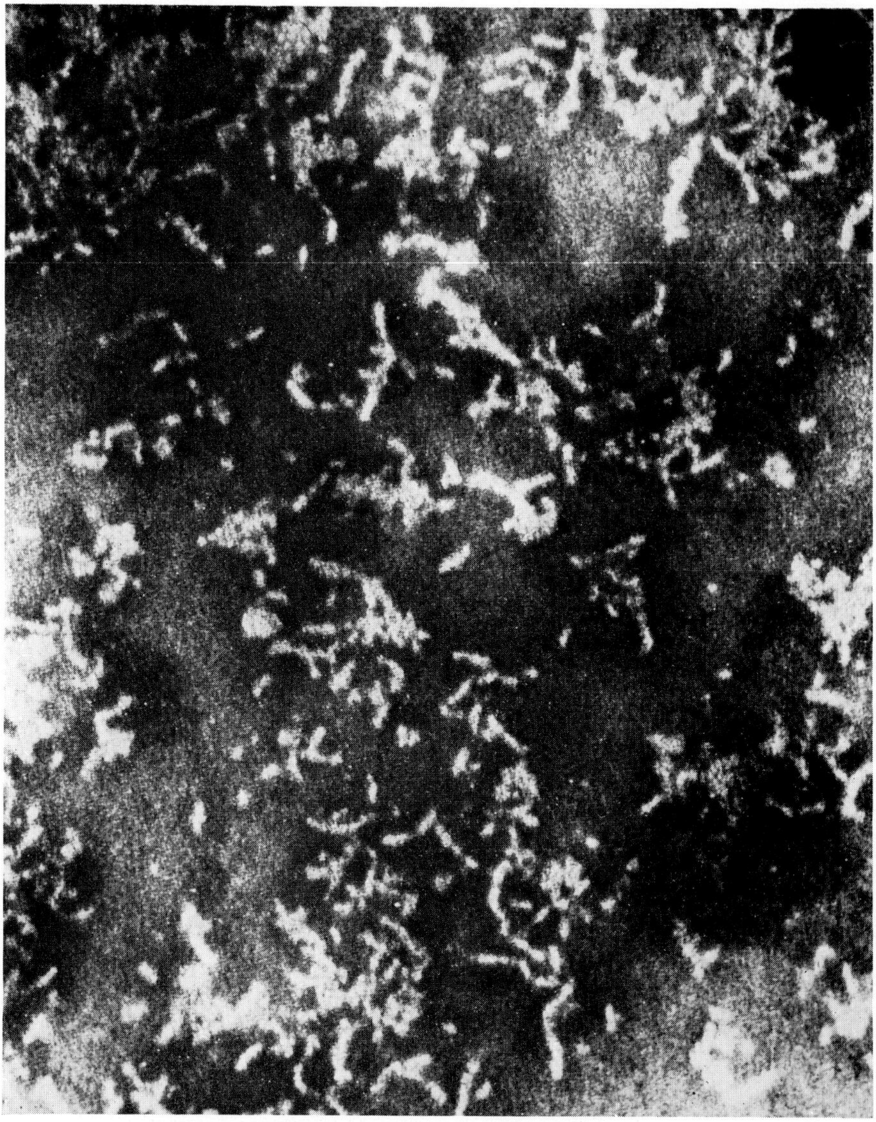

FIG. 1*a*. Lipid-depleted, detergent-free cytochrome oxidase. The small circles and rods are 90 Å wide. Negative stain.

Membrane structure

Structure of the cytochrome oxidase in association with increasing concentrations of lipid was examined by negative staining, cross sections and freeze fracture.

The lipid depleted, detergent free cytochrome oxidase with about two gram atoms of phosphorus and from 0 to 3 moles Triton per mole of cytochrome oxidase (assuming a molecular weight for the oxidase of 72,000) does not have a membranous structure. By negative staining we find 90 Å wide rods or 90 Å diameter spheres (Fig. 1*a*). Cross sections show short segments 90 Å wide or black spots (Fig. 1*b*). The segments show dark-light-dark staining similar to that of a unit membrane, but the lack of evidence for membranes by negative staining indicates that the trilamellar regions are in fact longitudinal sections of rods and that the oxidase protein clusters show darker staining on the outside than inside. Sections of oxidase after extraction of lipid by acetone have also been shown by Stoeckenius (1971) to reveal the 90 Å trilamellar structure and we feel that this should also be interpreted as sections of rod-like clusters of oxidase protein rather than a membrane.

With addition of 10 to 20 g atoms phospholipid phosphorus per mole of oxidase, larger rods (200 Å diameter) are found along with small sheet like regions covered with white spots 50–60 Å in diameter (Fig. 2).

Above 20 g atoms phosphorus per mole of oxidase negative staining shows rounded sheets with 50–60 Å white spots (Fig. 3). Sections show vesicles of trilamellar membranes. The membranes show good unit membrane structure but are only 50 Å thick. The vesicles are often concentric (Fig. 4).

If high levels of phospholipid are mixed with the oxidase (over 65 g atom P/mole oxidase) the membrane sheets are surrounded with white rings which look like pure phospholipid micelles (Fig. 5). Repeated washing by centrifugation of these preparations removes phospholipid and the white rings disappear. After repeated washing the phospholipid to protein ratio of the membranes approaches 65 g atom P per mole of oxidase. Figure 6 shows that thoroughly washed membranes are saturated at about 65 g atom P/mole cytochrome oxidase.

The activity of the enzyme also increases with the addition of phospholipid up to 65 g atoms P per mole of cytochrome oxidase (Fig. 7). Detergents such as Emasol can also activate the enzyme without causing membrane formation.

Figure 8 shows the detail structure of the surface of negatively stained cytochrome oxidase with 23 g atom P per mole of oxidase. Freeze fracture of a similar preparation confirms the membranous and vesicular structure without the use of fixatives or drying involved in sectioning or negative staining (Fig. 9).

A determination of the ratio of area for white spots and black regions on the surface of negatively stained membranes (Table 1) with different ratios of

Table 1. *Ratios of black and white surface areas in membranes with different proportions of phospholipid and protein*

Phospholipid content of membrane g atom P/mole cytochrome a	Ratio of white regions to dark regions on membrane*	Calculated ratio of protein to lipid area
32	1·56	1·64
37	1·34	1·44
51	1·01	1·02

* Ratios determined by cutting out white and black regions of enlarged photographs and weighing the paper. Calculated areas based on 55 Å diameter circle for protein (2360 Å^2) and 45 Å^2 for surface area of tightly packed lipid molecules (Shah, 1970; Engelman, 1971) for one half of a bilayer.

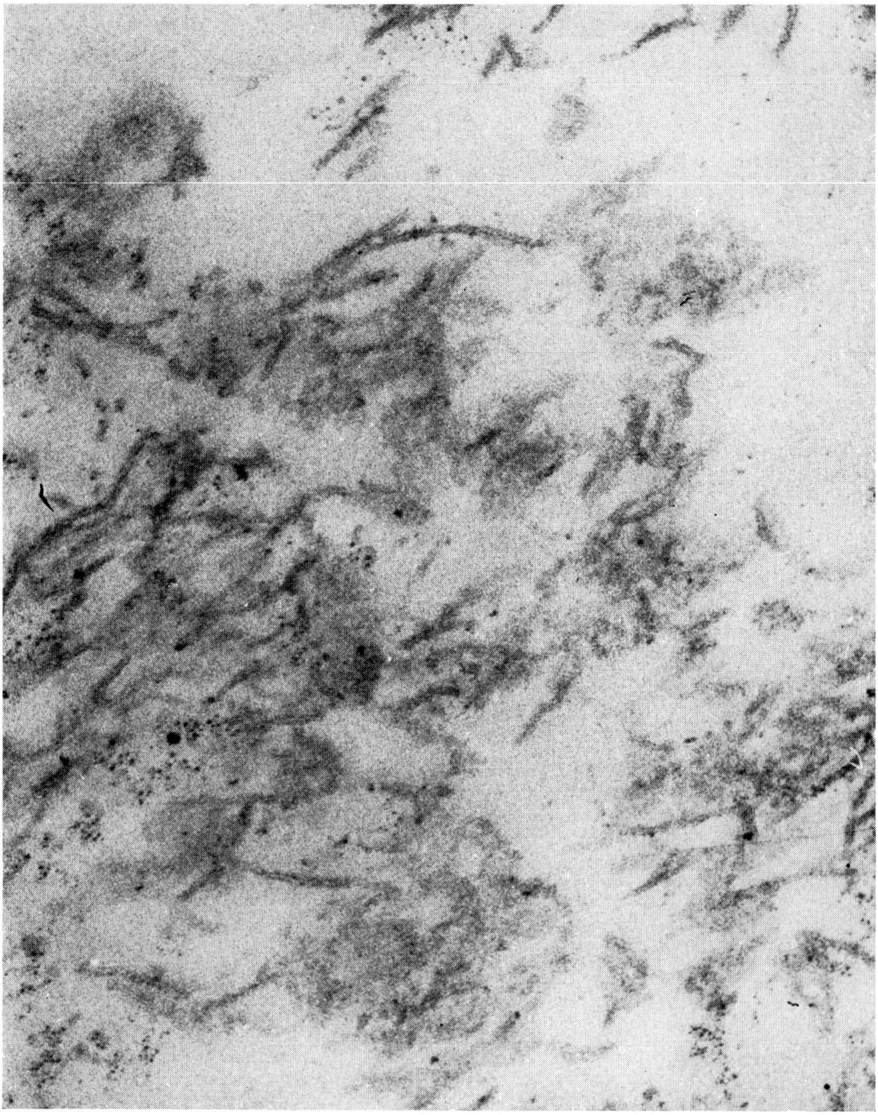

FIG. 1*b*. Cross section of lipid depleted, detergent free cytochrome oxidase. The trilamellar fragments are 90 Å wide and would represent longitudinal sections of the rods in Fig. 1*b*.

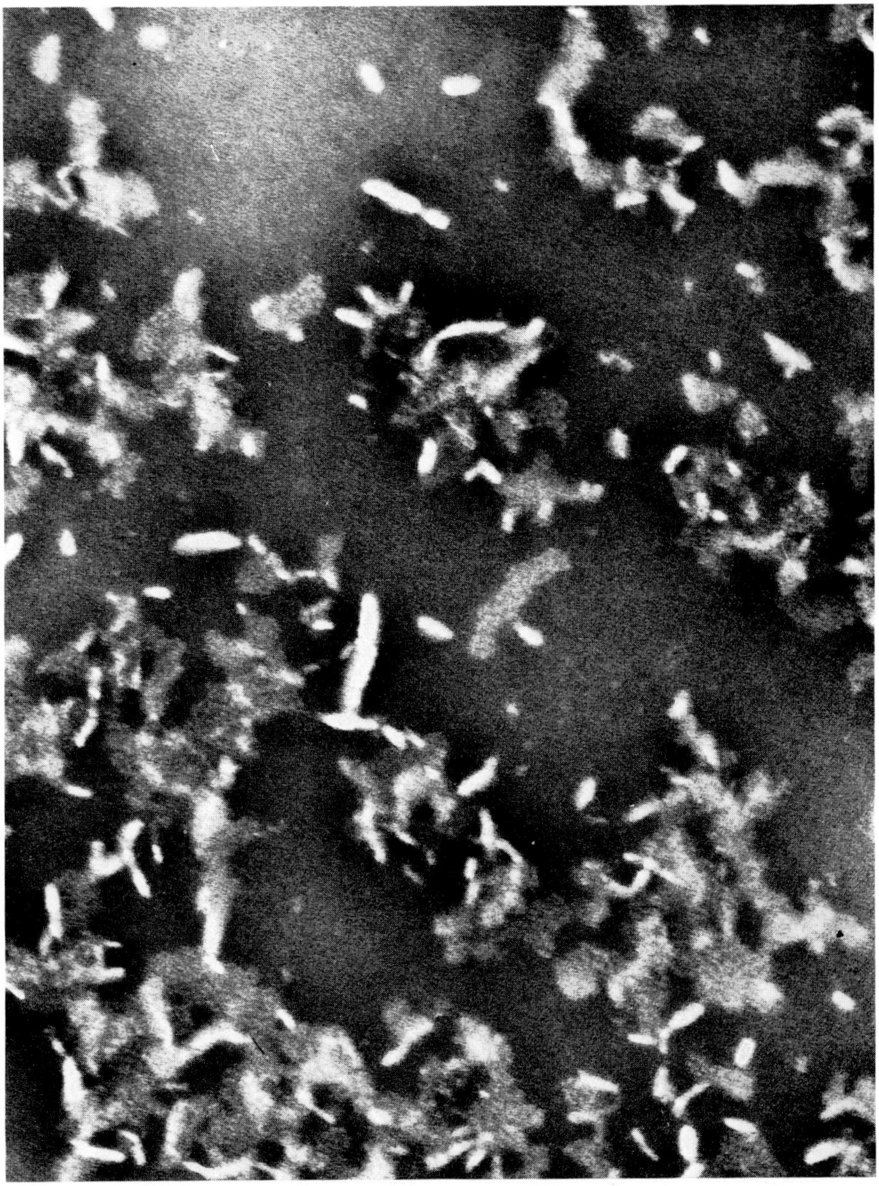

FIG. 2. Cytochrome oxidase with 10 g atoms phosphorus per mole of cytochrome oxidase. The large rods are 200 Å wide. A few small regions of membrane with 50–60 Å units can be seen near some rods. Negative stain.

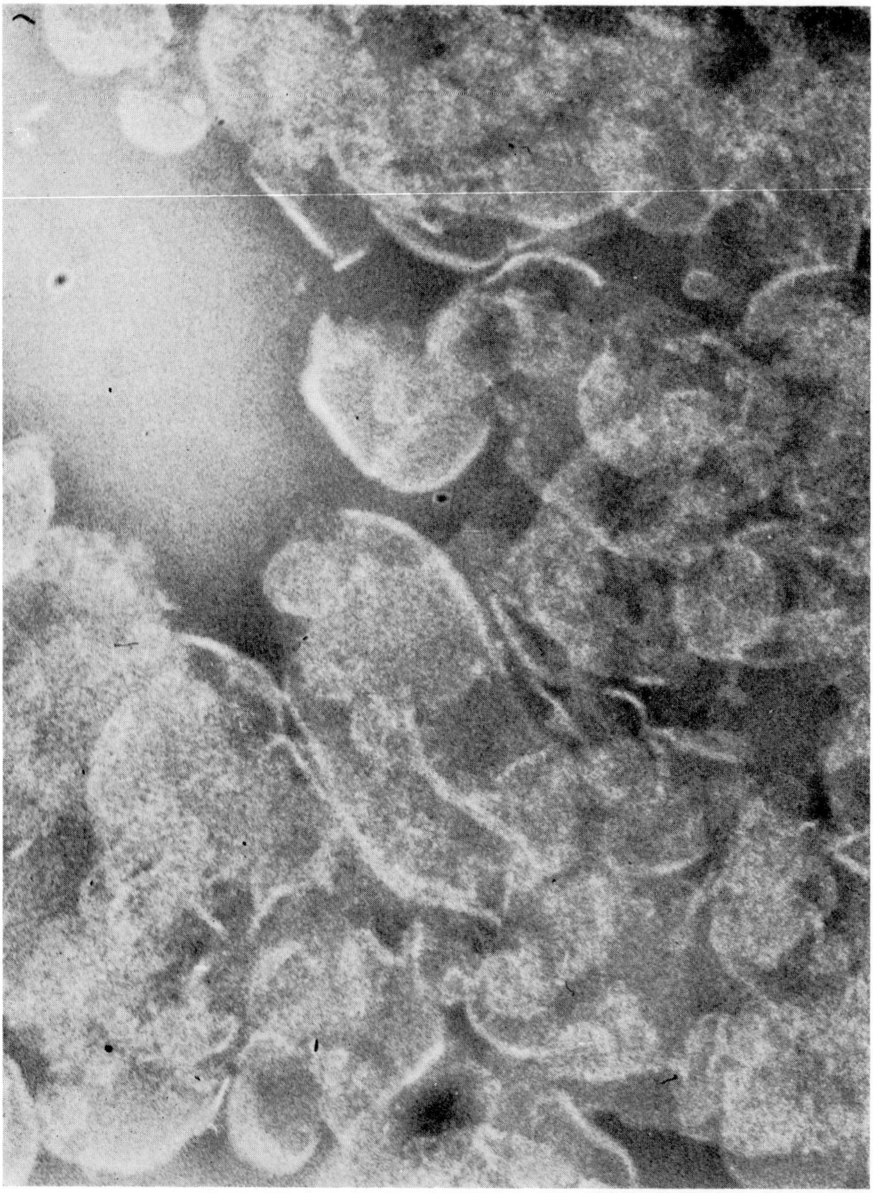

FIG. 3. Cytochrome oxidase with 31 g atoms phosphorus per mole of oxidase (25% phospholipid). Negative stain.

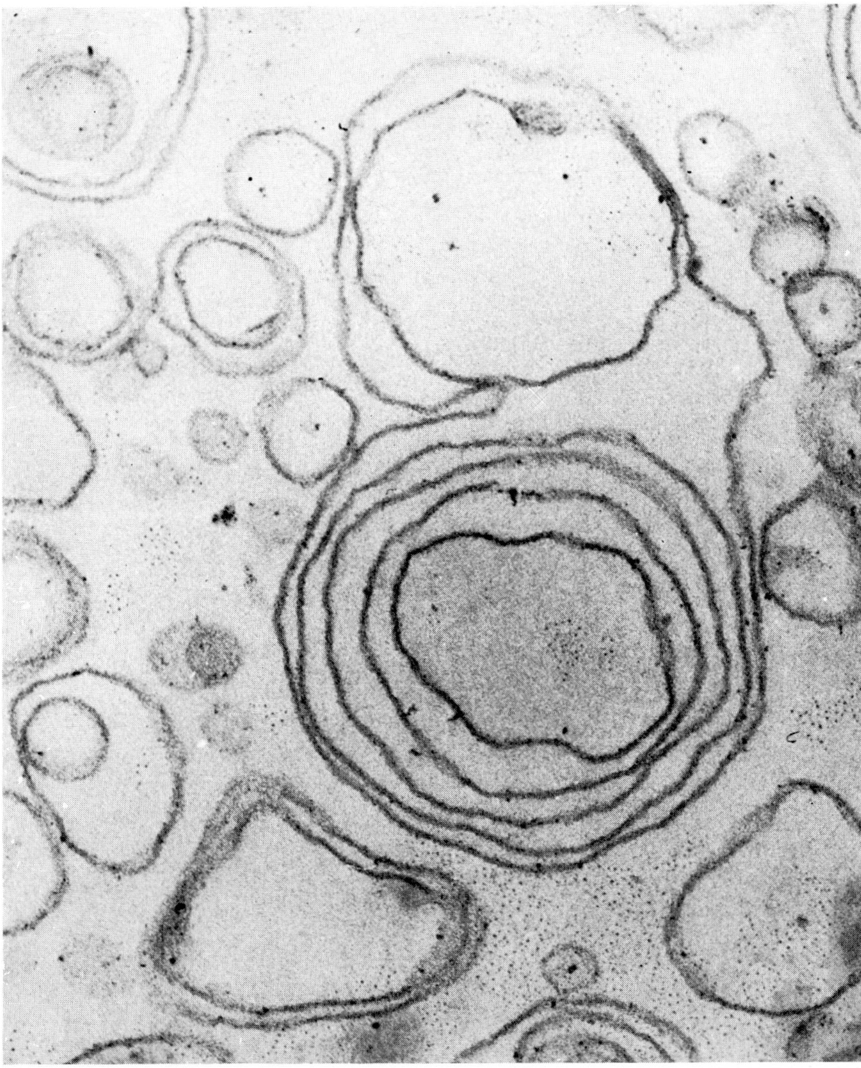

FIG. 4. Cross section of membranous cytochrome oxidase. Note the thin unit membranes (50 Å) and the concentric vesicles.

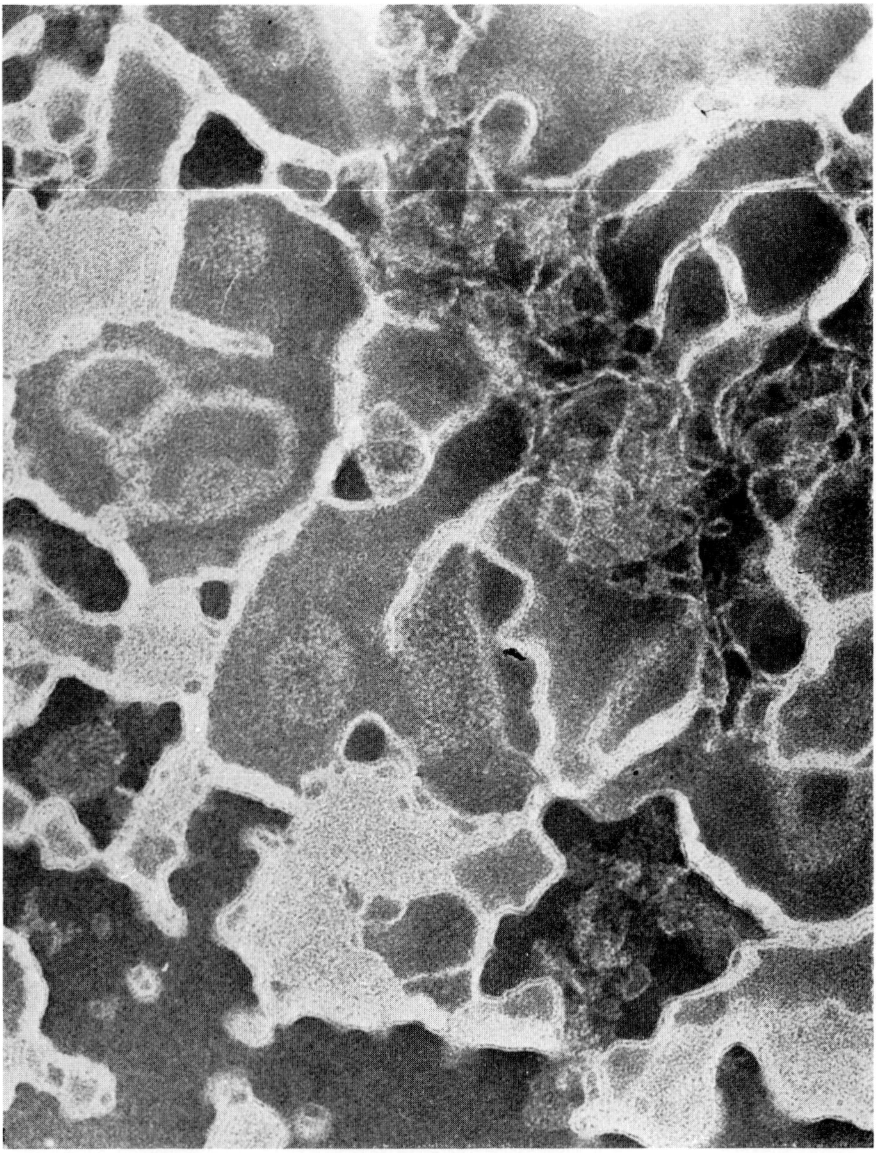

FIG. 5. Cytochrome oxidase with 235 g atoms phosphorus per mole of cytochrome oxidase (70% phospholipid). Negative stain. Note the excess phospholipid.

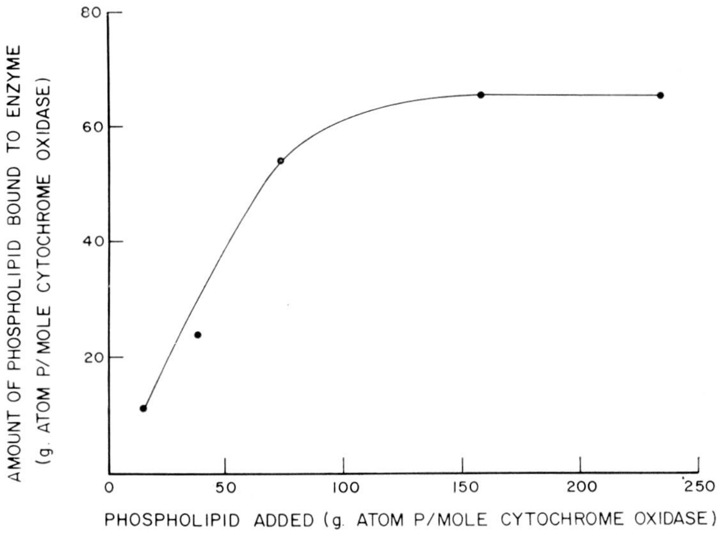

FIG. 6. Binding of phospholipid to cytochrome oxidase. After addition of phospholipid the membranes were washed to remove excess phospholipid or non-membranous cytochrome oxidase.

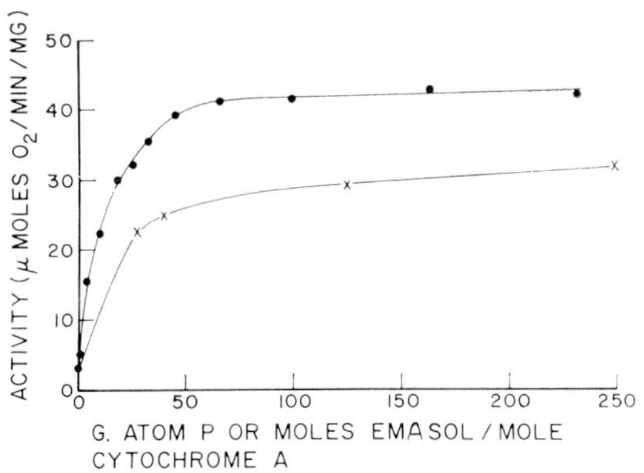

FIG. 7. Effect of phospholipid or detergent on cytochrome oxidase activity. —O— phospholipid. —x— Emasol concentration.

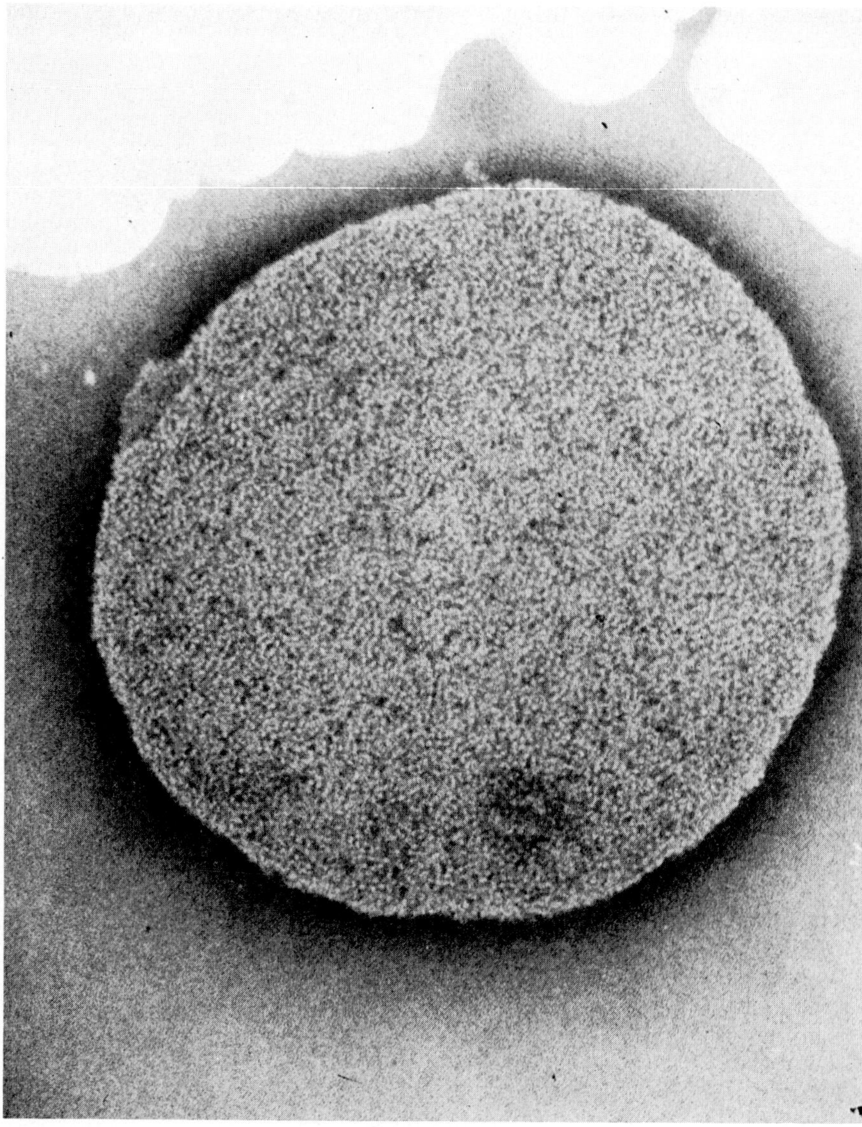

FIG. 8. A cytochrome oxidase–phospholipid vesicle showing the 50–60 Å globules spaced on the surface of the membrane. Negative stain.

FIG. 9. Freeze fracture surfaces of a series of cytochrome oxidase vesicles. Note globular units in the cross fractured membranes.

phospholipid to proteins shows that the dark regions increase as more phospholipid is added. This indicates that the white spots are in fact the oxidase molecules.

Stability of the membrane

Membrane formation occurs and the membrane is stable at ionic strengths up to 0·6. At higher ionic strength in the presence of higher levels of detergent the membrane structure may be reduced to small fragments (Sun *et al.*, 1968). Membranes form and are stable between pH 3·5 and 11. The stability of the membrane under these conditions indicates that ionic association between the protein and phospholipid is not the prime requisite for maintaining membrane structure. This stability is in marked contrast to removal of cytochrome c from association with phospholipids under conditions which reduce ionic association (Das, Haak & Crane, 1965).

Lipid specificity

The minimum attributes of a lipid molecule which can form membrane in association with cytochrome oxidase appear to be:

(1) a diglyceride structure;
(2) unsaturated fatty acid.

All of the major phospholipids from mitochondria such as lecithin, phosphatidylethanolamine or cardiolipin will form membrane with oxidase, although the size and shape of the vesicles will vary. On the other hand no detergent which we have tried will form membrane. These include Tritons, bile salts, fatty acids or Emasols.

If mitochondrial lecithin is hydrogenated then it will no longer form membrane. The negatively stained mixture of hydrogenated lecithin and cytochrome oxidase shows the same 90 Å rods and spheres as does the lipid depleted oxidase (Fig. 10) together with large white masses of lipid. Sections show short 90 Å trilamellar segments and black spots (Fig. 11). These would be consistent with sections of 90 Å thick rods cut at various angles to the long axis of the rod.

If membranes formed with lecithin and cytochrome oxidase are digested with phospholipase C to remove the phosphocholine from the lecithin, membrane structure with typical trilayer appearance 50 Å thick is retained. After this digestion concentric vesicles are no longer found but the membrane appears as irregular sheets which suggests that the charge on the phospholipid may influence the shape of the membrane vesicles (Chuang *et al.*, 1970).

To confirm this evidence that the charged groups on the phospholipid are not necessary for membrane formation one can form membrane sheets by a combination of diolein and cytochrome oxidase. This membrane in sections shows trilamellar unit membrane structure 50 Å thick and sheets with 50 Å globules by negative staining. A combination of oxidase and diolein does not restore cytochrome oxidase activity.

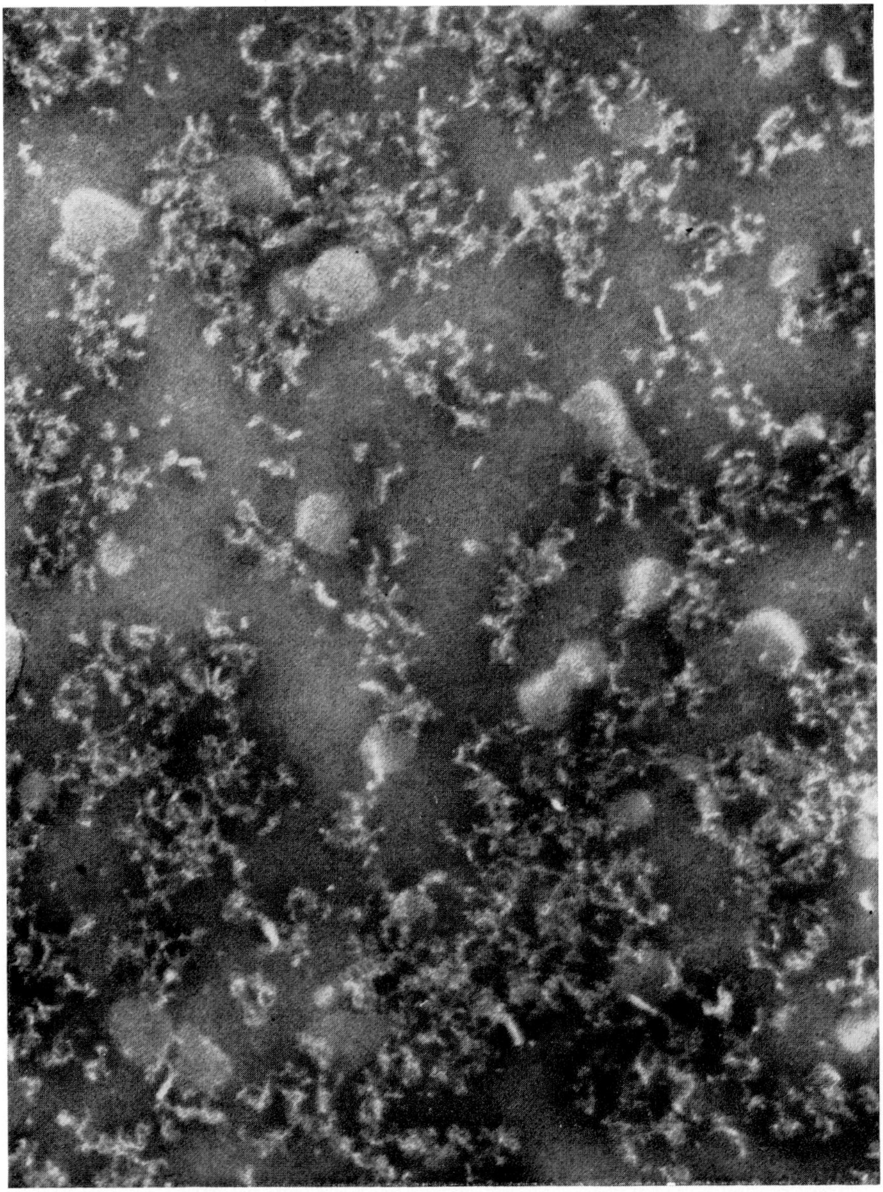

FIG. 10. Lipid depleted cytochrome oxidase plus hydrogenated lecithin. Negative stain.

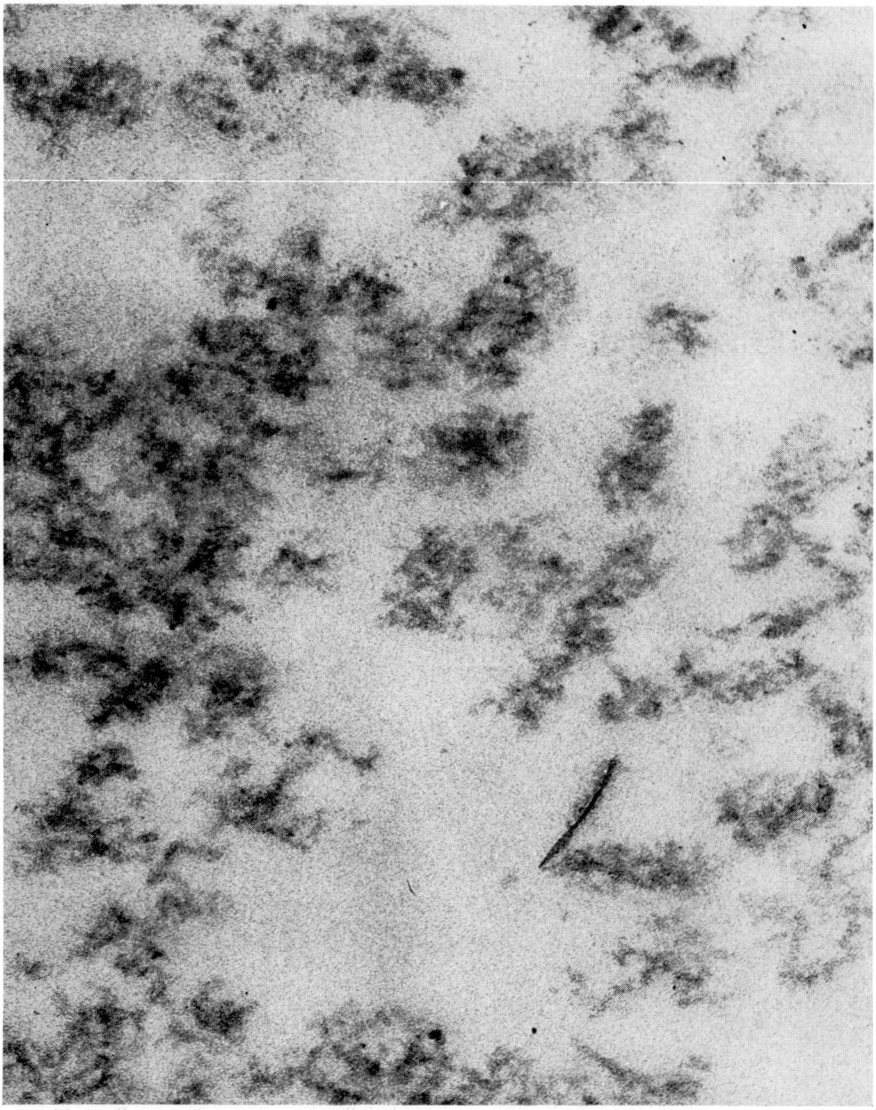

FIG. 11. Lipid depleted cytochrome oxidase plus hydrogenated lecithin. Cross section. Note the lack of any membranous structure except for very short trilamellar segments.

Physical properties

The visible spectrum of oxidized, reduced and reduced plus carbon monoxide cytochrome oxidase in the lipid depleted, detergent free state does not differ from that of the membranous oxidase. The site of the heme *a* in the protein is therefore not affected by association with lipids.

The circular dichroism of the membranous oxidase shows a strong negative maximum at 225 nm and a shoulder at 210 nm indicating considerable α helix in the protein (Fig. 12). The mitochondrial cristal membrane (ETP) shows a

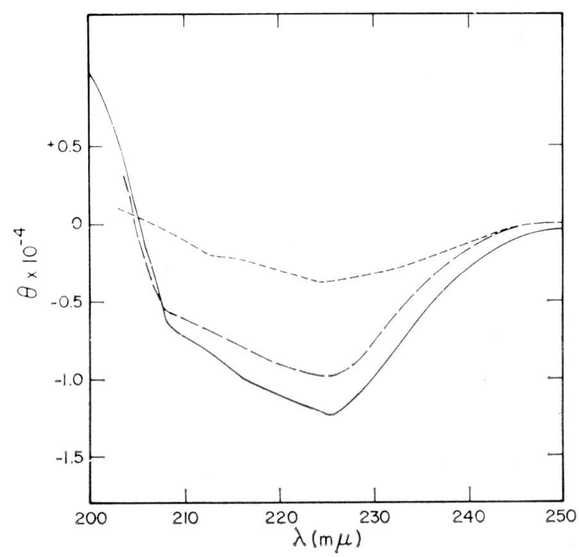

FIG. 12. Circular dichroism of cytochrome oxidase. – – – – detergent free lipid depleted cytochrome oxidase; — — — — cytochrome oxidase plus phospholipid; ———— mitochondrial cristae membranes (ETP).

similar spectrum. The lipid free, detergent free protein shows only a weak maximum but the general shape of the curve is appropriate for retention of significant α helix. The greater turbidity in the lipid free preparation may contribute to the decreased maxima (Wallach & Zahler, 1966).

The infra red spectrum of the lipid free and membranous cytochrome oxidase both show a strong peak at 1652 cm^{-1} characteristic of α helix or random configuration (Mizazawa & Blout, 1961) and little absorbance at 1630 cm^{-1} characteristic of β configuration (Figs. 13, 14).

A band at 720 cm^{-1} characteristic of methylene rocking is found in lipid bilayers and is present in phospholipid (Chapman, Kamat & Levene, 1968) (Fig. 15). Note, however, that the spectrum of the phospholipid cytochrome oxidase membranes in Fig. 14 does not show this band even at the equivalent level of pure phospholipid.

The dye, 8-anilino-1-naphthalene-sulphonic acid (ANS) can be used as an indicator of conformational change in protein where exposure of hydrophobic groups to the dye enhances the fluorescence (McClure & Edelman, 1965;

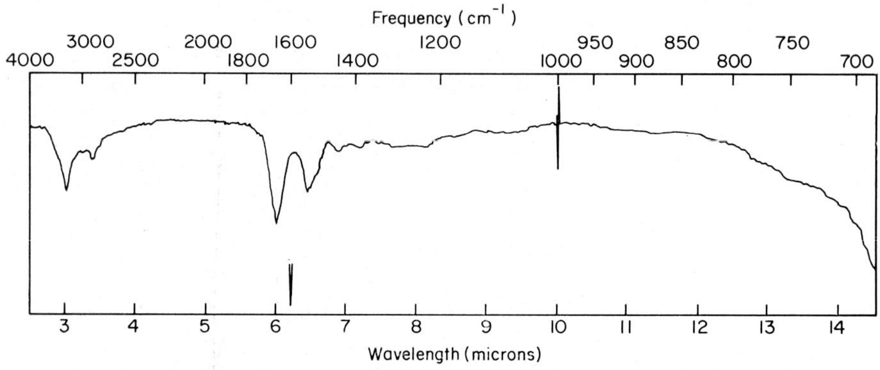

Fig. 13. Infra red spectrum of lipid free cytochrome oxidase.

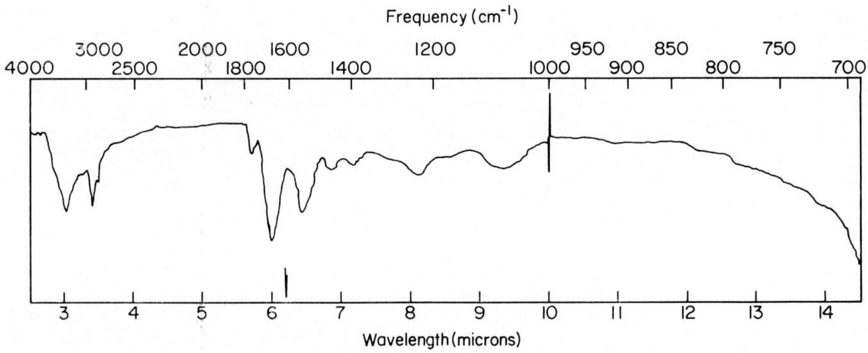

Fig. 14. Infra red spectrum of cytochrome oxidase plus phospholipid membranes.

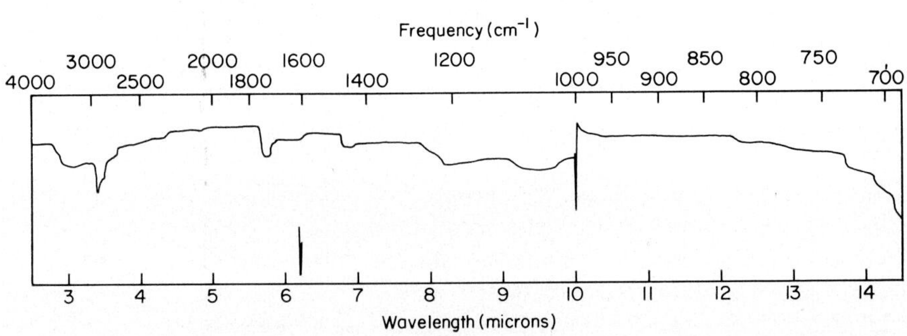

Fig. 15. Infra red spectrum of phospholipid preparation used for membrane formation.

Weber & Young, 1964). The lipid free cytochrome oxidase shows a high level of fluorescence when ANS is added indicating exposed hydrophobic groups. When phospholipid is added to the oxidase to form membrane, the fluorescence is decreased even though phospholipid alone would give increased fluorescence (Fig. 16). The loss of fluorescence during membrane formation indicates that

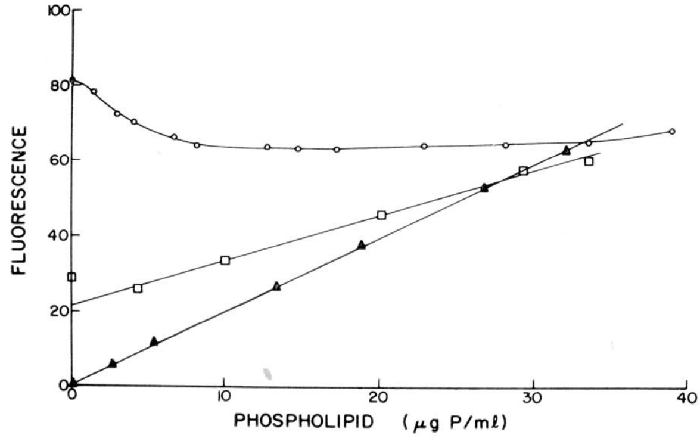

FIG. 16. Effect of increasing phospholipid concentration on fluorescence of the dye ANS.
—△— phospholipid only, 0·02 M TrisHCL, pH 7·4;
—□— phospholipid added to 0·3 mg Triton × 100 per ml;
—○— phospholipid added to 0·46 mg lipid-depleted oxidase per ml.

hydrophobic groups are less available to the dye in the membrane than they are in the lipid free oxidase or in phospholipid alone. It is likely that addition of phospholipid blocks hydrophobic sites on the surface of the oxidase protein and that hydrophobic sites on the phospholipid are not as readily available in the membrane as they are in a smectic bilayer.

Heat stability

Cytochrome oxidase activity is more resistant to heat in membrane form than in the lipid free state. As shown in Table 2 the oxidase activity in combination with phospholipid is quite resistant to exposure to 38° as it is in the mitochondrial membrane.

Table 2. *Effect of heat treatment on cytochrome oxidase activity*

State of cytochrome oxidase	Activity remaining after exposure to heat*					
	20 min			40 min		
	30°	38°	50°	30°	38°	50°
Lipid free cyt. oxidase	75%	45%	0	75%	25%	0
Membranous cyt. oxidase	95%	85%	20%	90%	85%	10%
Oxidase in mitochondrial membrane	—	98%	50%	—	98%	45%

* Cytochrome c oxidase supplemented with phospholipid after heat treatment in the case of the lipid-free enzyme. All assayed polarographically at 37°.

Protein specificity

The number and types of proteins in membranes which can form mosaic structures in association with phospholipids are yet to be determined. There is evidence that several of these membrane-directing proteins are present in mitochondrial cristae since each of the four functional complexes of the electron transport chain have been shown to form spotted membranes in association with phospholipid (Green & MacLennan, 1969). The directive nature of the protein can vary since the NADH dehydrogenase complex (I) forms small, stiff sheets of 90 Å thick membrane and the cytochrome b, c, complex (III) forms thin single vesicles in contrast to the thin concentric vesicles of cytochrome oxidase (Crane *et al.*, 1971). Certainly not all proteins are capable of directing this type of membrane formation. Bovine serum albumin, either native or denatured, does not form membrane. The F_1 ATPase from mitochondria remains in globular masses in presence of phospholipid. The major protein of cristae membranes which in addition to the denatured ATPase constitutes the major part of the structural protein does not form membrane, but rather forms a string-like structure in association with phospholipid which suggests a role in membranifibril formation (Stiles, Wilson & Crane, 1968; Hall & Crane, 1972).

Only one of the two proteins in the cytochrome oxidase complex is capable of membrane formation. We have previously shown that the cytochrome oxidase can be separated by gel electrophoresis into approximately equal amounts of two proteins which have molecular weights of 26,000 and 55,000 (Chuang & Crane, 1971). The heme *a* containing protein (MW26,000) can form membrane vesicles similar to the intact complex except that the white spots seen by negative staining on the surface of the membrane are only about 30 Å in diameter. This dimension is consistent with a smaller protein unit in the membrane. The other protein, which is devoid of prosthetic groups (MW55,000), does not form membrane with phospholipid and may be core protein IV.

CONCLUSIONS

We conclude that the cytochrome oxidase complex consists of 50–60 Å globules. In the lipid free inactive state these globules cluster together in pentameric or heximeric units 90–100 Å in diameter and these clusters may stack into long 90 Å rods (Fig. 17). As lipid is added the fatty acid chains penetrate between the protein globules so that the rods swell into 200 Å rods. With further lipid the globules are separated into a sheet with small regions of phospholipid bilayer between the protein units. As more lipid is added the sheet expands by increasing the lipid bilayer area and by spreading the protein apart until the mixed micelle structure can no longer be maintained or the protein is saturated. The mosaic membrane structure is diagrammed in Fig. 18. From 20 to 65 molecules of phospholipid are associated with each cytochrome oxidase unit (assuming 72,000 molecular weight for the unit) in the bilayer. One molecule of phospholipid is drawn inside each oxidase unit to indicate the one to two molecules of tightly bound cardiolipin which is present at all stages

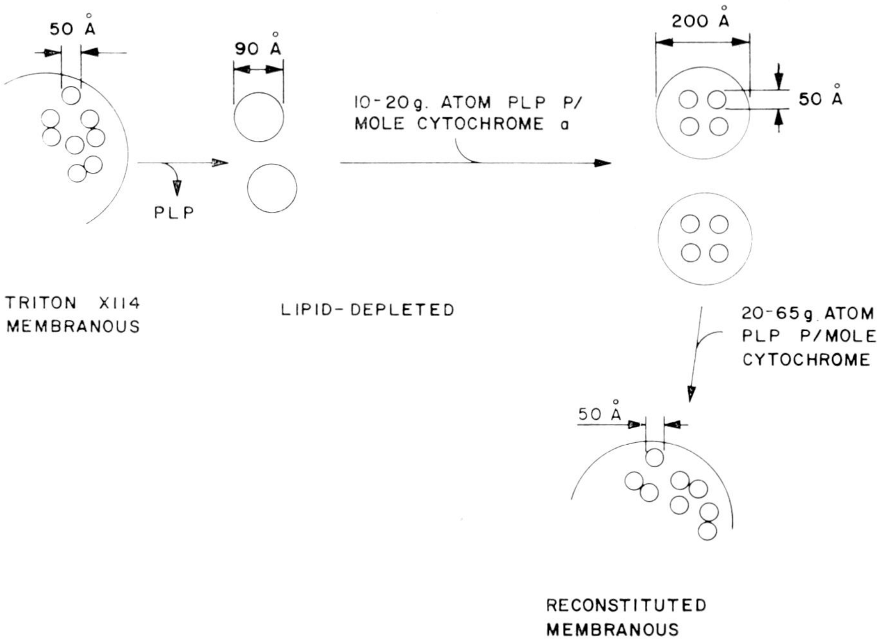

FIG. 17. Diagram of the organization of cytochrome oxidase in association with different levels of phospholipid.

(Awasthi *et al.*, 1971). This mosaic type membrane would be similar to that proposed by Wallach & Gordon (1968).

Both lipid and protein contribute to the stability of the membrane. The stabilizing effect of the protein is seen in the formation of membrane by

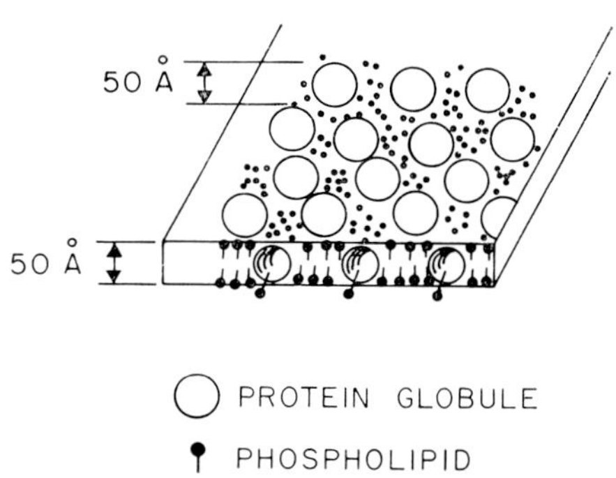

FIG. 18. Model mosaic cytochrome oxidase membrane. Large spheres, cytochrome oxidase; black beads with tails, phospholipid molecules.

diolein which does not form bilayer alone. Removal of lipid by detergent or solvent extraction also causes loss of membrane structure. A similar stabilizing effect of lipid can be seen in acetone-extracted mitochondrial membranes (Hall & Crane, 1971).

The structure of the oxidase complex remains to be determined. The smallest unit would appear to be composed of one heme *a* protein and one core protein

IV (MW81,000) which is appropriate for a 50 Å sphere. The heme *a* content of 9–10 nmoles per mg protein indicates a molecular weight of 100,000–110,000 per heme *a* (Tzagoloff *et al.*, 1965). Since a functional oxidase requires two heme *a* ($a + a_3$), the functional unit would be 200,000 daltons or would require the interaction of two 100,000 MW units. Of course the loss of some heme *a* during purification would give an apparent increase in the molecular weight of the oxidase. The combination of two 26,000 molecular weight heme *a* proteins plus a 55,000 MW core protein IV would constitute a functional unit of 107,000 daltons appropriate to 50 × 60 Å units in a membrane 50 Å thick. The loss of heme *a* which we have observed during purification of the smaller unit indicates that the definition of molecular weight on a heme *a* basis needs to be examined carefully. If no loss of heme *a* has occurred then association of two white spots in the membrane would apparently be necessary for oxidase activity.

The morphology of each of the membranous organelles of the cell must depend primarily on the directive influence of the membrane proteins. An influence for lipid in the structure can also be postulated on the basis of the differences in membrane structure when the lipid component is changed.

SUMMARY

The cytochrome oxidase complex forms membranes with phospholipids which are good models for mosaic membrane structure. Lipid free oxidase forms rod-like clusters, 90 Å in diameter, which swell and spread out on addition of phospholipid into membrane sheets and vesicles. The phospholipid binding capacity of the protein is 65 g atom P per mole of cytochrome oxidase. Evidence is presented that membrane is formed by hydrophobic association between fatty acid chains of the phospholipid and 50–60 Å diameter globules of oxidase inserted in a lipid bilayer. Infra red and CD spectra indicate retention of protein α helix in the membrane but that the 720 cm^{-1} methylene rocking of phospholipid micelles is lost in the membrane. Enhancement of ANS fluorescence by both protein and lipid is decreased in the membrane. Heat stability of oxidase activity is increased in the membrane. Hydrogenated phospholipid or detergents do not form membrane, but diolein does, indicating that a diglyceride structure with unsaturated fatty acid is the minimum requirement. Protein specificity for membrane formation is shown since purified heme *a* protein will form membrane but core protein IV, the other component of the oxidase complex, will not form membrane.

ACKNOWLEDGEMENTS

Supported under grant AM04663 from the National Institute for Arthritis and Metabolic Diseases and Career grant K6-21,839 from the National Institute of General Medical Science. We wish to thank Dr Judy Hall and Mrs Linda Funk for assistance in electron microscopy.

REFERENCES

Awasthi, Y. C., Chuang, T. F., Keenan, T. W. & Crane, F. L. (1971). *Biochim. biophys. Acta* **226**, 42.

Chapman, D., Kamat, V. B. & Levene, R. J. (1968). *Science* **160**, 314.
Chuang, T. F., Awasthi, Y. C. & Crane, F. L. (1970). *Proc. Indiana Acad. Sci.* **79**, 110.
Chuang, T. F. & Crane, F. L. (1971). *Biochem. biophys. Res. Commun.* **42**, 1076.
Chuang, T. F., Sun, F. F. & Crane, F. L. (1970). *Bioenergetics* **1**, 227.
Chuang, T. F., Awasthi, Y. C., Funk, L. & Crane, F. L. (1970). *Biochim. biophys. Acta* **211**, 599.
Crane, F. L., Arntzen, C. J., Hall, J. D., Ruzicka, F. J. & Dilley, R. A. (1971). In *Autonomy and Biogenesis of Mitochondria and Chloroplasts.* (Ed. by Boardman, N. K., Linnane, A. W. & Smellie, R. M.), p. 53. Amsterdam: North Holland.
Cunningham, W. P. & Crane, F. L. (1966). *Exptl. Cell Res.* **44**, 31.
Das, M. L., Haak, E. D. & Crane, F. L. (1965). *Biochemistry* **4**, 859.
Engelman, D. M. (1971). *J. molec. Biol.* **58**, 153.
Green, D. E. & MacLennan, D. H. (1969). *Bioscience* **19**, 213.
Green, D. E. & Perdue, J. F. (1966). *Proc. natn Acad. Sci. U.S.A.* **55**, 1295.
Hall, J. D. & Crane, F. L. (1972). *Biochim. biophys. Acta* **241**, 682.
Hall, J. D. & Crane, F. L. (1972). *Biochim. biophys. Acta* **255**, 602.
Jacobs, E. E., Andrews, E., Cunningham, W. P. & Crane, F. L. (1966). *Biochem. biophys. Res. Commun.* **25**, 87.
McClure, W. O. & Edelman, G. M. (1965). *Biochemistry* **5**, 1908.
McConnell, D. G., Tzagoloff, A., Maclennan, D. H. & Green, D. E. (1966). *J. biol. Chem.* **241**, 2373.
Mizazawa, T. & Blout, E. R. (1961). *J. Am. chem. Soc.* **83**, 712.
Seki, S. & Oda, T. (1970). *Archs. Biochem. Biophys.* **138**, 122.
Shah, D. O. (1970). *Adv. Lipid Res.* **8**, 347.
Stiles, J. W., Wilson, J. T. & Crane, F. L. (1968). *Biochim. biophys. Acta* **162**, 631.
Stoeckenius, W. (1971). In *Membranes of Mitochondria and Chloroplasts.* (Ed. by Racker, E.). New York: Academic Press.
Sun, F. F., Prezbindowski, K. S., Crane, F. L. & Jacobs, E. E. (1968). *Biochim. biophys. Acta* **153**, 804.
Tzagoloff, A., Yang, P. C., Wharton, D. C. & Rieske, J. S. (1965). *Biochim. biophys. Acta* **96**, 1.
Vanderkooi, G. (1972). *Ann. N.Y. Acad. Sci.* **195**, 6.
Wallach, D. F. H. & Gordon, A. (1968). In *Regulatory Functions of Biological Membranes.* (Ed. by Jarnefelt, J.), p. 64. Amsterdam: Elsevier.
Wallach, D. F. H. & Zahler, P. H. (1966). *Proc. natn Acad. Sci. U.S.A.* **56**, 1552.
Weber, G. & Young, L. B. (1964). *J. biol. Chem.* **239**, 1415.

CONCLUDING REMARKS

By J. GANGULY

Department of Biochemistry, Indian Institute of Science, Bangalore, India

THE meeting, of which this book is a record, was unusual, in the sense that a war was declared on the host country during it. However, in spite of the suspense and tension that usually accompany a war the meeting was successful.

Professor Lynen's significant contribution started the book on a very high plane. The next paper discussed the effects of relative rates of synthesis and degradation of the acetyl-CoA carboxylase in the ultimate regulation of fatty acid synthesis. The mechanism of biosynthesis of unsaturated fatty acids in plants has eluded us for a long time. Evidence has been presented here showing that oleoyl-CoA can be specifically desaturated to linoleoyl-CoA by a microsomal enzyme from maturing seeds and that the reaction is dependent on NADH and molecular oxygen.

The mode of action and possible active forms of all fat-soluble vitamins have always been challenging problems. It now appears that at last we are coming nearer to understanding those concerning vitamin D. We have read that vitamin D is hydroxylated in the liver to 25-hydroxycholecalciferol, which in turn is further hydroxylated in the kidney to 1,25-dihydroxycholecalciferol. It has been claimed that this dihydroxy compound might be the actual agent responsible for intestinal Ca^{2+} transport and bone mineral mobilization. Progress with vitamin A has not been so spectacular excepting the recent work on retinoic acid. But the transport mechanism of retinol appears clearer and we have seen that during vitamin A deficiency the RBP (retinol-binding protein) accumulates in the liver and falls in the blood, because in the absence of retinol the apo-protein remains immobilized in the liver.

Prostaglandins were discovered long ago. But only in recent years have we begun to understand their potential use, and their physiological and pharmacological functions. Our attention has been focused here on the possible role of these interesting compounds in the biogenesis of progesterone.

In the section on neurochemistry we have read about the metabolic defects in the two cerebral sphingolipidoses, namely metachromatic leukodystrophy and infantile amaurotic idiocy. We now have a better understanding of the types of glycosidases and the variants involved in these diseases. Newer concepts regarding the biogenesis of myelin membranes were presented in that, it now appears that myelin is not so stable metabolically as we have been led to believe, because cholesterol molecules are exchanged between myelin sheaths and oligodendroglial cells.

Although the first polyprenol, solanesol, was isolated several years ago, only after the systematic work at Liverpool did this group of compounds come into prominence. It is therefore most appropriate that the leader of the Liverpool school Professor Morton has given an excellent review of this subject emphasizing the role of undecaprenyl pyrophosphates as carriers in the synthesis of complex polysacharides, particularly in cell walls and that of dolichol metaphosphate in glucose transfer in liver.

The excellent section on the biogenesis of isoprenyl compounds has shown that it is now possible to demonstrate enzymic synthesis of polyprenyl derivatives of parahydroxy benzoate (potential precursors of ubiquinones) from pyrophosphate derivatives of polyprenols and not from their CoA derivatives. Apparently, to be successful in this reaction one has to be extremely careful and swift. It has also been interesting to learn that cyclization of lycopene to β-carotene takes place anaerobically and that desaturation of phytoene to lycopene proceeds without intermediate hydroxylation.

Another topic of considerable current interest, namely, metabolic regulation of isoprenoid and fatty acid synthesis receives a good deal of attention here. We have seen that in the obese mice the rate of synthesis of acetyl-CoA carboxylase is about five times that of its degradation, and also that, since on a fat-free diet the acetyl-CoA caroxylase and fatty acid synthetase show marked increase while HMG-CoA reductase decreases, there is a cellular mechanism by which acetyl-CoA can be shunted into the fatty acid synthesis pathway and away from the pathway for cholesterol synthesis. In this context the demonstration of inhibition of cholesterol synthesis by dietary ubiquinone is of interest.

Membrane biochemistry is becoming increasingly important for the proper understanding of the forces involved in the regulation of metabolism of the cell. This area has also been presented excellently. We have appreciated the heroic attempts that are currently being made to prepare artificial membranes which can simulate the permeability properties of natural membranes. It has been most interesting to learn that there has been some success in these efforts, because the loss in the activities of some enzymes caused by disaggregation of natural membranes can be partially restored on reconstituting the membranes. This has been shown particularly with cytochrome oxidase and also with the enzymes involved in phospholipid biosynthesis.

AUTHOR INDEX

Numbers in italics indicate pages where references are listed at the end of each article.

M

Q

R

Z

SUBJECT INDEX

B

E

F

R